Advances in
Biomedical
Measurement

Advances in Biomedical Measurement

Edited by

Ewart R. Carson
City University
London, England

Peter Kneppo
Slovak Academy of Sciences
Bratislava, Czechoslovakia

and

Ivan Krekule
Czechoslovak Academy of Sciences
Prague, Czechoslovakia

PLENUM PRESS • NEW YORK AND LONDON

Library of Congress Cataloging in Publication Data

IMEKO Conference on Advances in Biomedical Measurement (4th: 1987: Bratislava, Czechoslovakia)
Advances in biomedical measurement.

"Proceedings of the Fourth IMEKO Conference on Advances in Biomedical Measurement, held May 1987, in Bratislava, Czechoslovakia" — T.p. verso.
    Includes bibliographies and index.
    1. Medical electronics — Congresses. 2. Human physiology — Measurement — Congresses. 3. Imaging systems in medicine — Congresses. 4. Expert systems (Computer science) — Congresses. 5. Biomedical engineering — Instruments — Congresses. I. Carson, Ewart R. II. Kneppo, Peter. III. Krekule, Ivan. IV. International Measurement Confederation. V. Title. [DNLM: 1. Biomedical Engineering — congresses. 2. Electronics, Medical — congresses. W3 IM309 4th 1987a / QT 34 I32 1987a]
R895.A2I43   1987   610'.28   88-12533
ISBN-13:978-1-4612-8298-3   e-ISBN-13:978-1-4613-1025-9
DOI: 10.1007/978-1-4613-1025-9

Proceedings of the Fourth IMEKO Conference on Advances in Biomedical Measurement, held May 1987, in Bratislava, Czechoslovakia

© 1988 Plenum Press, New York
Softcover reprint of the hardcover 1st edition 1988

A Division of Plenum Publishing Corporation
233 Spring Street, New York, N.Y. 10013

The current scientific and technical literature in the fields of medicine and engineering, in addition to learned society journals, embraces
textbooks, monographs and conference proceedings. The last of these categories is potentially of prime importance given the increasing pace at which
knowledge is advancing. However, traditional conference proceedings often
tend to be indigestable, both due to their excessive volume and to the un-
evenness of the ingredients. Here we have attempted to overcome these
deficiencies by selecting a set of contributions which, in our opinion, in
the best sense convey the flavour of the 4th IMEKO Conference on "Advances
in Biomedical Measurement". This meeting, which took place in Bratislava
in May, 1987, was held concurrently with the 1st Regional Conference of the
Socialist European Countries on Biomedical Engineering and the 2nd Czechoslovak Conference on Biomedical Engineering.

From more than 200 papers presented in 14 sessions, 56 contributions
have been selected to represent four major areas within which advances in
biomedical measurement are occurring. These are: measurement and instrumentation (including prosthetics); signal and image processing; modelling
and simulation; and decision support.

The process of deciding which authors should be invited to contribute
to this volume was not always easy. What we have tried to do is to achieve
a mix which provides an overview of the state of the art across this broad
spectrum of endeavour. In doing this we have taken into account first of
all the scientific value of the contribution, but have also attempted to
provide a picture of the level of activity in the various countries represented. Clearly the volume does not claim to be exhaustive in its coverage,
as would be a textbook or a monograph, but rather it constitutes a colourful
and clear mosaic of important developments and trends in the subject.

The first section, comprising 19 contributions, is devoted to measurement, instrumentation and prosthetics, and opens with a consideration of the
philosophy of microcomputer application to biomedical instrumentation
which is an issue of current importance. This is followed by a series of
chapters illustrating measurements of variables and parameters which are of
clinical importance. Amongst the highlights of this section are a number
of new methods applied in cardiography. There are also examples of contemporary issues in prosthetics at both the organ and tissue levels. The
final part of this section is devoted to new types of measurement made
possible by technological advances in the application of biomagnetic fields.

The second section, with 12 chapters, deals with signal and image processing in the classical sense embracing ECG and EEG signals as well as
issues of medical images and imaging. However, all the important trends
in these domains are illustrated by, for example, ECG late potential detection, non-linear analysis applied to the EEG as well as the mapping of these
signals. Amongst the contributions covering two-dimensional signals, the

presentation of dynamic effects is discussed together with practical and theoretical problems relating to computer tomography.

The 7 chapters forming the section devoted to modelling and simulation cover a range of methodological issues in modelling including those of minimal representation and non-linear identification. The examples of modelling and simulation presented show the importance of such approaches in clinical practice as well as in biomedical research. Some of the philosophical issues involved are also discussed.

The large number of contributions (18) belonging to the section concerned with decision support correctly reflects the increasing extent to which techniques drawn from artificial intelligence are finding application in biomedical measurement systems. This section is introduced by a chapter which illustrates the way in which the information yielded by clinical instrumentation to the decision maker can be enhanced by the appropriate incorporation of knowledge-based systems and dynamic modelling. The increasing role of expert systems is illustrated by a number of chapters dealing with both practical and theoretical issues. Also treated are clinical database systems which are pre-requisites for optimal treatment as well as for the acquisition of knowledge necessary for the building of expert systems.

We should like to thank all the authors who have responded to our invitation in a positive and timely manner. We are also very much indebted to Audrey Cackett for her patience and for the masterly way in which she transformed the original manuscripts into the final camera-ready pages. Our thanks also go to Madeleine Carter, Ken Derham and their colleagues at Plenum who have encouraged us in this project and have exercised the necessary tolerance and forbearance vital for such an international venture to succeed.

January, 1988   Ewart Carson
                                                                 Peter Kneppo
                                                                 Ivan Krekule

CONTENTS

PART 2 SIGNAL AND IMAGE PROCESSING

PART 1

MEASUREMENT AND INSTRUMENTATION

A FAMILY OF MICROCOMPUTER ELECTROCARDIOGRAPHS

I. K. Daskalov

Institute of Biomedical Engineering
Medical Academy
Sofia, Bulgaria

INTRODUCTION

Since the development of our first fully computerised electrocardiograph in 1980-81, a family of four types has been designed and put into
production now reaching more than 5000 units. The concept adopted at that
time and published later (Dotsinsky et al., 1985) involved a multichannel
input amplifier in order to obtain synchronous signals from all the 12 ECG
leads, an A/D converter/multiplexer and a dedicated microcomputer system
with its microprocessor, RAM, EPROM, peripheral adaptors, keyboard and a
thermal-dot printer-plotter.

The selection of a high-resolution thermal-dot printer as a recorder
for the electrocardiographs proved to be a very important decision. It
permits the easy mixing of graphics with text and facilitates high-fidelity
recordings. Thus the standard heated-stylus recorders with their inherent
distortions, fragility and need of frequent servicing became outclassed.

The use of a microcomputer system naturally leads to the desire to
implement more and more functions in addition to the basic functions of
signal acquisition, filtering and recording. The solution could be either
to use a very fast and powerful µC system to deal with all the tasks, or to
adopt the concept of a multiprocessor system. It was decided that the
first solution leads to more complicated programming, especially in view
of the need for continuous improvements and the adding of new functions.
Therefore, we implemented the multiprocessor concept in the family of ECG
machines, where an input module deals with signal acquisition and filtering,
a central module takes over with data processing and analysis and an output
module is used for data buffering and controlling of the printer/plotter.
The modules are interconnected using serial lines.

The family has now 4 members (some with several versions), all of
them built according to the general concept described above. They acquire
8 input signals synchronously and simultaneously and use them to derive
the 12 routine ECG leads (I, II, III, aVR, aVL, aVF, V1-V6). This makes
these ECG machines 12-channel ones from the point of view of signal acquisition and memory, meaning that all the ECG waves acquired represent the
same cardiac cycles, as it would be with a full 12-channel recorder. The
four types of electrocardiographs differ mainly by the means of recording
and output data presentation, as well as by the software for analysis,
parameter measurement and automatic interpretation.

The first member of the family, started as a "single-channel electrocardiograph", now makes three-channel records of the peripheral leads and
two-channel ones of the precordial leads, using a 60 mm paper-width thermal
printer; therefore we call it "3/2-channel" ECG. The A/D conversion is
with 250 Hz and 8-bit. Its software includes (in several versions): ECG
wave detection, identification and measurement, basic parameter measurement
and tabulation, full parameter set tabulation, ECG-screening interpretation,
Frank leads computer derivation from the 12-leads and automatic morphological ECG interpretation. The hardware versions include rechargeable-accumu-
lator and mains-powered ones, flat keyboard simple push-button controls,
automatic trace-centring, signal quality control including patient electrode-
lead impedance, defibrillator protection, and so on.

The second electrocardiograph has the same possibilities, except for
the use of a 110 mm paper-width thermal printer (8/10 pts per mm). It can
accept direct Frank-leads or derive them from the 12-leads and produce
vectorcardiographic loops (normal and expanded) or polarcardiographic
tracings. There is a stationary version with scope outputs, experimental
inputs, stress-test software and a portable one which is the size of an
ordinary single-channel electrocardiograph. The A/D conversion is with
400 Hz and 10-bit.

The third member of the family is a true 12- or 15-channel ECG machine.
It makes use of a 215 mm paper-width thermal printer with a resolution of
8/10 points per mm. The recordings are made across the paper so that all
the 12 or 15 leads (50 mm paper speed) are presented in a one-page format.
The software functions are identical to those mentioned above.

The fourth instrument is a much larger electrocardiology computer
system, using the same input module as the other machines but with A/D conversion of 400 Hz and 12-bit. It is based on the well-known IBM-PC/XT/AT
professional personal computers and their compatibles. Its possibilities
include those of the family-members with some refinements, as well as the
storage of ECG databanks on disk, comparison with old ECG records and a
multitude of new functions developed, in progress and planned.

INPUT MODULE

The input signals are taken from the following electrodes: right
arm (R), left arm (L), chest (precordial positions) C1, C2, C3, C4, C5 and
C6, all with respect to the left leg (L) electrode signal. A "driven
right-leg" circuit is used, according to Levkov (1982), allowing the use
of non-differential amplifiers for each of the 8 input channels (Fig. 1).
These input buffers (A1 - A8) have an amplification factor usually in the
range 15 to 25.

Low-noise operational amplifiers should be used in the input stages
(A1 - A8). Concerning the choice, LM308 has about 10μVpp of noise referred
to the input, for the circuit shown in Fig. 1, with a bandwidth of up to
150 Hz, and TL072 could go down to 5μV. Better noise figures can be reached
with more expensive operational amplifiers. A9 should be a wideband stage,
in order to be able to drive the right leg (N) potential adequately over
the entire frequency band, so that the left leg potential always equals that
of the floating reference zero voltage. LF356 can be a good selection. It
should be powered by a voltage of about ±8V or more, if there is a need for
use of high-value resistors in the F and N leads. This requirement is not
so strong when other measures for electrical safety are taken, for example
making the whole module "floating" by powering it from an isolated transformer and using optocouplers for the communication lines with the central
module, as is the case with our input module.

4

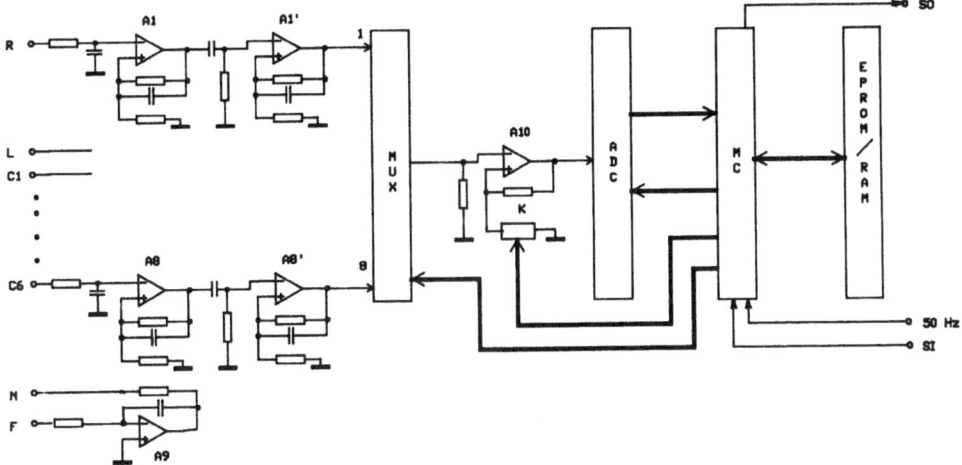

Fig. 1.   Block diagram of the isolated input module.

The input stages are followed by the traditional RC time constant
which should be not less than 3.2 s for ECG diagnostic units.   This relatively large value requires a follower, due to the need of a higher resistor.   The followers could also be wired as amplifiers (A1' - A8') in order
to avoid any multiplexer noise problems, with an amplification factor in
the range 35 - 65.

Another solution could be the use of very high capacitances (for instance tantalum capacitors) and no more than 1kΩ resistors in the RC circuit, allowing a standard multiplexer to be used, such as a 4051.   In this
case the A10 amplifier should be a low-drift one and of a relatively wideband type (OP-07, OP-05, LHO044A).   This stage can also be used for step-
gain selection in some versions.   If an 8- or 10-bit A/D converter is
selected (for economy only), its input dynamic range is not sufficient for
the acquisition of all possible input ECG signals.   The gain can be varied
in one or several "x ½" steps by accurate and stable resistor dividers and
switches (K in Fig. 1), driven by the microcomputer when it senses input
signal saturation.

The functioning of the input module is entirely controlled by the
microcomputer (MC in Fig. 1).   A single-chip HD6303 is normally used, connected to the central unit by serial input (SI) and serial output (SO) lines
(Fig. 1).   The module has its own EPROM and RAM.   The latter serves as a
buffer memory, as well as for storing some programs transferred from the
central unit.

The entire module as shown in Fig. 1 is isolated from the main system
by the use of a powering transverter and 4 kV optocouplers in the SI, SO
and 50 Hz connecting lines.

The basic functions of the module are: A/D conversion and input signal
dynamic range adaptation, 50 Hz interference elimination, drift filtering
and communication with the main unit.

The A/D conversion is with different sampling rate and resolution,
depending on the type of electrocardiograph: 250 Hz 8-bit, 400 Hz 10-bit
and 400 Hz 12-bit.   One important point here is that the conversion frequency has to be synchronised with the mains frequency, which is necessary
for accurate 50 Hz interference elimination.   Another point is the completion
of the A/D conversion and multiplexing procedure for all the 8 input

5

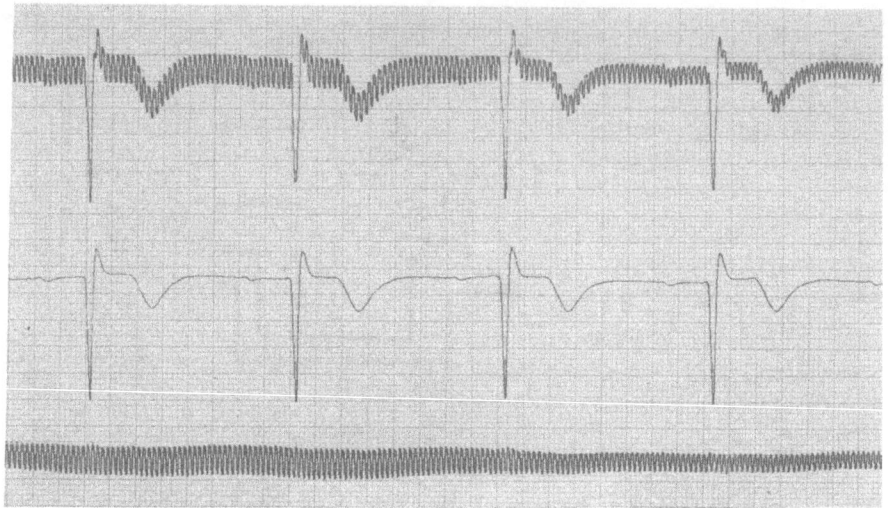

Fig. 2. An example of strong 50 Hz interference eliminated by the special
subtraction procedure. The baseline drift is filtered too.

channels within less than 0.4 ms per cycle, thus avoiding the use of sample
and hold circuits. Our experiments showed that phase errors due to multiplexing without sample and hold circuits began to lead to distortions of
the recorded signals with phase differences in the range of 0.7 - 1.0 ms.
Such minimal distortions can be visually detected only in the vectorcardiographic loops.

The 50 Hz interference elimination procedure has been described in
detail elsewhere (Levkov et al., 1984). It implements a very powerful
method, preserving entirely the original signal shape, based on the measurement of the interference and its subsequent subtraction from the mixed
signal. An illustration of the possibilities of this method is given in
Fig. 2. Baseline drift suppression is also shown. Recently the speed of
the procedure was increased several times by Christov and Dotsinsky (1988)
so that a real-time implementation with a 6303 or 6809 microprocessor
became possible.

The baseline drift filtering has been implemented by a first-order
high-pass digital filter, applied forward and backward on the sampled
signal data. An example of the filtering is shown in Fig. 3. Another
approach to this problem is to reduce the traditional value of the time-

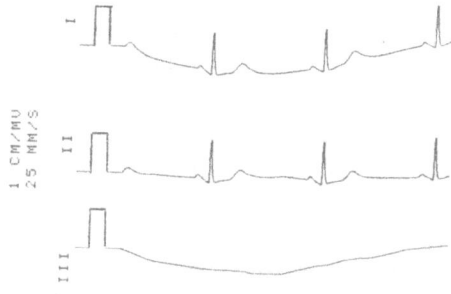

Fig. 3. Possibilities of the baseline
filtering procedure.

6

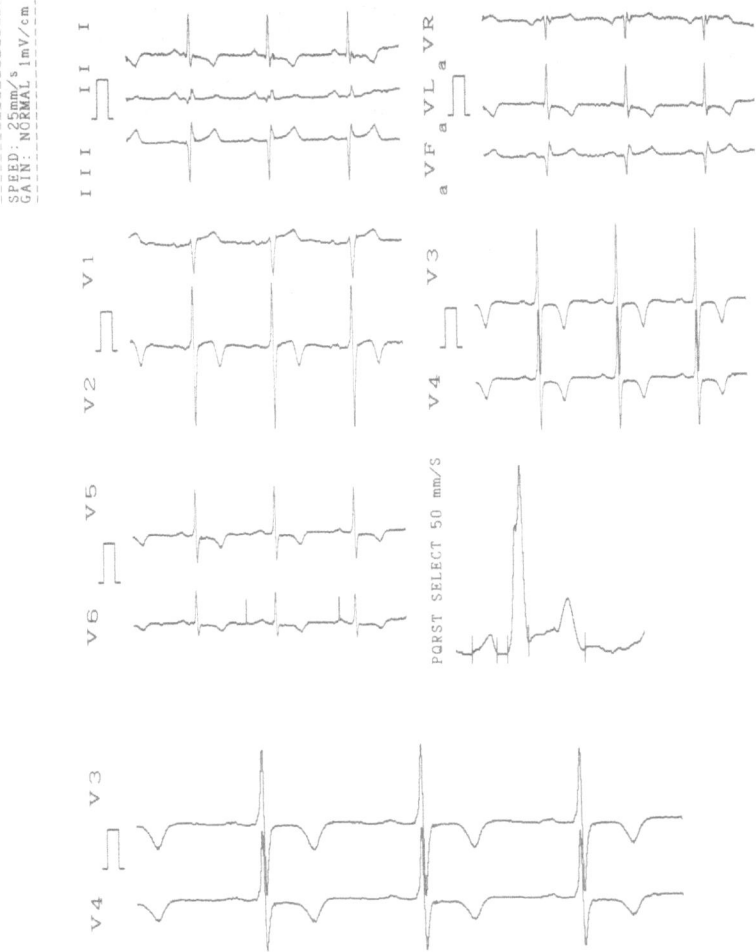

Fig. 4. A record of the 3/2-channel electrocardiograph on 60 mm
wide paper. The record is cut and arranged in 4 rows.
The synthesised lead is on the 3rd row. The V3 - V4
record is repeated on the 4th row with a paper speed
of 50 mm s^{-1} for better observation of possible trace
crossings.

constant from 3.2 s to about 0.3 - 0.4 s and then to apply the digital filter
backwards only, in order to compensate for the phase error from the analog
RC filter. In this case, attention should be paid to ensuring that the
exact RC time constant value corresponds to the respective digital filter
coefficients. Such a solution has the advantage of leading to much
shorter amplifier transients with input overload, occurring, for example,
after defibrillation or in some monitoring situations.

THE MAIN PROCESSING AND OUTPUT FUNCTIONS

The basic mode of functioning of the electrocardiographs is not a
real-time one. In our opinion, it is preferable to take an epoch of the
input signals and, making good use of the possibilities of the microcomputer system, to process it in a few seconds and then plot an extremely high

7

quality multichannel synchronous electrocardiogram, rather than to write in a real-time mode and obtain an asynchronous record with no guarantee of quality. Of course, a real-time mode is available, to be used in emergencies (defibrillation, rhythm monitoring, and so on), but not as a basic one. The epoch lengths normally used by us are between 3 and 6 s, except in rhythm recording/analysis modes, where 20 to 50 s epochs are acquired and stored. In the real-time mode, up to 3 synchronous leads can be recorded. In the store mode, a single lead or two leads can be recorded in two or three traces.

The data transferred from the input module to the central one are arranged into 8 RAM buffers, storing about 3 to 6 s (depending on the type of electrocardiograph) epochs of input signal samples. These data are used to compute the 12-lead signals from the 8 buffered ones, as described earlier (Dotsinsky et al., 1985). This simple procedure avoids the use of analog summing and subtracting circuitry with limited accuracy and at the same time significantly reduces the quantity of input data. Further, it was decided that the "standard" paper widths used in the 1, 3 and 6-channel ECG machines, respectively 50 – 60 mm, 100 – 120 mm and about 200 mm, could take more than the 1, 3 and 6 traditional traces, if the microcomputer system was used to arrange the records according to their amplitudes. Following this strategy, our "3/2-channel" electrocardiograph plots the peripheral lead traces in two groups of three and the precordial lead traces in three groups of two (Fig. 4), on 60 mm paper. Each trace occupies a part of the paper width depending on its amplitude. Also, having in mind that the input signals are synchronously acquired, the records can be arranged one below the other and thus a full 12-channel record obtained.

High-amplitude signals can cross each other without loss of information, as can be seen in Fig. 4 and still better in its bottom traces, where the V3 – V4 leads were repeated with a paper speed of 50 mm per second.

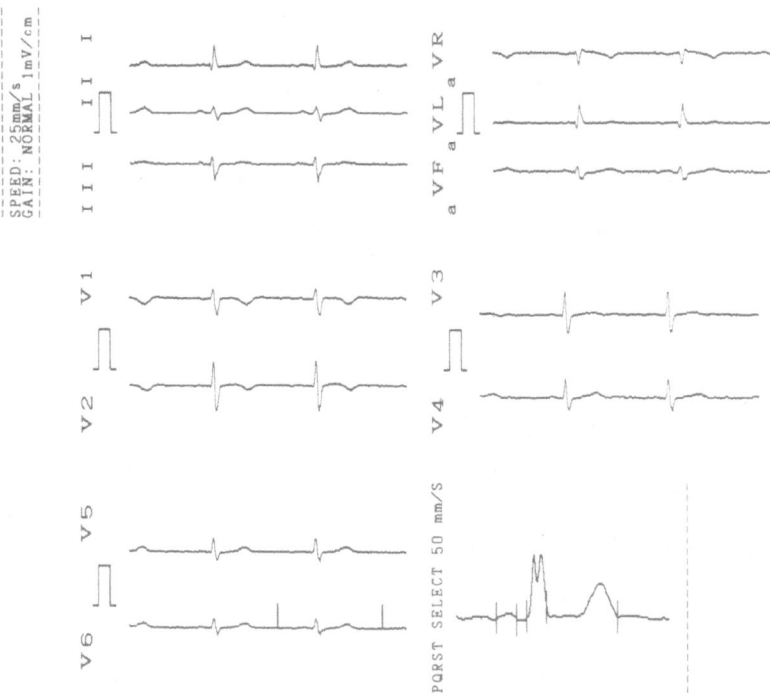

Fig. 5.   Enhancement of P and T waves in the synthesised lead.

In some cases the ECG amplitudes in the precordial leads could even reach 10 mV and then the microcomputer switches down the gain to 2 mV per cm.

The program then proceeds to the selection of a representative PQRST-complex for subsequent measurements, having previously detected all the valid QRS-complexes. The PQRST-complex selection is based on analysis of the signal quality in all the 8 input buffers. Then two types of special "leads" are synthesised by my colleague Pundjev from samples around the selected complex. One of them is a sum of several input lead data and their derivatives, designed for better delineation of the beginning and end of the QRS complex. The other is composed of weighted input leads and is designed for the enhancement of the P and T waves. Its effect can be seen from the record in Fig. 5, where practically no P waves can be seen in most of the leads, but a P wave appears in the composed lead. Some special algorithms are applied on these two synthesised leads in order to locate accurately the beginnings and ends of the P wave, the QRS complex and the end of the T wave. The P and T enhancement lead is shown on the records in order that the operator can verify the accuracy of the detection and decide whether or not to rely on the data further computed from the detected points (Figs. 4 - 7).

Using the detected points data, the program computes and prints out a "data report", containing the basic ECG parameters such as the heart rate, the PQ interval, the P duration, the QRS duration, the QT interval and the corrected QT interval, the QRS vector axis and the T vector axis in the frontal plane (Fig. 6).

The software package could include a program for the screening analysis of the electrocardiogram. It takes into account the wave amplitudes in II, III, V1 and V5 leads, as well as the Q/R, T/R and S/R ratios, the parameters from the data report and the Sokoloff-Lion criteria for left and right ventricular hypertrophy (Tomov et al., 1977; Matveev et al., 1977). The output of this program is either a statement "no abnormal findings", a table of the parameters found to exceed their respective normal limits or statements related to ratios and other combined parameters (Fig. 6).

Another program of the package makes the measurement and tabulation of all the wave amplitudes in all the 12 leads, taking into account negative P and T waves too (Fig. 7). These accurate measurements proved to be very useful in the interpretation of borderline or difficult cases.

Having in mind a certain interest in the orthogonal leads and the eventual need for vectorcardiographic and polarcardiographic recordings, we included in our ECG instruments the acquisition and recording of the Frank leads by relocation of the electrodes according to the well-known Frank positions. Moreover, we decided to try and develop a method for computed derivation of the Frank leads from the 12 leads. Thus we could obtain automatically the orthogonal leads without relocation of the electrodes, avoiding additional positioning errors and saving examination time.

The derivation was accomplished using the technique of Levkov (1987) and very good results have been obtained with comparison of direct and computed Frank leads. Now the program for the automatic computation of the Frank leads is included in all the ECGs of our family. In the 3 and 12-channel ECGs the computed Frank leads are also used for obtaining the vectorcardiograms and polarcardiograms, preserving, of course, the possibility to take-off the signals in the classical way. An example of original and computed Frank lead recordings can be seen in Fig. 8.

The good accuracy of the synthesised Frank leads stimulated us to try

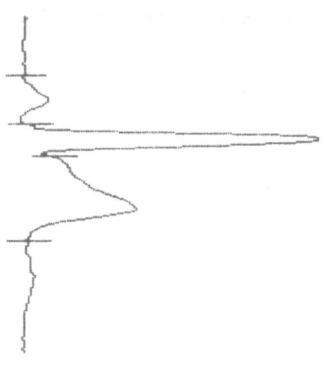

PQRST SELECT 50 mm/s

```
-------------------
DATA   REPORT
-------------------

HEART RATE          59 bPm

R-R INTERVAL    1016 mS
P-Q INTERVAL     156 mS
  P DURATION       -
QRS DURATION     104 mS
Q-T INTERVAL     384 mS

QTr              0.96

QRS AXIS         -20 deg
  T AXIS         120 deg

---------------------
SCREENING REPORT
---------------------
```

	II	III	V1	V5
Q	--	--	--	--
R	--	--	--	--
S	--	12	23	9
		40	40	
ST	--	--	--	--
T+	--	--	--	0
				20
T-	--	--	--	4
				40
Q/R	--	--	--	--
T/R	--	--	--	0
				01
S/R	--	--	--	--

```
RV5+SV1        39.60 mm
   T AXIS        120 deg
```

Fig. 6. Example of a data
report, containing
the basic ECG para-
meters and followed
by a screening report
table. The amplitudes
are in mV.10 units.

and implement an automatic interpretation system, using the orthogonal
leads and based on the work of Pipberger et al. (1982). The system is
attractive on account of its reduced number and simply taken parameters
and a well-defined set of most important electrocardiographic diagnostic
statements, including myocardial infarction with its localisation, left
ventricular hypertrophy, right ventricular hypertrophy and obstructive
pulmonary disease, ventricular conduction defects, fascicular blocks, QRS
axis deviations, J, ST, T and P abnormalities, and so on. Its diagnostic
accuracy is better than the more complicated Minnesota-code system, based
on the 12-lead electrocardiogram.

```
        -----------------
        DATA   REPORT
        -----------------

  HEART RATE          64 bPm

  R-R INTERVAL       937 mS
  P-Q INTERVAL       172 mS
     P DURATION      112 mS
  QRS DURATION        92 mS
  Q-T INTERVAL       384 mS

  QTr                0.99

  QRS AXIS            75 deg
    T AXIS            45 deg


        ----------------------
        WAVE   AMPLITUDES
             ( MV/10 )
        ----------------------

          I        II       III
  Q      .0        .0        .0
  R     3.6      10.6       8.0
  S     1.2        .8        .0
  ST     .6        .8        .2
  T+    4.2       5.6       1.6
  T-     .0        .0        .0
  P+     .8       1.8       1.8
  P-     .0        .0        .0


         AVR       AVL       AVF
  Q     6.4        .0        .0
  R     1.0        .0       9.4
  S      .4       2.4        .4
  ST   -0.8        .2        .8
  T+     .0       1.6       3.6
  T-    4.8        .0        .0
  P+     .0        .2       2.0
  P-    1.2        .8        .0


          V1        V2        V3
  Q      .0        .0        .0
  R     2.6       5.2       6.2
  S    12.0      16.0       7.6
  ST     .8       2.6       2.8
  T+    2.4      10.6      13.0
  T-     .0        .0        .0
  P+     .6        .6        .6
  P-     .0        .0        .0


          V4        V5        V6
  Q      .0        .0        .0
  R    19.2      17.2      11.4
  S     2.6        .8        .0
  ST    2.4       1.4        .8
  T+   13.2       9.0       5.4
  T-     .0        .0        .0
  P+     .6        .8        .8
  P-     .0        .0        .0
```

Fig. 7.   A table of amplitude
          parameters in the 12
          leads, given in mV.10
          units.

The interpretation system was implemented in an initial set of 10 ECG units and 300 ECG records have been made. 220 of them were examined independently by four leading electrocardiologists of the National Centre of Cardiovascular Diseases. After comparison of the results and a subsequent analysis of the discrepancies, slight changes to the original system were made, including correction of the QRS duration limit (from 120 to 128 for males and from 108 to 116 for females, which might be due to our method of duration measurement), discarding of the Rz amplitude as a parameter for right ventricular hypertrophy and excluding some of the overriding of several diagnostic statements by other findings.

This corrected system was included in the program package of the 3/2 channel electrocardiograph. An example of a diagnostic statement is shown in Fig. 9, the 12-lead ECG of this patient having already been given in Fig. 4.

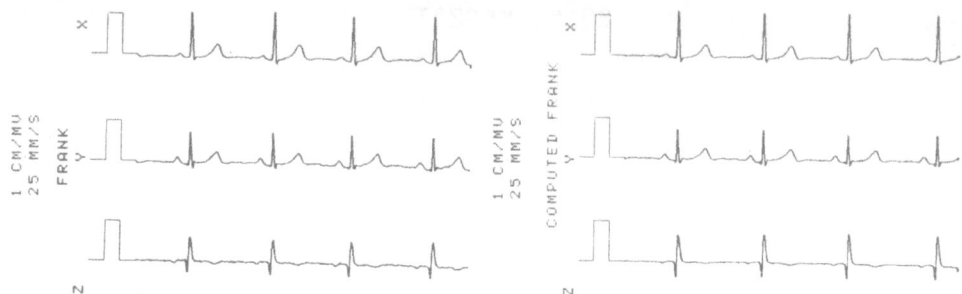

Fig. 8. An example of original and computed Frank leads, taken with the 3-channel electrocardiograph.

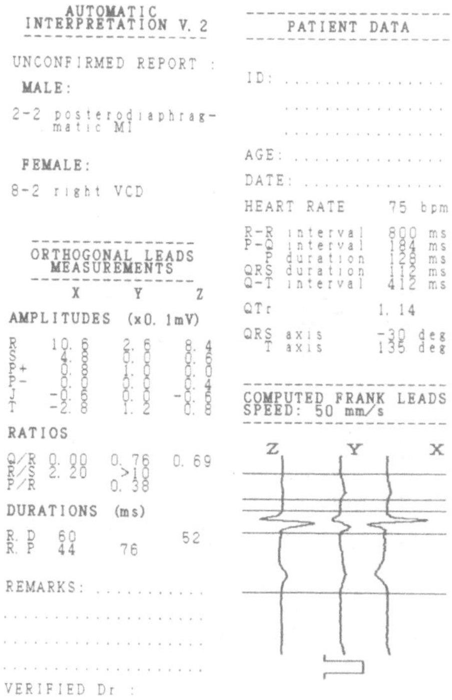

Fig. 9. ECG-diagnostic statements of
         the 3/2 channel instrument,
         with the computed Frank leads
         and their parameter table.
         This is a continuation of the
         record shown in Fig. 4.

The statement has a heading "unconfirmed report", meaning that a doctor's confirmation is always desirable. Therefore, a space for remarks and for the doctor's signature is left at the end of the record.

An example of a compressed (single PQRST) record is shown in Fig. 10, with a "no abnormal findings" statement.

Other interpretation systems are under consideration for an eventual implementation in our ECG family.

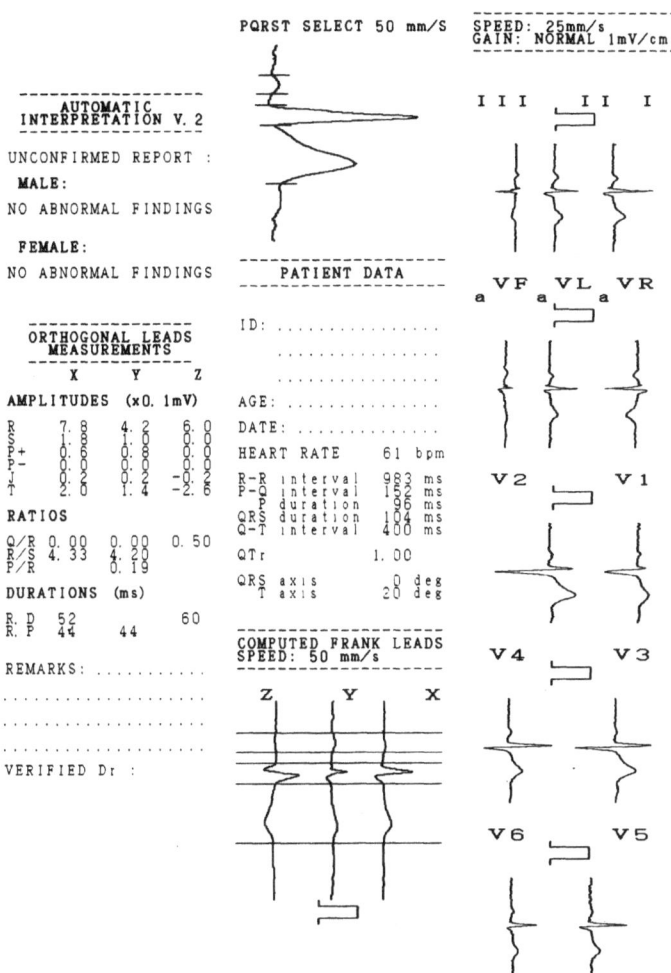

Fig. 10. A compressed (single-PQRST) record of
the 3/2-channel machine with the synthesised lead, wave identification,
data report, computed Frank leads and
diagnostic interpretation.

Some examples of records from the different ECG machines are presented
in the following figures. A polarcardiogram recording from the 3-channel
electrocardiograph is shown in Fig. 11, beginning with the magnitude of the
spatial vector and the subsequent frontal, horizontal and saggital plane
tracings of the plane vector and its respective longitude and latitude
angles. The vector loop tracings shown in Fig. 12 are taken with the same
machine.

A one-page 12-channel record from the 12/15-channel electrocardiograph
is shown in Fig. 13. Its continuation includes a data report, wave amplitudes table and the computed Frank leads (Fig. 14).

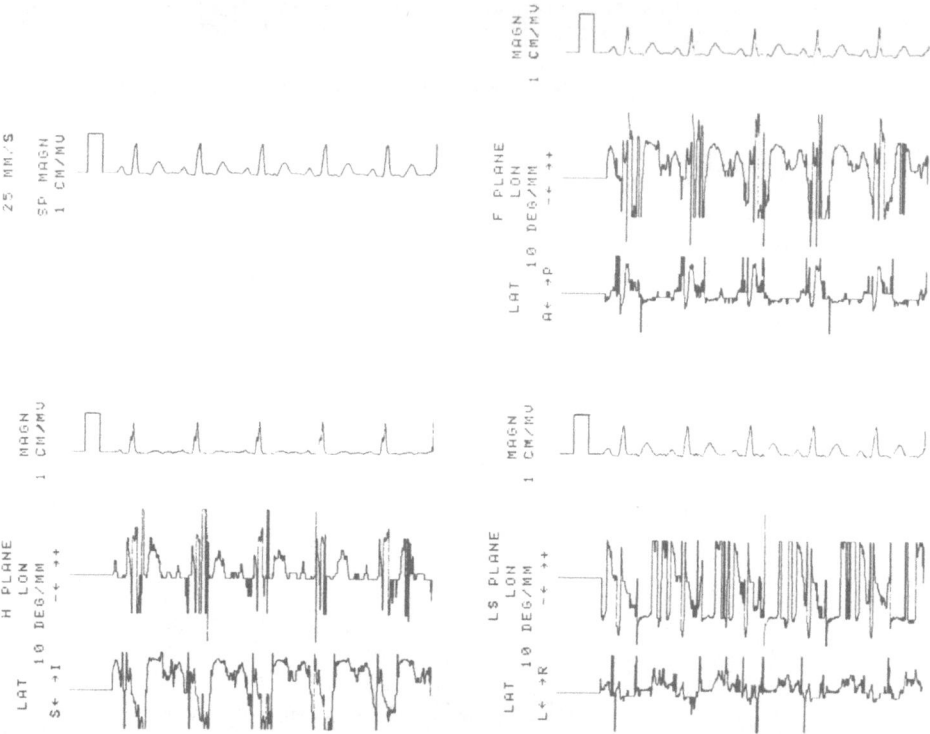

Fig. 11. A polarcardiogram, containing the spatial vector magnitude and the plane vectors with their latitude and longitude angles in the frontal, horizontal and saggital planes.

ECG COMPUTER SYSTEM

The last and most recent member of the family is an ECG computer system, designed for use in specialised cardiological departments, combining high-precision electrocardiography including ECG databanks, comparison with old records, polarcardiography and vectorcardiography, stress-test ECG, and so on. Of course, the system can combine sophisticated and routine investigations. It can also be used as a routine 12/15-channel high-performance electro/vector/polarcardiograph.

This system, called "ECG computer " is based on the most popular professional personal computers, IBM-PC/XT/AT types and their compatibles. It has its own floating-input module as described above, implemented as a PC board and inserted in a slot of the PC. The A/D conversion rate is 400 Hz with a resolution of 12-bit, where the LSB weight is 4.88μV. The PC system should have a RAM with a minimum of 512K, 2 floppy-disk drives or 1 floppy and 1 hard disk drive, a "Hercules" graphic (or compatible) add-on printed-circuit board and a high-resolution dot-matrix printer (1920 x 256 points) or a thermal dot printer (minimum 8 dots mm^{-1}).

The software has been written in assembly and FORTH languages by my colleagues Gramatikov and Lolov. It includes all the functions described above, with some improvements. The main menu is shown in Fig. 15. A part of a record can be seen in Fig. 16, with detailed wave-amplitude measurements. A Frank orthogonal ECG with its wave measurements is shown in Fig. 17.

14

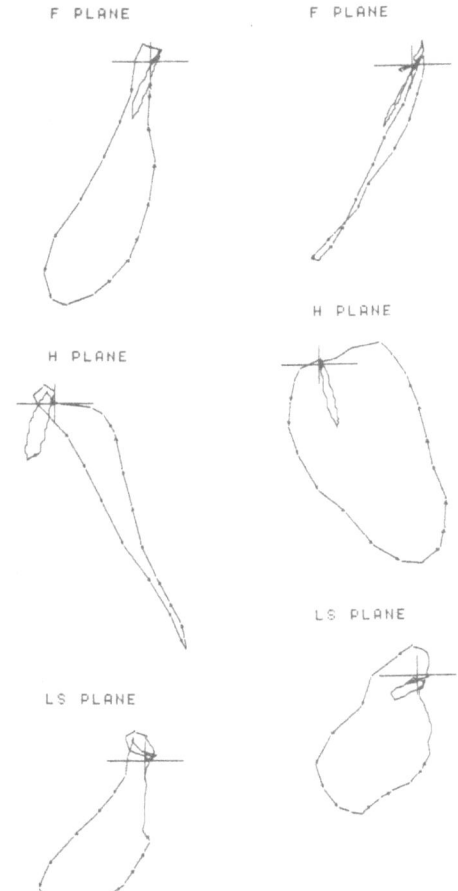

VECTOR LOOPS 4 CM/MV   VECTOR LOOPS 4 CM/MV

F PLANE   F PLANE

H PLANE

H PLANE

LS PLANE

LS PLANE

Fig. 12.   Examples of vectorcardiograms
           of the 3-channel machine. The
           loops should be observed with
           the axis across the paper,
           parallel to the headings, cor-
           responding to Y, Z and Y, res-
           pectively.

Vector loops with a gain of 12 cm per mV are shown in Fig. 18. The
P and T wave loops can be easily observed, as in some of our earlier works
(Daskalov, 1977). The most important VCG parameters are measured and a
table is printed below the loops.

An important feature of the system is its digitised ECG records databank, which is continually accumulated and updated. The records can be
easily retrieved, edited, deleted, compared with other records and restored.
The databank is an important basis for further electrocardiological investigations and improvement of the system itself and of all the family of electrocardiographs.

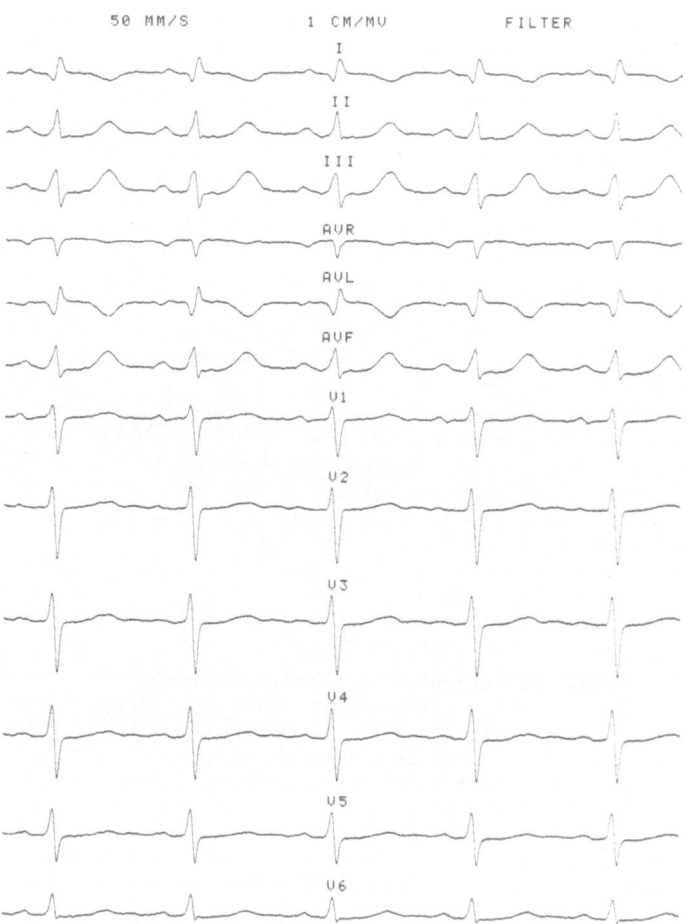

NAME . . . . ID . . . . DATE . . . . .

50 MM/S   1 CM/MV   FILTER

Fig. 13.   A 12-lead record of the 12/15-channel ECG in a
one-page format.

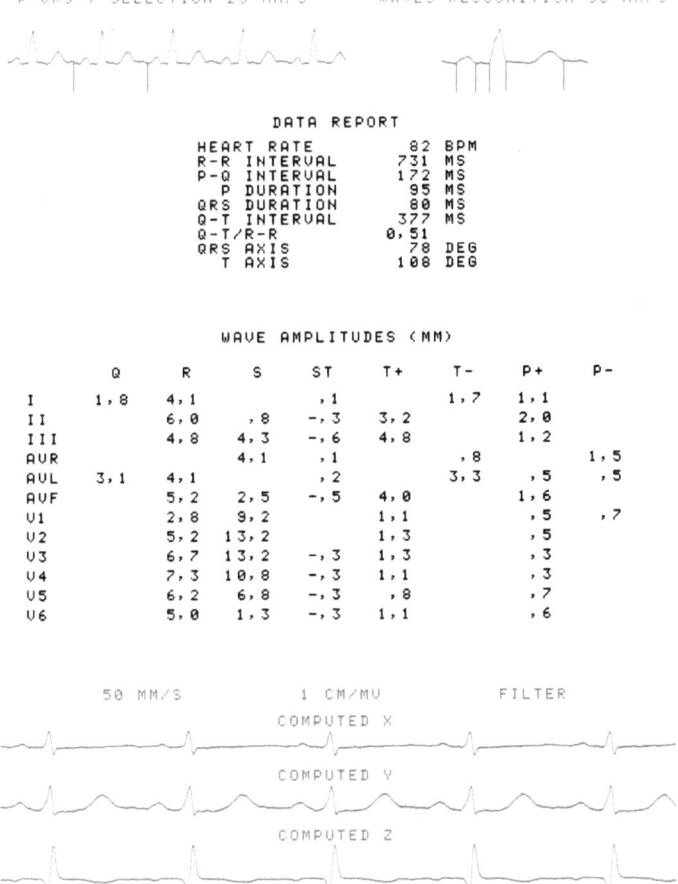

DATA REPORT

HEART RATE	82 BPM
R-R INTERVAL	731 MS
P-Q INTERVAL	172 MS
P DURATION	95 MS
QRS DURATION	80 MS
Q-T INTERVAL	377 MS
Q-T/R-R	0,51
QRS AXIS	78 DEG
T AXIS	108 DEG

WAVE AMPLITUDES (MM)

	Q	R	S	ST	T+	T-	P+	P-
I	1,8	4,1		,1		1,7	1,1	
II		6,0	,8	-,3	3,2		2,0	
III		4,8	4,3	-,6	4,8		1,2	
AVR		4,1		,1		,8		1,5
AVL	3,1	4,1		,2		3,3	,5	,5
AVF		5,2	2,5	-,5	4,0		1,6	
V1		2,8	9,2		1,1		,5	,7
V2		5,2	13,2		1,3		,5	
V3		6,7	13,2	-,3	1,3		,3	
V4		7,3	10,8	-,3	1,1		,3	
V5		6,2	6,8	-,3	,8		,7	
V6		5,0	1,3	-,3	1,1		,6	

50 MM/S   1 CM/MV   FILTER

COMPUTED X

COMPUTED Y

COMPUTED Z

Fig. 14.   The continuation of the record of Fig. 13 including the data
report, the wave-amplitude table and the computed Frank leads.

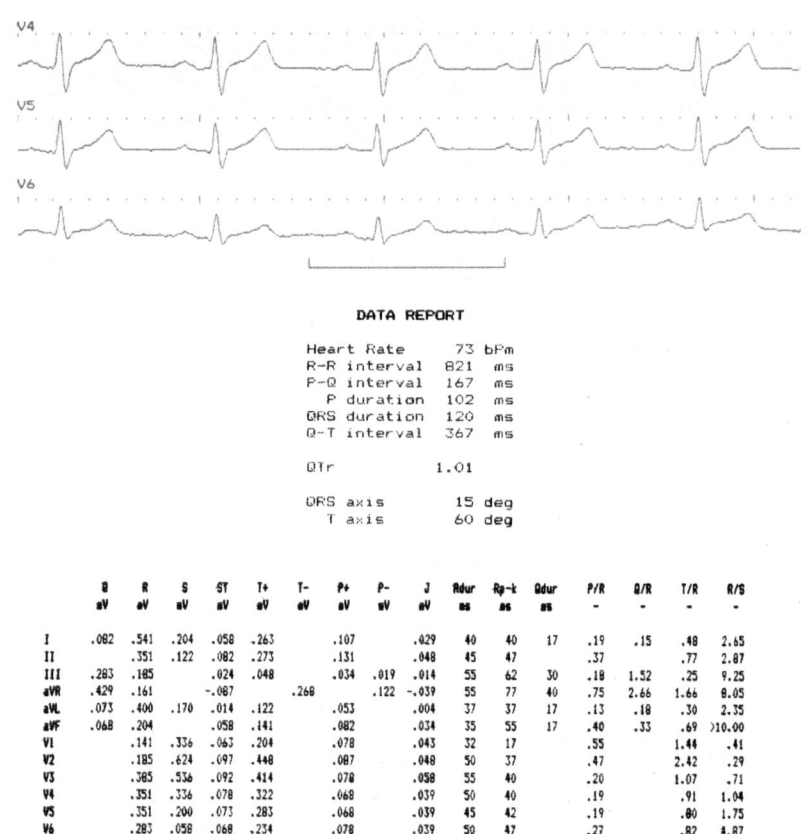

```
                    ECG - VCG  Computer System
           Institute of Biomedical Engineering, Med.Academy, Sofia

  New PATIENT Data     Typical QRS        PROGRAM 1        SENSIT.:  10  mm/mV
  Change PAT. Data     Waves RECOGNITION  PROGRAM 2                  20  mm/mV
  DATA Acquisition     12 Standard LEADS  PROGRAM 3                  40  mm/mV
  8 Raw Chann.DRAW     Computed FRANK     PROGRAM 4                 100  mm/mV
  SAVE Data on Disk    VCG                PROGRAM 5        SPEED  :  25  mm/s
  50 Hz Elimination                                                 50  mm/s
  30 Hz Filter                                                     100  mm/s
  DRIFT Filter         Disk UTILITIES     Continue PROGRAM


  Time 00:54   Date  23jun1987      Sensitivity  10  mm/mV   Speed  25  mm/ms

  Name:Rusy Panchev
  Age:37    Sex(M/F):m   Date:15APR1987     Time:14:35    ID No:8
  Remarks:Wet electrodes, chest 1 cm down


  Select desired function with ↑  and press ↵ to EXECUTE or h for HELP.
```

Fig. 15.   The main menu of the ECG-computer.

DATA REPORT

Heart Rate   73 bPm
R-R interval   821 ms
P-Q interval   167 ms
 P duration   102 ms
QRS duration   120 ms
Q-T interval   367 ms

QTr   1.01

QRS axis   15 deg
  T axis   60 deg

	Q aV	R aV	S aV	ST aV	T+ aV	T- aV	P+ aV	P- aV	J aV	Rdur ms	Rp-k ms	Qdur ms	P/R -	Q/R -	T/R -	R/S -
I	.082	.541	.204	.058	.263		.107		.029	40	40	17	.19	.15	.48	2.65
II		.351	.122	.082	.273		.131		.048	45	47		.37		.77	2.87
III	.283	.185		.024	.048		.034	.019	.014	55	62	30	.18	1.52	.25	9.25
aVR	.429	.161		-.087		.268		.122	-.039	55	77	40	.75	2.66	1.66	8.05
aVL	.073	.400	.170	.014	.122		.053		.004	37	37	17	.13	.18	.30	2.35
aVF	.068	.204		.058	.141		.082		.034	35	55	17	.40	.33	.69	>10.00
V1		.141	.336	.063	.204		.078		.043	32	17		.55		1.44	.41
V2		.185	.624	.097	.448		.087		.048	50	37		.47		2.42	.29
V3		.385	.536	.092	.414		.078		.058	55	40		.20		1.07	.71
V4		.351	.336	.078	.322		.068		.039	50	40		.19		.91	1.04
V5		.351	.200	.073	.283		.068		.039	45	42		.19		.80	1.75
V6		.283	.058	.068	.234		.078		.039	50	47		.27		.82	4.87

Fig. 16.   Part of an ECG record with a detailed parameter table.

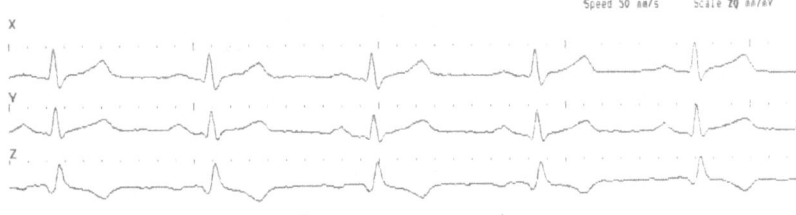

COMPUTED FRANK

Speed 50 mm/s   Scale 20 mm/mV

X

Y

Z

DATA REPORT

Heart Rate   73 bPm
R-R interval   821 ms
P-Q interval   167 ms
P duration   102 ms
QRS duration   120 ms
Q-T interval   367 ms

QTr   1.01

QRS axis   15 deg
T axis   60 deg

	Q mV	R mV	S mV	ST mV	T+ mV	T- mV	P+ mV	P- mV	J mV	Rdur ms	Rp-k ms	Qdur ms	P/R	Q/R	T/R	R/S
X		.331	.131	.058	.239	.068			.034	45	42		.20		.72	2.52
Y		.292	.068	.073	.219	.117			.043	42	52		.40		.75	4.29
Z	.087	.302		-.053		.165		.039	-.024	75	67	37	.12	.28	.54	>10.00

Fig. 17. Computed Frank leads and their parameter table.

VCG of: P Q R S T

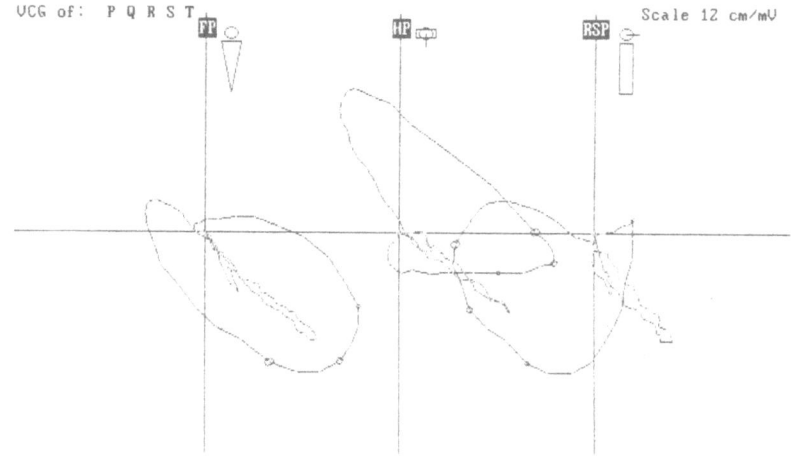

Scale 12 cm/mV

FP   HP   RSP

	FRONTAL		HORIZONTAL		SAGITAL	
	Modul mV	Angle deg	Modul mV	Angle deg	Modul mV	Angle deg
QR Vector ms						
10	.024	-169	.039	126	.034	-8
20	.029	-81	.087	87	.087	-18
30	.161	-5	.180	27	.082	-9
40	.366	25	.336	10	.165	68
50	.380	49	.248	-10	.292	99
60	.200	83	.229	-84	.302	139
Max Vector	.395	42	.336	10	.312	149

Fig. 18. Amplified vectorcardiograms, showing the T and
          P loops and a table with the initial QRS-vectors
          and the maximal vector measurements.

19

REFERENCES

Christov, I., and Dotsinsky, I., 1988, A new approach to the digital elimination of 50 Hz interference from the electrocardiogram, Med. Biol. Eng. Comput., 26:(in press).

Daskalov, I., 1977, Signal processing for vectorcardiograph, in: "Digest 1st Mediter. Conf. on Med. and Biol. Eng., Sorrento, Italy", vol. II: 8.

Dotsinsky, I., Christov, I., Levkov, C., and Daskalov, I., 1985, A microprocessor electrocardiograph, Med. Biol. Eng. Comput., 23 : 209.

Levkov, C., 1982, Amplification of biosignals by body potential driving, Med. Biol. Eng. Comput., 20 : 248.

Levkov, C., 1987, Orthogonal electrocardiogram derived from the limb and chest electrodes of the conventional 12-lead system, Med. Biol Eng. Comput., 25 : 155.

Levkov, C., Michov, G., Ivanov, R., and Daskalov, I., 1984, Subtraction of 50 Hz interference from the electrocardiogram, Med. Biol. Eng. Comput., 22 : 371.

Matveev, M., Tomov, L., Tomov, I., and Daskalov, I., 1977, Lead selection for ECG screening, Adv. Cardiol., 19 : 147.

Pipberger, H. V., Simonson, E., Lopez, E., Araoye, M., and Pipberger, H. A., 1982, Circulation, 65 : 1456.

Tomov, L., Tomov, I., Matveev, M., and Daskalov, I., 1977, ECG parameters for screening analysis, Adv. Cardiol., 19 : 150.

A PERSONAL COMPUTER-BASED SYSTEM FOR CARDIAC ELECTRIC FIELD INVESTIGATION

IN THE CLINIC AND IN RESEARCH

M. Tyšler, V. Rosík and M. Turzová

Institute of Measurement and Measuring Technique
Electrophysical Research Centre
Slovak Academy of Sciences
Bratislava, Czechoslovakia

INTRODUCTION

The results of using a special minicomputer-based system EKKG-80 for measuring and modelling the cardiac electrical field (Tyšler et al., 1984) showed that it would be desirable to have some more convenient system for clinical studies and, at the same time, for additional studies in electro-cardiological research. After some analysis of the clinical needs we decided to develop a personal computer-based system which enables different methods of the heart state analysis to be used, such as classical ECG and

Fig. 1. The KARDIOMAT system.

VCG recording methods, computerised ECG and VCG classifications and also some newer methods based on body surface potential mapping.

The KARDIOMAT system is based on a PDP-11 compatible professional personal computer with an ECG measuring sub-system specially designed to fulfil the safety requirements for measurements on human subjects. The system enables measurements to be made of 12-lead ECG, Frank VCG and 24-lead ECG for mapping. The modular software includes packages for the input of patient personal data, ECG measurements and body surface potential mapping. Additional PDP-11 software for ECG analysis can also be run.

TECHNICAL DESCRIPTION

The system consists of two basic parts: an analog multichannel measuring unit and the microcomputer.

The measuring unit amplifies electric signals of the human heart activity (ECG) picked up from the body surface with the required accuracy for their digitisation and processing by the microcomputer. A block diagram of the analog multichannel measuring unit is given in Fig. 2.

Electronic circuits of the unit are segmented into several pluggable modules. This structure enables effective adjustment of the hardware to the demands of particular measurements.

The limb leads module KZ-1 picks up signals R, L, F, (N) from the limbs and implements standard limb leads I, II, III, aVR, aVL, aVF, along with the signal of the Wilson central terminal W. In addition, the module contains a circuit for active neutralisation through the right leg (N) with a voltage gain $A = -100$ and active shielding of lead wires of the patient cable ($A = 0.99$).

The chest leads module HZ-1 picks up and processes signals from the chest leads. Each module contains 8 measuring channels. Parallel combination of these modules enables the system to be extended up to 64 channels (leads).

Each measuring channel of the modules KZ-1 or HZ-1 is realised as an input buffer amplifier with a voltage gain of 40, followed by a differential 3-OPA instrumentation amplifier with a voltage gain of 25, so the resultant voltage gain is 1000. These modules also contain electronic circuits indicating bad contacts of the sensing electrodes.

Signals from the KZ-1 and HZ-1 modules are fed to several controlled amplifier/multiplexer modules MP-1. Each MP-1 module contains a 16-channel amplifier with digitally controlled gain in 4 steps (0.5, 1, 2, 3) and a sample/hold circuit. The outputs of all channels are connected to the 16-channel analog multiplexer, which forms the output of the module. The outputs of all MP-1 modules are connected and fed to the A/D converter. The analog multichannel measuring unit is characterised by the following main technical parameters:

number of measuring channels	32 (8 + 24)
extension	up to 64 channels
input resistance for any lead	minimum 1000 MΩ
discrimination factor (without active neutralisation)	minimum 100dB
frequency range (selectable)	0.05 - 1000 Hz
	0.05 - 100 Hz.

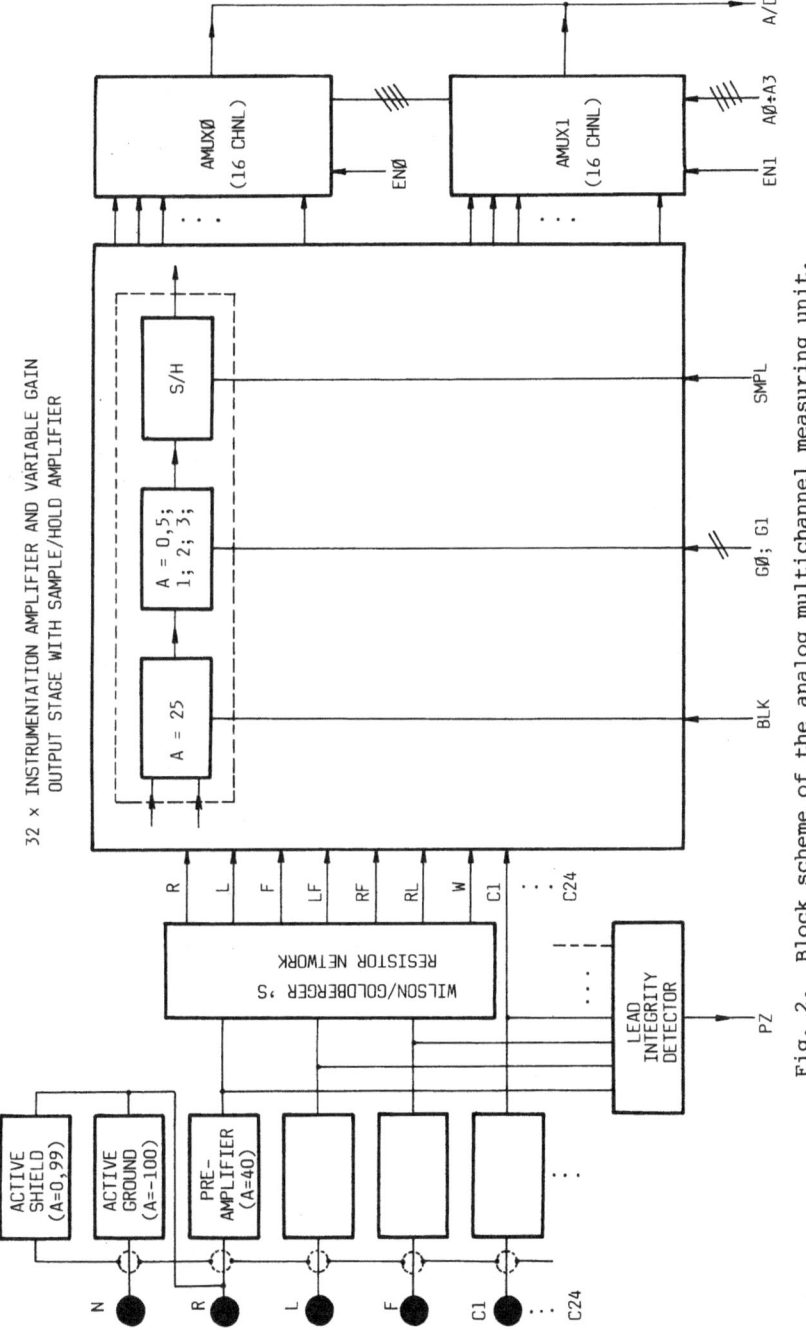

Fig. 2. Block scheme of the analog multichannel measuring unit.

The microcomputer is based on a PDP-11 compatible 16-bit processor, with memory up to 248 kB, cache memory and optional floating point processor. The peripheral devices include a keyboard with 82 keys (some of them are used as functional keys), a 12-inch colour monitor with a raster of 256 x 256 points and 8 colours, two 8-inch floppy disk drives (512 kB each) and a matrix printer with several character sets and 480, 640 or 960 graphic points per line.

The analog measuring unit is connected to the microcomputer using a standard laboratory peripheral system interface card. The microcomputer is based on modules connected to a Unibus-like bus. This also makes it possible to add other useful peripherals.

Mechanically, the device is designed as a compact mobile table, with the display, keyboard and printer on the table and all other modules under the table in two cabinets (the computer unit and the analog measuring unit). The inputs are connected to the measuring unit at the right side of the table.

FUNCTIONS IMPLEMENTED IN THE SOFTWARE

The user-oriented characteristics of the device are determined to a large extent by the application software implemented in the system. The application software running under the RT-11 operating system is modular and additional software packages can be implemented or developed. The following programs are already implemented in the device:

Measurement of the Standard 12-lead ECG

The program enables the input of patient personal data and simultaneous 12-lead calibrated ECG record 2.5 s long with a sampling rate of 2 ms (as required by the analysing program-Minnesota code). A basic check on the patient data and the status of the measuring sub-system is performed and visual checking of all the measured signals is also possible before continuing to the analysing programs.

Measurement of the Frank Vectorcardiogram

The program reads the patient personal data and calibrated three-channel VCG record (X, Y, Z) 10 s long with a sampling rate of 2 ms, as required by the modified Pipberger analysing program. A basic check on the patient data and the status of the measuring sub-system is performed and visual checking of all the measured signals is also possible.

Measurement of the 24-lead ECG for Mapping

The program enables the input of patient personal data and simultaneous 25-lead calibrated ECG record (one standard limb lead and 24 selected chest leads). The record is 1.2 or 2.4 s long using a sampling rate 2 or 4 ms. An elementary check on the patient data and the status of the measuring sub-system is performed and consecutive visual checking of all measured signals is also possible. This program also automatically preprocesses the measured signals (it determines the time instants of the QRS complex and the baseline, makes calibrations and estimates the noise level in the signals).

Graphical Output of the Measured Signals

This set of programs facilitates the display or graphical print-out of the measured 12-lead ECG, VCG and 24-lead ECG signals in selected scales and time intervals.

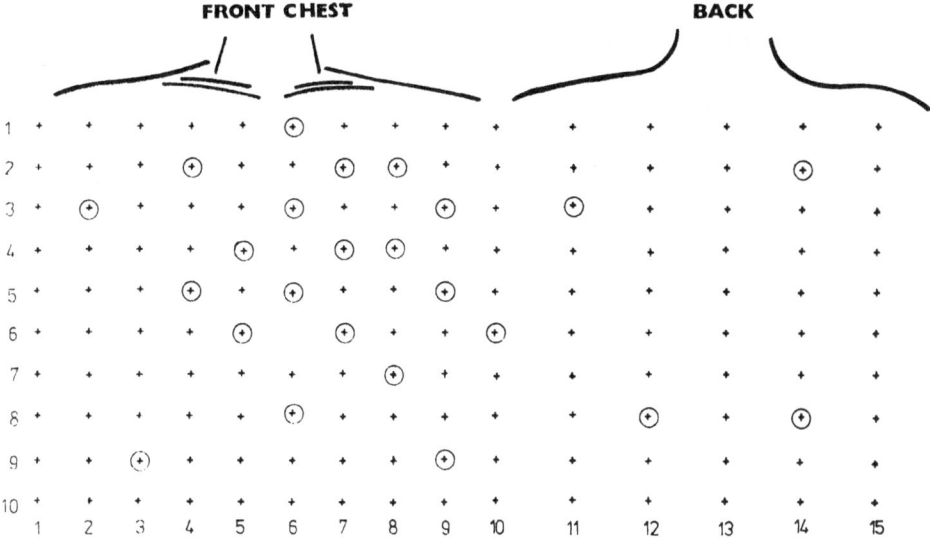

Fig. 3.   150 mapped positions and 24-lead set for mapping.

Computation of Body Surface Potential Maps

The use of a limited number of leads for estimating the total body surface potential distribution was suggested as a practical solution to the problems associated with the impracticability of the usual 100 - 300 lead mapping. In the KARDIOMAT system the 24-lead system for estimation of 150 body surface potentials, reported by Barr et al. (1971), is used. The positions of the 24 measured leads and 150 mapped points on the torso are shown in Fig. 3.

Body surface potential map frames can be computed for any selected time instant within the cardiac cycle according to the matrix equation

$$\underline{W} = \underline{H} \cdot \underline{U}_L \tag{1}$$

where $\underline{W}$ is the vector of 150 estimated potentials (optimally in the least square sense), $\underline{U}_L$ is the vector of 24 measured chest potentials, and $\underline{H}$ is the 150 x 24 transformation matrix common for all individuals and all time instants. $\underline{H}$ was determined on a large group of normals and patients (more than 1000). An example is shown in Fig. 4.

Integral Body Surface Potential Mapping

A quantity of detailed information about the cardiac electrical field included in the set of instantaneous maps makes medical interpretation difficult. Recently, the integral potential distribution on the body surface has been investigated for global evaluation of the heart state.

Integrals p_i of potentials $u_i(t)$ are evaluated over a selected time interval $<T_1, T_2>$ at each mapped site $i = 1, \ldots, N$, according to the formula:

$$p_i = \int_{T_1}^{T_2} u_i(t) \, dt \tag{2}$$

The program computes p_i by means of a trapezoidal rule, using the

25

MAPOVANIE NAPATI NA POVRCHU HRUDNIKA

kod merania: MP630L
rodne cislo: 63-10-11/0008
datum a cas: 12.12.86 00:00

komentar: DIAFRAG. IM (OKT. 86)

pocet meranych bodov: 24
mapovane body: 10 x 15
zvisly mapovany rozsah: 270 mm
pociatok casu voci max. R: -50.0 ms

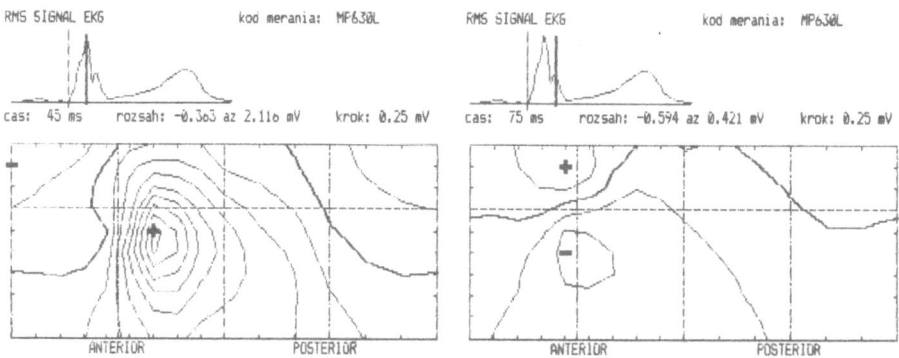

Fig. 4. An example of a body surface potential map printout.

IZOINTEGRALOVE MAPY POVRCHU HRUDNIKA

kod merania: MP630L
rodne cislo: 63-10-11/0008
datum a cas: 12.12.86 00:00

komentar: DIAFRAG. IM (OKT. 86)

pocet meranych bodov: 24
mapovane body: 10 x 15
zvisly mapovany rozsah: 270 mm
pociatok casu voci max. R: -50.0 ms

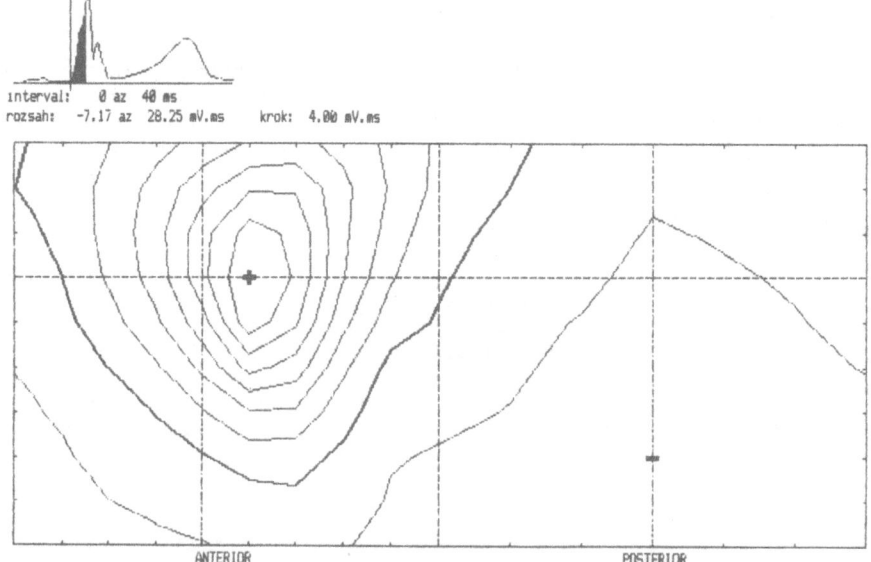

Fig. 5. An example of an integral body surface potential map.

150 site map frames, estimated from 24-lead ECG measurements (see Fig. 5).

Programs for Graphical Output of the Maps

These provide the display or graphical printout of the resultant potential and integral map frames in several modes (single map frame or group of them; using isolines or colour surfaces). All the maps are supplemented with basic identification text and parameter description.

Other Supporting Programs

These enable the measurement protocol, computed parameters and map tables to be printed out and also patient personal data to be stored in a small database, and the system to be calibrated and tested.

TESTING OF THE 24-LEAD MAPPING SYSTEM

Acceptability of the 24-lead mapping system for our population was tested on a group of subjects before implementation into clinical system for widespread use. Measured 150-lead and 24-lead estimated potential distributions were compared using two criteria - the correlation coefficient ρ between measured and estimated map frames and the relative mean square error ε (see Figs. 6 and 7).

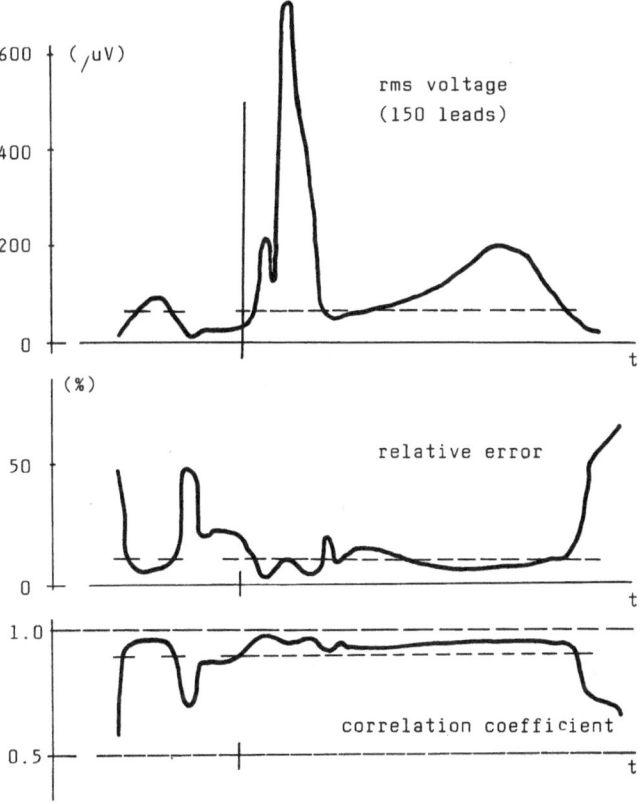

Fig. 6.   The variation of the 24-lead testing criteria during the heart cycle.

27

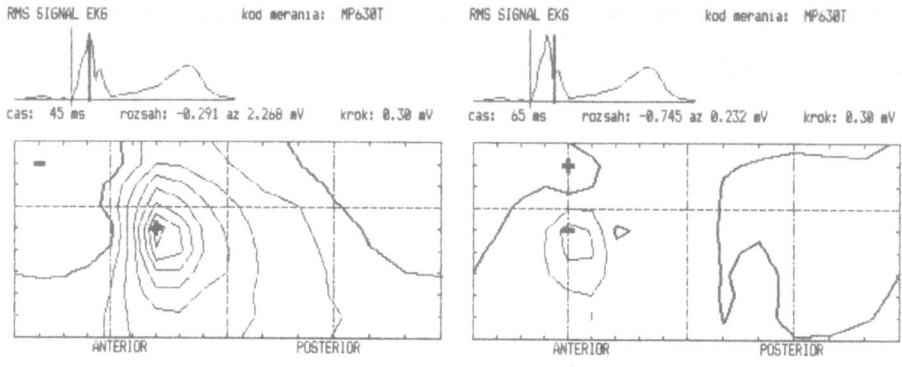

150 - LEAD MEASUREMENTS

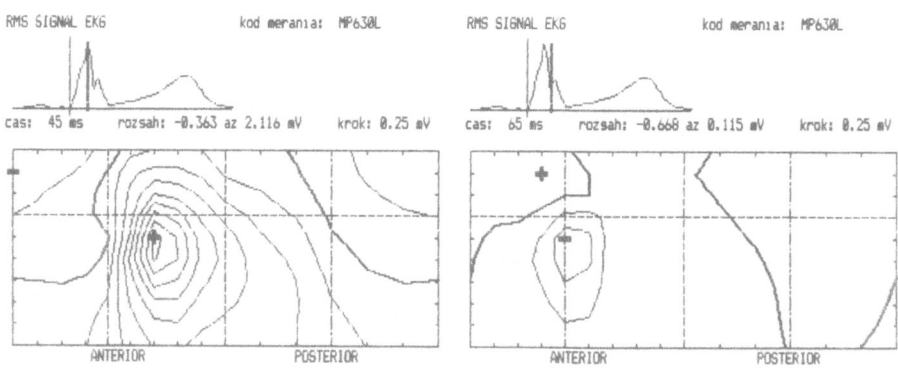

24 - LEAD MEASUREMENTS

Fig. 7. Comparison of isopotential map frames measured at l50 points
and estimated from 24 points.

Mean values of the criteria obtained for the QRS interval (ε = 5.7%,
ρ = 0.97) are comparable with values reported by Barr et al. (1971)
(ε = 5%, ρ = 0.97). For the T-wave interval the estimation is acceptable,
but the error is greater. The visual comparison showed good agreement between estimated and totally measured map frames.

DISCUSSION AND CONCLUSIONS

The aim of the work was to develop a practical measuring and processing
system for clinical electrocardiology and for additional studies in electrocardiological research.

The concept of the KARDIOMAT device based on a professional personal
computer, with the measuring unit specially designed for medical purposes
and completed with appropriate application software, makes the device a
versatile tool for electrocardiological research and clinical application.
Using the same hardware, other useful methods for patient examination or
monitoring might be added according to the demands of the users. Software
for modelling the electrical heart generator and the heart surface excitation
sequence based on the multipole model (reported by Tyšler et al. (1984)) is
under development. Implementation of the Minnesota code and the Pipberger

program for ECG and VCG classification is expected soon. The possibility of using the KARDIOMAT system in magnetocardiographic research is also being considered.

REFERENCES

Tyšler, M., Kužmová, J., Kneppo, P., Rosík, V., and Kričfaluši, M., 1984, Body surface mapping and analysis of human cardiac electric field by the automated system EKKG-80, in: "Electrocardiology 83. Proceedings of the 10th International Congress on Electrocardiology", Excerpta Medica, Amsterdam : 160.

Barr, R. C., Spach, M. S., and Herman-Giddens, G. S., 1971, Selection of the number and positions of measuring locations for electrocardiography, IEEE Trans. Biomed. Eng., BME-18 : 125.

TECHNICAL MEANS FOR CLINICAL INVESTIGATIONS OF BRAIN BIOELECTRICAL PROCESSES

S. G. Danko and Y. L. Kaminsky

Institute for Experimental Medicine AMS USSR
Leningrad
USSR

INTRODUCTION

It is one of the most difficult methodological problems to investigate real activities and state mechanisms in the living human brain. The reasons arise not only from its complexity, but also from severe medico-ethical limitations. The technical means for such investigations can be effective only if they are adequate to meet the numerous demands which follow from the medical and physiological tasks, taking into consideration real clinical conditions. The demands are particularly severe when dealing with patients who are treated with intracerebral electrodes implanted for diagnosis and treatment in accordance with proper medical considerations. Taking diagnosis and treatment as the main and most important goals, one should use the opportunities available for ethically admissible researches in pathology and normal physiology of the human brain (Bechtereva, 1978; 1980; Bechtereva et al., 1969). The relative importance of concrete features is different, as a rule, for a clinical instrument compared to a general scientific instrument. Nevertheless, one can obtain an acceptable compromise trying to achieve both relative initial simplicity of the means and the possibilities of increasing their computing and controlling potential when proper knowledge and experience are acquired.

Current methods of human brain neurophysiological investigation are based, first of all, on the consideration of all types of brain bioelectrical process (BBP) - the well-known electroencephalogram, its cortical and subcortical equivalents, slow and infraslow electrical processes down to steady potential, impulse neuron activity. One needs to analyse great quantities of data and to do so within a time that permits the use of the results in the treatment of this particular patient. It is also important to have the following possibilities: to know BBP in many brain sites (structures) at the same instant and over a long period of time; to have electrophysiological apparatus connected on-line to a computer; to apply adaptive control through real-time processing; to obtain and store BBP while a patient is in a number of different situations and locations; to display and record both primary processes and the results of their processing; and to have the means of achieving this with equipment which is sufficiently compact, simple and convenient in use, and not too expensive. In our attempts to compromise these somewhat contradictory demands in an acceptable manner, we approached every piece of the necessary equipment, not as an isolated one significant in its own right, but as a part of a complete system, providing

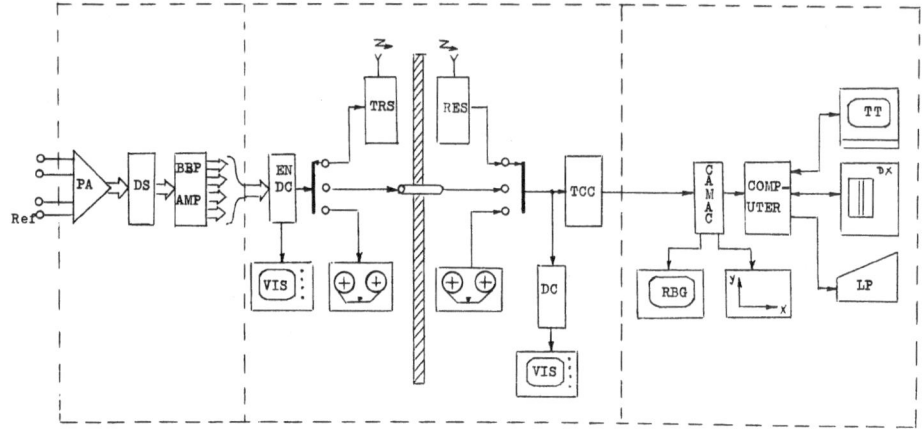

Fig. 1. Structural diagram for BBP acquisition and measurement.

the functions depicted in Fig. 1. First of all, it meant examining the technical specification of each element of the system, taking into consideration the capabilities and limitations of other system elements. This approach was applied to both hardware and software and it was found that for a number of the system components it appeared most beneficial to develop and produce special equipment instead of aggregating commercially available ones. In the following sections we describe the system developed, one which is widely adopted in clinical investigations sponsored by the Human Neurophysiology Department of the Institute.

POLYELECTRONEUROGRAPH - MULTICHANNEL MODULAR AMPLIFIER COMPLEX FOR SIMULTANEOUS REGISTRATION OF ALL TYPES OF BRAIN BIOELECTRICAL PROCESSES

Brain bioelectrical processes (BBP) are still the main source of information on the states and activities of the living brain, and neurophysiologists are eager to know as much as possible about all the types of BBP that it is possible to record by means of surface and intracerebral electrodes.

For long-term intracerebral implantation, electrodes are used which are made of gold wire 0.1 mm in diameter isolated with teflon (ftoroplast) (Anitchkov et al., 1978). The electrode-brain contact sites are of the order 0.05 - 0.2 mm² and BBP can be derived using the electrodes as follows (see Fig. 2): impulse neuron population activity - random impulse sequences with amplitudes up to 100 mkV; stereoelectroencephalosignals (SEES) - analogous random processes with amplitudes up to 200 mkV; infraslow electrical processes (ISEP) - analogous random processes with wave amplitudes up to 10 mV and wave durations from 2 to 120 s; steady electrical voltages (steady or d.c. potential) - usually measured in discrete values between +100 and -100 mV The physiological meaning of the BBP, as well as other problems and the results of using intracerebral electrodes clinically, is discussed in books and articles published by N. P. Bechtereva and her colleagues (for example, Bechtereva, 1978; 1980; Bechtereva et al., 1969; 1978; 1980; 1983; 1985).

If using Ag-AgCl electrodes with proper electrolyte bridges in scalp derivations, one can record not only traditional EEG infraslow electrical processes (whose amplitudes are lower than for intracerebral ISEP - lower than 100 mkV), but also steady potentials (ω - potentials in the terminology of Ilyukhina (1981)) in the range +80 - -20 mV. These types of BBP are also clinically significant (Ilyukhina, 1981; Ilyukhina et al., 1986).

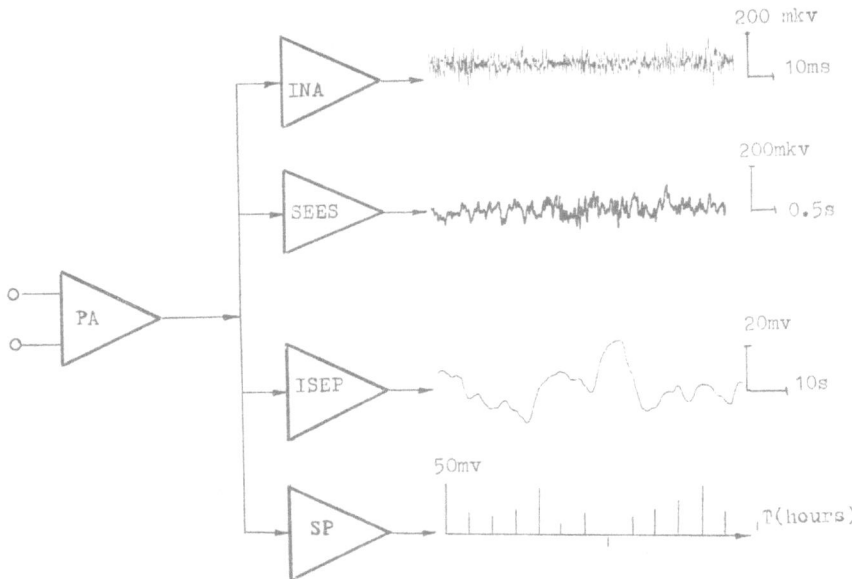

Fig. 2. Polyelectroneurography principle. Examples of BBP components
are shown. The nomenclature is explained in the text.

In the past amplifiers for different types of BBP were developed independently. In a single investigation one type of BBP was generally used,
although sometimes different types of BBP were derived from different electrodes. This put certain limitations on the structure of investigations
and interpretation of the results. It was practically impossible to overcome the difficulties in accordance with the requirements mentioned above
combining different existing amplifiers. The solution was to design a
special functionally-integrated modular apparatus for simultaneous amplification of all types of BBP picked up with the same electrode and with many
electrodes at the same time (Danko and Kaminskij, 1982; 1984).

Polyelectroneurography is based on the principle of preliminary wide
band amplification of all types of BBP combined with consecutive separate
filtering and necessary further amplification of every component of the
BBP (Fig. 2).

The functional circuit diagram of the complex is presented in Fig. 3.
Electrodes are connected to a multichannel pre-amplifier. Its frequency
response is flat from d.c. up to 20 kHz. Its gain is sufficiently small,
for example, 20, so that steady electrode voltage does not influence its
linearity. The multichannel pre-amplifier is not a common set of one-
channel devices, but rather a special circuit providing in every channel
amplification of the voltage difference between the electrode connected to
the channel input and the electrode connected to the reference input. Output signals of the pre-amplifier may be treated as monopolar derivatives
because voltages between any of the electrodes and an electrode common to
all channels are amplified. At the same time the derivatives are equivalent
to bipolar ones in rejection of interference, as can be shown.

Considering that every electrode connected to a non-inverting input
of the pre-amplifier (Fig. 3) is influenced by the sum of the signal voltage
(let this be on the reference input, Er) and interference voltage, En (the
latter considered to be the same on every input since the electrodes are of

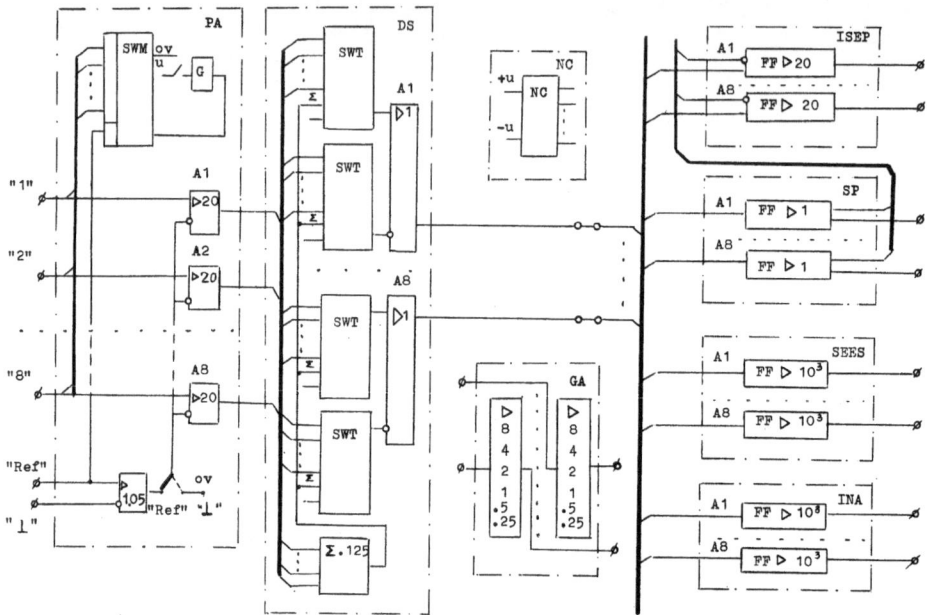

Fig. 3. A polyelectroneurographical amplifier complex block diagram.
The nomenclature is explained in the text.

the same type and their loads are equal) and using

$$Kp = 1 + Km \tag{1}$$

for an operational amplifier with proper feedback one can write for the
output voltage, Ur

$$Ur = (Er + En)*(1 + Km') \tag{2}$$

and for any other operational amplifier output voltage, for example Ui,

$$Ui = (Ei + En)*(1 + Km) - Ur*Km = (Ei - Er)*(1 + Km) \tag{3}$$

if the interference rejection condition

$$Km*Km' = 1 \tag{4}$$

is fulfilled.

The pre-amplifier is situated as near to the electrodes as possible
and its input impedance is high. So we have the best conditions for interference rejection and for small high frequency distortions despite the
high signal source impedance. A low pre-amplifier output resistance provides small distortions for all signal frequencies, although the cables to
the other modules are long (up to 10 m) and the input capacitances of the
modules are high, especially the montage capacitance of the derivative
selector.

The pre-amplifier developed provides opportunities for derivative
selection just as a traditional EEG-apparatus, but for any kind of BBP.
It is possible by means of the differential amplifiers and the switches
in the derivative selector (DS in Fig. 3) to form differences between BBP

at any pair of electrodes (bipolar derivatives), between BBP at an electrode and the mean BBP over all connected electrodes except the, referent
(the proper signal is fed to switches from the summator-divider designated
in Fig. 3) or to preserve monopolar derivatives (either with the same
polarity or inverted).

Furthermore, the integrated BBP are separated into traditional components - steady potentials, infraslow electrical processes, electroencephalo-
signals, impulse neuron activity - by means of proper filter-amplifiers
(SP, ISEP, SEEG, INA in Fig. 3). In ISEP amplifiers, steady electrode
voltage components can be compensated by means of a separate null corrector
(NC) - the source of calibrated steady voltages set by an operator with
proper polarity and value. Another way to compensate the voltage is to
connect the output of the SP filter with proper polarity to the input of
ISEP instead of NC. The steady voltages will be eliminated in ISEP automatically, but its frequency band will be changed (Danko and Kaminskij,
1984).

The amplified component signals can be fed to suitable devices for
display, recording and processing. If necessary, wideband modules (GA)
can be inserted between the main amplifiers and the devices for simultaneous
amplification or attenuation, the signals being in groups of eight channels.

Polyelectroneurographic amplifier complexes are constructed as a set
of interchangeable modules including ones for commutation. The basic set
inserted in one crate provides simultaneous amplification of 64 processes
from 16 electrodes.

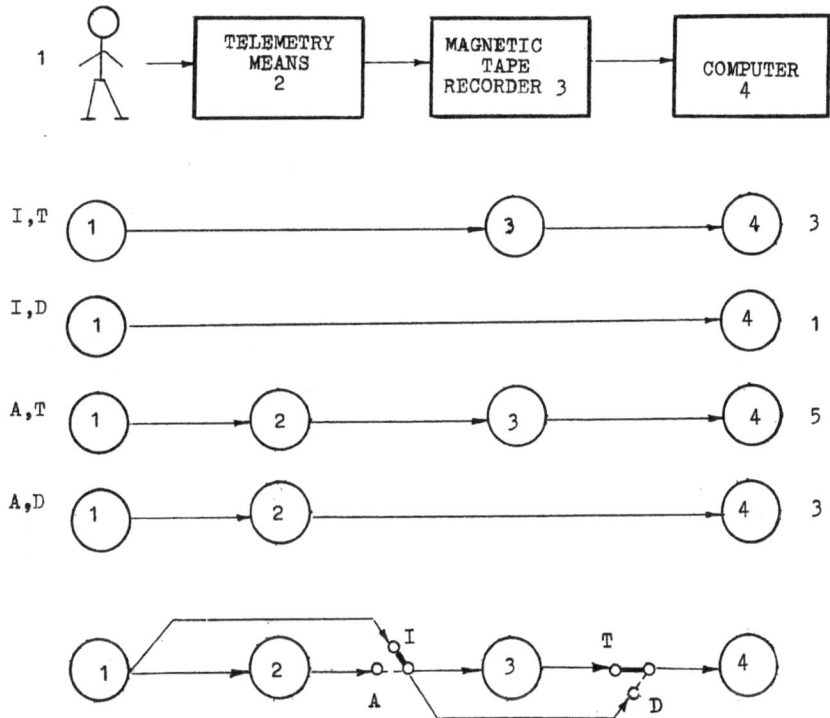

Fig. 4. Commonly used structures for signal transfer from a patient
to a computer. The nomenclature is explained in the text.

The problem of checking the multichannel unit is solved with a special circuit built in a pre-amplifier (G on Fig. 3) generating a mixture of proper signals fed simultaneously to all pre-amplifier inputs and separated in the main filter-amplifiers.

Experience gained with a number of polyelectroneurographic complexes in years of BBP clinical investigations confirms the following possibilities: to have a direct view of different BBP components taken at the same time at the same sites; to decrease the duration (or quantity) of BBP investigations and so benefit the patients (this is especially important during surgery, diagnostic trials, and during sleep research); to eliminate technological limitations while choosing a proper set of processes for a specific research task; and to use interference resistive monopolar derivatives for obtaining optional derivatives (for example, different bipolar ones) while processing a posteriori.

UNIFIED MEANS FOR BBP CONVERSION, TRANSMISSION, STORAGE AND FEEDING INTO THE COMPUTER

The need for automatic processing of multichannel BBP of a patient, not only in a special laboratory but also elsewhere in a hospital, forced us to pay much attention to the transfer of BBP to a computer according to the requirements mentioned above.

In common practice, the structure which provides the means of transfer (TM) is defined for particular conditions of investigation and for the available equipment (see Fig. 4). A subject may be immobilised (I) or active (A). BBP may be fed into a computer directly (D) or it is possible to register it on a magnetic tape (T) beforehand. A computer may be distant or nearby. Analysis of the possible combinations shows that a rational TM structure should include telemetry (TL), a magnetic tape recorder (TR) and an analog-digital converter.

In TM design it is the usual practice to provide consecutive conversions: to transmit BBP via telemetry channels and to restore them at the receiver; to write them down on TR and to restore them during replay; to digitise them while feeding them into a computer. The multiple conversions accumulate errors and increase the complexity and cost of the TM in proportion to its capabilities. These drawbacks are minimised in the variant which has been developed (Danko and Kaminskij, 1984). It is based on a time-pulse method of analog-digital conversion combined with time multiplexing of BBP channels. The method permits separation of its stages in time and space if necessary. The first stage is realised in an encoder (EN) which includes an analog multiplexor and pulse-width modulator, the second stage in a time-code converter (TCC) which is connected to a computer (Fig. 5). The type of transmission channels adopted depends on the conditions attached to the investigation, and the TM developed permits the use of radio-channel, telephone line, simple commercial tape recorder which can be used in any combination, for example direct BBP feeding into a computer is usually carried out in parallel with tape recording. EN and PCC characteristics define the TM precision in the case of direct input and this precision is not influenced by the characteristics of a decoder which is used only to check the display of the transmission. So the decoder circuit may be quite simple.

The EN structure is shown in Fig. 6. There are two blocks (marked with broken lines) providing both fast (SEEG) and slow (ISEP) signals transmission at the same time in EN. The slow signals may be sub-multiplexed instead of one fast signal.

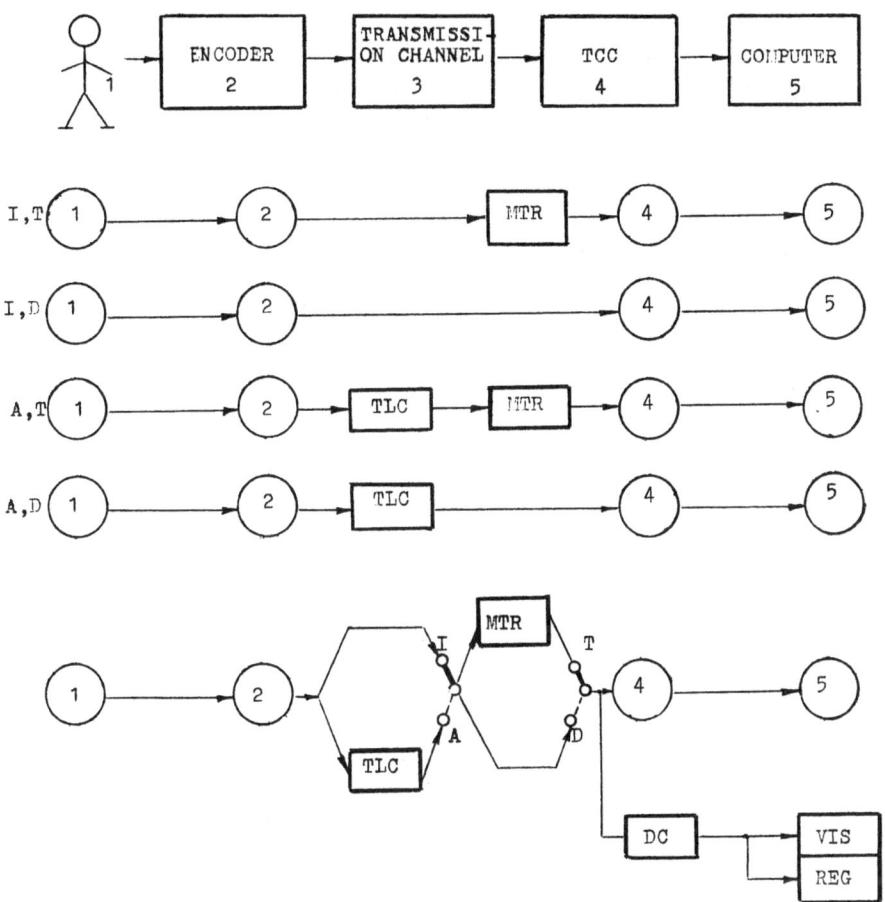

Fig. 5.   The unified structure developed for BBP transfer from a patient.
The nomenclature is explained in the text.

The main specification details of the TM which has been developed are:

primary modulation - PW;
channels number - 15 (7 fast + 8 slow) or 8 fast;
discretisation frequency for fast channels - 128 Hz;
modulation depth - 50%;
resulting relative TM errors in case of
     radio   - 0.4%
     cable   - 0.2%
     tape recording - 1.5%

Tape recording errors can be reduced by means of software demodulation
in accordance with the relation

$$A = (T1 - T2)/2*(T1 + T2) \qquad (5)$$

where T1 is the pulse duration in a PWM signal, and T2 is the duration of
the pause which follows the pulse. As the main source of errors is detonation with relatively low frequencies, then as a first approximation the
detonation influences T1 and T2 equally and does not influence A.

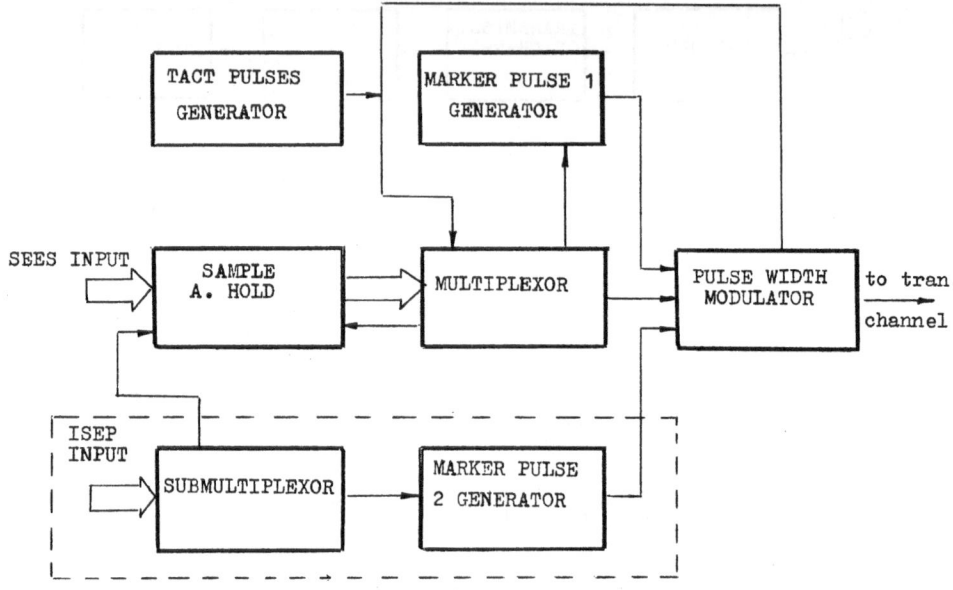

Fig. 6. Block diagram of a TM encoder.

Although nowadays one witnesses amazing successes in the development of computer memories with great capacities, our experience with BBP multi-channel recording on one- and two-channel tape recorders confirms that it is still quite a useful alternative in creating a versatile, inexpensive and reliable network for many laboratories and laboratory stations. The network permits the storage of large quantities of data and the easy transfer between different means of processing them.

BBP PARAMETER MEASUREMENT USING MICROCOMPUTERS

Technically, the system for processing BBP is implemented using 3 identical sub-systems with "Elektronika 60" microcomputers. They provide second stage conversion of the transferred BBP signals, with intermediate and final results being displayed in colour graphics and with a numerical hard copy.

The problem-oriented software developed provides: synchronous averaging of evoked activity in 8 channels at a time; spontaneous analog processes (SEES, ISEP) halfwave distribution evaluation in certain duration bands ("rhythm analysis"); spectral and correlation analysis of BBP; statistical batch processing of numerical data. The processing software is written in ASSEMBLER and FORTRAN, the statistical portion in BASIC.

The synchronous averaging is implemented with a mean recursive calculation algorithm so one always has the current result on a screen and with a "cyclic buffer" for prestimulus interval averaging. There is a process realisations superposition on one half of the screen, thus enabling scatter and artefacts in the realisations to be seen. When the averaging is over one can scale and smooth the results and measure the co-ordinates of extrema automatically or in an interactive mode. To measure the parameters of the integral evoked potentials, dispersion of the averaged processes around their mean values are calculated at the prestimulus time and over two poststimulus intervals. Total form coefficients are also calculated

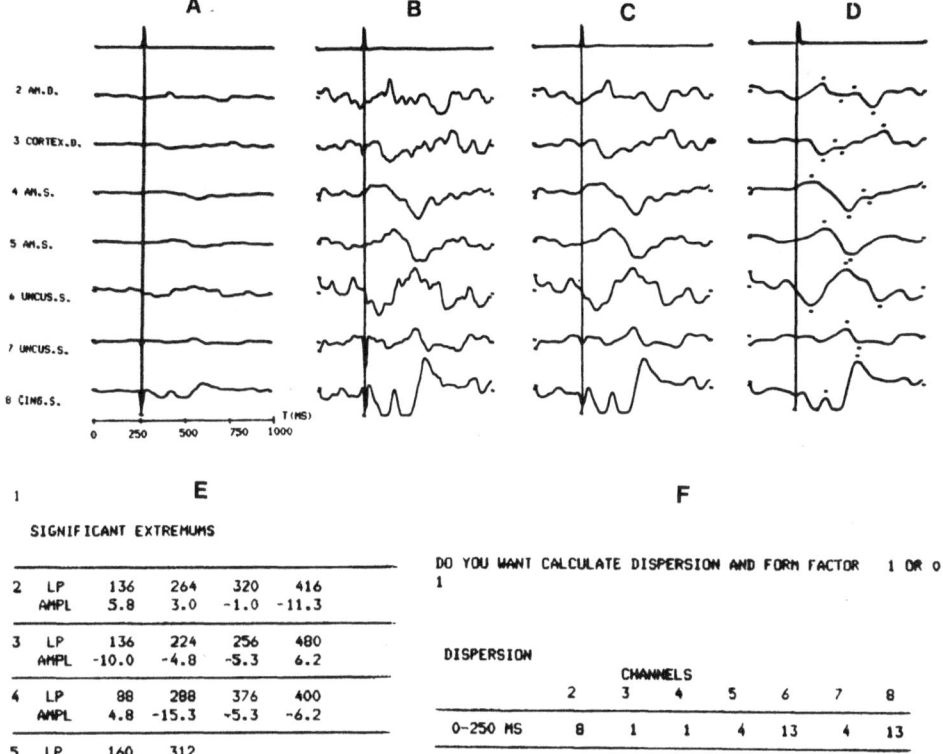

E

SIGNIFICANT EXTREMUMS

2	LP	136	264	320	416
	AMPL	5.8	3.0	-1.0	-11.3
3	LP	136	224	256	480
	AMPL	-10.0	-4.8	-5.3	6.2
4	LP	88	288	376	400
	AMPL	4.8	-15.3	-5.3	-6.2
5	LP	160	312		
	AMPL	6.5	11.5		
6	LP	80	272	456	
	AMPL	-12.5	12.3	-9.5	
7	LP	272	352		
	AMPL	5.5	-4.3		
8	LP	104	160	224	336
	AMPL	-16.0	-7.5	-18.7	16.5

F

DO YOU WANT CALCULATE DISPERSION AND FORM FACTOR   1 OR 0
1

DISPERSION

	CHANNELS						
	2	3	4	5	6	7	8
0-250 MS	8	1	1	4	13	4	13
250-750 MS	21	21	45	30	61	9	119
750-1000MS	10	6	10	5	41	5	9
KF	1	3	8	35	4	3	5

Fig. 7.   Results of multichannel evoked potential processing. Plotted
averaged signals are scaled (A, B) and smoothed (B, C, D).
General means are marked with points at the beginnings and at the
ends of the lines. Significant extrema are marked with points at
(D). Measured extremum parameters (E) and measured evoked potential integral parameters (F) are printed.

as the result of dividing the maximal halfwave amplitudes by the mean amplitudes of the other halfwaves for every result. An example of the processing
of evoked potentials is shown in Fig. 7.

The developed software also permits the sorting of records in accordance
with the situations in terms of "type of stimulus - type of response" if one
signal channel is used for optional service marks. One can exclude records
contaminated by artefacts from the averaging by delivering their serial numbers in an interactive mode after a preliminary inspection.

In the measurement of the parameters of spontaneous analog BBP dynamics
programs are used which divide the processes into sequential halfwaves,
define halfwave amplitudes and durations, distribute them according to the
preset duration limits and calculate mean amplitude and total relative

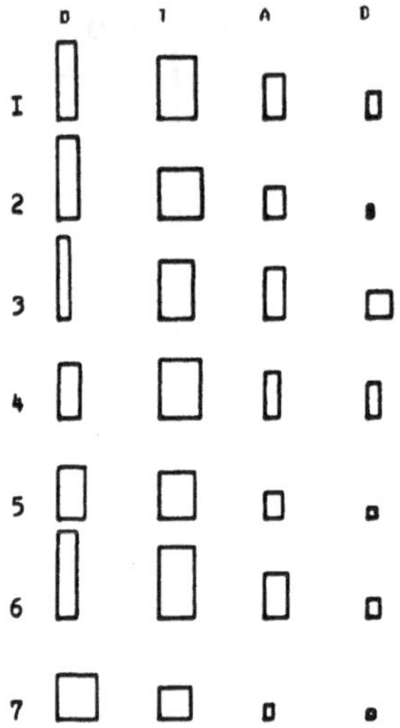

	DELTA		TETA		ALFA		BETA	
N	A(MKV)	K	A(MKV)	K	A(MKV)	K	A(MKV)	K
1	23.2	0.23	19.0	0.39	14.0	0.23	8.7	0.15
2	24.1	0.25	15.0	0.45	9.8	0.22	4.7	0.08
3	24.2	0.16	17.4	0.35	15.4	0.23	8.7	0.26
4	16.5	0.25	17.8	0.43	14.0	0.17	11.1	0.15
5	15.9	0.30	14.3	0.37	8.4	0.20	4.4	0.13
6	25.7	0.22	21.3	0.37	13.7	0.26	6.5	0.16
7	13.9	0.43	10.0	0.34	5.5	0.13	3.8	0.10

MEAN VALUE

DO YOU WANT OUTPUT ON PLOTTER?(1 OR 0)

Fig. 8.   Results of spontaneous SEES processing (an example). Mean amplitudes and indices of presence are printed according to generally used EEG-rhythms. In graphical presentations the vertical bars are proportional to the mean amplitudes and the horizontal bars to the indices.

duration of the half waves. As a result an extremum found earlier is verified only if the segment between the earlier and the nearest newly found extrema has co-ordinate projections greater than the duration and amplitude thresholds. The latter is proportional to the mean half-wave amplitude in the analysis of the preceding epoch.

The duration of an epoch being analysed can be varied in an interactive manner from 1 to 20 s for SEEG and from 2 to 16 min for ISEP. 20 epochs are processed with a pause between each of them of up to 1 s before the result output is mandatory. The start of the processing can be synchronised with a service mark in the transferred signals. The output is presented both numerically and graphically, the latter for the convenience of preliminary review (Fig. 8).

In sleep research the processing is implemented in two stages. During sleep, the current parameters of the patient are stored on floppy-disks only. Mean parameter calculations and the output of results, including the presentation of parameters as functions of time, are executed a posteriori.

Programs developed for spectral and correlation analysis calculate the functions for 8 signals at a time, both for primary BBP and for primary processing results stored on floppy-disks. Power spectra are calculated in 8 sequential epochs and stability is achieved with the averaging of these 8 evaluations.

Statistical analysis includes calculation of means, dispersions, confidence intervals, enthropia, linear and range correlation coefficients and histograms. The calculations can be executed in data of up to 80 x 80 elements.

Thus the sub-systems, although modest enough in their computer hardware capabilities, have successfully met real primary needs in automated BBP processing and have confirmed their effectiveness in clinical situations.

REFERENCES

Anitchkov, A. D., Beljaev, V. V., and Usov, V. V., 1978, Konstruktsija mnozhestvennyh elektrodov i sposoby ich vvedenija v golovnoj mozg tcheloveka, Fiziologija Tcheloveka, 4 : 371.

Bechtereva, N. P., 1978, "The Neurophysiological Aspects of Human Mental Activity", 2nd edn., Oxford Univ. Press, New York.

Bechtereva, N. P., 1980, "Zdorovyi i Bolnoi Mozg Tcheloveka", Nauka, Leningrad.

Bechtereva, N. P., Bondartchuk, A. N., Smirnov, V. M., and Trochatchev, A. I., 1969, "Physiology and Pathophysiology of the Human Deep Brain Structures", Volk und Gesundheit, Berlin.

Bechtereva, N. P., Kambarova, D. K., and Pozdeev, V. K., 1978, "Ustoitchivoe Patologitcheskoe Sostojanie pri Boleznjach Golovnogo Mozga", Medicina, Leningrad.

Bechtereva, N. P., Bundzen, P. V., Gogolitsin, Y. L., and Medvedev, S. V., 1980, Physiological correlates of states and activities in the central nervous system, Adv. Physiol. Sci./Brain Behav., 17 : 395.

Bechtereva, N. P., Gogolitsin, Y. L., Ilyukhina, V. A., and Pakhomov, S. V., 1983, Dynamic neurophysiological correlates of mental processes, Int. J. Psychophysiol., 1 : 49.

Bechtereva, N. P., Gogolitsin, Y. L., Kropotov, Y. D., and Medvedev, S. V., 1985, "Neirofiziologitcheskie Mechanizmy Myshlenija", Nauka, Leningrad.

Danko, S. G., and Kaminskij, Y. L., 1982, "Sistema Technitcheskyh Sredstv Neirofiziologitcheskyh Issledovanij Mozga Tcheloveka", Nauka, Leningrad.

Danko, S. G., and Kaminskij, Y. L., 1984, Polielektroneirograf - modulnaja usilitelnaja sistema dlja kompleksnogo issledovanija bioelektritcheskoj aktivnosti mozga, Fiziol. Zhurnal SSSR, 57 : 1061.

Danko, S. G., and Kaminskij, Y. L., 1986, Avtomatizatsia klinitcheskyh neirofiziologitcheskyh issledovanij na baze mikroEVM, Avtometrija, 3 : 42.

Ilyukhina, V. A., 1981, Sverchmedlennye protsessy mozga (terminologija i utochnenie nekotoryh ponjatij). Soobshenie 1. Spontannaja dinamika sverchmedlennyh protsessov kory i podkorkovyh struktur v kliniko-fiziologitcheskich issledovanijach, Fiziologija Tcheloveka, 7 : 512.

Ilyukhina, V. A., Habaeva, Z. G., Nikitina, L. I., Medvedeva, T. G., Movsisjants, S. A., Minitcheva, T. V., Kozhushko, N. Y., and Orlov, V. V., 1986, "Sverchmedlennye Fiziologitcheskie Protsessy i Mezhsis-temnye Vzaimodejstvija v Organizme", Nauka, Leningrad.

MICROCOMPUTER-BASED OBJECTIVE VISUAL FIELD DIAGNOSIS

G. Henning and W. Müller

Department of Biomedical Engineering, Institute of Tech-
nology, Ilmenau; and
Clinic for Eye Diseases, Medical Academy, Erfurt,
GDR

INTRODUCTION

The purpose of a perimetric investigation is to determine the outer
limits of the visual field and to examine the functionality of the total
region within the limits, as well as to determine the enlargement of the
physiological blind spot, the so-called papilla nervi optici. Hence, peri-
metry represents an important differential diagnostic method not only for
ophthalmology, but also for neurology and neurosurgery. Conventional perimetric methods are completely subjective measurements in the sense that the
patient whose visual system is the object of measurement is, at the same
time, the most important element of the measuring system. The minimum
necessary pre-conditions for achieving reliable results are the ability and
readiness of the person tested to co-operate. If these pre-conditions are
lacking the method will fail.

THE MEASUREMENT SYSTEM

On the basis of these facts, our inter-disciplinary working group
sought to develop an objective measuring procedure in addition to the subjective methodology which does not have these restrictions. The visually
evoked cortical potentials (VECP) offer themselves as the objective measured
value, because they contain information on the total course of the visual
pathways up to the visual cortex. The conceptual basis of the procedure
adopted as the solution is represented schematically in Fig. 1.

The following main requirements must be met for the investigations to
be carried out:

(i) Localised retina stimulation. This is a complex issue. The most
     serious problem is posed by scattered light. To eliminate its in-
     fluence, we have to establish an optimum relationship between the
     intensity of the stimulus and the brightness of the surrounding
     field. Clinical proof that localised stimulation has been success-
     ful is provided by the evidence that we may manage to give for the
     existence of the physiological blind spot of the retina. If a stimu-
     lation of that area yields no or only a negligible potential, the
     conditions of localised stimulation have been successfully estab-
     lished.

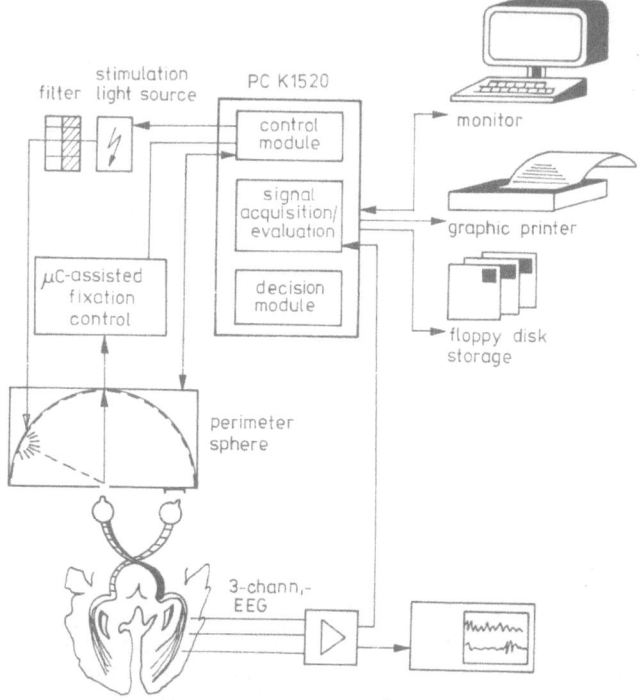

Fig. 1.   Microcomputer-based objective perimeter
IEP 3 - Block diagram.

(ii)   It must be possible to test any selected section of the retina.
         This is realised by programme-controlled positioning of the stimu-
         lating light source.

(iii)   In cases of repeated stimulation, there must be a guarantee that
         exactly the same point of the retina is stimulated in every stimula-
         tion. The patient's fixation must therefore be under automatic con-
         trol. If there are deviations from a pre-set tolerance range, the
         recording must be interrupted. We managed to do this by using a
         modified corneal reflection method in the infra-red light range,
         measuring the position of the eye every 40 ms and evaluating it on-
         line by a slave microcomputer.

(iv)   The system of automatic signal evaluation must operate under on-
         line conditions. It is only in this way that the physician can
         assess the latest evidence and start making additional checks
         without any delay.

        The evaluation of the measurement data up to the computer-assisted
diagnosis proposal therefore represents a priority objective for the
methodological work. The main purpose of signal evaluation in the case of
electroperimetry is not primarily the determination and quantitative evaluation of parameters, but rather the decision as to the presence or absence
of the stimulus response which is made even more difficult by the unfavourable signal-to-noise ratio of the signal embedded in the spontaneous EEG
activity.

        As a basis for a computer-assisted decision, 87 characteristic para-
meters of the following groups were determined from the single responses
and the averaged signal:

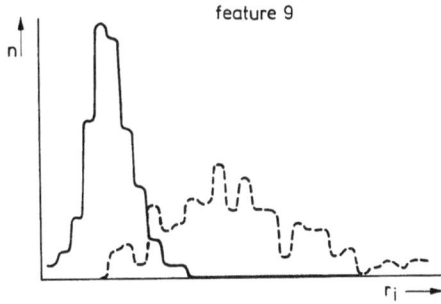

feature 9

Fig. 2.   Distribution of criterion 9 -
          correlation coefficient r_i -
          for perimetric EEG - data with
          VECP (broken line) and without
          VECP (solid line). Classes
          are well separated; a low
          error univariate diagnostic
          decision seems to be possible.

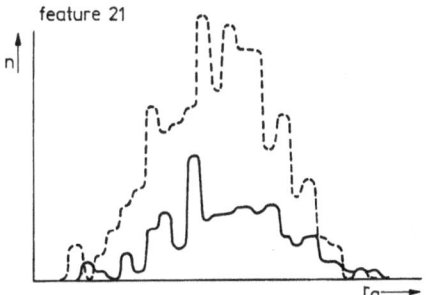

feature 21

Fig. 3.   Distribution of criterion 21 -
          correlation coefficient r_a
          for perimetric EEG - data with
          VECP (broken line) and without
          VECP (solid line). Classes are
          totally overlapping; univariate
          separation is impossible.

   (i)   outer correlation coefficients;
  (ii)   amplitude criteria;
(iii)   amplitude and time parameters of the extrema;
  (iv)   area criteria;
   (v)   internal correlation coefficients; and
  (vi)   power density spectrum parameters.

EVALUATION

The evaluation of the univariate effectiveness of the characteristic
parameters for separating the diagnostic classes was made on the basis of
an extensive series of investigations of probands by evaluating the distributions for the characteristic features by means of the bunching parameter
F and the minimum average decision error DE (see Figs. 2 and 3).

45

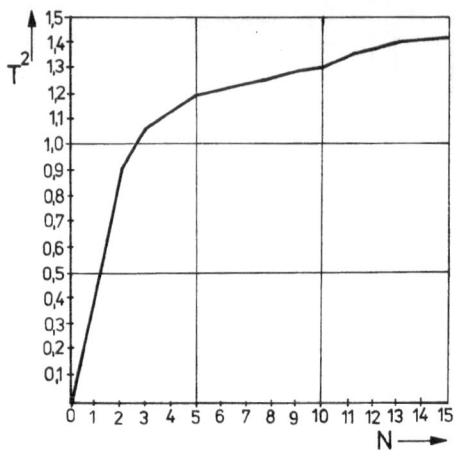

Fig. 4. Multivariate separation factor T^2 as a function of the
number N of included parameters.

The decision certainty which can be reached with the most effective univariate characteristic parameter is around 75%. A further improvement of the separation efficiency was to be attained by multivariate intercon-nection of the characteristic parameters using linear discriminant analysis. The most selective univariate characteristic criterion was then the star-ting parameter for the determination of a sub-optimum parameter set by means of the bottom-up method. The optimisation of the set of parameters was realised by a combination of bottom-up and top-down methods. The multivariate separation factor T^2 according to Ahrens and Laeuter (1981) served as an optimisation criterion (Fig. 4).

After evaluation on the basis of the rate of false classification and taking into consideration the shortest possible computing time, we decided in favour of a classifier having 9 characteristic parameters. On this basis an evaluation system was realised for clinical application which has by now been found to be effective in routine operation for one year. The error classification rate determined for each channel by way of π-type error estimation on the basis of extensive proband examination results is about 15%.

For comparison purposes, the following figures show:
- a visual field subjectively determined with a co-operative patient
  (Fig. 5); and
- the visual field of the same patient measured under nearly the same
  measuring conditions by the use of our computer-assisted objective
  procedure (Fig. 6).

The routine evaluation system has been implemented into the microcom-puter-assisted on-line perimeter type IEP 3, which also realises automatic control of the whole measuring procedure, including all the important para-meters of the test. The diagnostic equipment has now been in clinical use for one year.

Based on the clinical experience gathered to date we can say, in summary, that the application of a computer-assisted decision procedure represents an important contribution towards improvement of the diagnostic

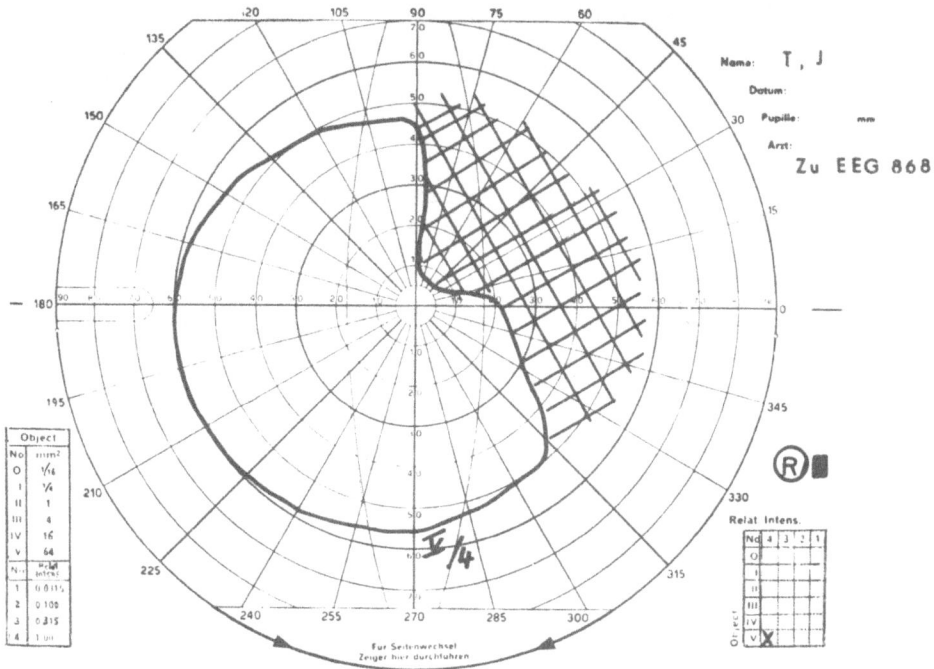

Fig. 5. Subjectively-determined visual field of a co-operative patient.

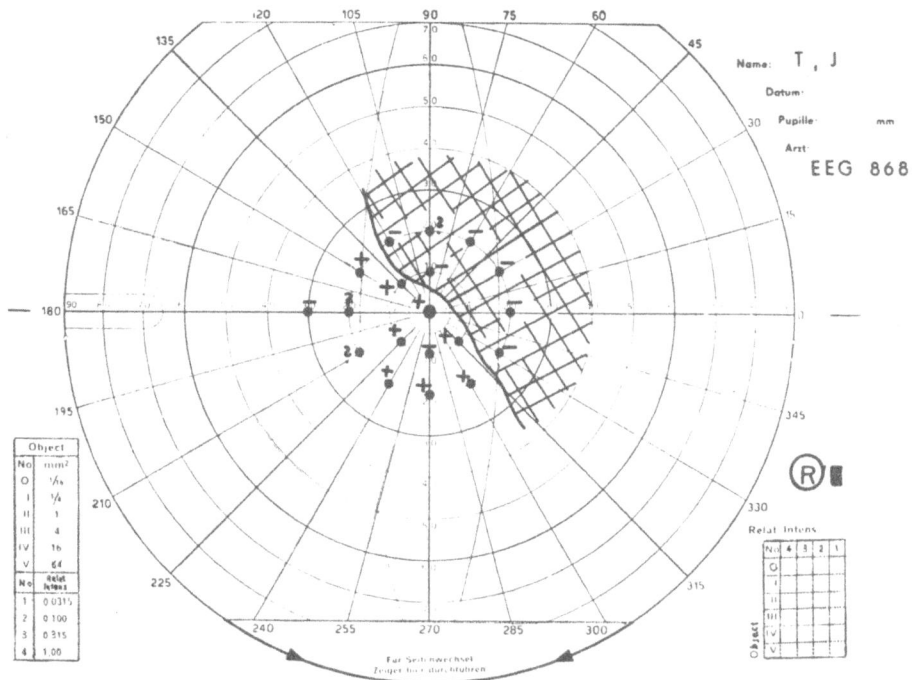

Fig. 6. Visual field of the same patient determined by the use of the method described.

certainty and thus of the clinical relevancy of the method. Furthermore, what cannot be underestimated is the fact that the physician is relieved of time-consuming manual and graphical evaluation.

REFERENCE

Ahrens, H., and Laeuter, J., 1981, "Mehrdimensionale Varianzanalyse –
    Hypothesenpruefung, Dimensionserniedrigung, Diskrimination", Akademie-
    Verlag, Berlin.

A TEMPERATURE SENSITIVE MICROELECTRODE FOR MEASUREMENTS IN SOFT TISSUES

AT THE CELLULAR LEVEL

I. Dittert and F. Rech

Institute of Physiology
Czechoslovak Academy of Sciences
Prague, Czechoslovakia

INTRODUCTION

This chapter describes the design and construction of a temperature sensitive microelectrode suitable for measurements in soft tissues at both the cellular and sub-cellular levels. The microelectrode technique is currently used to measure potentials, currents and ion fluxes. A functional temperature sensitive electrode would extend the scope of quantities to be measured. Knowledge of the temperature in micro-objects is useful in general, and in particular in the study of the behaviour of a simpler structure in relation to the overall features of living systems. Physiological and biomedical applications include, for instance, studies of the energy balance of photosynthetic processes, metabolic processes or the onset of malignant growth as affected by carcinogens and chemotherapeutics, and so on.

DESIGN REQUIREMENTS

The requirements posed on the physical properties of the probe can be illustrated in an experiment designed to measure the thermophysical properties of a leaf tissue for the purpose of modelling. The apparatus is shown schematically in Fig. 1. The leaf was illuminated on its upper side by a focused light source - an electric bulb; the illuminated area was 1 mm in diameter. The temperature was monitored by thermocells placed on the reverse side of the leaf at random along concentric circles 2 and 10 mm in diameter. The recorded time courses of temperature are shown in Fig. 2. Course A was expected and was obtained in most cases. Courses B, C and D, however, were unexpected; they signified marked temperature drops below the steady state initial temperature of the environment that had taken place in the immediate vicinity of the damaged tissue owing to the heating effect of radiation. Thermocells with weld diameter of about 90 μm proved to be too big to be used for a more detailed investigation of the monitored phenomena.

The main demands for a more suitable method of measuring temperature by a miniature probe are as follows:

(i) the dominant dimension of the active part of the probe should not
     exceed units of μm;

(ii)   effective suppression of deformation of naturally formed temperature field in the studied structure, caused by the probe itself as a result of its geometry and the thermophysical properties of the construction materials, or brought about by the electric effects of the necessary electronic circuits; and

(iii)   a sampling frequency of at least 1 kHz, the highest possible resolution, minimum resolution of 0.1 K.

PROBE DESIGN

The probe design is shown in Figs. 3 and 4. It consists of a modification of a conventional glass micropipette. Amorphous semiconductive material with suitable mechanical and electrical properties is placed in the micropipette tip, giving rise to a temperature sensitive microelectrode

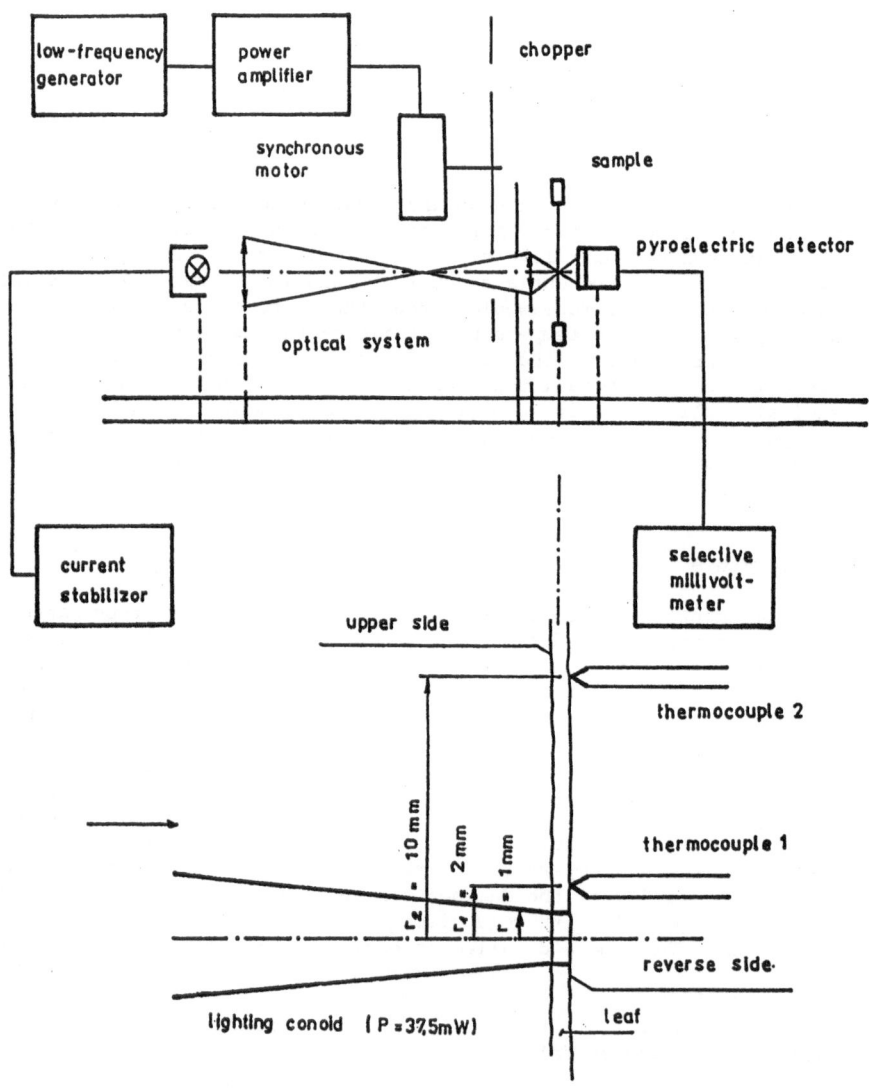

Fig. 1.   Schematic representation of the measurement system.

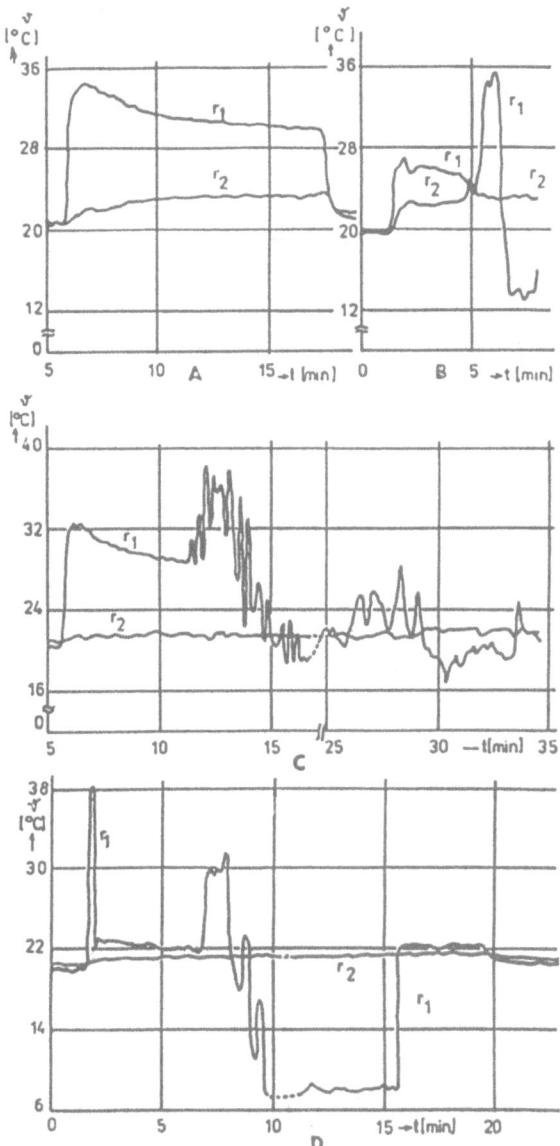

Fig. 2. Recorded time courses of measured temperature.

(TSM) functionally resembling a miniature thermistor. The thermophysical properties of water and glass are very similar and the TSM, on insertion into the cell, will thus cause no marked deformation of the temperature field inside the cell. The least suitable parameters of measurement with the TSM are found on the surface of the cell. To avoid errors caused by heat losses through the TSM body higher than 1%, half of the tip angle of a cone-shaped approximation of the actual TSM tip should not exceed 7°. This condition is easy to satisfy. Index 1 in Fig. 3 relates to quantities concerning the semi-infinite environment inside the cell, while index 2 is related to the TSM. The nomenclature adopted is:

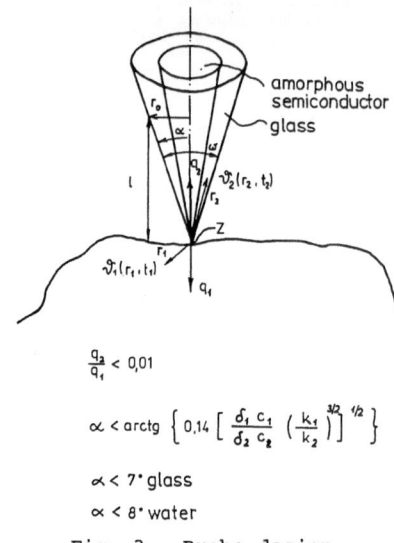

$$\frac{q_2}{q_1} < 0.01$$

$$\alpha < \text{arctg} \left\{ 0.14 \left[\frac{\delta_1 c_1}{\delta_2 c_2} \left(\frac{k_1}{k_2} \right)^{3/2} \right]^{1/2} \right\}$$

$\alpha < 7°$ glass

$\alpha < 8°$ water

Fig. 3.   Probe design.

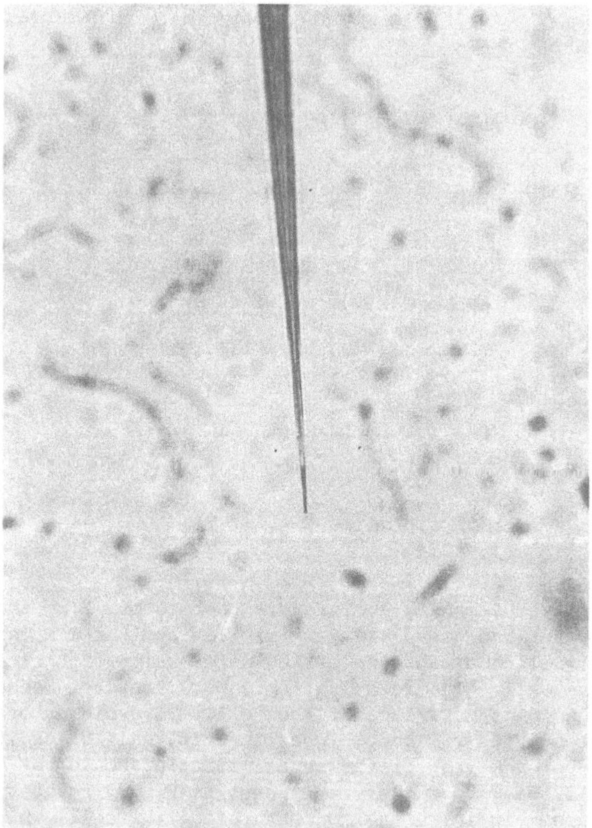

Fig. 4.   Magnification of the probe tip.

q $(W\,m^{-2})$ —   heat flow from point z

$\theta(r,t)$ (K) —   temperature as a function of the radius vector and time

α $(°)$ —   half of the tip angle of a cone-shaped approximation of the actual TSM tip shape

$\omega(sr)$ —   space angle delimited by the TSM cone

δ $(kg\,m^{-3})$ —   density

c $(J\,(kg\,K)^{-1})$ —   specific heat

k $(m^2\,s^{-1})$ —   heat conductivity

The microelectrode tip in Fig. 4 is filled with a semiconductive amorphous material to a height of $18\,\mu m$. The impedance of the TSM is for the most part concentrated in the tip, whereas the temperature sensitivity is localised at a certain distance from the tip. If the height of the temperature sensitive zone is sufficiently small, of the order of μm, the semiconductive material need not be confined solely to the tip and the whole TSM can be filled with the material. Analysis of static and dynamic parameters of the TSM can thus lead to simplification of the technology of TSM construction. The contact with the semiconductive material in a partially filled electrode can be secured both from the inside and from the outside via electrolytes. In an electrode filled to a height of several millimetres, internal contact can be ensured by a metal conductor without the risk of appreciable deformation of the thermal field near the active zone of the TSM.

ELECTRICAL PARAMETERS

Electrical static and dynamic parameters of the TSM were derived by generalisation of the actual tip geometry in a sub-micron region. Fig. 5 shows an equivalent diagram of a conductive element in the TSM tip filled with a .semiconductor. The symbols used are as follows:

$r(x)$ (Ω/m) —   elementary resistor

c' (F/m) —   elementary transverse capacitor

The tip shape was gradually replaced by a linear and parabolic approximation with the condition of passage through the measured points (inside and outside) at site $x = 0$. For both approximations

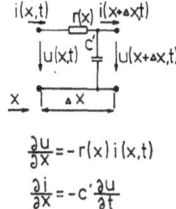

$$\frac{\partial u}{\partial x} = -r(x)\,i(x,t)$$
$$\frac{\partial i}{\partial x} = -c'\frac{\partial u}{\partial t}$$

Fig. 5.   Equivalent circuit of a conductive element in the TSM tip filled with a semiconductor.

$$r(x) = \frac{\rho}{\pi\, y_i^2} \; ; \quad R_c(x) = \int_0^x r(x)\,dx; \tag{1}$$

$$C' = \frac{2\pi\varepsilon_r\varepsilon_o}{\ln(y_i/y_o)} \tag{2}$$

where ρ (Ωm) is the specific resistance of the semiconductor whose temperature dependence forms the basis of the method

R_c (Ω)	–	total d.c. resistance from TSM tip ($x = 0$) to height x
y_i (m)	–	perpendicular distance between the TSM axis at point x and the inner surface of the glass TSM wall
y_o (m)	–	perpendicular distance between the TSM axis at point x and the outer rim
ε_r (·/·)	–	relative permittivity of the micropipette glass
ε_o (F m^{-1})	–	permittivity of vacuum.

Fig. 6 illustrates the dependence of elementary and total impedance of the TSM on the x(m) co-ordinate. It can be seen that the maximal d.c. impedance is reached for lower values of x when the microcone has a linear shape than when it is parabolic. As documented in Fig. 7, an increasing value of the tip angle is accompanied by a marked shift of the temperature sensitive region of TSM towards the origin. The actual value of the tip angle represents a compromise between the height of the temperature sensitive zone and the measurement error caused by heat losses via the TSM body (see also Fig. 3). Consequently, the shape of the microcone has a strong effect on static parameters of the TSM. The temperature sensitive zone can be shortened even for very small tip angles by introducing inhomogeneity in the specific resistivity of the semiconductor. This is accomplished by placing in the tip a material that exhibits a higher specific resistivity at a constant ambient temperature relative to the material filling the remaining part of the TSM.

The equations in Fig. 5 describe fundamental formulae for voltage and current in a conductor with distributed parameters. Their modification yields a linear approximation of the TSM microcone shape in the form of a

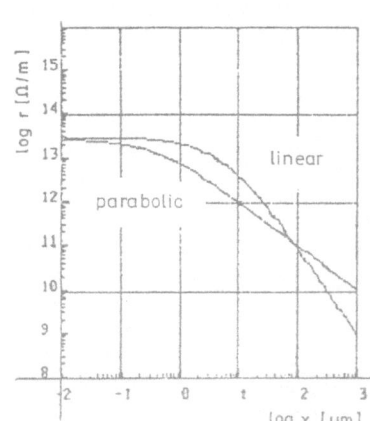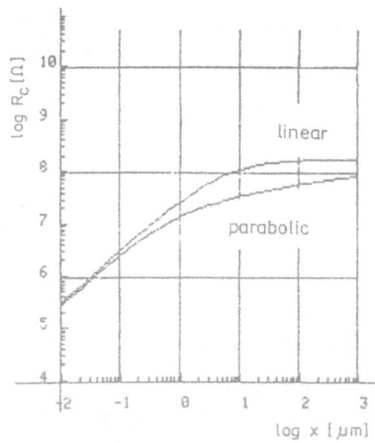

Fig. 6. Dependency of elementary and total impedance of the TSM on the x(m) co-ordinate for various approximations to the microcone shape. $\rho = 1\ \Omega$m; $r_o = 0.1\ \mu$m; $\alpha = 1°$.

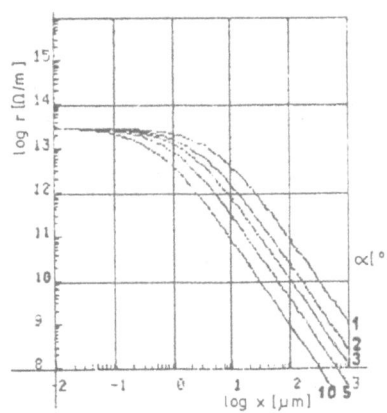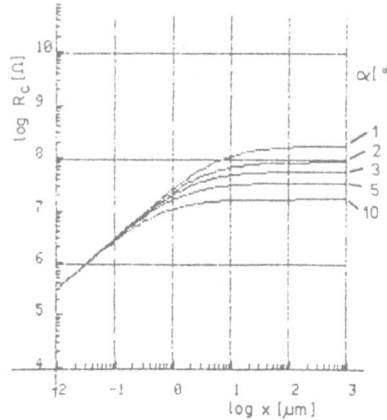

Fig. 7.   Log/log plot showing how an increase in the value of the tip
angle (α) is accompanied by a marked shift of the temperature
sensitive region of the TSM to the origin. $\rho = 1\,\Omega\,m$;
$r_o = 0.1\,\mu m$; α-par.

differential equation of Euler's type:

$$U''(b + x)^2 + 2U'(b + x) - p\tau U = 0 \tag{3}$$

$$\tau = aC'\,[s] \tag{4}$$

$a\,[\Omega m]$; $p\,[s^{-1}]$ Laplace operator

$b\,[m]$

The parabolic approximation results in a differential equation of
Bessel's type

$$U''(b + x) + U' - p\tau U = 0 \tag{5}$$

$$\tau = aC' \tag{6}$$

$a\,[\Omega]$

$b\,[m]$

where a, b are parameters characterising the geometry of the tip and the
electrical properties of the semiconductor.   Euler's equation can be solved
exactly and the impedance has the form

$$Z_{VEp} = \frac{a}{x + b} \cdot \frac{\sinh \gamma s}{\gamma \cosh \gamma s - \frac{1}{2} \sinh \gamma s} \tag{7}$$

$$\gamma = \frac{1}{2} \sqrt{1 + 4p\tau} \tag{8}$$

$$s = \ln\left(1 + \frac{x}{b}\right). \tag{9}$$

The Bessel equation was solved in the form of a succession in which the
first two and three terms were taken into consideration.   For two terms,
the impedance is of the form

$$Z_{VB} = R_c(x, \rho)\,\frac{1 + pA}{1 + pT}\,; \quad A < T \tag{10}$$

$$A = \tau b\left(2 + \frac{x}{b}\right) \tag{11}$$

$$R_c(x,\rho) = a \ln\left(1 + \frac{x}{b}\right) \tag{12}$$

$$T = \tau b\{1 + \left(1 + \frac{x}{b}\right)[\ln\left(1 + \frac{x}{b}\right) - 1]\} \tag{13}$$

When the expression for impedance obtained from the exact solution of Euler's equation is transformed into a series and the first two terms are considered, the approximated impedance has a form which is formally identical to that obtained from the Bessel equation:

$$Z_{VE_A} = R_c(x,\rho)\,\frac{1 + pA}{1 + pT} ; \quad A < T \tag{14}$$

$$A = 2\tau \tag{15}$$

$$R_c(x,\rho) = \frac{a}{b} \cdot \frac{x}{x + b} \tag{16}$$

$$T = \tau\left(2 + \frac{x}{b}\right). \tag{17}$$

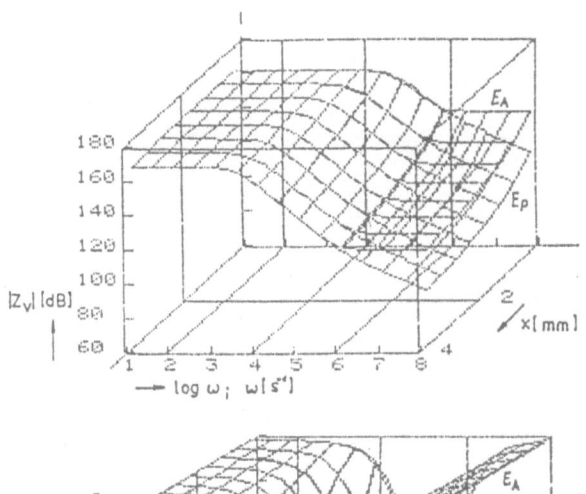

Fig. 8. Comparison of both forms of impedance calculated from Euler's equation with the dependence of amplitude and phase on frequency and depth of TSM immersion also shown.

DISCUSSION

Fig. 8 compares both forms of impedance calculated from Euler's equation, with the dependence of amplitude and phase on frequency and the depth of TSM immersion also being given. The approximate form is identical with the exact solution up to frequencies of about 10^4 Hz. For two terms of the series we found an acceptable equivalent circuit of the TSM with concerted (lumped) parameters. For three terms an equivalent circuit can no longer be found.

The frequency independent part of the impedance $R_C(x,\rho)$ is proportional to the measured temperature. Its measurement was performed by the impulse method. Two basic circuits were constructed for the compensation of transmitting properties of the TSM by means of a negative or positive feedback. Both circuits have their advantages and the choice of a given circuit depends on the type of application. Compensation by negative feedback makes it possible to accomplish the measuring cycle within a period shorter than 100 µs, with the electrical time constant of the TSM including input amplifier capacity leads reaching the range of hundreds of ms. This type of connection also has a more suitable signal-to-noise ratio.

The upper part of Fig. 9 shows the calculated time courses of the current through the microelectrode. The lower part shows the course of generated power output per unit length of the TSM. In both cases the co-ordinate x serves as the parameter. Both graphs were obtained from the

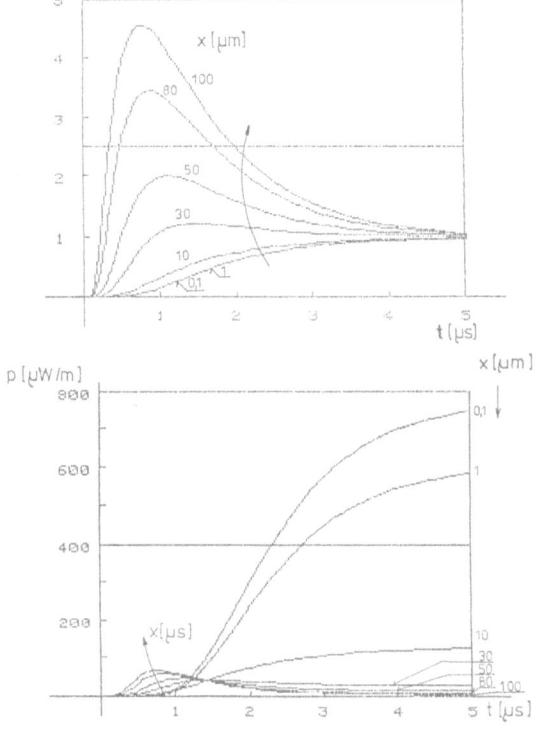

Fig. 9. Upper panel - calculated time course
of current flow through microelectrode; lower panel - time course
of generated power output per unit
length of the TSM.

exact solution for the linear approximation of the microcone. The current was calculated from the derived expressions:

$$i*(x',t) = R_c(x',\rho) \cdot i(x',t) \tag{18}$$

$$i*(x',t) = 1 + e'\sqrt{\frac{1+x'}{1+e'}} \sum_{k=1}^{\infty} \frac{\frac{(k\pi)^2}{S_e} \cos\left(k\pi\frac{S_x}{S_e}\right) - k\pi\sin\left(k\pi\frac{S_x}{S_e}\right)}{[S_e^2 + (k\pi)^2] \cos k\pi} \exp\left\{-\frac{t}{4\tau}\left[1+\left(\frac{k\pi}{S_e}\right)^2\right]\right\} \tag{19}$$

$$x' = \frac{x}{b} \quad ; \quad S_x = \ln\sqrt{1+x'} \tag{20}$$

$$e' = \frac{e}{b} \quad ; \quad S_e = \ln\sqrt{1+e'} \quad ; \quad x \leqslant e \tag{21}$$

where e (m) – height of semiconductive material filling the TSM

t (s) – time since a unit jump in the measuring impulse.

The power output is given by:

$$p(x',t) = \left[\frac{i*(x',t)}{R_c(e')}\right]^2 \cdot r(x') \tag{22}$$

The pulse amplitude on the output of the compensated amplifier is proportional to the measured temperature. Evaluation of the amplitude was performed by a microcomputer. The timing, control and data signals were separated from the computer by opto-couplers. The TSM itself with the sample was shielded.

First TSM samples are being prepared and the theoretically calculated parameters will be compared with experimental data. The laboratory TSM measurement requires not only measuring equipment but also microelectrode preparation and calibration devices.

ELECTRICAL MEASUREMENT OF FLUID DISTRIBUTION IN HUMAN LEGS AND ARMS AND

ITS CLINICAL APPLICATION

H. Kanai, M. Haeno, H. Tagawa and K. Sakamoto

Department of Electrical and Electronic Engineering
Sophia University, Tokyo, Japan; and
Department of Internal Medicine
Mitsui Memorial Hospital
Tokyo, Japan

INTRODUCTION

Since the distribution of fluid in living tissue is greatly affected
by physiological variables such as those of blood circulation, tissue metabolism and electrolyte concentrations in the intra- and extracellular
fluids, information concerning fluid distribution is of some importance in
the diagnosis of various diseases, the monitoring of seriously ill patients
and in treatments such as artificial dialysis. It is, however, quite difficult to measure the distribution of fluid in the tissues in vivo.

The frequency characteristics of the electrical impedance of the living
tissues show that there are three types of frequency dispersions, α, β and
γ, due to three different mechanisms of relaxation (Ackman et al., 1976;
Cole and Cole, 1941; Geddes and Baker, 1967; Schwan, 1957; 1959). The β
dispersion is well-known as a structural relaxation, because it is mainly
affected by the structure of tissues composed of the membranes of cells and
intra- and extracellular fluid. Therefore, the electrical impedance around
the β dispersion frequency gives us information concerning the cell membrane
and intra- and extracellular fluid (Ackman et al., 1976).

In this chapter a new method for the measurement of fluid distribution
in the tissues is proposed using electrical impedance. This method is based
on the β dispersion theory mentioned above. Some clinical applications of
the resulting measurements are also shown (Haeno et al., 1985; Kanai et
al., 1983; Sakamoto et al., 1979).

THEORY

β dispersion of living tissue occurs at radio frequencies between 10kHz
and 10MHz. Fig. 1 shows a schematic diagram of the tissues. Fig. 2a shows
an equivalent electrical circuit of the tissues. The electrical properties
of the cell membrane can be represented by a large capacitance which lies
between 1 and 10 $F\,cm^{-2}$ because the membrane resistance R_m (see Fig. 2a) is
very large and the membrane is very thin. When the frequency of the electrical voltage applied to the tissues is lower than the β dispersion frequency
F_o, as shown by the solid lines in Fig. 1, electrical current flows mainly

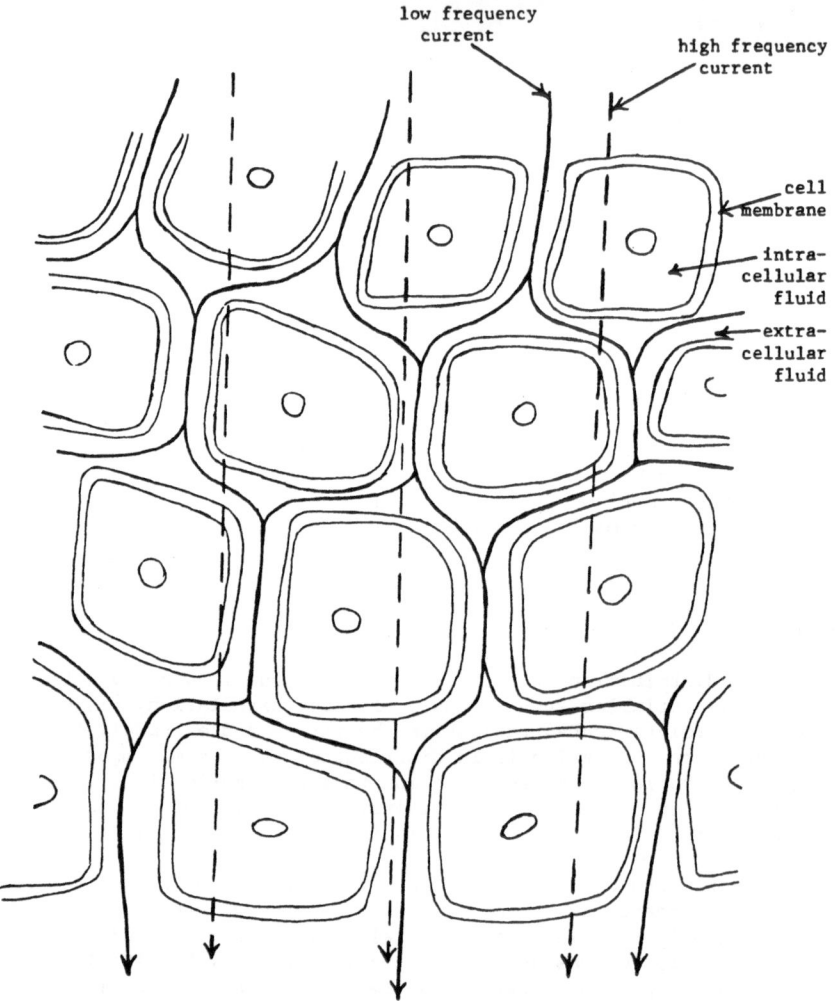

Fig. 1.   Schematic diagram of tissue and current flow.

through the extracellular fluid since the impedance of the cell membrane
is very large at low frequencies. With increase of frequency, the impedance
of the membrane decreases so that the current can flow through the membrane.
Therefore, at a frequency higher than F_o, the current flows through both
intra- and extracellular fluids, as shown by the broken lines in Fig. 1.
Finally, the impedance of tissue at low frequency gives us information about
extracellular fluid, whilst that at high frequency provides information
about the sum of the intra- and extracellular fluids. From these facts
the equivalent electrical circuit of living tissues shown in Fig. 2a can be
reduced to the simplified circuit shown in Fig. 2b in the radio frequency
range. R_i, R_e and C_m in Fig. 2b are the intra- and extracellular fluid
resistances and capacitance of the cell membrane, respectively. The admittance locus of the simplified circuit shown in Fig. 2b is a semicircle, as
shown by the solid line in Fig. 3. The time constant T_o of β dispersion is
given by $C_m R_i$, and the dispersion frequency F_o is $\frac{1}{2}\pi T_o$. The time constants
of cells in tissue such as muscle are not the same but are distributed. The
average dispersion frequency is about 70 kHz for skeletal muscle and 3 MHz
for blood. The measured admittance loci of living tissues usually constitute

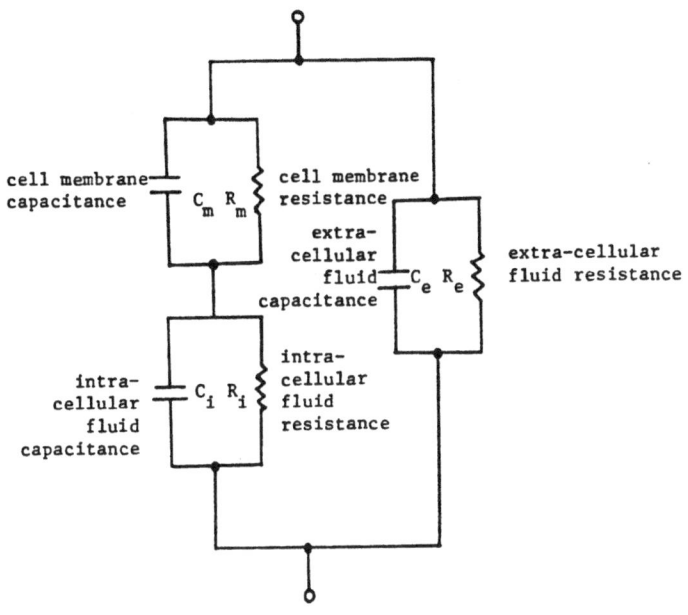

(a) Equivalent circuit for a cell

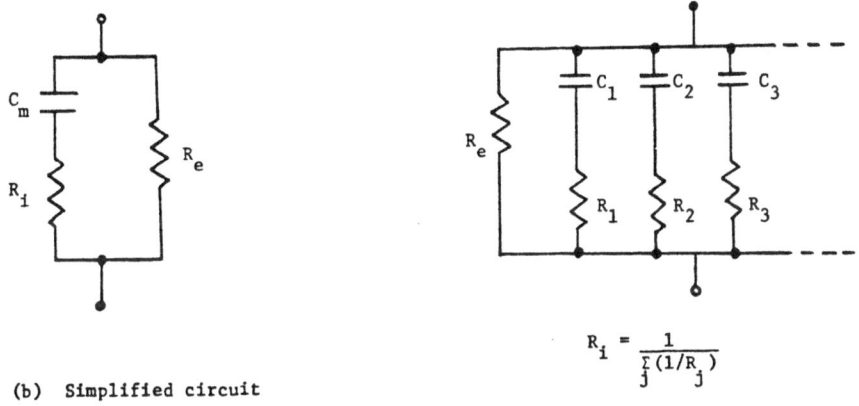

(b) Simplified circuit

$$R_i = \frac{1}{\sum\limits_{j}(1/R_j)}$$

(c) Distributed time constant circuit

Fig. 2. Equivalent circuits of a cell or a tissue.

only a part of a semicircle with the centre below the real axis. This is shown by a broken line in Fig. 3. The distribution function of the time constant of living tissue should be a normally distributed one because the size of cells is statistically distributed. The nature of the admittance loci of the normally distributed time constant circuit has been discussed in a number of papers (Cole and Cole, 1941; Haeno et al., 1985; Kanai et al., 1983; Schwan, 1957). The real equivalent circuit of tissues is given by the distributed time constant circuit shown in Fig. 2c. The admittance Y of the distributed circuit is usually represented by the Cole-Cole equation:

$$Y = \frac{1}{R_e} + \frac{1}{R_i} - \frac{1}{R_i\left(1 + (j\omega T_o)\right)^{1-\alpha}} = \frac{1}{R_e} + \frac{1}{R_i} - \frac{1}{R_i}\int_0^\infty \frac{f(t)}{(1 + j\omega T_o)}\,dt \qquad (1)$$

61

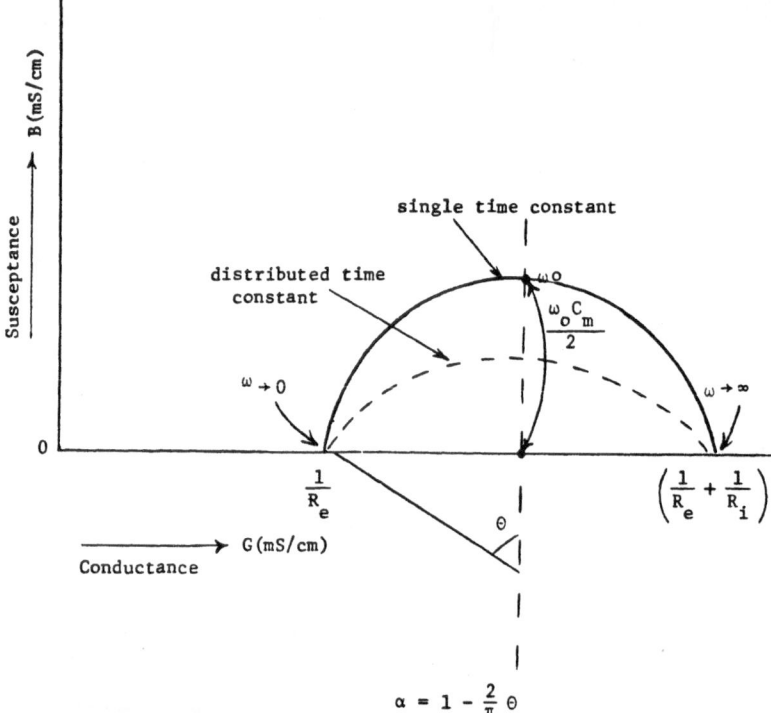

Fig. 3. Admittance loci for single time constant circuit and
         distributed time constant circuit.

where f(t) is the Cole-Cole distribution function given by the following
equation (Cole and Cole, 1941; Schwan, 1957):

$$f(t) = \tfrac{1}{2} \sin\alpha\pi/\left(\cos h\{(1 - \alpha)\log T/T_0\} - \cos\alpha\pi\right) \qquad (2)$$

T_0 is the average value of the distributed time constant and the R_i of the
whole tissue is given by the following equation:

$$R_i = \frac{1}{\sum\limits_{k=1}^{\infty}\left(\dfrac{1}{R_{ik}}\right)} \qquad (3)$$

R_i and R_e are related not only to the resistivities and volumes of the
intra- and extracellular fluids, but also to the shape factors of the tissue
s_i and s_e, respectively. The shape factors are the effective ratios of
fluid to other elements of intra- and extracellular substances. Therefore,
they are affected by the shape and structure of the cells and the volume of
intra- and extracellular fluids. Since it can be assumed that the resistivities of intra- and extracellular fluid and the shape and structure of the
cells do not change when the fluid volume changes, R_i and R_e are mainly
affected by the intra- and extracellular fluid volumes.

    To discuss the results it is necessary to normalise the data obtained
from subjects of various shapes and sizes. Assuming that the legs and arms
between two voltage electrodes are a cylinder with cross-section S and
length L, normalised resistivities RE and RI are shown as:

$$RE = R_e \cdot S/L \qquad (4)$$
$$RI = R_i \cdot S/L \qquad (5)$$

On the other hand, RE and RI are represented by

$$RE = \rho_e \cdot (S/S_e) \qquad\qquad (6)$$

and $RI = \rho_i \cdot (S/S_i),$ (7)

respectively, where ρ_e and ρ_i are the real resistivities of extra- and intracellular fluid. S_e and S_i are the effective areas of extra- and intracellular fluid; S_e and S_i depend on the shape factor s_e and s_i, respectively.

Since the real resistivities ρ_i and ρ_e are almost the same for human beings, RE and RI are mainly affected by the ratios S/S_e and S/S_i, respectively. Therefore, the effective intra- and extracellular fluid volumes can be estimated from RE and RI.

EXPERIMENTAL METHOD

      Fig. 4 shows the block diagram of the experimental arrangement. Impedance is measured by a 4-electrode method to reduce the effects of electrode impedance, skin impedance and spread impedance. Current electrodes of 2 cm x 6 cm area and voltage electrodes 2 cm x 3 cm area are placed on a leg or an arm, as shown in Fig. 4. In order to measure the frequency characteristics of living tissue impedance, it is required to vary the frequency of the applied current around the dispersion frequency. Our measurement system is computer-controlled and sweeps the frequency band from 2 kHz to 500 kHz automatically within one second. A constant current of 100 µA amplitude is supplied to the current electrodes. The magnitude and phase of the impedance are calculated from the supplied current and the potential differences between the two voltage electrodes. The constant current source and the detecting circuit are floated from ground and isolated from the computer by photocouplers in order to prevent any electrical hazard. All input and output circuits are shielded by the shield wire whose potential is kept at the same value as the inner wire to reduce the effects of stray capacitance. The admittance locus and the parameters α, T_o, R_i, R_e and C_m are calculated from the measured values by the least square curve fitting method. Measurement errors are ±1% in amplitude and ±1 degree in phase. For discussion of the results, the resistivities of whole blood and plasma sampled from the subjects are measured because they are closely related to the real resistivities of intra- and extracellular fluid.

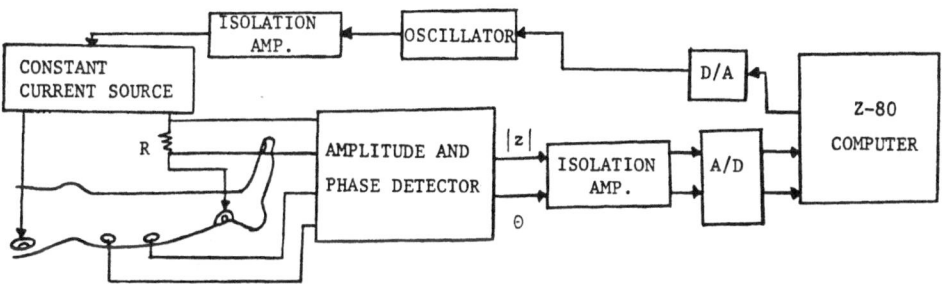

Fig. 4. Block diagram of experimental arrangement.

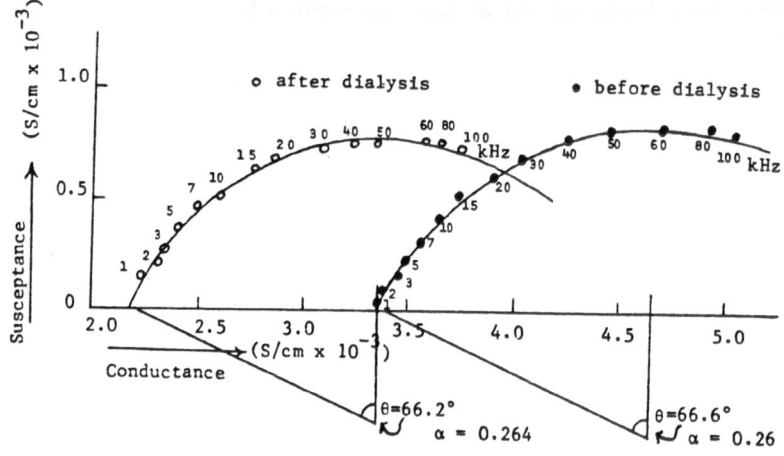

(a) Admittance loci

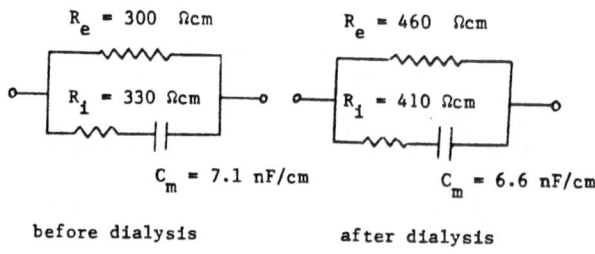

(b) Simple equivalent circuits

Fig. 5. Admittance locus and simple equivalent circuits for a
patient before and after artificial dialysis.

RESULTS

The frequency characteristics of electrical properties of the human
arms and legs are measured before, during and after artificial dialysis
and various exercises. The changes in fluid distribution are discussed
on the basis of the measured values.

Impedance Change by Artificial Dialysis

Typical admittance loci and equivalent circuits of a patient's leg
before and after artificial dialysis are shown in Fig. 5. In this case,
the weight of the patient decreased from 64.5 to 63.1 kg as a result of
the artificial dialysis. The extracellular fluid resistivity RE increased
remarkably from 300 to 465 Ωcm, whilst the intracellular fluid resistivity
RI decreased slightly from 390 to 384 Ωcm.

Fig. 6 shows some of the results both before and after 5 hours' regular dialysis. The regular dialysis was repeated 5 times over a period of
12 days. In the figure, the filled and open circles show the results obtained from a leg and an arm, respectively. The solid lines show the
differences of RE, RI, C_m, α and weight over the period of the dialysis.
From these results we can see that RE increases substantially and RI

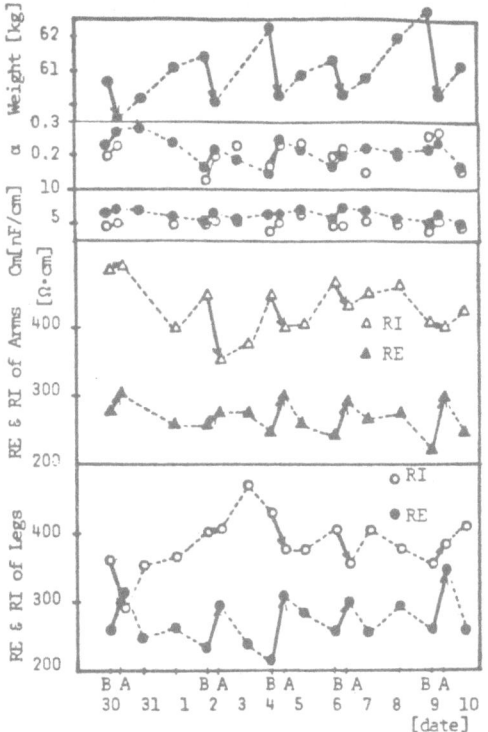

Fig. 6. Change of parameters due to dialysis, B - before dialysis, A -
after dialysis.

decreases slightly due to artificial dialysis. This means that regular
artificial dialysis brings about a decrease in extracellular fluid volume
and an increase in intracellular fluid volume. From a physiological point
of view, it is very interesting that the intracellular fluid volume increases despite the decrease of body weight due to the dialysis. Therefore,
even after dialysis the patient is in abnormal physiological condition.
It is, however, very difficult to obtain precise information about the exact
fluid volume from RE and RI, because these parameters are greatly dependent
on the cross-sectional structure of the tissues at the measurement site.
After dialysis, the value of RE of both arm and leg is always about 300 Ωcm.

We also periodically measured the change of leg impedance of a patient
both during and between each artificial dialysis over a period of 4 days.

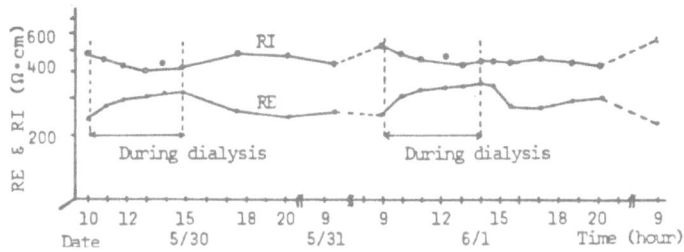

Fig. 7. Change of RE and RI for a haemodialysed patient.

The measurement results are shown in Fig. 7. Most of them indicate that
RE increases substantially and RI decreases due to regular artificial dialysis.

The chemical and electrical properties of blood were measured before,
during and after artificial dialysis. These results can be summarised as:

  (i)   the haematocrit value and the concentration of the protein in plasma
        increase by about 10% due to artificial dialysis;
  (ii) the osmotic pressure of plasma decreases by 5 - 10%; and
 (iii) the electrical properties do not change.

From these results it can be seen that by regular artificial dialysis fluid
in the plasma and the extracellular space is extracted at first, and then
the extracellular fluid moves into the cells due to the increase of osmotic
pressure of the extracellular fluid; and as a result of fluid movement, the
intracellular fluid volume increases.

Results of High Sodium Dialysis (HSD)

In order to discuss the fact that extracellular fluid moves into the
cells due to artificial dialysis, it should be noted that the electrical
impedance changes for some patients, due to high sodium artificial dialysis,
because HSD keeps the osmotic pressure of the plasma almost constant during
dialysis. Typical changes in RE and RI are shown in Fig. 8. During HSD,
RE gradually increases, whilst RI either does not change at all or else
shows a slight increase. On the other hand, RE increases and RI obviously
decreases during regular sodium artificial dialysis (RSD). From these results it might be concluded that both intra- and extracellular fluid is
gradually extracted during HSD and that extracellular fluid moves into cells
during RSD like the results shown in Figs. 6 and 7 (van Stone et al., 1980).

Analytical data on the chemical properties of the patient's blood are
shown in Table 1. From these results it can be concluded that the increase
of osmotic pressure as a result of HSD causes the decrease of intracellular
fluid volume. The slight decrease in total protein during HSD treatment

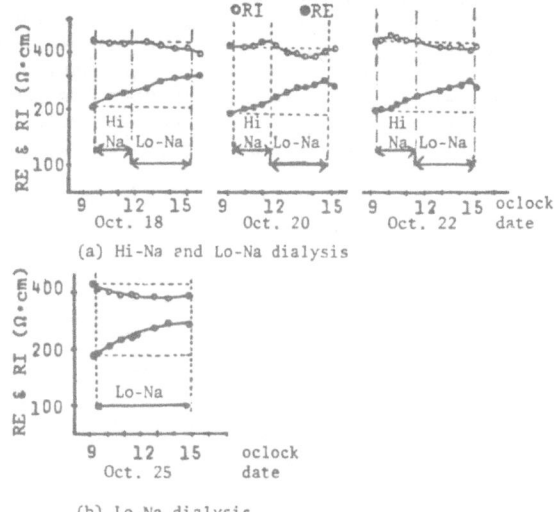

Fig. 8.   Change of RE (solid circle) and RI
          (open circle) by high sodium (Hi-Na),
          regular sodium (Lo-Na) dialysis.

Table 1. Change in Laboratory Data with High and Regular Sodium Haemodialysis

			(a) High Sodium			(b) Regular Sodium		
			before	during	after	before	during	after
Blood	TP	$[g\ dl^{-1}]$	6.6	6.4	6.7	7.1	7.6	8.4
	BUN	$[mg\ dl^{-1}]$	110	81	43	107	69	44
	Na	$[mEq\ l^{-1}]$	137	141	138	136	137	136
	Posm	$[mOsm\ kg^{-1}]$	318	311	292	319	303	295
	Posm'	$[mOsm\ kg^{-1}]$	279	282	277	281	278	279
Dialysate	Na	$[mEq\ l^{-1}]$	152	153	138	137	-	-
	Posm	$[mOsm\ kg^{-1}]$	311	309	276	-	-	-

leads to the slight increase in total plasma volume which is observed.
These results agree well with the changes in RE and RI.

Fluid Distribution Change Due to Various Exercises

It is very important to know the change of the fluid distribution
brought about by exercise since the change of intracellular fluid depends
on both the metabolic status and the potential for recovery of the human
subject. We measured the change of the frequency characteristics of the
electrical impedance of the legs and arms due to ergometer, running and
swimming exercise. The filled and open circles in Fig. 9 show the results
obtained before and after ergometer exercise, respectively, for a well-

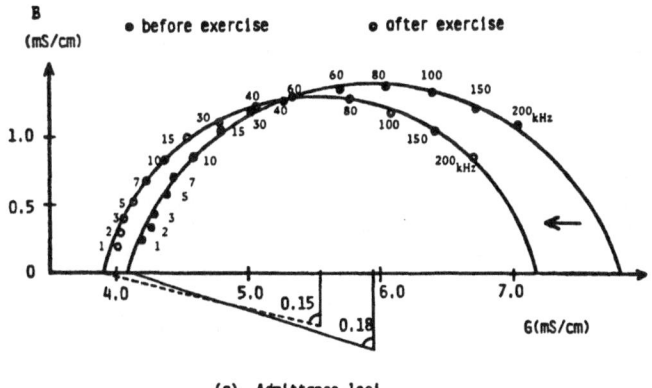

(a) Admittance loci

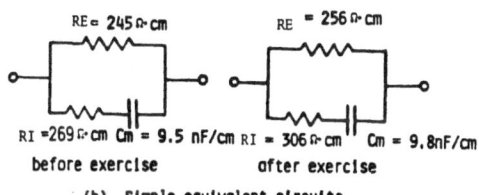

(b) Simple equivalent circuits

Fig. 9. Admittance loci and simple equivalent circuits for a trained subject: the change due
to ergometer exercise.

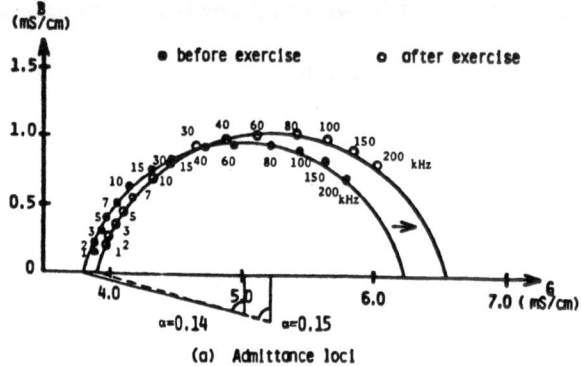

(a) Admittance loci

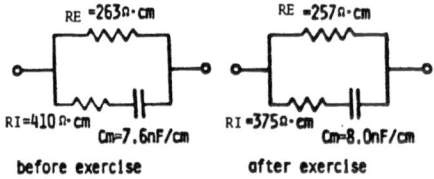

RE =263Ω·cm   RE =257Ω·cm

RI=410 Ω·cm   Cm=7.6nF/cm   RI=375Ω·cm   Cm=8.0nF/cm

before exercise   after exercise

(b) Simple equivalent circuit

Fig. 10.   Admittance loci and simple equivalent
circuits for an untrained subject:
the change due to ergometer exercise.

trained athlete. The weight of the subject decreased from 56.0 to 55.2 kg
as a result of the exercise. There was a slight increase in resistivity.
The characteristic frequency of structural relaxation also increased
slightly. The calculated values of RE, RI and C_m are shown in Fig. 9b.
RI and RE increased by 12% and 5% respectively. On the other hand, for
untrained subjects RE exhibits only a slight decrease of about 1%, whilst
RI decreases by about 9%, as shown in Fig. 10.   This phenomenon is not yet
well understood physiologically. However, it is of interest that the
changes of fluid distribution in untrained subjects, as a result of exercise, are greatly different from those of well-trained subjects. This
phenomenon might be useful for the monitoring of patients and the measurement of athletic ability because the change of fluid distribution depends
on the metabolic status and the potential for recovery of the subject.

The changes in RI and RE of the arms and legs as a result of 3500 m
or 7000 m running exercise or 800 m swimming exercise are shown in Fig. 11.
RI and RE depend not only on the fluid volume, but also on the resistivity
of the fluid. We therefore also measured the resistivity of whole blood
(Rb) and plasma (Rp), the haematocrit value (Ht), heart rate (HR) and body
weight (W) before and after exercise, as shown in Fig. 11. Subjects A, B,
C and D are untrained males, E, F, G, H and I are well-trained male athletes,
and J and K are well-trained female athletes. Changes in body weight for
all exercise patterns are less than 1 kg. Rb and Ht increased slightly
after exercise. RE in the leg decreased remarkably for the untrained subjects, even though Rb increased and W decreased. On the other hand, RE for
well-trained subjects usually increased after exercise, as shown in Fig. 11.
This divergence may be caused by the difference in metabolic rate and recovery rate of the individuals. The recovery rate depends mainly on the
capability of the circulatory system. From these experimental results, it
is concluded that the changes in RI and RE can be used for the diagnosis
of circulatory disease and also for the investigation of athletic ability.

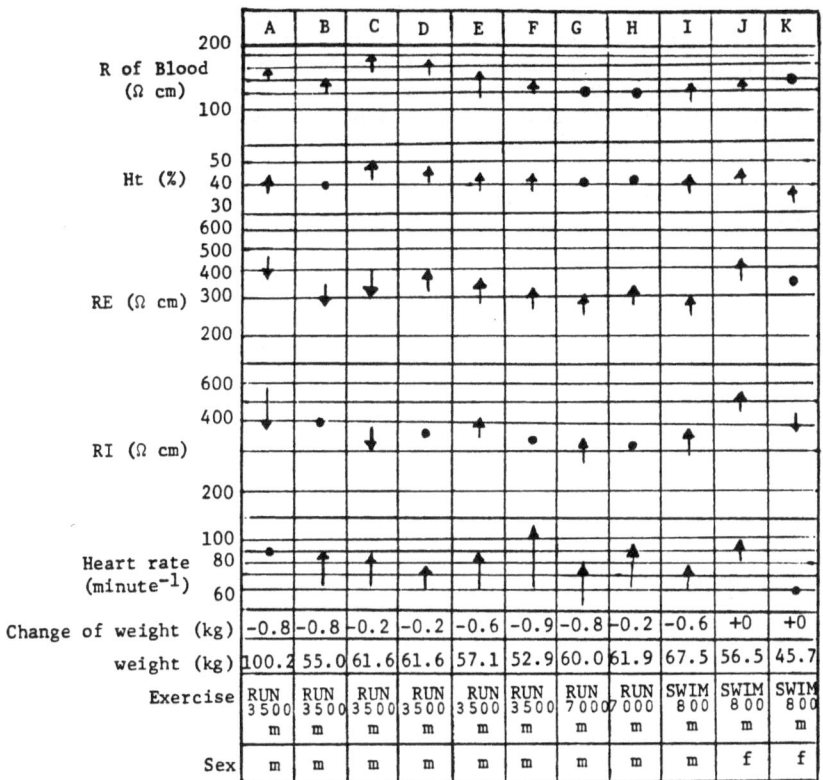

Fig. 11. Change of various values in 11 subjects (A - K) due to
running or swimming exercise. Arrows show the values
and direction of the changes. Closed circles mean no
change.

CONCLUSIONS

It has been shown, theoretically and experimentally, that the electrical characteristics of the human arms and legs over frequencies between
1 kHz and 500 kHz can be represented by a simple equivalent circuit which
consists of an equivalent resistivity of the extracellular fluid RE, and
equivalent resistivity of the intracellular fluid RI and an equivalent
capacitance of the cell membrane C_m. These parameters give us rough information about the intra- and extracellular fluid volumes and their distribution.

The frequency characteristics of the electrical properties of human
arms and legs have been measured with sufficient accuracy within 1 second.
In this chapter measured values obtained before, during and after artificial
dialysis and various patterns of exercise have been presented. From the
measured values, RE, RI, C_m and the characteristic frequency of the β dispersion were calculated; and from these results the intra- and extracellular fluid volumes and their distribution were estimated.

It can be concluded that R_e and R_i are very useful for the monitoring
of artificial dialysis and the estimation of athletic ability. More useful
information will be obtained from the measured values, such as the equivalent membrane capacitance, the characteristic frequency of the β dispersion
and the Cole-Cole distribution function. RE, RI and other parameters are

quite useful for the measurement of oedema, the estimation of metabolic rate and recovery potential, the diagnosis of various circulatory diseases, the monitoring of seriously ill patients and the effect of medicines, such as diuretics (Haeno et al., 1985; Kanai et al., 1983; Sakamoto et al., 1979; Thomasset et al., 1983; van Stone et al., 1980). These clinical applications will be discussed elsewhere.

ACKNOWLEDGEMENT

We are very grateful to Mr. T. Higashiizumi, Yomogawa Medical System, Ltd., Mr. M. Tanishima, Nihon Koden, Ltd., and Dr. N. Ono and Mr. Kano, Mitsui Memorial Hospital, for their valuable assistance in these experiments. We are also very grateful to Professor K. Nakayama, Sophia University, for his valuable discussion. A part of this work was supported by the Suzuken Memorial Foundation.

REFERENCES

Ackman, J. J., Seitz, M. A., Dawson, C., and Sances, A., 1976, Complex bio-electric impedance measurement for the detection of pulmonary edema, in: "Digest of 11th ICMBE, Ottawa, Canada" : 288.

Cole, K. S., and Cole, R. H., 1941, Dispersion and adsorption in dielectrics, J. Chem. Rev., 9 : 341.

Geddes, L. A., and Baker, L. E., 1967, The specific resistance of biological material, Med. Biol. Eng., 5 : 271.

Haeno, M., Sakamoto, K., and Kanai, H., 1985, Estimation of fluid distribution by impedance method, Jap. J. Med. Electron. Biol. Eng., 23 : 354.

Kanai, H., Sakamoto, K., and Haeno, M., 1983, Electrical measurement of fluid distribution in human legs: estimation of extra- and intra-cellular fluid volume, J. Microwave Power, 18 : 233.

Sakamoto, K., Higashiizumi, T., and Kanai, H., 1979, Frequency characteristics of the electrical impedance of legs, Jap. J. Med. Electron. Biol. Eng., 17 : 264.

Schwan, H. P., 1957, Electrical property of tissues and cells, Adv. Biol. Med. Phys., 5 : 147.

Schwan, H. P., 1959, Alternating current spectroscopy of biological substances, Proc. IRE, 10 : 1845.

Thomasset, A. L., Lenoir, J., Roullet, C., Jenin, P., Beruard, M., Bernard, C. H., and Baur, F., 1983, The physiological surveillance of hemodialysis session by the continuous measurement of L.F. impedance of the circulating blood, Clin. Exper. Dialysis Apheresis, 7 : 235.

van Stone, J. C., Bauer, J., and Carey, J., 1980, The effect of dialysate sodium concentration on body fluid distribution during hemodialysis, Trans. Am. Soc. Artif. Intern. Organs, 26 : 383.

A NEW METHOD FOR ELIMINATING MICRO-BUBBLES IN THE EXTRACORPORIAL BLOOD

CIRCULATION SYSTEM

K. Sakamoto, T. Yagimuma, M. Fujii and H. Kanai

Department of Electrical and Electronic Engineering
Sophia University
Tokyo, Japan

INTRODUCTION

Recently, in the medical field, especially in surgery, remarkable progress has been made in the area of artificial organs such as the artificial lung. Unfortunately, these organs sometimes have a major and most serious problem which occurs when air-bubbles intermix into blood. Usually, air-bubbles in the extracorporial circulation system produced by various sources such as the artificial heart-lung system are eliminated by various types of mechanical blood filters before the blood recirculates into the human vein. However, air-bubbles which are only a little larger or smaller than the mesh size of the filter can pass through it and these air-bubbles could sometimes be micro-embolic in the peripheral circulation bed and hence cause tissue damage.

Air-bubbles in liquid grow or diminish in size due to resonance phenomena, or else they coalesce to form a large air-bubble after being drawn towards the pressure node or anti-node by the force due to the pressure gradient, or else they approach each other as a consequence of mutual attraction in the stationary weak sound field.

Here, a new method is proposed based on these theories. In this chapter we explain, theoretically and experimentally, these phenomena and show some results of the experimental and clinical application of our new method. Small air-bubbles in the artificial lung grow in size and can be trapped and can be easily and effectively eliminated by the usual blood filter without any damage to the blood.

THEORY

Resonance Phenomena

It can be assumed that a small bubble in the blood whose radius is less than a few hundred micrometres is spherical because of its very large surface tension. When such small bubbles are in an ultrasound field, they vibrate with three modes which are volume vibration, translational motion and surface vibration. Here, we are interested in volume vibration.

If the applied acoustic ultrasound pressure $P(t) = P_0 + a.\exp(j\omega t)$ and the wavelength of this ultrasound is much larger than the radius of the bubble, the spontaneous average radius R of a bubble which vibrates volumetrically is given by (1):

$$R = \bar{R} + P.B/\left(\rho_o \bar{R} \omega_r^2\right) \exp[j(\omega t - \theta)] \tag{1}$$

where

$$\theta = \tan^{-1}[\delta\omega.\omega_r/(\omega_r^2 - \omega^2)] \tag{2}$$

$$B = 1\Big/\sqrt{[1 - (\omega/\omega_r)^2]^2 + (\delta\omega/\omega_r)^2} \; , \tag{3}$$

$\bar{R}$ is the average bubble radius, ω_r is the resonant angular frequency, ρ_o is the density of the liquid, and δ is the damping factor. When the damping factor δ is zero, the relation between the radius of the bubble and the resonance frequency F_o is given by:

$$F_o = 1/R_o\sqrt{3\gamma\bar{P}_i/\rho_o} \tag{4}$$

where R_o is the bubble radius (resonance radius), $\bar{P}_i$ is the average pressure inside the bubble, and γ is the specific heat of gas.

When the bubble radius R is less than the resonance radius R_o, a small bubble absorbs air which has melted into the liquid around the bubble and its radius increases to a certain stable value R_C which is almost the same as the resonance radius R_o. On the other hand, when R is larger than R_o, air in the bubble melts into the liquid, and the air-bubble diminishes in size to R_C.

In our case, the growth or contraction rate is important and is given by (5) (Higashiizumi, 1979):

$$\frac{d\bar{R}}{dt} = \frac{\alpha_B \kappa \rho_2}{\rho_{20}(\bar{P}_o\bar{R} + 2T)}\left[S\bar{P}_o - \frac{2T}{R} + \frac{2\bar{R}^2\rho_1\omega^2P^2B^2}{9\gamma^2\omega_r^4(P_o + 2T/\bar{R})^2}\right] \tag{5}$$

where α_B is the Bunsen absorption coefficient, T is surface tension, κ the diffusion constant of gas in liquid, ρ_2 the density of gas under standard conditions, ρ_{20} the actual density of gas, S the degree of the supersaturation in the liquid, and f the frequency of the applied ultrasound. The relation between the irradiation time and the radius of an air-bubble in liquid can be calculated under various conditions from (5).

The Effects of a Non-uniform Sound Field

When the ultrasound field is not uniform, a force (the Bjerkness force) caused by the pressure gradient is exerted on air-bubbles. Now let us assume that the ultrasound field, which consists of the plane standing acoustic wave P_S as a non-uniform ultrasound field, is symmetric about the z-axis and is given by (6):

$$P_S = P\sqrt{1 + K^2 + 2K\cos(4\Pi Z/\lambda)}\exp j(\omega t + \alpha) \tag{6}$$

where

$$\alpha = \tan^{-1}(K_1\tan 2\Pi Z/\lambda) \tag{7}$$

$$K_1 = (1 - K)/(1 + K) \tag{8}$$

where λ is the wavelength in the Z direction and K is the reflection coefficient.

The average acoustic force F_g resulting from the pressure gradient in the Z direction acting on the air-bubble is given by (9):

72

$$F_g = \frac{4\Pi^2 R P^2 B \sqrt{(1 + K^2)^2 - 4K^2 \cos(4\Pi Z/\lambda)}}{\rho_1 \omega_r^2 \lambda} \cos(\theta + \alpha - \beta - \Pi/2) \tag{9}$$

where

$$\beta = \tan^{-1}[(\tan 2\Pi Z/\lambda)/K_1] \tag{10}$$

When a bubble moves as a result of the force given by (9), by taking into consideration the viscous drag F_v and inertia F_i of the liquid around the bubble and the buoyancy F_b, the moving velocity V_o of the bubble is given by (11):

$$V_o = (4R^3 \Pi \rho_1 g/3 + Fg)/6R\Pi\mu \tag{11}$$

where μ is the viscosity of the liquid around the bubbles and g is the acceleration due to gravity. From this equation it can be seen that when the bubble radius R is smaller than R_o, the bubble moves to a certain stable location L_o near the pressure anti-node of the standing acoustic wave at which F_g is equal to F_b. On the other hand, if R is larger than R_o, then it moves to L_o near the pressure node.

From these results it can also be understood that if there are many small bubbles in the same field, they could move to L_o near the node or anti-node and coalesce into a larger bubble.

Mutual Attraction

The theory concerning the mutual attraction force F_m has been discussed in detail by Crum (1969). Equation (12) defines this attraction force as calculated by him:

$$F_m = -2\Pi\rho_1 \omega^2 d_1 \cdot d_2 \cdot \bar{R}_1^2 \bar{R}_2^2 \cos(\theta_1 - \theta_2)/r^2 \tag{12}$$

where $\bar{R}_1$ and $\bar{R}_2$ are the radii of bubbles which vibrate with amplitudes d_1 and d_2, respectively. θ_1 and θ_2 are the phase differences between the applied ultrasound and the vibrations of the bubbles and r is the distance between the two bubbles. d_1 and d_2 are the amplitudes of the bubble vibrations as calculated from the second term in (1). Since the mutual attraction force is inversely proportional to the square of the distance between the air-bubbles, when such bubbles are close to each other, they rapidly approach one another and coalesce to form a larger bubble.

THEORETICAL AND EXPERIMENTAL RESULTS

The radius of the air-bubble was determined from the modified Stokes equation using its measured buoyant speed. The ultrasound field is provided by (ferrite) magnetostrictors. The number and size of the air-bubbles in the extracorporial circulation system are counted just at the outlet from the artificial lung and before and after the mechanical blood filter by the micro-bubble activity monitor which is a 2MHz ultrasound type detector. In order to measure the velocity of the moving bubbles, a very high-speed video camera with a 10 ms exposure light (LED) is used.

Fig. 1 shows one set of calculated and experimental results for the resonance phenomena. The vertical and horizontal axes are the spontaneous radius of the air-bubble and the ultrasound irradiation time, respectively. The solid lines and the open and filled circles correspond to the theoretical and experimental results. The pressure amplitude and frequency of applied ultrasound are 0.5 and 0.77 atm and 28.7 kHZ, respectively. The resonance radius R_o of the air-bubble is 134 µm and 137 µm, corresponding to 0.5 and 0.77 atm. From these results it can be seen that experimental

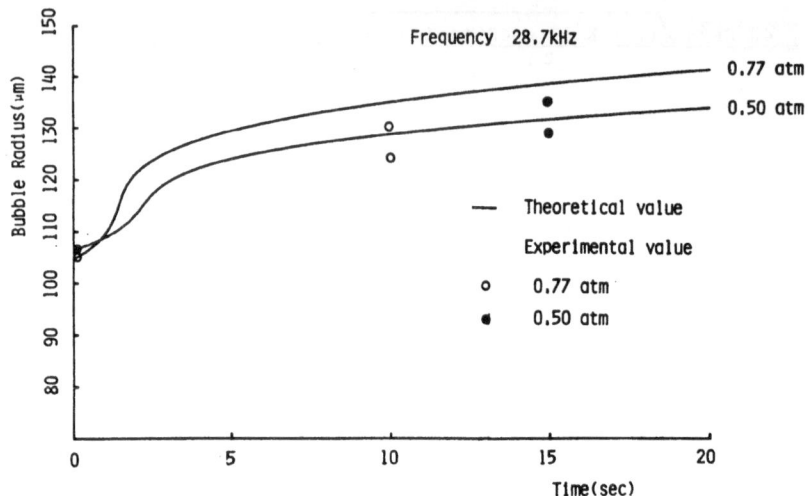

Fig. 1. The relation between spontaneous average bubble radius
and the irradiation time of the ultrasound.

values agree well with the theoretical predictions. Even if the applied
ultrasonic field is very weak, when the radius of the air-bubble is smaller
than the critical value, the air-bubble expands and approaches a certain
stable size R_C. From the theoretical and experimental results it is concluded that even though small bubbles, the radii of which are less than
10 μm, can pass through the usual mechanical blood filter, they can be absorbed in the blood by their own surface tension. For small bubbles which
can pass through the usual mechanical blood filter but are not absorbed in
the blood, it takes about 30 s to approach the stable magnitude under a weak
ultrasound field whose frequency is 50 kHz and pressure amplitude is about
0.5 atm.

Fig. 2 shows the movements of bubbles in an acoustic standing wave
whose frequency, standing wave ratio and maximum amplitude are 28.7 kHz,

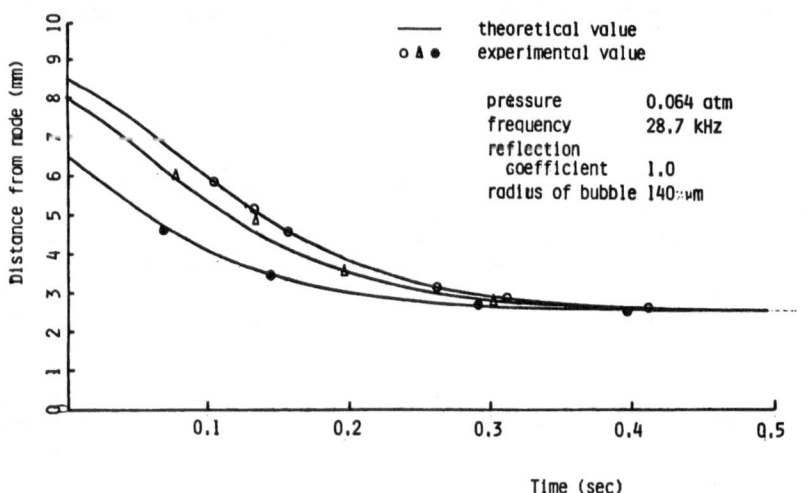

Fig. 2. The movement of the bubble due to the force caused by
the pressure gradient of the acoustic standing wave.

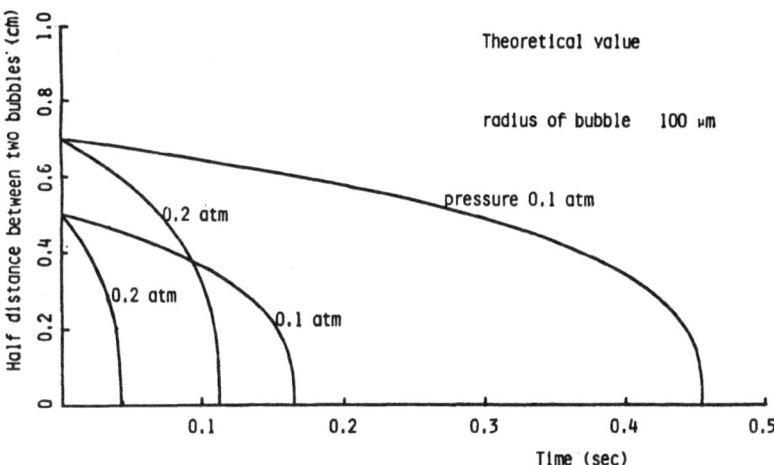

Fig. 3. The change of the half distance between two bubbles
resulting from the mutual attraction.

approximately 2, and 0.06 atm, respectively. In this figure the solid
lines, filled and open circles and triangles correspond to the theoretical
values and experimental results. Experimental results accord well with the
theoretical predictions. In this case it takes about 0.3 s for a 140 μm
radius air-bubble to reach the pressure node. From these theoretical and
experimental results it can be concluded that if there are many small
bubbles in the same field, they could rapidly move to near the pressure
node or anti-node and coalesce to form a single large bubble. Small bubbles
move to near the pressure anti-node and coalesce to form a large bubble
which then starts to move towards the pressure node. Therefore, finally
all the small bubbles rapidly congregate near the pressure node.

Fig. 3 shows the effect of the mutual attraction on small bubbles, as
calculated by Crum. The vertical and horizontal axes correspond to the
spontaneous distance between two bubbles and the irradiation time of the
ultrasound, respectively.

Next, we show some results of clinical application of our new device.
A transducer which provides ultrasound of a pressure amplitude 0.2 atm and
frequency 50 kHz is set on to the bottom of the bubble-type artificial lung.
The resonance bubble radius for 50 kHz is approximately 70 μm.

Fig. 4 shows the ultrasound pressure distribution inside the blood
reservoir of the artificial lung. The pressure node is located in the
middle part of the blood reservoir.

Fig. 5 comprises photographs which show the bubble condition inside
the blood reservoir as measured by ultrasono-tomography. The upper photograph corresponds to the condition before the irradiating ultrasound is
applied. There are many air-bubbles in the lower part of the blood reservoir. The lower photograph shows the condition during ultrasound irradiation. The bubbles in the lower part of the blood reservoir disappear and
no bubble flows through the outlet of the reservoir. Bubbles are trapped
in the middle part of the blood reservoir where the pressure node is
located.

Before the irradiating ultrasound is applied there are many bubbles
at the outlet of the artificial lung and just before the blood filter, as

Fig. 4. Pressure distribution in the blood
reservoir (x 10^{-2} atm).

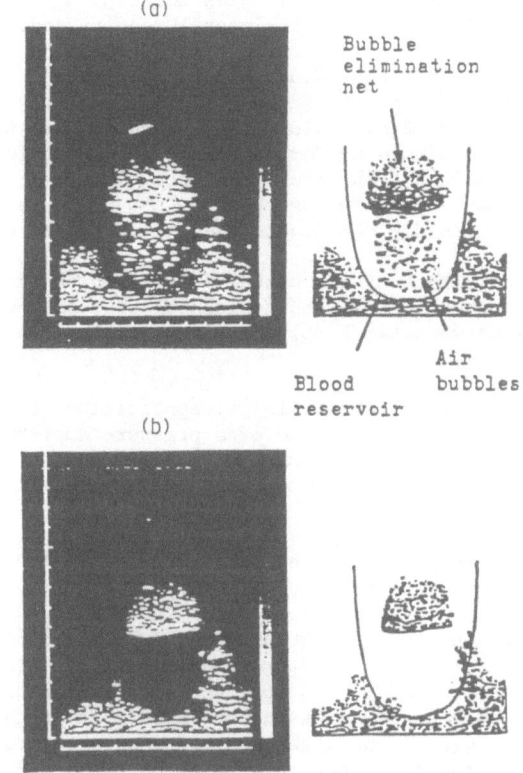

Fig. 5. Photographs of bubbles inside the
blood reservoir showing the condition (a) before and (b) during irradiation of ultrasound.

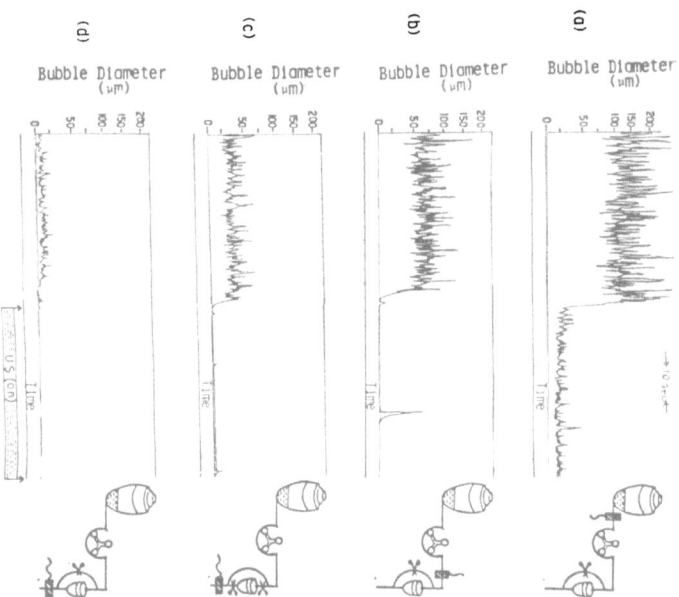

Fig. 6. Clinical results of the bubble elimination
effect of the ultrasound air-bubble remover.

shown in Fig. 6(a) and (b). Even though the number of bubbles decreases
remarkably after the blood filter whose mesh size is 40 μm, bubbles between
20 μm and 40 μm still exist, as shown in Fig. 6(c). During the irradiating
ultrasound, the number and size of the air-bubbles decreases substantially,
and in particular no bubbles are detected after the blood filter, as shown
in Fig. 6(d).

From these results it can be concluded that small air-bubbles in the
blood reservoir of the artificial lung move rapidly to the pressure node
and coalesce to form a large bubble by the net force caused by the pressure
gradient and the mutual attraction force. Then this large bubble is
trapped near the pressure node or else moves to the upper part of the blood
reservoir due to buoyancy.

Finally, Fig. 7(a) and (b) shows the change in haemoglobin contents
in the plasma of resting and flowing blood due to the irradiation of ultrasound. The solid and broken lines in Fig. 7(a) show the results when ultrasound of 60 mW cm^{-1} is irradiated and when it is not, respectively.

There is no remarkable difference between them. Also there is no
remarkable difference in the increasing rate of haemoglobin contents during
ultrasound and non-ultrasound irradiation. In addition, it is found that
there is no change in the oxygen content in blood by irradiating ultrasound
of 60 mW cm^{-1}. From these theoretical and experimental results, it is concluded that the effect of ultrasound on haemolysis is quite small.

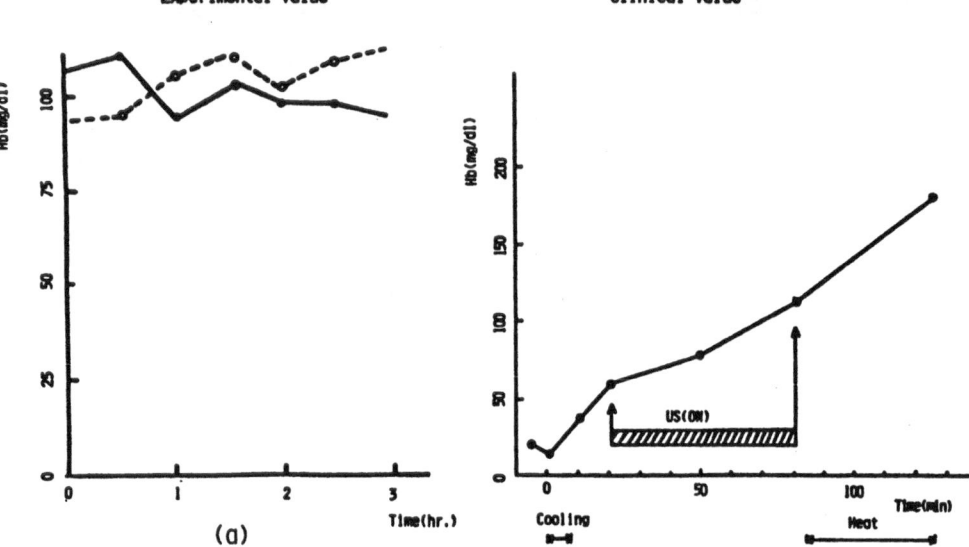

Fig. 7. Haemolysis change in the extracorporial circulation before and after ultrasound irradiation.

CONCLUSIONS

From these theoretical and experimental studies, our new method can effectively eliminate small bubbles in the extracorporial circulatory system which the usual mechanical blood filters cannot eliminate. Such small bubbles would sometimes be micro-embolic in the peripheral circulatory bed and thus cause tissue damage. Therefore, this ultrasonic bubble elimination method has considerable clinical applicability.

ACKNOWLEDGEMENT

We thank Professor K. Nakayama, Sophia University, Tokyo, for his useful advice and discussion, Mr. T. Higashiizumi, Yokogawa Medical System, Ltd., Tokyo, for his excellent theoretical analysis, discussions and experimental work, and Dr. N. Ono, Mitsui Memorial Hospital, Tokyo, for his valuable assistance in the experiments. Part of this work was supported by grant-in-aid from The Mochida Memorial Foundation for Medical and Pharmaceutical Research.

APPENDIX

The values of the parameters used in the theoretical calculations are:
$\alpha_B = 0.021$, $C_p = 1.01 \times 10^3$ [J(kgK)$^{-1}$], $C_1 = 4.19 \times 10^3$ [J(kgK)$^{-1}$],
$C = 1.47 \times 10^3$ [m s^{-1}], $K = 2.2 \times 10$ [m^2 s^{-1}], $K_g = 2.52 \times 10^{-2}$ [W mK^{-1}],
$K_1 = 5.88 \times 10^{-1}$ [W mK^{-1}], $\gamma = 1.4$, $\rho_1 = 1.0 \times 10^3$ [kg m^{-3}], $\rho_{20} = 1.29$
[kg m^{-3}], $\rho_2' = 1.23$ [kg m^{-3}], $\rho_{20} = 1.23$ [kg m^{-3}], $P_O = 1.0 \times 10^5$ [N m^{-2}],
$S = 0$, $T = 73 \times 10^{-3}$ [N m^{-1}], $\mu = 1.14 \times 10^{-3}$ [Ns m^{-2}].

BIBLIOGRAPHY

Crum, L. A., 1969, Motion of bubbles in a stationary sound field, J. Acoustic. Soc. Am., 48 : 181.

Crum, L. A., 1975, Bjerkness forces on bubbles in a stationary sound field, J. Acoustic. Soc. Am., 57 : 1363.

Devin, Jr., C., 1959, Survey of thermal, radiation and viscous damping of pulsating air bubbles in water, J. Acoustic. Soc. Am., 31 : 1654.

Eller, A., 1968, Force of a bubble in a standing acoustic wave, J. Acoustic Soc. Am., 43 : 170.

Higashiizumi, T., 1978, "Physical Properties of Micro-bubbles in Blood in a Stationary Ultrasound Field and Application", Master's Thesis, Sophia University, Tokyo.

Saneyoshi, J., 1953, Growth and disappearance of bubble which absorbs ultrasonic wave in liquid, Bull. Tokyo Inst. Technol., (Series-B) 6 : 13.

William, M., 1977, A new noninvasive technique for cardiac pressure measurement; resonant scattering of ultrasound from bubbles, IEEE Trans. Biomed. Eng., BME-242 : 102.

MEASUREMENT OF LEG BLOOD VOLUME CHANGE BY IMPEDANCE PLETHYSMOGRAPHY

WITH SPECIAL REFERENCE TO MICROGRAVITY SIMULATION TESTING

K. Makie, K. Nakayama, S. Yagi, A. Miyamoto and K. Yajima

Department of Electrical & Electronic Engineering
Sophia University, Tokyo; and
Department of Hygiene, Nihon University
Tokyo, Japan

INTRODUCTION

Exposure of the human circulatory system to microgravity causes a headward shift of blood and leads to an increase in venous return, cardiac output, heart rate and systolic and mean pressures, together with a decrease in diastolic pressure. In the resultant adjustment, the total blood volume is decreased. The adaptation of the circulatory system to microgravity proceeds fairly smoothly, but some symptomatic cardiovascular dysfunctions are reported to occur when astronauts return to earth. The causes which provoke these symptoms are considered to be not only blood volume reduction, but also the change of cardiovascular control mechanisms, such as modification of the baroreceptor reflex. Orthostatic intolerance is known as one of the marked symptoms and this is considered to be provoked primarily by the blood shift to the lower extremities upon standing in the gravitational state. Thus, in order to analyse the adaptation and the re-adaptation of the circulatory system to microgravity and to normal gravity, it is important to study the dynamics of the blood shift to the lower extremities (Gaffey, 1985; Thornton et al., 1977). A multi-voltage-channel impedance plethysmograph has been developed for this purpose and applied in an orthostatic intolerance test following a microgravity simulation experiment.

MULTI-VOLTAGE-CHANNEL IMPEDANCE PLETHYSMOGRAPHY

Impedance plethysmography is a non-invasive method which is conventionally and widely used to measure blood volume change in the upper or lower limbs. A four-electrode system with a pair of voltage electrodes and a pair of current-supplying electrodes is normally used. However, if one intends to measure the blood volume change of the whole leg, it is desirable to use more voltage electrodes so as to minimise the error resulting from the non-uniformity of the tissue structure of the leg. In this section, we shall present a simplified analysis of this error, and give the rationale for the multi-voltage-channel system.

If the leg can be regarded as a uniform conductive cylinder, and if the blood volume change can be considered as a simple parallel addition of a uniform blood column to that cylinder, then the blood volume change can be estimated from the admittance change of this cylinder as follows:

$$\Delta V = \Delta Y L^2 / \sigma_b \tag{1}$$

where ΔV, ΔY, L and σ_b denote blood volume change, admittance change, segment length and the electrical conductivity of blood, respectively.

However, equation (1) is not fully applicable to the whole leg for the following reasons. First of all, the anatomical difference leads to the inhomogeneity of the cross-sectional area and the electrical conductivity of tissue. Secondly, the difference of vascularisation causes the inhomogeneous distribution of the amount of possible change of stored blood mass.

Let us now consider the leg as an inhomogeneous conductive cylinder along the x-axis. If the increase in the admittance in an infinitely small section can be modelled as a simple adding of a uniform blood column parallel to that infinitesimal section of the leg, the increase in admittance in the section from x_1 to x_2 can be expressed by the following equation:

$$
\begin{aligned}
\Delta Y &= Y - Y_o \\
&= \left(\int_{x_1}^{x_2} \frac{dx}{\sigma S + \sigma_b \Delta S} \right)^{-1} - \left(\int_{x_1}^{x_2} \frac{dx}{\sigma S} \right)^{-1} \\
&= \frac{\displaystyle\int_{x_1}^{x_2} \frac{\sigma_b S dx}{\sigma S (\sigma S + \sigma_b \Delta S)}}{\displaystyle\int_{x_1}^{x_2} \frac{dx}{\sigma S} \cdot \int_{x_1}^{x_2} \frac{dx}{\sigma S + \sigma_b \Delta S}}
\end{aligned}
\tag{2}
$$

where $\sigma(x)$, $S(x)$ and $\Delta S(x)$ denote the electrical conductivity of tissue, the initial cross-sectional area and the cross-sectional area change caused by the stored blood mass change, respectively.

If we use a one-voltage-channel system, that is the conventional four-electrode system in which two voltage electrodes are on the edge of the segment of interest, blood volume change would be estimated by (1).

However, the true value of blood volume change ΔV_{true} is the integration of the piece-wise volume change along the x direction. Thus the error in the estimated blood volume change can be calculated as follows:

$$
\begin{aligned}
\text{ERROR} &= \Delta V_{est} - \Delta V_{true} \\
&= \Delta Y \cdot L^2 / \sigma_b - \int_L \Delta S dx \\
&= \frac{\displaystyle\int_L \frac{\Delta S dx}{\sigma S (\sigma S + \sigma_b \Delta S)} \cdot \left(\int_L dx \right)^2}{\displaystyle\int_L \frac{dx}{\sigma S} \cdot \int_L \frac{dx}{\sigma S + \sigma_b \Delta S}} - \int_L \Delta S dx
\end{aligned}
\tag{3}
$$

If the segment of interest is divided into N sections and the admittances are measured in each section separately, we shall obtain a better estimate for the total blood volume change. The error will thus be as follows:

$$
\begin{aligned}
\text{ERROR} &= \Sigma \Delta V_{est}(i) - \Delta V_{true} \\
&= (L/N)^2 \cdot \Sigma \Delta Y_i / \sigma_b - \int_L \Delta S dx
\end{aligned}
$$

$$
\Delta Y_i = \frac{\displaystyle\int_{L(i)} \frac{\sigma_b S dx}{\sigma S (\sigma S + \sigma_b \Delta S)}}{\displaystyle\int_{L(i)} \frac{dx}{\sigma S} \cdot \int_{L(i)} \frac{dx}{\sigma S + \sigma_b \Delta S}}
\tag{4}
$$

where ΔY_i denotes the admittance change in the section from x_i to x_{i+1}.

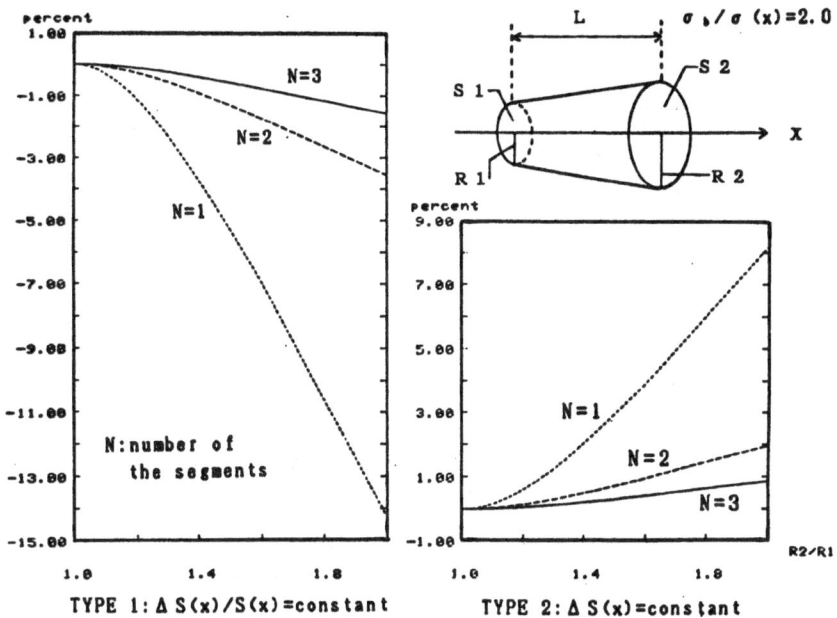

Fig. 1. Inhomogeneous error estimation.

We shall estimate this error based upon the following assumptions:

(i) the conductivity of the tissue is constant along the x-axis;
(ii) the conductivity of blood is twice as large as that of tissue;
(iii) the linear dimension of the cross-sectional area $S(x)$ is increasing linearly with x; in other words, the leg is assumed to be a truncated cone;
(iv) as for the cross-sectional area change $\Delta S(x)$, two types of inhomogeneities are examined:

Type I: $\Delta S(x)$ is proportional to the initial cross-sectional area $S(x)$

Type II: $\Delta S(x)$ is constant along x.

Fig. 1 shows the result of the theoretical estimation of inhomogeneity error. In the graph, the horizontal axis represents the radius ratio of the cross-section of the right edge to the cross-section of the left edge. If R_2/R_1 is equal to 2, S_2 is four times as large as S_1 and this seems reasonable to simulate the leg from the thigh to the ankle. In both graphs there are three traces according to N, the numbers of divided sections. As the number of segments increases, the error becomes smaller naturally, and approaches zero as N tends to infinity. If the measured segment is divided into three sections, the error of the inhomogeneity is estimated to be almost within 1% in this simulation model. Our newly developed system has three voltage channels to measure the whole leg and seems to have enough accuracy from this standpoint.

SYSTEM CONFIGURATION

Fig. 2 shows the schematic diagram of the newly developed three-voltage-channel impedance plethysmograph. A frequency of 20 kHZ was selected since we are concerned with the extra-cellular fluid measurement and it is reported (Kanai et al., 1983) that the β-dispersion of human calf tissue occurs at around 50 kHz. A floating constant current source is

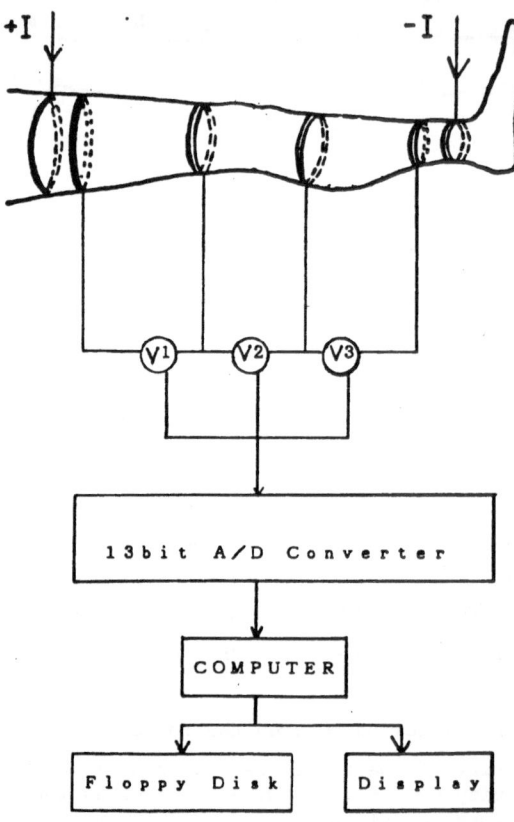

Fig. 2. Schematic diagram of three-voltage-
channel impedance plethysmograph.

realised by two independent single-ended constant current sources which
have balanced opposite-signed current outputs of 200μA r.m.s. All the
cables to and from the current and voltage electrodes are driven-shielded.
Three amplified voltage channel signals, as well as the current monitoring
signal and zero reference, are multiplexed by an analogue switch and then
fed to the synchronous detector followed by a D.C. amplifier. The final
analogue output is A/D converted with a 13-bit resolution and fed to the
personal computer which calculates the blood volume change in real time,
and also saves the raw data on floppy disk by means of an interrupt scheme.

EXPERIMENTAL RESULTS

The Blood Volume Change of the Control State

    A pair of current-supplying band electrodes were set around the ankle
and thigh, and four voltage-detecting band electrodes were set around the
ankle, the under knee, the upper knee and the thigh of the left leg of a
healthy 22 year-old male subject. He lay down in a supine position for
about 30 minutes, then stood up for about 25 minutes, and then lay down
again. He basically stood on the right leg on which the electrodes were
not set.

    Fig. 3 shows the blood volume change per unit length of each segment.
The blood volume change per unit length was largest in the calf section and

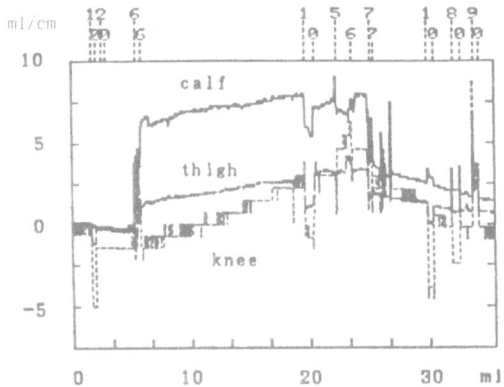

Fig. 3. Blood volume change per unit length of
each segment of the control state.

smallest in the knee section. These differences can be considered to result
from the differences of vascularisation among the sections.

Fig. 4 shows the comparison of the estimated total blood volume change
by the three-voltage-channel method with the one-voltage-channel method.
The solid line represents the result of the three-voltage-channel method, that
is, the total blood volume change calculated by multiplying the unit length
volume change by the measured section length and summing them up. The
dotted line represents the results of the one-voltage-channel system.
Actually, the three-channel-voltage data were directly summed to obtain the
potential difference between the outermost voltage electrodes, and the total
blood volume change of the whole segment was calculated from these overall
admittance data. The difference between the two traces is considered to be
the error caused by the inhomogeneity. This difference seems quite plausible since the cross-sectional areas of the three sections were 183 cm^2,
111 cm^2 and 54 cm^2, and the initial admittance values per unit length were
0.0455 S, 0.0145 S and 0.0173 S, respectively. From the graph it could be
seen that as soon as the subject stood up, the leg blood volume increased
abruptly and then it gradually increased at an almost constant rate while
standing.

The numbers above the traces represent the events described above the
graph. Event number 1 represents the instruction to constrict the muscles
of the left leg. As soon as the leg was constricted, the blood volume decreased. When the leg was relaxed, the blood volume increased and quickly
returned to the former level. It is considered that the blood volume
decrease here resulted from the pumping action of the muscles which promotes
venous return, and the rate of blood volume increase on muscle relaxation
represents the blood inflow from the arteries since there would be a small
amount of venous outflow at this time. In Fig. 5, the part corresponding
to the event of constricting and relaxing the leg muscle is magnified. As
the ascending part shown by the leg muscle relaxation took 10 s, we consider
that this ascending rate would not be a motion artefact but rather one good
figure expressing the arterial blood inflow rate.

The Blood Volume Change of Post-microgravity Simulation

Fig. 6 shows the blood volume change of the whole leg during the

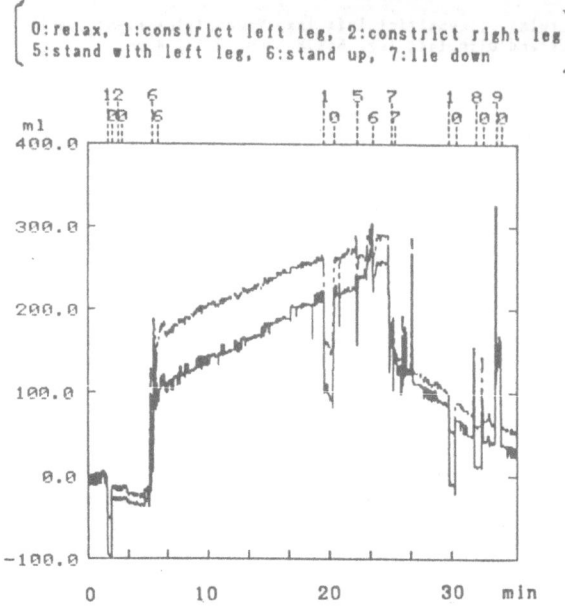

```
┌ 0:relax, 1:constrict left leg, 2:constrict right leg ┐
└ 5:stand with left leg, 6:stand up, 7:lie down        ┘
```

Fig. 4. Comparison of the estimated total blood
volume change of control state by the
three-voltage-channel method (dotted
line) and the one-voltage-channel method
(continuous line).

orthostatic intolerance test (70 degrees head-up tilting) of post-microgravity simulation. Microgravity simulation is performed by lying down for
6 days in the state of -6 degrees head-down tilting. After this simulation
the circulatory system is considered to show a change similar to that corresponding to a condition of microgravity such as a space flight. Comparing
the blood volume change after the microgravity simulation (Fig. 6) to the
control (Fig. 4), the rate of the blood volume increase while standing is
almost the same (control state: $0.16 \, \mathrm{ml \, s^{-1}}$, post-microgravity simulation:

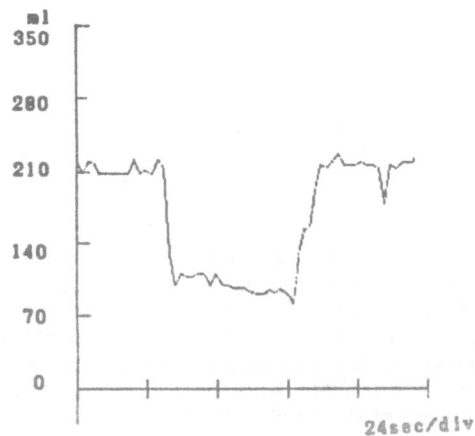

Fig. 5. The blood volume change on the
event of constricting and relaxing the leg muscle.

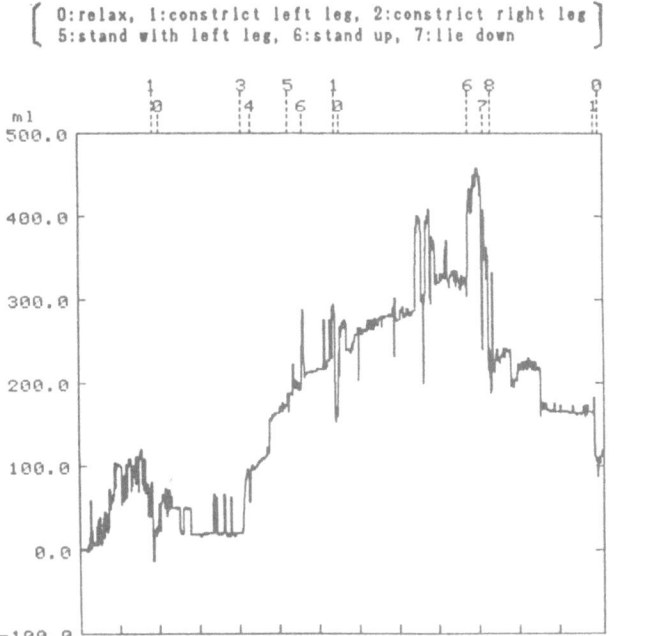

Fig. 6. Blood volume change of the whole leg during orthostatic intolerance test of post-microgravity simulation.

$0.17 \, ml \, s^{-1}$). After standing for about 25 min, the leg blood volume increased by about 290 ml (control state) and by about 330 ml (post-microgravity simulation). From these preliminary results, no significant difference was found in the amount of increase of leg blood mass upon standing in the post-microgravity simulation and the control.

CONCLUSION

A three-voltage-channel impedance plethysmograph to measure the blood volume change of the leg has been developed, and applied in an orthostatic intolerance test after a microgravity simulation experiment. The system worked satisfactorily. This equipment should become one of the important components of the measurement system in the study of the adaptation and re-adaptation of the human circulatory system to microgravity and normogravity.

REFERENCES

Gaffney, F. A., 1985, Space life sciences flight experiments: an integrated approach to the study of cardiovascular deconditioning and orthostatic hypotension, in: "Preprints of the 36th Congress of the International Astronautical Federation", Pergamon, Oxford:316.

Kanai, H., Sakamoto, K., Haeno, M., 1983, Electrical measurement of fluid distribution in human leg: estimation of extra- and intra-cellular fluid volume, J. Microwave Power, 18 : 233.

Thornton, W. E., Hoffler, G. W., and Rummel, J. A., 1977, Anthropometric changes and fluid shifts, in: "Biomedical Results from Skylab", NASA Report SP-377, R. S. Johnson and L. F. Dietlein, eds., NASA, Washington, DC : 330.

A METHOD FOR CONTINUOUS MEASUREMENT OF MUSCLE FIBRE CONDUCTION VELOCITY

R. Mucke and G. Küchler

Central Institute of Occupational Medicine of the GDR
Berlin
GDR

INTRODUCTION

In work and in sports physiology, parameters of the electromyogram (EMG) are often used to assess muscular performance, muscle fatigue, and the effects of training. While in clinical neurology the EMG is in many cases recorded invasively by needle electrodes, in the fields of work physiology commonly estimated parameters of the surface electromyogram are detected by skin electrodes over the investigated muscle.

Here changes can be observed in the amplitudes as well as in the spectral distributions of the myoelectric signal, related to the state of the contracting muscle and the form of contraction. In brief non-fatiguing isometric muscle contractions there is a nearly linear relation between muscle force developed and the mean amplitudes of the EMG (iEMG). With increasing duration of contraction at constant force the iEMG rises to higher values and the power of lower frequencies in the EMG spectral distribution goes up, that is, a shift of the power density spectra (PDS) to lower frequencies and so a decrease in the mean frequency of PDS (MPF) is observed (De Luca, 1985; Petrofsky et al., 1982). The reasons for the spectral changes are not completely clear. Central and peripheral factors have been discussed as possible mechanisms (Bigland-Ritchie et al., 1986). Central processes influencing the electromyogram are the synchronisation of excitation of the motor units, their firing frequency and the recruitment order of different motor units.

Local processes leading to shifts in EMG spectral distributions are changes in motor unit action potential shape attributed to alterations of muscle fibre conduction velocity (CV) caused by influences on the electrical membrane properties by accumulation of metabolic byproducts in the intercellular space (Gatev et al., 1981). Theoretical models for the closed relations between CV and EMG spectral distributions have been described by Lindström et al. (1970) and between CV and different moments of the EMG power density spectra by Stulen and De Luca (1981).

Measurements of CV in human muscles in situ by using invasive methods (needle electrodes) were reported by Buchthal et al (1955) and by Stalberg (1960). The estimation of conduction velocity by non-invasive methods from the power density spectrum of the surface EMG was described by Lindström et al. (1970). Zipp (1978) discovered in EMG power density spectra characteristic frequencies (f_{dip}) with pronounced diminution of power. Because of the transversal filter configuration of the surface electrodes, with the muscle fibres acting as a delay line, these frequencies are directly connected to the conduction velocity of motor units by the relation:

$$v = f_{dip} \cdot d \tag{1}$$

where d is the electrode distance.

A useful method for measuring conduction velocity in human skeletal muscles was introduced by Lynn (1979). The basis was the assumption that two surface electromyograms recorded in the fibre direction between the region of motor endplates and the tendon are time delayed in relation to the mean conduction velocity of the motor units involved and the inter-electrode distance. The estimation of conduction velocity was realised on a computer measuring the time delays continuously.

On the basis of the results of Lynn, other investigators proposed procedures for calculation of the cross-correlation function to determine the time lag between EMG1 and EMG2 (Naeije and Zorn, 1983; Yaar et al., 1984). Masuda et al. (1982) simulated a method for direct measurement of conduction velocity without A/D conversion by estimating the time lag between characteristic events (zero crossings) of the electromyograms. A hardware realisation of this computer simulation was suggested. An essential improvement of the procedures for the estimation of muscular conduction velocity was the recording of three surface electromyograms in the muscle fibre direction and the differentiation of the amplified signals in a second layer of differential amplifiers (Broman et al., 1985). By means of this configuration, non-delayed activities leading to errors are eliminated.

With regard to the papers of Lynn (1979) and Masuda et al. (1982), as well as the proposal of Broman et al. (1985), hardware for continuous measurement of muscle fibre conduction velocity was developed. The aim of the design was to provide a method useful in laboratory investigations and under field conditions to investigate long-lasting muscle contractions including an effective artefact reduction. A detailed description of the method has been given by Mucke (1986).

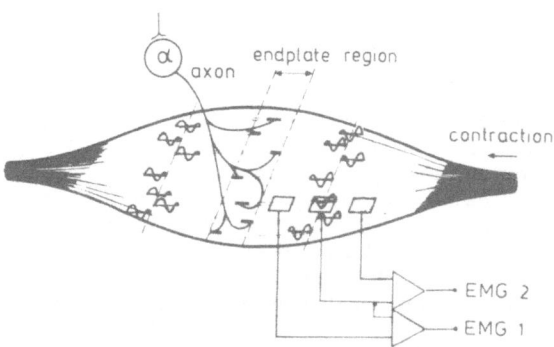

Fig. 1. Position of surface electrodes in relation to the muscle fibres. The delay between EMG1 and EMG2 is the conduction time.

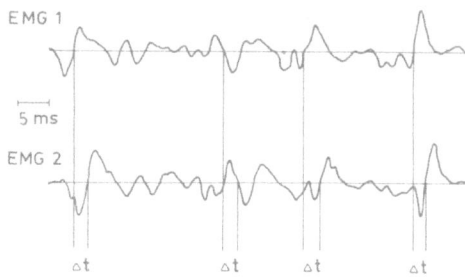

Fig. 2.   Registration of time delayed
electromyograms for the estimation of conduction velocity.
Δt - selected characteristic
time delays.

METHODS

The conduction velocity of muscle fibres was estimated from the delay
of two electromyograms recorded using a bipolar double electrode configuration.   The electrodes are located between the region of the endplates and
the tendon of the muscle (Fig. 1).   Motor unit action potentials move from
the endplates of the muscle fibres in both directions to the tendons.
Thus a time delay is observed between EMG1 and EMG2 related to the electrode distance and the mean conduction velocity of the muscle fibres.   An
original recording of electromyograms with characteristic time delays Δt
between EMG1 and EMG2 is shown in Fig. 2.

Previous investigations were carried out by recording the electromyograms using a double surface electrode configuration (Fig. 3B).   In the

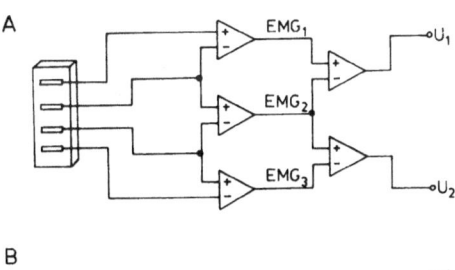

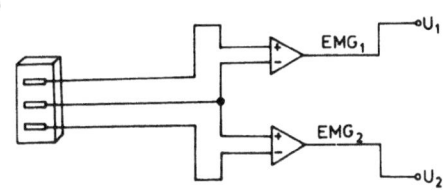

Fig. 3.   Electrode configurations for
the recording of myoelectric
activity.   A: triple bipolar
electrode with differentiation
of the signals by a second
layer of amplifiers.   B: double
bipolar electrode for the
recording of two delayed
electromyograms.

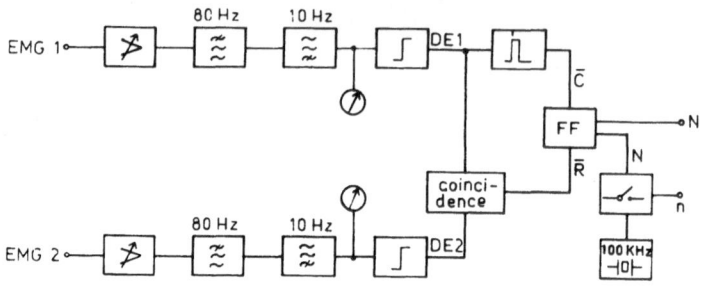

Fig. 4. Illustration of the technique for continuous
measurement of conduction time from surface
electromyograms. Scheme of hardware solution.

current realisation of this method, a bipolar triple electrode and the
differentiation of amplified electromyograms in a second layer of differential amplifiers is used (Fig. 3A). The individual electrodes consist of
silver bars of width 1.5 mm and length 10 mm. The silver bars and their
connections are mounted at distances of 10 mm in epoxy resin. The lag of
U1 to U2 - derived from the electromyograms by amplification or differentiation - is the mean conduction time.

A device was developed for estimation of conduction velocity and
measurement of conduction time. The function of the technique is demonstrated by a raw scheme shown in Fig. 4. The signal EMG1 denotes the
electromyogram picked up near the endplate region. EMG2 is delayed relative to EMG1. Both signals are amplified to comparable amplitude levels,
indicated by light-emitting diodes connected to the output of the filters.
To eliminate baseline alterations and noise of higher frequency, the signals are filtered by 80 Hz low pass and 10 Hz high pass filters (Bessel-
characteristic, 4th order). The cut-off frequencies were estimated from

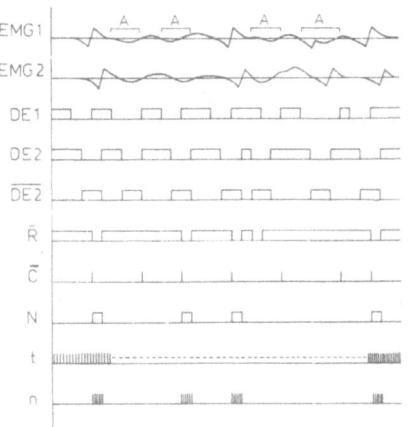

Fig. 5. Diagram of the signals
during EMG processing.
EMG1 is recorded near the
endplate region, EMG2 is
delayed to EMG1. n - bursts
of 10 µs impulses with the
duration of the conduction
time samples.

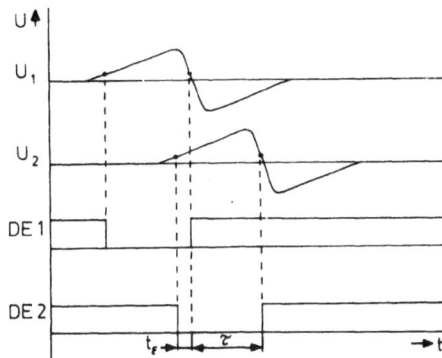

Fig. 6. Time conditions for the digital artefact rejection. U_1,
U_2 – output signals of the
filters; DE1, DE2 – digital
signals derived from U_1 and
U_2; t_ϵ – coincidence interval
of U_1 and U_2; τ – conduction
time.

experiments and seem to be optimal for electrode distances of 10 mm.

For further processing, the analog signals were digitised by comparators and schmitt triggers to the signals DE1 and DE2 (Fig. 5). The trigger
threshold of the comparators was chosen to be +0.3V. This represents 10%
of the internally defined standard levels of EMG amplitudes of ±3V. By this
means portions of the EMG at lower amplitudes, which would be involved by
triggering at zero crossing, are suppressed. The time lag is measured between the positive slopes of DE1 and DE2 corresponding to the steepest portions of the electromyograms. In order to suppress artefacts (EMG portions
from different sources, time delays < 0) the following conditions must be
fulfilled: the positive slope of DE1 has to arrive prior to the positive
slope of DE2 and there has to be coincidence of DE1 and DE2 for a certain
time. To realise these conditions DE2 is inverted to $\overline{DE2}$, and DE1 and $\overline{DE2}$
are connected to $\overline{R}$ by a NAND-gate. A 1 μs impulse $\overline{C}$ is derived by means of
a monoflop from the positive slope of DE1. $\overline{C}$ is connected to the carry
input and $\overline{R}$ to the reset input of a flip-flop (FF). Time delays are only
accepted when impulse $\overline{C}$ and delay impulse $\overline{R}$ coincide. The principle of
digital artefact reduction is demonstrated by an ideal motor unit action
potential (Fig. 6). U_1 and U_2 are the input signals of the comparators
with the trigger thresholds for digitisation (phase shifts of filters are
neglected). Delay intervals τ are only measured when $t_\epsilon > 0$, that is, the
potentials in U_1 and U_2 have the same source. In the case of artefacts or
confusion of electrodes occurring, an event in U_2 arriving before an event
in U_1 becomes $\tau = 0$ and the events are not accepted as a time delay. The
flip-flop (Fig. 4) cannot change state because there is no coincidence
between $\overline{R}$ and $\overline{C}$.

When the time delay has fulfilled the conditions for artefact rejection, a 100 kHz generator is switched on by the output impulse N of the
flip-flop. This provides bursts of 10 μs impulses with the duration of
the time delay denoted by n. The number of time delayed events is denoted
by capital N. The mean conduction velocity is calculated for a time interval with respect to the electrode distance by the formula:

$$v = \frac{100 \text{ s } N}{n} \qquad (2)$$

93

with:

      v = conduction velocity (m/s)
      s = electrode distance (mm)
      N = number of time delayed events per time unit
      n = total of time delays per time unit (μs 10).

The condition for the calculation of conduction velocity from the mean of
several single time delays is their symmetric distribution.

RESULTS

    In Fig. 7 distributions of measured time delays are shown in biceps
contractions sustained for 2 minutes at 15, 30, 45, 60 and 75% of maximum
voluntary force for one subject. After decrease of the force below the
required level, contractions had to be continued at maximum effort.

    Each of the histograms contains 512 time intervals, the class width
being 200 μs. Samples were taken at the beginning and after 30, 60 and 90 s.
The duration of one sample was about 10 s. The histograms are symmetrically
distributed, so the calculation of the mean conduction velocity is reliable.
The symmetry of the distributions was proved by a χ^2-test with the hypothesis of equal class frequencies below and above the mean value. The
hypothesis was confirmed by the test.

    The figures at the top of the histograms are the calculated mean con-
duction velocity for one sample. With the onset of fatigue the conduction

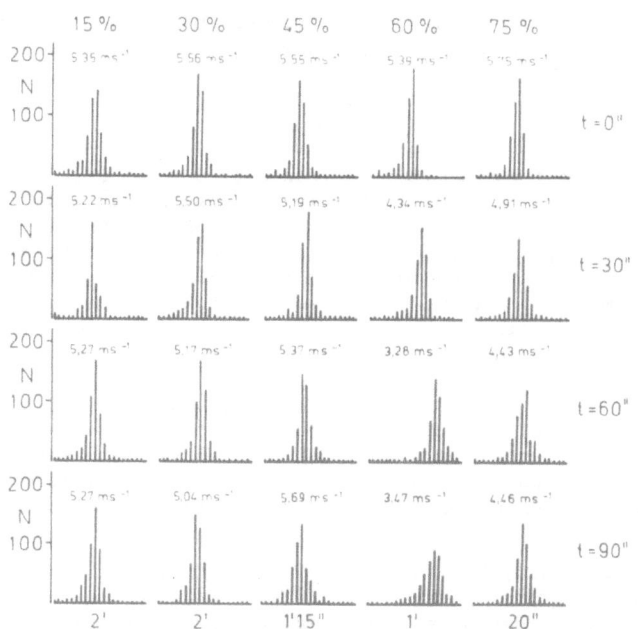

Fig. 7. Histograms of measured conduction times Δt
(class width 200 μs, N = 512 time intervals).
Duration of contractions 2 min, samples
taken at the beginning and 30, 60 and 90 s
later. Figures at bottom – time of contraction with required force.

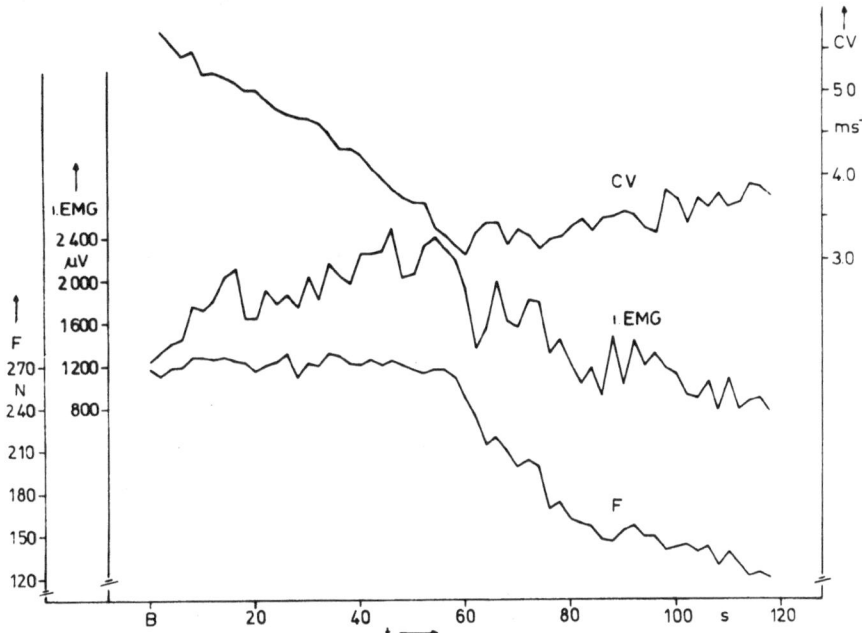

Fig. 8.   Example of continuous registration of force (F), integrated
electromyogram (iEMG), and conduction velocity (CV) during
a 2 min long fatiguing arm contraction at an initial force
of 60% MVC in one subject.

velocity decreases during contractions at higher force. After diminution
of force a partial recovery of change in conduction velocity is observed.
The continuous recording of conduction velocity, integrated electromyogram
(iEMG) and force during a 2-minute long contraction at an initial force of
60% MVC is shown in Fig. 8.   The force required was kept constant for 60 s.
The iEMG rose during the first minute of contraction and the conduction
velocity slowed down continuously. During the second minute the force of
contraction decreased as well as the iEMG. No further decline in conduction
velocity was observed. With improvement of muscular circulation a slight
increase in conduction velocity was measured so that a depletion of metabolic by-products in the intercellular space could be assumed.

The last example of the application of the method described shows the
results of an experimental investigation to enquire into the effect of different muscle temperatures on EMG parameters and conduction velocity. In
these experiments a triple electrode configuration was used. The five subjects had during one experiment to perform three brief biceps contractions
at maximum voluntary force, interspersed by pauses of one minute, followed
by a rest period of three minutes and a test contraction at 30% of maximum
voluntary contraction for three seconds under normal conditions. After
this contraction, constant ambient temperatures were provided by a cuff
closely applied to the upper arm. To achieve the constant temperature conditions, water at controlled temperatures of 0, 10, 20, 30 and 40°C was run
through the cuff. Immediately after application of the cuff a further test
contraction was performed at 30% of maximum force and repeated every five
minutes for an hour. The 17th test contraction was followed by a 2-minute
contraction at instantaneous maximum voluntary force.

The behaviour of conduction velocity and mean frequency of the EMG
power density spectra during experiments at different temperatures is

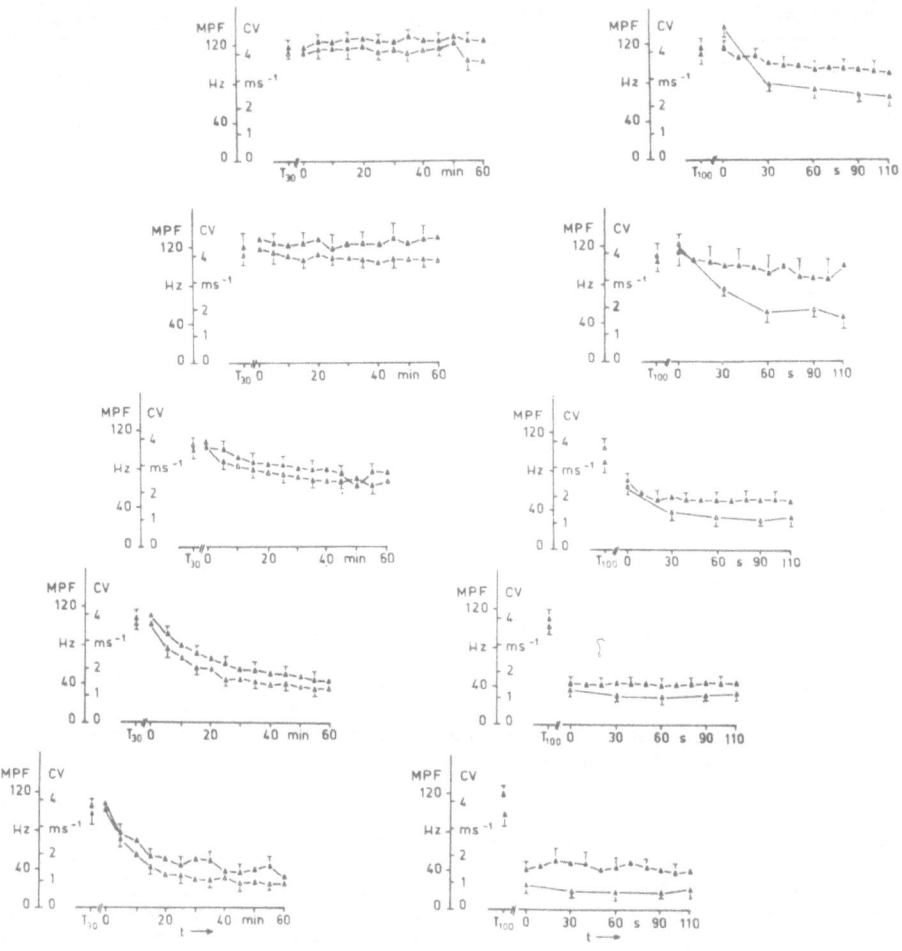

Fig. 9. Behaviour of EMG mean power frequency (MPF) Δ——Δ and muscle
fibre conduction velocity (CV) ▲——▲ in non-fatiguing and
fatiguing biceps contractions at different temperatures.

presented in Fig. 9. The diagrams on the left demonstrate the values of
test contractions at 30% MVC during the 60 minutes of temperature application related to the test contractions under normal conditions before application of the cuff. On the right side of Fig. 9 the results for the two-
minute contractions at instant maximum force are shown related to the maximum contractions at the beginning of the experiments. Temperature variations
are presented from 0°C - the bottom diagrams to 40°C - the top diagrams.

     With the decrease in temperature a similar diminution of conduction
velocity and mean frequency of EMG spectra in the test contractions can be
observed. During maximum contractions a slight increase was found for the
initial values of MPF and CV at 30 and 40°C, but in the process of progressive muscular fatigue the mean frequency decreased to lower values than the
conduction velocity. We assume, with respect to these results, that changes
in mean frequency of EMG caused by alteration of muscle temperature are
directly related to changes in muscle fibre conduction velocity, whereas
after the onset of muscle fatigue the EMG spectral distribution and the
conduction velocity seem to be affected in a different way by other

influences also. Further results obtained by this method have been described by Mucke and Zöllner (1986).

DISCUSSION

     All the methods developed in recent years to estimate muscular conduc-
tion velocity from surface electromyograms are founded on the assumption
that the recorded action potentials are conducted at the muscle fibres in
one direction and delay related to conduction velocity occurs (Graham et
al., 1984; Masuda et al., 1982; Naeije and Zorn, 1983; and Yaar et al.,
1984). Pre-requisites for this assumption are, on the recording site, the
localisation of the endplates outside the muscle tissue region below the
electrodes, the parallel direction of electrode axis and muscle fibres, and
homogeneity and isotropy of the volume conductor. Despite certain advantages and disadvantages in the different technical signal processing techniques (cross-correlation function, direct lag measurements, polarity correlation), in principle, identical results can be expected under ideal conditions.

     The human skeletal muscle fibres are not ordered in parallel struc-
tures and the conductive tissue between the bioelectrical source and the
surface electrode is inhomogeneous and anisotropic. These real conditions
could lead to errors in conduction velocity values (Sollie et al., 1985).
If the electrodes are not arranged in a direction parallel to the muscle
fibres, lower values for delay time are obtained (Sadoyama et al., 1985).

     Endplates are located predominantly in the middle of the biceps
muscle, but a small number may be distributed over the whole dimension of
the muscle. When they appear below the recording electrodes action potentials of muscle fibres excited by these endplates cause non-delayed and
phase altered EMG signals (Sadoyama et al., 1985). The non-linear properties of the tissue influence the myoelectric signal in its amplitude and
phase, in particular those of motor units far from the surface. These
effects may provide higher values for the estimated conduction velocity
and less pronounced changes in physiological investigations (Roy et al.,
1986).

     An essential improvement in the techniques for non-invasive measure-
ment of conduction velocity was the recording of three electromyograms and
the double differentiation of the signals to reduce the presence of non-
delayed activity (Broman et al., 1985). Techniques for the estimation of
muscle fibre conduction velocity known from the literature can be classified
into two groups: calculation of the cross-correlation function of the
analog signals including A/D conversion of the electromyograms (Naeije and
Zorn, 1983; Yaar et al., 1984); and measurements of delay time between
characteristic points of the electromyograms (zero-crossing, triggering at
the potential slopes) (Graham et al., 1984; Lynn, 1979; Masuda et al.,
1982; Mucke, 1986). The advantage of analog correlation of myoelectric
signals is the independence of amplitudes over a wide range. However, the
calculation of the cross-correlation function is associated with a greater
demand on the processing technique. Single events cannot be separated and
investigation of conduction time distributions is impossible.

     With the aid of direct measurement of delay time and pre-processing of
data during the investigation, trials of long duration can be registered.
Results obtained are delay histograms as well as mean values of the conduction velocity for certain time intervals chosen by the investigator. The
problem is the definition of the thresholds for triggering the signals.
Measuring the delay between zero crossings might produce errors caused by
low amplitude activity of motor units far from the electrode or by noise.

Triggering at a threshold on the slopes of the action potentials as in the method described in this paper demands identical amplitude of EMG signals. Differences lead to constant errors in time measurements due to the finite rise times of the potentials.

The method described was developed for investigations in work physiology. It is possible to estimate continuously muscular conduction velocity in long-lasting contractions or muscular work. With careful electrode localisation and application, the required identical myoelectric signal amplitudes can be recorded without problem. The calculated conduction velocities are comparable with results obtained by other methods applied in situ as well as in isolated muscle fibres.

REFERENCES

Bigland-Ritchie, B., Furbush, F., and Woods, J. J., 1986, Fatigue of inter-mitted submaximal voluntary contractions: central and peripheral factors, J. Appl. Physiol., 61 : 421.
Buchtal, F., Guld, C., and Rosenfalck, P., 1955, Propagation velocity in electrically activated muscle fibres in man, Acta Physiol. Scand., 34 : 75.
Broman, H., Bilotto, G., and De Luca, C. J., 1985, A note on the non-invasive estimation of muscle fiber conduction velocity, IEEE Trans. Biomed. Eng., BME-32 : 341.
De Luca, C. J., 1985, Myoelectrical manifestations of localized muscular fatigue in humans, CRC Crit. Rev. Biomed. Eng., 11 : 251.
Gatev, P., Dimitrov, G. V., Gydikov, A., and Gerilovsky, L., 1981, Effect of ischemia on the potentials of human single muscle fibres, Acta Physiol. Pharmacol. Bulg., 7(2) : 3.
Graham, A. J., Hudgins, B. S., and Parker, P. A., 1984, Polarity correlator for conduction velocity measurement, IEEE Trans. Biomed. Eng., BME-31 : 675.
Lindström, L., Magnusson, L. R., and Petersen, I., 1970, Muscular fatigue and action potential conduction velocity changes studied with frequency analysis of EMG signals, Electromyography, 4 : 341.
Lynn, P. A., 1979, Direct on-line estimation of muscle fiber conduction velocity by surface electromyography, IEEE Trans. Biomed. Eng., BME-26 : 564.
Masuda, T., Miyano, H., and Sadoyama, T., 1982, The measurement of muscle fiber conduction velocity using a gradient threshold zerocrossing method, IEEE Trans. Biomed. Eng., BME-29 : 673.
Mucke, R., 1986, "Eine Methode zur fortlaufenden Messung der Erregungs-leitungsgeschwindigkeit im Muskel aus Oberflächenelektromyogrammen", Thesis, Ingenieurhochschule, Dresden.
Mucke, R., and Zöllner, I., 1986, Muscle fibre conduction velocity during fatiguing and nonfatiguing isometric arm contractions, Biomed. Biochim. Acta, 45 : 77.
Naeije, M., and Zorn, H., 1983, Estimation of the action potential conduction velocity in human skeletal muscle using the surface EMG cross-correlation technique, Electromyogr. Clin. Neurophysiol., 23 : 73.
Petrofsky, J. S., Glaser, R. M., Lind, A. R., Williams, C., and Phillips, C. A., 1982, Evaluation of the amplitude and frequency component of the surface EMG as an index of muscle fatigue, Ergonomics, 25 : 213.
Roy, S. H., De Luca, C. J., and Schneider, J., 1986, Effects of electrode location on myoelectric conduction velocity and median frequency estimates, J. Appl. Physiol., 61 : 1510.
Sadoyama, T., Masuda, T., and Miyano, H., 1985, Optimal conditions for the measurement of muscle fibre conduction velocity using surface electrode arrays, Med. Biol. Eng. Comput., 23 : 339.

Sollie, G., Hermens, H. J., Boon, K. L., Wallinga-De Jonge, W., and Zilvold, G., 1985, The boundary conditions for measurement of the conduction velocity of muscle fibres with surface EMG, Electromyogr. Clin. Neurophysiol., 25 : 45.

Stalberg, E., 1960, Propagation velocity in human muscle fibers in situ, Acta Physiol. Scand., 70 (suppl. 287).

Stulen, F. B., and De Luca, C. J., 1981, Frequency parameters of the myoelectric signal as a measure of muscle conduction velocity, IEEE Trans. Biomed. Eng., BME-28 : 515.

Yaar, I., Shapiro, M. B., Mitz, A. R., and Pottala, E. W., 1984, A new technique for measuring muscle fiber conduction velocities in full interference patterns, Electroencephalogr. Clin. Neurophysiol., 57 : 427.

Zipp, P., 1978, Effect of electrode parameters on the bandwidth of the surface EMG power density spectrum, Med. Biol. Eng. Comput., 17 : 537.

A SYSTEM FOR INTRACRANIAL PRESSURE PROCESSING AND INTERPRETATION IN

INTENSIVE CARE

M. Czosnyka, W.Zaworski, P. Wollk-Laniewski and L. Batorski

Institute of Electronic Fundamentals,
Warsaw University of Technology; and
Department of Anaesthesiology and Intensive Care and
Clinic of Neurosurgery,
Child's Health Centre,
Warsaw, Poland

INTRODUCTION

An opinion that continuous monitoring of intracranial pressure (ICP) improves management of patients with certain severe brain disorders can be found in many papers and reports (Avezaat and van Eijndhoven, 1984; Marmarou, 1973; Sliwka, 1980; Snyder et al., 1980; Sullivan et al., 1980). These cases form an important group in intensive care in neurosurgery, including brain tumours, head injuries, intracranial aneurysms, encephalitis, Reye syndrome, and so on. Almost all of them can produce symptoms of brain oedema and cerebral blood flow disorders. This is one main reason for the poor outcome of treatment in patients after cerebral injuries and neurosurgical operations. Appreciation of the patient's status, based only on clinical guides and level of consciousness, is commonly inexact and too late. Information on pathological processes occurring within the intracranial space can be achieved on the basis of ICP signal analysis. Such techniques of ICP monitoring are now widely accepted in intensive care (Snyder et al., 1980; Sullivan and Becker, 1980).

In the early work of Langfitt (see, for example, Langfitt et al., 1964) the problem of pressure response to external volume addition was considered. It was shown that the so-called pressure-volume characteristic, expressing the increase of ICP level due to addition of cerebrospinal fluid, was a rising exponential function (Avezaat and van Eijndhoven, 1984; Langfitt et al., 1964; Marmarou, 1973; Sliwka, 1980). Interpretation of such characteristics required an adequate theoretical model of intracranial space. Due to the exponential shape of the pressure-volume characteristic, the pressure response to blood volume addition depends almost linearly on the mean ICP level, starting from a certain pressure level (see Fig. 1). For lower levels of ICP the amplitude remains constant (Avezaat and van Eijndhoven, 1984; Sliwka, 1980). Measurement of this characteristic can be performed using an infusion test. An increase in ICP is forced by the infusion of a definite volume of cerebrospinal fluid. The slope of the amplitude-pressure characteristic (AMP/p) contains important information on features of the cerebrospinal system, being the product of the portion of blood volume ΔV_T flowing in a pulsatile manner into the cerebrospinal vascular bed and the so-called intracranial space elasticity. This elasticity (α)

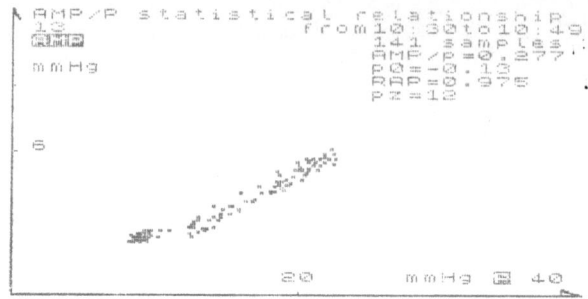

Fig. 1. Amplitude-pressure characteristic measured during infusion test. Abscissa:
ICP in mmHg, ordinate: amplitude of
pulse wave in mmHg. pO denotes inter-
cept of regression line and pressure
axis.

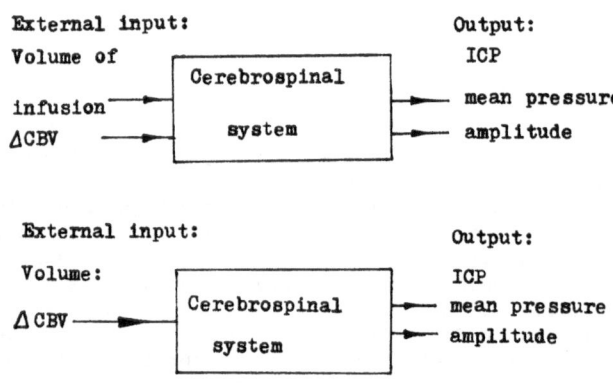

Fig. 2. General scheme of cerebrospinal system
being identified using infusion test
(a - two input system) and in steady
state (b - one input system).

describes the ability of the intracranial space to compensate for changes
in volume.

The cerebrospinal system being analysed using the infusion test is depicted as a black box in Fig. 2a. It is considered as a system of unknown
features, of two volume inputs - a portion of the blood ΔV_T, added after
each heart beat, and a definite volume, added externally to force an increase in ICP level. This system has subsequently two "outputs" (this is,
in fact, a single pressure output, but the signal consists of two components
of different origin): the mean ICP level and the amplitude of the pulse
wave. Application of this infusion test is, however, severely limited in
intensive care. For continuous ICP monitoring the scheme can be reduced
to that shown in Fig. 2b. In this case it is not the external but the internal sources of change of intracranial volume which are considered. We
obviously cannot control changes of this volume, but information on them
is included in spontaneous changes of ICP level. A model, in this case,
can be identified by means of analysis of correlated changes of ICP mean
level and the amplitude of the pulse wave.

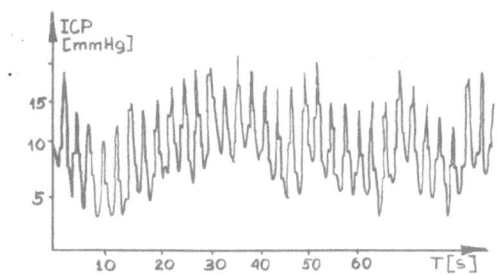

Fig. 3.   Fragment of typical ICP signal regis-
            tered during post-operative care.
            Ordinate: pressure in mmHg; abscissa:
            time in seconds. Pulse, respiratory
            and slow components can be easily ob-
            served; however, they interfere in
            time.

THE METHOD OF ICP PROCESSING

     Typically, the ICP signal consists of several components. The most
specific is the pulse wave. Its fundamental frequency is equal to heart
rate and its amplitude is due to the blood portion V_T and cerebrospinal
system elasticity. Commonly, a respiratory component can be observed. Its
amplitude can be even larger than the amplitude of the pulse wave, especially
in the case of controlled ventilation (Czosnyka, 1985). There are also
several rhythmic or stochastic changes of ICP mean level - the so-called
Lundberg slow waves (Lundberg et al., 1965; Snyder et al., 1980; Sullivan
and Becker, 1980) and others which cannot be classified so perfectly (see
Fig. 3).

     Our basic aim was to measure the changes of the amplitude of pulse
wave (denoted a or AMP) correlated with the changes of the ICP mean level p.
The measurement should be precise, because the variance of the estimator of
the slope of amplitude-pressure characteristic, calculated as:

$$\alpha\Delta V_T = \frac{\sigma_a}{\sigma_p}\ r_{ap},$$ (1)

(where σ_a is the r.m.s. of amplitude from the chosen time period, σ_p is the
r.m.s. of mean ICP and r_{ap} is the correlation coefficient between a and p)
is proportional to:

$$\frac{\sigma_a^2}{\sigma_p^2}\ (1 - r_{ap}^2)$$ (2)

In practice, however, both the amplitude and the ICP mean level, interpreted
as realisations of stochastic processes, consist of two components:

$$a = a_n + a_s$$

$$p = p_n + p_s$$ (3)

where a_s and p_s represent components perfectly correlated to each other
(thus r_{aps} is equal to 1), and a_n and p_n are uncorrelated. The correlation
coefficient r_{ap} is, on average, equal to 0.4 in long-term monitored patients
after operations for a tumour of the fossa posterior (Czosnyka, 1985).
Amplitude and pressure are not correlated as strongly as in the case of the
infusion test (compare Figs. 1 and 8).

103

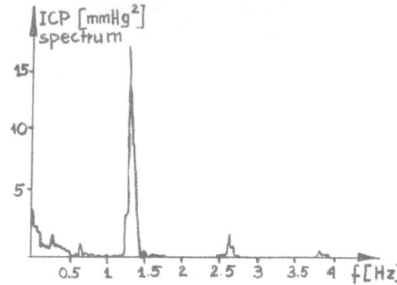

Fig. 4. Spectrum of the ICP signal from Fig. 3.
Abscissa: frequency scale (in Hz), ordinate: energy of pressure signal (in
$mmHg^2$ - note that d.c. level was sub-
tracted before analysis). Components
are undoubtedly better separated in the
frequency domain.

From (3) the following relationship can be derived:

$$\sigma^2_{\alpha \Delta V_T} = K \frac{\sigma^2_{as} + \sigma^2_{an}}{\sigma^2_{ps} + \sigma^2_{pn}} \left(1 - r^2_{ap}\right) \tag{4}$$

where K is a constant. To calculate the slope of the amplitude-pressure
characteristic accurately, the power of a_n and p_n should be filtered out
from a and p. The most influential source of disturbing component a_n is
other components of the ICP signal - that is the respiratory and slow waves.
Thus, selective filtering of the ICP waveform should be performed. The
current value of the heart rate should be known in order to solve this
problem accurately.

The efficient means of selective filtering, without prior knowledge of
heart rate, is frequency analysis. All the components of ICP interfere in
the time domain, but commonly they can be well separated in frequency domain
(see Figs. 3 and 4). The Discrete Fourier Transform of the sequence of ICP
samples is used to estimate the discrete power spectrum of the signal. The
sampling frequency is automatically chosen to be 5 times greater than the
heart rate. As a result of this, the peak of the discrete spectrum, corresponding to the fundamental frequency of the pulse wave, is placed near the
middle of the frequency scale. It allows the highest accuracy of interpolation algorithm to be reached when a short DFT (that is 32 or 64 point) is
calculated (Czosnyka, 1985). Interpolation improves the accuracy of the
analysis of the pulse component more than 10 times. Such a short DFT allows
calculations to be performed between two subsequent heart beats. The total
improvement, expressed as a reduction of the power of component a_n with respect to the direct calculation of amplitude from the ICP signal, is approximately equal to 2.5 (Czosnyka, 1985).

A further reduction of power of the uncorrelated components is performed by means of processing sequences of samples a and p, collected
during repetitive spectral analysis. Each of these sequences represents
the changes in measured parameters. It can clearly be seen (Fig. 5a)
that they contain components which are correlated and uncorrelated to each
other.

Typically, component a_n in the sequence (a) lies higher in the frequency domain than a_s. The sequence of ICP mean level (p) commonly has the

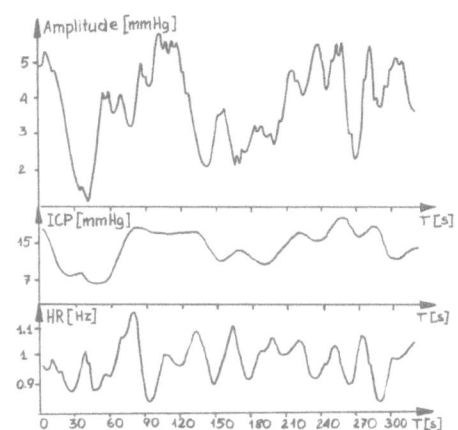

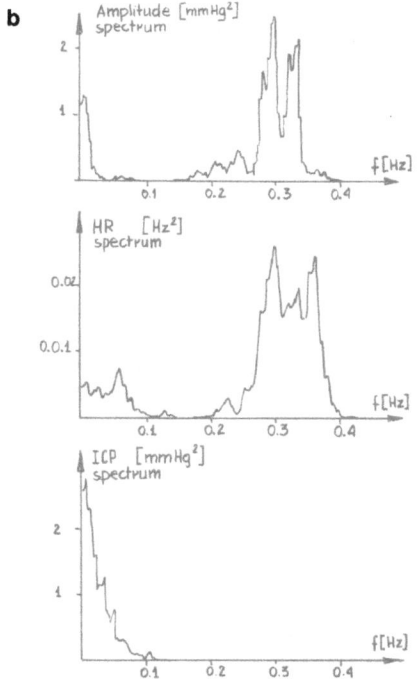

Fig. 5. (a) Example of three sequences: amplitude
of pulse wave (a), mean ICP level (p), and heart
rate (f_{HR}) registered in intensive care. Note
some components of sequence of amplitudes which
are correlated to changes of mean pressure p and
others correlated to heart rate. Common abscissa:
time in seconds.   (b) Example of power spectra
of three sequences from Fig. 5a. Note that the
spectrum of amplitudes A contains components
lying outside the bandwidth of the spectrum of
mean pressure level P.   Common abscissa:
frequency in Hz.

```
Program ICP processing;
VAR
    break: BOOLEAN(interrupt);
    k:INTEGER (index of sequences);
    a,p,f_hr: ARRAY (1..512) OF REAL;

BEGIN
Initialization (all parameters);
 REPEAT
 Signal recognition and finding the proper sampling
 frequency;
 k:=0;
  REPEAT
  Fast Fourier Transform of ICP samples;
  Interpolation;
  Calculation of a[k],p[k], f_hr[k];
  k:=k+1
  UNTIL k=512;
 Automatic artefacts extractions from a,p,f_hr;
 Matched filtering of a;
 Linear regression between a and p;
 Calculations of other parameters **
 UNTIL break;
END.
```

Fig. 6.   Scheme of the program for ICP proces-
         sing in intensive care, denoted in
         pseudo-Pascal notation.

character of a low bandwidth Gaussian stochastic process. It can be shown,
performing a spectral analysis of sequence (a) and (p) (see Fig. 5b), that
the spectrum of (a) consists of peaks both lying in the bandwidth of (p)
and outside this bandwidth. This second peak, commonly positioned higher
in the frequency domain, is undoubtedly related to changes of heart rate,
known as a breathing arrhythmia. These changes of (a) should be extracted,
since they form the second important source of the disturbing component a_n.

    The low pass matched filtering of sequence (a) was carried out. The
finite impulse response filter for a Gaussian shape of impulse response,
matched to the bandwidth of changes of (p) calculated from its spectrum,
was used. The frequency characteristic of the filter and the z-transform
of its transfer function can be expressed as:

$$F(\omega\Delta t) = \{\cos(\omega\Delta t/2)\}^k$$

$$F(z) = \frac{1}{4^k}\{1 + z^{-1}\}^{2k} \tag{5}$$

where k is the parameter chosen to achieve a proper bandwidth match.

    Finally, the correlation coefficient (r_{ap}) between sequences (a) and
(p) is calculated. If the coefficient is positive and differs statistically
from zero, the slope of amplitude-pressure characteristic is computed from
(1).

    A complete scheme of the program for performing the analysis is shown
in a pseudo-Pascal notation in Fig. 5. Not only the correlation coefficient
and the slope of amplitude-pressure characteristic, but also some morphological parameters of the signal are computed. They help to achieve a proper
interpretation of the intracranial processes, either pathological or normal,
being observed in ICP fluctuations. They are:
- mean ICP level;
- power (r.m.s. value) of slow waves;
- mean amplitude of a pulse wave;
- mean amplitude of the respiratory wave;
- mean heart rate.

106

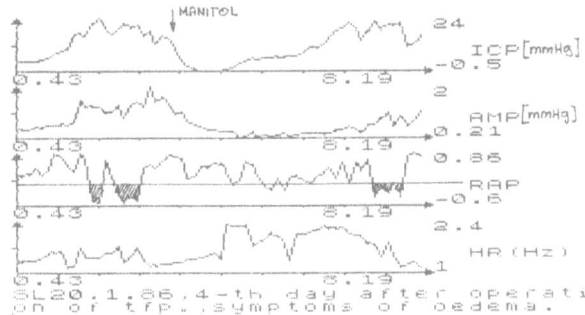

Fig. 7. Trends of parameters registered over 7
hours on the 3rd day after operation
for tumour of fossa posterior, ICP -
mean pressure level [mmHg], AMP - amplitude of pulse wave [mmHg], RAP - short-
term correlation coefficient between
amplitude of pulse wave and mean pressure
level, HR - heart rate in Hz. Note
periods of negative RAP and recovery to
equilibrium state after mannitol administration (75 ml). Common abscissa:
time in hours and minutes.

All the parameters are presented on the screen of the system as time trends.
They can be calculated continuously with the maximum rate of one sample per
2 minutes and collected in the memory.

The program described was written in assembler language of Intel 8080
for the microprocessor system built especially for ICP monitoring at Warsaw
University of Technology.

RESULTS AND THEIR INTERPRETATION

A group of 73 patients, comprising 64 following surgery for a tumour of
the fossa posterior, 4 with encephalitis, 3 with head injuries and 2 with
SAH, was monitored in the Intensive Care Unit of the Child Health Centre in
Warsaw by means of the system described above. The most important of our
clinical findings can be listed as follows:

(1) The mean pressure level varies in all cases in the range 0 to 35 mmHg.
     Variations of ICP reflect the existence of compensated or uncompensated
     volume processes, introduced either externally (drugs, CSF drainage)
     or internally (see Fig. 7);

(2) In many cases variations of amplitude of pulse wave and ICP remain
     proportional to each other. The so-called amplitude-pressure charac-
     teristic is a straight line (Fig. 8). Shifts of this characteristic
     reflect changes of cerebrospinal system features - its elasticity and/
     or cerebral blood volume (see Fig. 10);

(3) RAP reflects the general state of the cerebrospinal system. If it is
     relaxed, RAP is close to 0. RAP increases to 1 if an uncompensated
     volume process is introduced externally or internally, and becomes
     negative in cases of impairment of autoregulation of CBF (see Fig. 7).
     In these cases the upper breakpoint of the amplitude-pressure charac-
     teristic is commonly observed. The ICP level of the breakpoint can be
     interpreted as the critical level for a given patient (see Fig. 9);

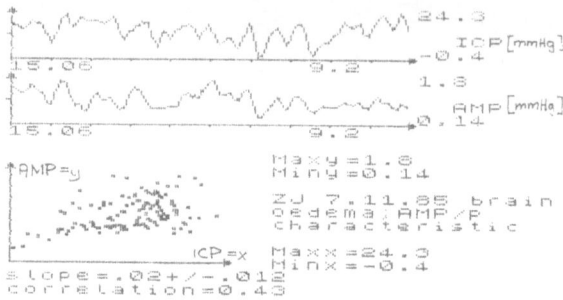

Fig. 8. Long-term amplitude-pressure characteristic registered in state after
operation for tumour of fossa posterior
(lower section). Note rather poor correlation between trends of ICP mean
level (upper trend and abscissa of
amp/p plot) and amplitude of pulse
wave AMP (lower trend and ordinate
of amp/p plot).

(4) The method allows for continuous monitoring of each ICP component
separately. The influence of drugs (for example, mannitol, see Fig. 7)
or $PaCO_2$ on ICP (see Fig. 10) can be more clearly appreciated in this
way.

The RAP can be interpreted as an index of the global state of intracranial space. Several typical patterns can be observed in ICP analysis in
intensive care:

(1) RAP = 0, ICP < 14 mmHg (equilibrium state) then ICP increases and RAP
becomes close to 1 (the loss of equilibrium), then ICP increases and
RAP becomes close to -1 (impairment of autoregulation of CBF; commonly
AMP and HR decreases) and then after mannitol, hyperventilation or CSF
drainage, ICP decreases and RAP becomes close to 0. This pattern is
typical for cases of brain oedema or CBF insufficiency;

(2) RAP = 0, ICP low, then ICP increases and RAP becomes close to 1, then
either after drainage of CSF or spontaneously, RAP recovers to zero
and ICP decreases, and so on. This can be interpreted as the state
of variations from equilibrium to the loss of equilibrium of the intracranial system;

(3) RAP remains close to 0 in spite of variations of mean ICP. This means
a permanent equilibrium state of the cerebrospinal system.

CONCLUSIONS

The method presented for intracranial pressure monitoring and processing has been used in the practice of intensive care. It seems to be useful, providing precise information of the state of the patient's cerebrospinal system. The results presented suggest that the correlation coefficient between amplitude and pressure can be interpreted as a steady state
index. It is close to 0 in the steady state, and differs significantly
from 0 in the case of loss of this equilibrium state. Its negative sign
indicates cerebral blood flow insufficiency.

The method and its interpretation is a new proposition for intensive
care. It needs, of course, more clinical evaluation. The authors hope that

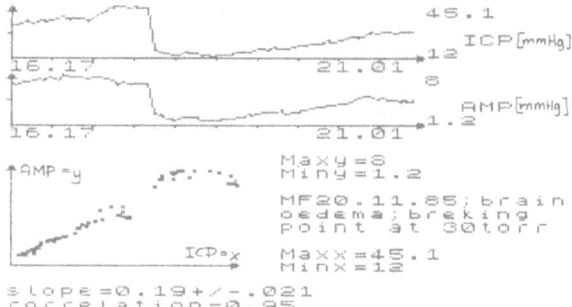

Fig. 9. Upper breakpoint of amplitude-pressure
characteristic, interpreted as a sign
of cerebral blood flow autoregulation
impairment (lower section). Note that
above a pressure level of about 35 mmHg,
amplitude (AMP) decreases gradually in
spite of further increase in ICP.

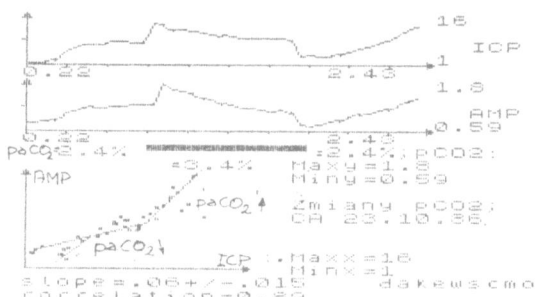

Fig. 10. Example of analysis of amplitude-
pressure characteristic in case of
$PaCO_2$ reactivity test. Change of
end-tidal CO_2 from 2.4 to 3.4 vol%
forces increase of mean ICP level
and increase of slope of amplitude-
pressure characteristic.

some clinical centres will be interested in introducing it into their practice. They would also be grateful for comments and opinions.

ACKNOWLEDGEMENT

Two of the authors (M.C. and W.Z.) were supported by Polish Research
Project 11.9. The authors would also like to express their gratitude to
Simonsen & Weel Medico Teknik A/S from Denmark for technical support.

REFERENCES

Avezaat, C. J. J., and van Eijndhoven, J. H. M., 1984, "Cerebrospinal Fluid
    Pulse Pressure and Craniospinal Dynamics", PhD Thesis, Academic
    Hospital and Erasmus University, Rotterdam.
Czosnyka, M., 1985, "Digital Spectral Analysis of Intracranial Pressure",
    Doctoral Thesis, Warsaw University of Technology.

Langfitt, T. W., Weinstein, J. D., and Kassell, N. F., 1964, Transmission of increased intracranial pressure, J. Neurosurg., 21 : 989.

Lundberg, N., Troupp, H., and Lorino, H., 1965, Continuous recording of ventricular fluid pressure in patients with severe acute traumatic head injuries; preliminary report, J. Neurosurg., 22 : 581.

Marmarou, A., 1973, "A Theoretical and Experimental Evaluation of Cerebrospinal Fluid System", PhD Thesis, Drexel University.

Sliwka, S., 1980, "The Clinical System for Examination of Dynamic Features of Intracranial Space", PhD Thesis, Polish Academy of Sciences.

Snyder, J. V., Posner, D. J., and Grenvik, A., 1980, Neurologic intensive care, in: "Anaesthesia in Neurosurgery", W. Cotrell, ed., C. V. Mosby, St. Louis.

Sullivan, H. G., and Becker, D. P., 1980, Intracranial pressure monitoring and interpretation, in: "Anaesthesia in Neurosurgery", W. Cotrell, ed., C. V. Mosby, St. Louis.

ELECTROMAGNETIC KINETOCARDIOGRAPHY: THEORY, CLINICAL AND TECHNICAL

ASPECTS

M. Kocí, M. Odehnal, V. Král and J. Hönig

Institute of Physiological Regulation and Institute of Physics
Czechoslovak Academy of Sciences
Prague, Czechoslovakia

INTRODUCTION

The motion of the ventricular walls is a result of the contractile effort of myocardial fibres and their structures. Under normal conditions the contraction of the fibres is said to be synergic; it stands for a co-ordinated synchronous upstroke of tension in the muscular elements during the isovolaemic phase and a shortening in the ejection phase. The kinetics of the ventricular walls then may be characterised as a gradual motion of hypothetical myocardial points towards the centre of the cardiac geometry. Any impairment in the contractility of any part of the myocardial fibres leads to altered local kinetics. The motion of the cardiac walls is a direct response to an actual state of contractility and it is an important indication of myocardial disorders if it is changed.

Well-known methods for ventricular wall motion detection are as fol- lows: cine-ventriculocardiography, radionuclide ventriculography and 2-D echocardiography. They signify   progress in the imaging of the anatomy and the kinetics of the ventricular structures. These methods either use some projections of the endocardial contours or cross-sectional views from certain standard directions. The end-diastolic and end-systolic phases are usually selected for the segment motion description. The step-by-step motion of all surfaces of both ventricles are, for these methods, difficult to image, because many projections must be performed and synthetised.

We have used other physical principles allowing us to visualise the kinetics of the endocardial surfaces in the form of dynamic maps. The method, entitled "Electromagnetic kinetocardiography" (EM KCG), is based on a new approach to the electromagnetic field properties.

PHYSICAL ASPECTS OF EM KCG

Primary Electromagnetic Field

A high frequency electromagnetic field (EMF) is used to transform the motion of the heart ventricles into electrical quantities measurable on the body surface. The simplified model of the interactions taking place in a human torso exposed to an EMF is shown in Fig. 1. The thorax of the person under examination is situated inside the inductor loop (L). The loop is

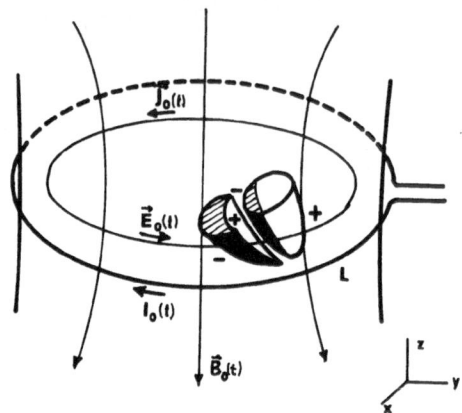

Fig. 1. Primary EMF and charge layers
at endocardial surfaces.
These charge layers generate
secondary EMF.

fed by the RF current $I_0(t)$. This current generates a magnetic field with magnetic induction $\vec{B}_0(t)$ and magnetic flux $\vec{\Phi}_0(t) = \vec{B}_0(t) \cdot \vec{S}$, where $\vec{S}$ is the area of the loop. According to Faraday's induction law, the time variation of Φ_0 induces the electric field $\vec{E}_0(t)$ and the Foucault current density $\vec{J}_0(t)$ in the conductive medium of the thorax. All indices, denoted by a small zero subscript, refer to the so-called primary EMF.

Electrical Parameters Inside the Thorax

The intra-thoracic medium is electrically non-homogeneous. Of the fundamental parameters σ, ε and μ which appear in Maxwell's equations, only σ (conductivity) and ε (permittivity) should be considered since μ refers to the magnetic medium and the body is non-magnetic. The mean values of σ (at f = 20 MHz) are, for blood, $0.6 (\Omega^{-1}m^{-1})$ and, for cardiac muscle, $0.4 (\Omega^{-1}m^{-1})$ along the fibres and $0.17 (\Omega^{-1}m^{-1})$ across the fibres. The fat tissue has only $0.05 (\Omega^{-1}m^{-1})$. We can see that the difference in σ between fat tissue and blood is an order of magnitude. Also σ of blood is 3 to 4 times greater than that of cardiac muscle taken across the fibres. The average value of the relative permittivity ε_r at a frequency of 20 MHz and T = 37°C is about 150 for muscle and 70 for blood.

Two types of current density are present inside the thorax: polarisation current density

$$|\vec{J}_P| = |i\omega\varepsilon_r\varepsilon_0 \vec{E}_0| \tag{1}$$

and conductivity current density

$$|\vec{J}_C| = \sigma|\vec{E}_0| . \tag{2}$$

When we compare (1) and (2), we obtain

$$\frac{|\vec{J}_P|}{|\vec{J}_C|} = \left|\frac{i\omega\varepsilon_r\varepsilon_0}{\sigma}\right| \tag{3}$$

and if we put

$$\left| \frac{i\omega\varepsilon_r\varepsilon_o}{\sigma} \right| \ll 1 \tag{4}$$

we obtain one of the conditions of quasi-stationarity. If we solve (4) for blood $\left(\sigma = 0.6\,\Omega^{-1}m^{-1}; \quad \varepsilon_r = 70; \quad \varepsilon_o = 0.885 \times 10^{-11}\,F\,m^{-1} \text{ and } f = 20\,MHz \right)$ then $|\vec{J}_P|/|\vec{J}_C| = 0.078$. This fulfils the condition of quasi-stationarity rather well, but for fatty tissues the result is 0.9, which is at the limit of the condition. Fortunately, fatty tissues that are deep inside the thorax are rare. The fulfilment of the conditions of quasi-stationarity means that we can treat the intra-thoracic medium as being mainly a resistive material and so we can consider only vortex current densities and neglect polarisation currents. Then the problem is rather diffusive in nature and no longer one of the propagation of an electromagnetic wave.

The parameters of the primary EMF are as follows:

magnetic induction of the loop $|B_{oz}| = 1.5 \times 10^{-6}T$,

$|\vec{E}_o| = 20\,Vm^{-1}$ at the chest surface,

$|\vec{J}_A| = 19\,Am^{-2}$ (vortex current density near the chest surface)

σ is considered to be $1(\Omega^{-1}m^{-1})$, $f = 20\,MHz$,

$I_o = 0.5\,A$ (loop current) and $P = 100\,Wm^{-2}$ (power density).

Secondary Electromagnetic Field

The primary EMF acting in a conductive inhomogeneous volume conductor (represented by a human torso) generates two secondary effects:

(i) <u>Inductive effect</u>. The heart can be approximated by a conductive sphere of radius a. This sphere is placed in the RF magnetic field with induction $\vec{B}_o\,e^{j\omega t}$. In the sphere are created eddy currents whose current density is $\vec{J}_A = -\sigma\frac{\partial \vec{A}_o}{\partial t}$, where $\vec{A}_o$ is a vector potential. The magnetisation $\vec{M}$ (magnetic dipole moment of volume current distribution) is given by $\vec{M} = \frac{\pi}{10}\,j\sigma\vec{B}_o.a^5.\hat{z}$, where $\hat{z}$ is a unit vector in the z direction. During contraction the heart volume varies over time, as also does its magnetic dipole moment.

(ii) <u>Galvanic effect</u>. This second effect is more important for EM KCG and is, in fact, a galvanic coupling of the eddy currents flowing in the thorax through interfaces separating regions with different values of σ and ε. On these interfaces are created single layers of charge, which help to explain the refraction of the $\vec{E}_o$ lines when passing through boundaries separating two media with $\sigma_i \neq \sigma_j$ and $\varepsilon_i \neq \varepsilon_j$. The charged surfaces form microscopic and macroscopic sources and sinks of the secondary EMF in the body volume conductor. The endocardial surfaces (see Fig. 1) are dominant charge layers with different polarities. They form macroscopic dipoles or multipoles which vary in time and space with the heart motion and thus change or modulate the secondary EMF. The motion of these charge layers can be seen under a spherical angle by electrodes placed over the chest.

Measurement of Secondary EMF Variations

In order to reveal abnormal local ventricular kinetics, the planar distribution of the secondary EMF must be studied. It is well known that the electric field can be characterised by a potential Φ, intensity $\vec{E}$ or by higher spatial derivatives of the potential $\nabla^2\Phi$. For EM KCG it is very important to differentiate particular sources and sinks on the endocardial surfaces. In Fig. 1 we can see that the right ventricle has its source (+) on the septal endocardial plane and its sink (-) on the external ventricular

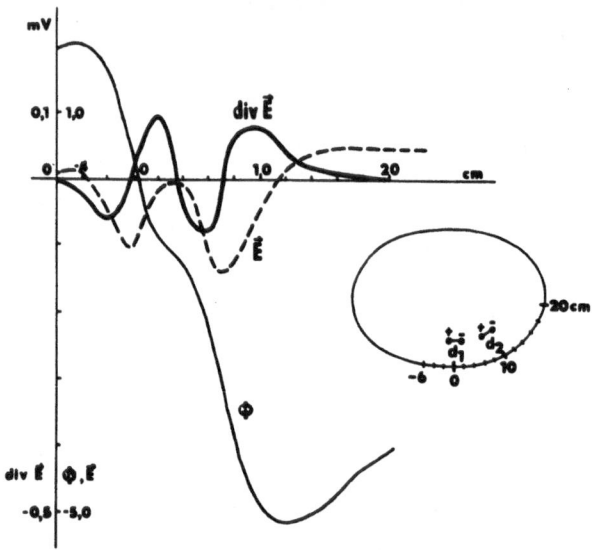

Fig. 2.   Distribution of Φ, $\vec{E}$ and div $\vec{E}$ of
two separated dipoles measured on
the thorax-shaped electrolytic tank.

plane. Similarly, the left ventricle has its source (+) on the anterior
and lateral endocardial surface and sink (-) on the left septal surface.
To distinguish the sources and sinks in the precordial distribution of the
field we use a special technique for direct measurement of $\nabla^2\Phi$ or divergence $\vec{E}$ (div $\vec{E}$). The div $\vec{E}$ is proportional to the sum of the diagonal components of the symmetrical part of the general second order tensor of
grad $\vec{E}$ (G_2). Off-diagonal components of this tensor relate to magnetic
sources, in our case to the magnetic dipole $\vec{M}$ generated by eddy currents
in the ventricular cavities. The reason for measurement of div $\vec{E}$ is explained by a model experiment. In a thorax-shaped electrolytic tank are
situated two dipoles, d_1 and d_2 (see Fig. 2). Measurements of the surface
distribution of the electric field were performed as potential Φ, and as
intensity $\vec{E}$ and div $\vec{E}$ as well. The distribution of potential Φ is a superposition of both fields without differentiating between particular sources
and sinks. The first derivative (in our case more correctly the difference)
of Φ is the intensity $\vec{E}$ which gives a distribution with two peaks differentiating two maxima of intensity. The second derivative (difference) of Φ
shows two pairs of sources and sinks in the distribution. This documents
the fact that the time and space behaviour of particular sources and sinks
in the heart can be detected very clearly by this method.

A PROBE FOR DIRECT DIV $\vec{E}$ MEASUREMENT

The principle of direct div $\vec{E}$ measurement lies in the analogue summation of the diagonal components of the tensor G_2. This leads to a Laplacian
type:

$$\left(P_1 + P_2 + P_3 + P_4 - 4P_5\right)/1^2 = \nabla^2\Phi \tag{5}$$

According to Fig. 3, the P_1, P_2 ..... P_5 are potentials (in fact normalised
to m^2) on equi-distant electrodes, arranged with a distance $1 = 1\,cm$. In
our case we get

114

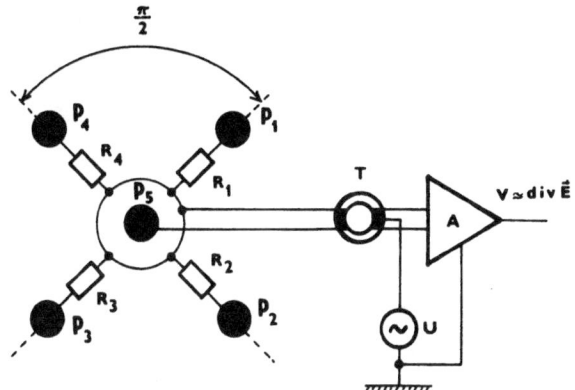

Fig. 3.　Electrode configuration and electronics of the probe for direct measurement of div $\vec{E}$.

$$\frac{1}{l^2}\left[\frac{P_1 + P_2 + P_3 + P_4}{4} - P_5\right] = \left(\frac{1}{4}\right)\nabla^2\Phi = \left(\frac{1}{4}\right)\text{div }\vec{E} \qquad (6)$$

since the summation with 4 resistors gives a quarter of the real value of
the sum $\left(P_1 + P_2 + P_3 + P_4\right)$. The toroidal transformer T has two functions.
The first is to separate galvanically the measuring circuits from the other
electronics lest the patient should be exposed to an electrical hazard.
The second is to transfer the measured voltage to the inputs of the differential amplifier A which is opened only during the positive half-period
of the cycle of the reference voltage U. We obtain, by such procedure, a
phase detection of the measured signals. The maximum voltage at the primary of T is 0.5 mV. The low frequency output voltage of amplifier A is
about 5 mV and requires additional amplification up to the level necessary
for processing. The special multi-electrode probe consists of 23 electrodes
(see Fig. 4). Every electrode has an independent telescopic mechanism to
ensure good contact with the curved precordial surface. The probe has 7
channels allowing simultaneous measurement of the distribution of the
div $\vec{E}$ response.

CLINICAL APPLICATION OF EM KCG

The induction loop L is tightened around the chest of a patient at the
level of the heart ventricles (Fig. 4). The loop resonates at the frequency
13.56 MHz of the exciting generator G. The probe for div $\vec{E}$ measurement is
attached to the loop. It can be freely shifted along the loop, thus

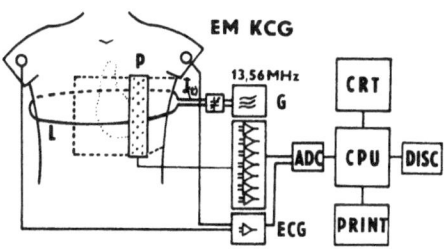

Fig. 4.　Schematic diagram of EM KCG
equipment.

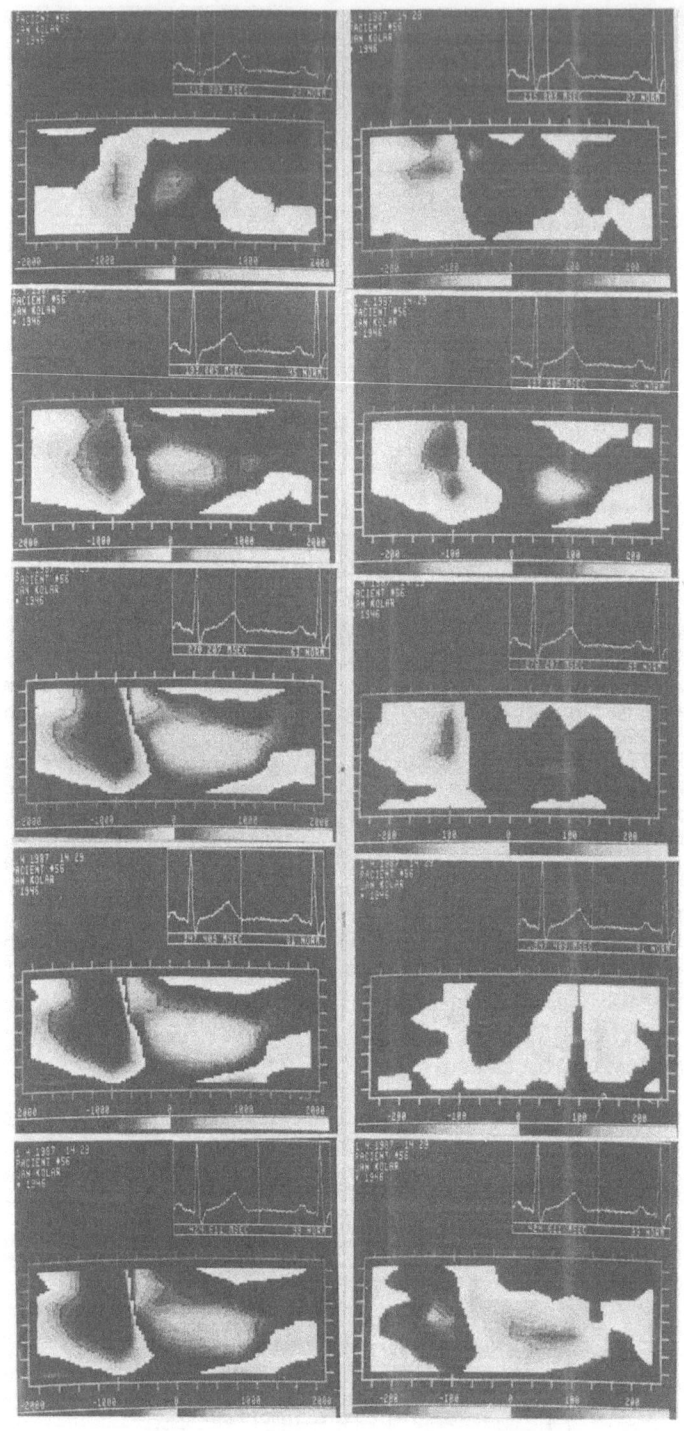

Fig. 5. Examples of dynamic maps of div $\vec{E}$ and div $\vec{J}$ in
the case of synergic contraction.

allowing 15 successive measurements to be performed in 2 cm steps. All the
measurements form a standard matrix of 7 x 15 measuring points, covering
the precordial area of 14 x 28 cm^2. The measurement is performed in the
supine position at the end of the expiratory phase of the breathing cycle.
One standard ECG lead is used during each successive measurement in order
to synchronise all the measured signals. These signals are calibrated by
1 mV. The procedure described is without any risk to the patient. The
investigation lasts about 10 - 15 minutes. No problems arise with electrode
contacts. The reproducibility of the measurements is very good if the
probe is precisely located.

COMPUTER-AIDED DATA EVALUATION

The information from all the measurements is synthetised in the form
of isoline maps. This task can be solved with any computer, so our aim was
to develop a program that could be easily modified for a particular target
machine.

The data processing can be divided into two phases:

(i) Data Acquisiton and Processing
As a result of two years' experience with EM KCG recording, we can
describe today's "standard" method.

Data are recorded sequentially in 15 probe positions, with a sampling
interval of 4 ms, the record length in each probe position being 1500
samples. Each record contains 3 - 6 systoles, from which only the best one
is stored. The selection is carried out automatically, the criterion being
that of minimal baseline shift (the data are rather sensitive to breathing).
The reliability of the detection algorithm is very high, but the operator
can, if necessary, change the results of the automatic procedure. The next
step is the base-line level subtraction and the time normalisation to minimise the unwanted influence of heart rate changes. The resulting record has
256 samples for each point on the chest surface. The position of two consecutive R points on the ECG are the 40th and 240th samples, respectively.

We have developed a "skeleton" program, to which only three modules
must be added for any computer: an A/D converter driver, which is totally
hardware dependent; graphics - only 3 procedures are necessary - draw
point, vector and character; and I/O disk operations, which may be dependent on the operating system.

(ii) Data Evaluation
This phase includes the display of the resultant EM KCG curves (time
courses) or maps on a graphic display or printer in order to determine the
diagnosis and the patient's file management.

Many procedures for the presentation of two-dimensional arrays have
been developed in our computer laboratory for EM KCG. The library includes
procedures for different modes of both black and white and colour display.
Some examples are presented in Figs. 5 and 6. The maps can be displayed
as a moving picture on a screen or can be printed out (Fig. 7). Fast
algorithms for map interpolation and isoline detection were specially
designed for small machines, so the system may run even on 8-bit CPUs.

Today, two versions of the software are in practical use. The first
runs on a PDP 11/34A under the RSX-11M operating system, the second on a
Kontron Cardio 2000 under the CP/M operating system. The first is written
in Pascal, the second in Fortran (Microsoft F 80), both with a small

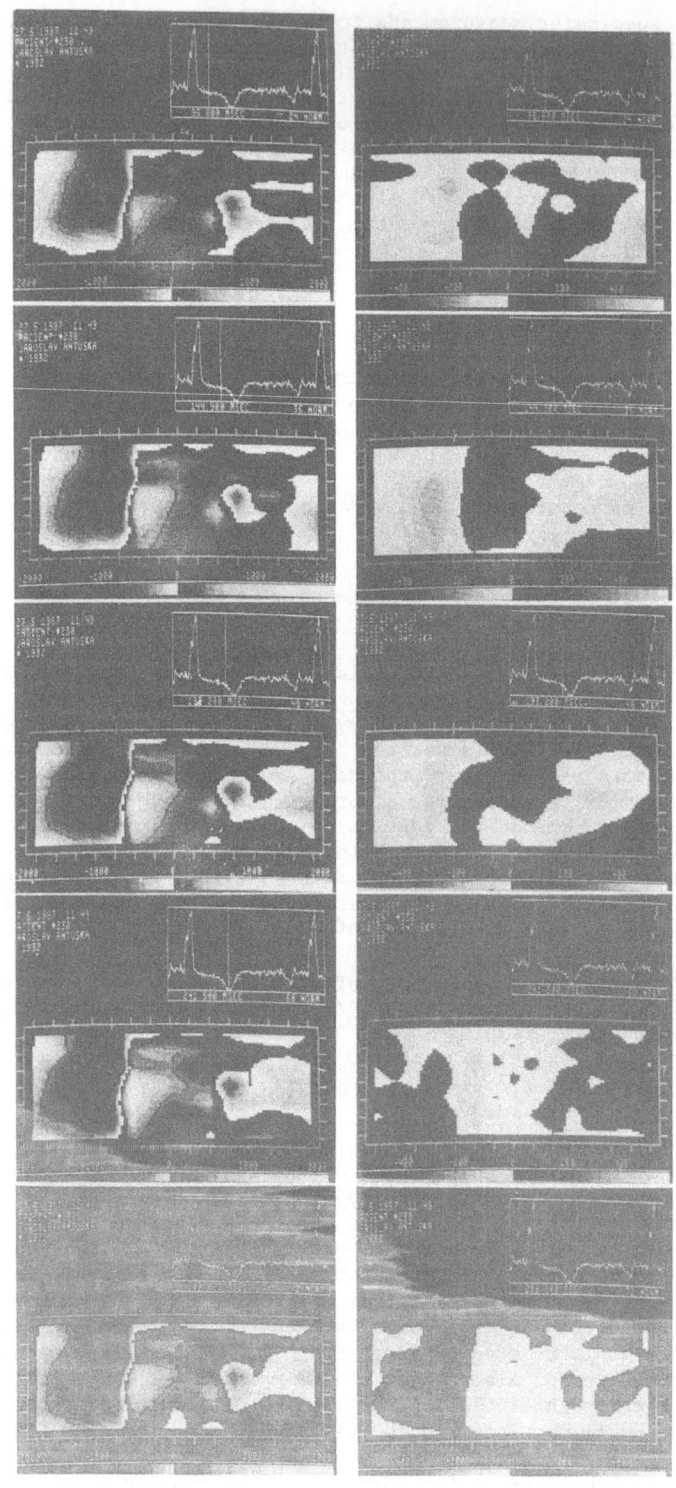

Fig. 6. Dynamic maps of div $\vec{E}$ and div $\vec{J}$ in the case of apical dyskinesia.

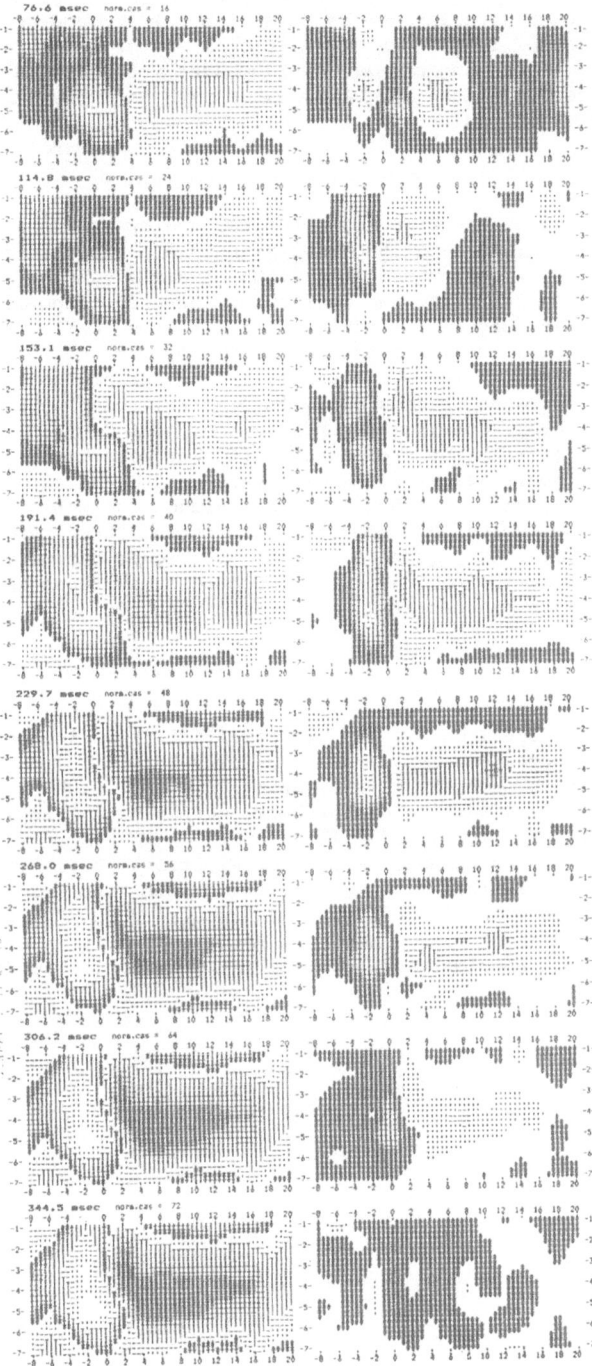

Fig. 7. Examples of printed maps of div $\vec{E}$ and div $\vec{J}$ during the systolic phase in the case of synergic contraction.

portion of the code written in assembly language. In contrast to the
PDP machine, the Cardio 2000 is not well-known and it needs a special com-
ment. This computer is delivered as an additional image processing unit
for an ATL ultrasound imaging system. It consists of a special array processor for image processing and a host Z80 based CPU. For EM KCG a simple
A/D unit was added. Because EM KCG can be a good complement to other cardiological investigations and the Cardio 2000 computer can be found in many
echo laboratories, we believe that our Cardio 2000 software can be useful
for future clinical application.

The topographical measurements described are synthesised in maps of
isodivergence $\vec{E}$ (isoline maps) and isoline maps of time differences of
div $\vec{E}$. The latter express the distribution of div $\vec{J}$, where $\vec{J}$ is the current density of the inflowing and outflowing charges. Maps of isodivergence
$\vec{E}$ express the topography of the ventricular kinetics, while the maps of
isodivergence $\vec{J}$ relate to the synchrony of the motion of different ventricular segments. In both types of isomap, the number of isolines increases
in systole and decreases in diastole. This correlates closely with the
change of distance of ventricular parts to the point of measurement. On
the maps there are different areas with different polarities. These correspond to different polarities of charge layers on the endocardium. The
polarity of some areas can change if local paradoxical motion appears.
Moving maps (dynamic maps) describe, at any instant, the state of the global
motion of the endocardial surfaces of both ventricles during the whole cardiac cycle. Imaging the motion of the endocardial surfaces is a new and
distinct approach to the detection of cardiac kinetics.

BIBLIOGRAPHY

Geselowitz, D. B., 1967, On bioelectric potential in an inhomogeneous
    volume conductor, Biophys. J., 7 : 1.
Kocí, M., Král, V. , Odehnal, M., and Hönig, J., 1985, High frequency elec-
    tromagnetic field mapping of the heart mechanic activity, Med. Biol.
    Eng. Comput., 23 (suppl. Pl) : 117.
Kocí, M., Král, V., and Odehnal, M., 1986, Electromagnetic kinetocardio-
    graphy, a new clinical method, in: "Proc. 8th Annual Conference of
    the IEEE Engineering in Medicine and Biology Society", G. V. Kondraske
    and C. J. Robinson, eds., IEEE, New York : 441.
Plonsey, R., 1969, "Bioelectric Phenomena", McGraw Hill, New York.

IDENTIFICATION OF AVERAGE NON-LINEAR PROPERTIES OF A TYPICAL ULTRASONIC

REAL-TIME SECTOR SCANNER UNDER MORE REALISTIC CONDITIONS

J. Jan, R. Kubák and M. Knotek

Department of Medical Electronics
Technical University
Purkynova, Brno, Czechoslovakia

INTRODUCTION

Improving lateral resolution in ultrasonic tomograms is still a challenging problem as the published attempts to date, based on linear image processing techniques either in the video or the r.f.-domain (Ardouin, 1985; Demoment et al., 1984; Robinson and Wing, 1984), have shown only a little, if any, improvement under realistic conditions. It seems that only non-linear algorithms could provide practically useful results.

The design of such an algorithm requires the identification of the properties of a typical scanner as a non-linear system. In Jan et al. (1986) the property of isoplanarity of the image forming process has been shown to be valid for a greater part of the original non-converted image (in polar co-ordinates), by means of measuring the responses to single-"point" targets. In the same work, the complexity of the image forming process was proved to exclude the use of simple models for the design of a restitution algorithm. It has been shown experimentally, by measuring responses to a series of two-point targets, that the qualitative behaviour of two-target imaging differs substantially, not only from the ideal sum of individual responses, but also from what would be expected on the basis of a simple far-field and point-target model. The exact form of a measured pattern was very sensitive to small differences in the relative position of both targets, though the results were well reproducible. The overall structure of the image could be described as quite complex, which may look discouraging with regard to the possibilities of restoration based on precise modelling. On the other hand, it may allow, in a real complicated situation, consideration of the imaging process as a stochastic one.

The paper by Jan (1986) thus analyses the statistical properties of the relation between measured patterns of couples and patterns computed as the sum of the individual responses to single-point targets. Surprisingly enough, there was a good qualitative agreement between both groups in a substantial number of cases when evaluated visually. This led to a trial to quantify the degree of correlation more conclusively by means of two-dimensional histograms which will be described in more detail later. The histograms showed the non-linear dependence between measured and computed pixel values, and that the results for different cases were of reasonable stability, enabling a concept of "average" non-linearity to be introduced.

In this present contribution the experimental analysis is generalised
to multiple- (namely quadruple-) point targets in order to determine whether
the previously found non-linearity holds in even more complicated cases that
correspond better to the real situation in the imaging of biological objects.

MEASUREMENTS

The experimental set-up consisted of the real-time scanner ADR 4000 S/L
with a moving crystal sector transducer (3 MHz, 13 mm), connected by means of
a special interface to an 8-bit microcomputer providing for data acquisition
and processing. The measured object - a quadruple of linear targets perpendicular to the tomographic plane, realised as copper wires with a diameter
of 0.08 mm - was fixed on a mechanical support enabling fine adjustments to
be made of the relative position of the two fixed couples of targets
mutually and with respect to the transducer in both the required directions.
The target arrangement was submerged in a water tank, the walls of which
were covered by ultrasound-absorbing material.

The original image as provided by a sector transducer has the form of
data in a matrix, the two indices of which correspond to the radial distance and azimuth angle, respectively. The displayed image, on the other
hand, must be represented by a matrix in cartesian co-ordinates; the format
conversion included in every scanner complicates the image properties to
the extent of making the analysis of imaging properties substantially more
difficult. In particular, the isoplanarity of the transfer function, referred to in Jan et al. (1986), does not apply to the converted image. For
this reason our analysis concerns the primary envelope-detected image data
in polar co-ordinates as loaded to the input of the format converter.

The attention of this study was focused on the interference phenomena
which are present when four-point targets are close to each other; the
interference causes significant differences from the ideal sum of four individual responses of separate targets. The measurement, the geometry of
which is shown in Fig. 1, was made at an axial depth of approximately
100 mm, which is usually in the region of practical interest. The quadruple
of targets was formed by two identical couples. A series of measurements
has been made for different horizontal distances of the couples.

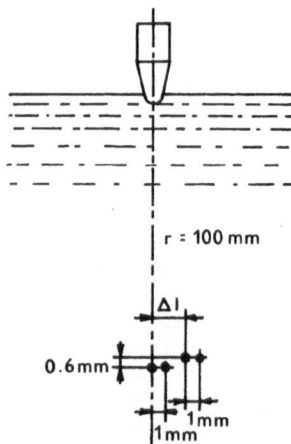

Fig. 1. Geometry of mea-
surement.

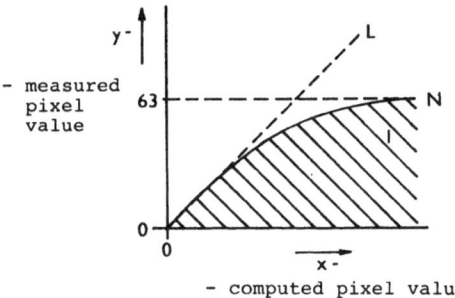

Fig. 2. Explanatory outline of the two-
dimensional histogram.

ANALYSIS AND DISCUSSION

Similarly, as in the case of comparing the measured and computed
responses of couples (Jan, 1986), there was again a good qualitative agreement between both types of response, even in the case of a quadruple target
when evaluated visually. Therefore, the same histogram technique was used
in order to find the relation between measured and additively-computed
pixel values.

The form of the histogram adopted is basically as shown in Fig. 2;
$H(x,y)$ means here the frequency of combinations (x,y) in corresponding
pixels in both compared patterns. Were both patterns identical, only the
elements $H(x,x)$ - on the line L - would be non-zero, while in the case of
totally uncorrelated (for example noise) images, the non-zero values would
be scattered evenly in the whole area of possible combinations, which is
obviously

$$A = \{(x,y): \quad x = 0, 1, \ldots, 252, y = 0, 1, \ldots, 63\} \tag{1}$$

Even if all the individual r.f. responses are totally in phase, hence the
envelope-detected responses are ideally added, the system has to make a
non-linear transform (obviously by log. amplification) in order to keep the
pixel values y less than 64 as the maximum value representable in the system
is 63. Thus it is clear that the points corresponding to the ideal sum
would then fill the histogram locations on some curve similar to N. Any
phase mismatch would lower y for the same x, thus filling the histogram
counters in the area I below the curve N.

The histograms computed for all the measurements of the above-mentioned
series demonstrated a similarity with the results obtained for couples of
point targets (Jan, 1986) showing a reasonable correlation in most cases.
As there is always interference present to some extent, due to closely
spaced targets in the individual couples, even in the case of a large distance between couples the values follow a curve from the very beginning
rather than the line L. When the couples are closer, a longer part of the
curve is active, but its form remains close to the results of Jan (1986).
Again, there seems to be only a slight dependence on the phase conditions,
so that summing all the histograms and approximating the relation $y = y(x)$
by a curve, while taking into account the weights $H(x,y)$, can produce the
desired description of the "average" non-linearity which is only an extension of the dependence following from the results of Jan (1986). Examples
of the results of analysis are shown in Fig. 3 and of the sum of histograms
in Fig. 4.

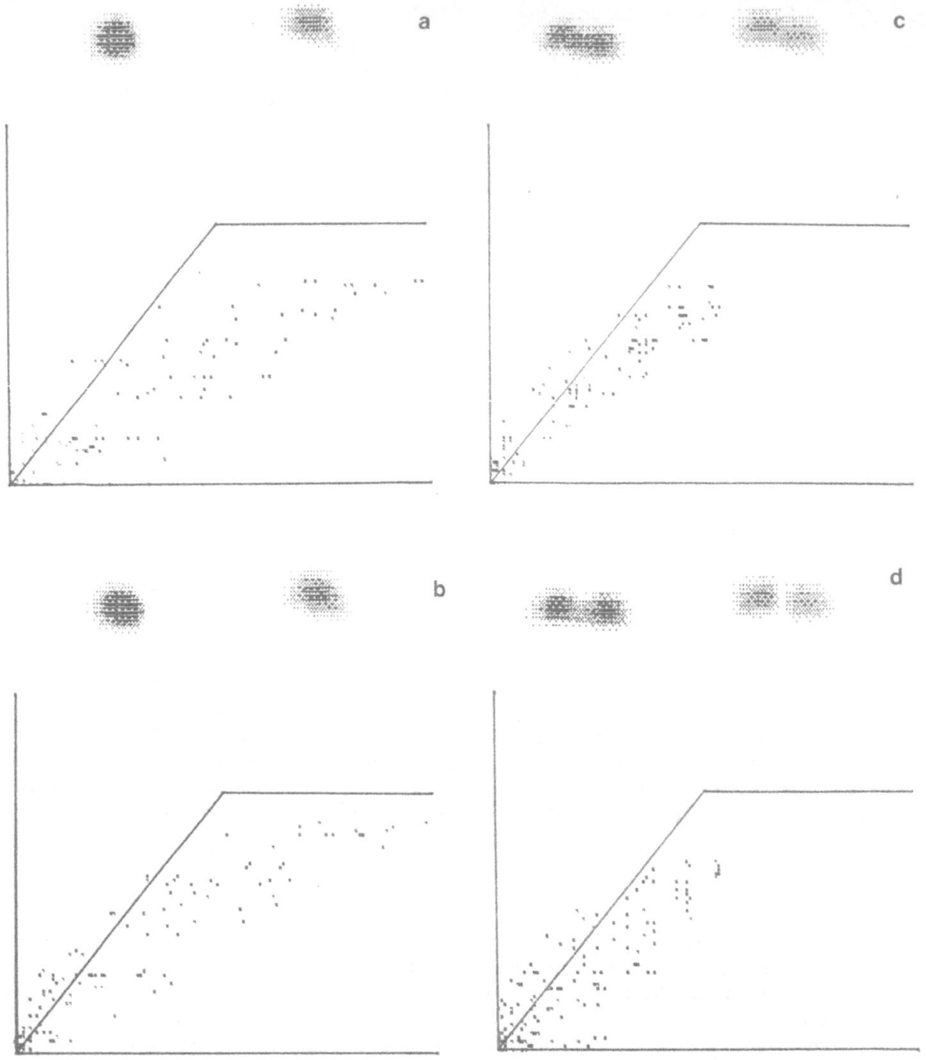

Fig. 3.   Examples of results of analysis. Above: patterns of four-point
response (left - calculated, right - measured); below: corresponding histograms a - 1 = 0.2 mm, b - 1 mm, c - 7 mm, d - 9 mm.

CONCLUSIONS

The measurement of the four-target arrangement proved that the conclusions of Jan (1986) were valid even in the case of a more complicated image
which is closer to the real conditions in a complex image. As the N-curve
seems to describe the non-linearities of the system regardless of wave
interference, and because the properties are rather isoplanar, there is
certain prospect that the attempt to design a restitution algorithm on
this basis need not be hopeless.

REFERENCES

Ardouin, J. P., 1985, "A Model for an Ultrasonic Phased-array B-scan
    Imaging System and its Application to Image Restoration", MSc Thesis,
    University of Toronto.

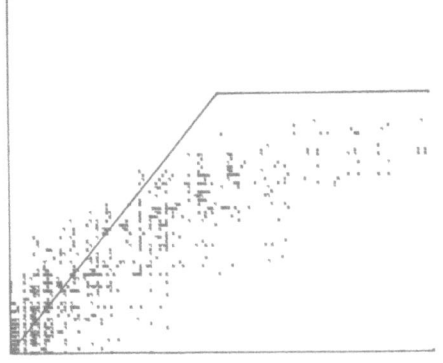

Fig. 4. Sum of histograms for l =
10 mm, 9 mm, 7 mm, 5 mm, 3 mm,
1 mm, 0.2 mm and 0 mm.

Demoment, G., Reynaud, R., and Herment, A., 1984, Range resolution improvement by a fast deconvolution method, Ultrasonic Imaging, 6 : 435.
Jan, J., 1986, Degree of nonlinearity in ultrasonographic image forming,
in: "Proc. 8th IEEE-EMBS Conference", G. V. Kondraske and C. J.
Robinson, eds., IEEE, New York : 1043.
Jan, J., Kubák, R., and Knotek, M., 1986, Point spread function of ultrasonic real time scanners, in: "Proc. Biosignal '86", DT CSVTS, Brno :
12.
Robinson, D. E., and Wing, H., 1984, Lateral deconvolution of ultrasonic
beams, Ultrasonic Imaging, 6 : 1.

AN EXPERIMENTAL APPROACH TO THE PROBLEM OF POSSIBLE EMBRYOTOXICITY DUE

TO BIOMEDICAL ULTRASOUND

I. Hrazdira, M. Doskočil, M. Dvořák and J. Šťastná

Department of Biophysics, Faculty of Medicine, Purkyně
University, Brno, Czechoslovakia;
Department of Anatomy, Faculty of General Medicine, Charles
University, Prague, Czechoslovakia; and
Department of Histology and Embryology, Faculty of Medicine,
Purkyně University, Brno, Czechoslovakia

INTRODUCTION

The extensive employment of ultrasonic methods in medicine is, in
general, based on the presumption of their harmlessness. Nevertheless, the
expansion of ultrasonic diagnostic methods in obstetrics has once more
raised the question of their safety. It has been demonstrated in several
studies that, under certain conditions, biological tissues may be
especially susceptible to damage induced by ultrasound. Embryonic tissues
are more sensitive to the influences of different physical and chemical
factors than differentiated tissues. That is why embryos at different
stages of development have frequently been used as experimental objects
for the assessment of the biological effectiveness of ultrasound.

Our approach to the problem of possible embryotoxicity of ultrasound
is based on experimental results obtained by ultrasonic irradiation of
both chicken and mouse embryos, as indicated in Table 1.

The ultrasonic intensity was measured in a water bath by means of a
microbalance and a calibrated hydrophone.

Table 1. Ultrasound Exposure Conditions

Continuous wave ultrasound	
Frequency:	0.8 MHz non-focused transducer
Intensity - SA:	100 mW cm^{-2}
Exposure time:	5 min
Pulsed ultrasound	
Frequency:	6 MHz non-focused transducer
Pulse width:	2 μs
Repetition frequency:	800 Hz
Intensity - SATA:	24.5 mW cm^{-2}
SPTP:	31 W cm^{-2}
Exposure time:	10 min

Table 2. Mortality, Body Mass and Number of Malformations in both Treated and Control Chick Embryos after 8 Days of Incubation.

Series	Prior incubation (h)	Number of eggs	Dead embryos No.	(%)	Body mass of living embryos (g)	(±SD)	Malformations No.	(%)
Control	0	37	6	(16)	1.21	0.196	3	(8)
Irradiated	0	30	6	(20)	1.14	0.221	5	(17)
Control	24	26	4	(15)	1.25	0.155	0	(0)
Irradiated	24	15	3	(20)	1.22	0.130	1	(7)
Control	48	18	3	(17)	1.33	0.239	0	(0)
Irradiated	48	17	3	(18)	1.29	0.247	1	(6)
Control	72	22	2	(9)	1.38	0.364	1	(5)
Irradiated	72	16	2	(13)	1.28	0.275	1	(6)

Repeated experiment

Series	Prior incubation (h)	Number of eggs	Dead embryos No.	(%)	Body mass of living embryos (g)	(±SD)	Malformations No.	(%)
Control	0	33	2	(6)	1.21	0.216	3	(9)
Irradiated	0	31	2	(6)	1.15	0.182	1	(3)

CHICK EMBRYOS AS THE EXPERIMENTAL MODEL

Bird embryos make it possible to investigate the direct effects of both physical and chemical factors without modifications due to the maternal organism. In chick embryos, continuous wave ultrasound of therapeutic intensities applied in the early stages of organogenesis produces an increased number of malformations (Taylor and Dyson, 1973). This teratogenic effect is most probably due to elevated temperature and cannot be considered as a specific action of ultrasound.

Irradiation of fertile eggs by pulsed ultrasound before and after 18 hours of incubation did not give rise to any deleterious effect on the embryonic development (Barnett, 1983).

In our experiments fertile eggs were irradiated in ovo through an opening in the shell under conditions corresponding to a static diagnostic procedure prior to incubation and after 24, 48 and 72 hours of incubation. The results are summarised in Table 2.

Irradiation of chick embryos during organogenesis does not result in any deleterious effect which could be ascribed to a direct action of ultrasound. The irradiation of fertile eggs prior to incubation, however, resulted in a higher proportion of malformation. This finding was interpreted as a non-specific effect at the period of maximal susceptibility of the embryonic target to external stimuli (Hrazdira and Doskočil, 1986). To verify this result we carried out a second experimental series under the same conditions. In this experiment we did not confirm our previous result.

The average body mass of all irradiated embryos was lower than the controls. The differences were not statistically significant. The finding, however, gives evidence for a great susceptibility of chick embryos in the early developmental stage to mechanical intervention.

EFFECTS ON MAMMALIAN EMBRYOS AND FOETUSES

Most of the studies which provide data on foetal effects in mammals have been performed in small laboratory rodents (mice, rats). In these experiments the entire foetus and a great part of the maternal body were exposed to the ultrasonic beam. The results indicated that exposure to sufficiently high intensities of ultrasound was able to produce death of the foetus, reduction in its weight, and a variety of malformations (Shoji et al., 1975; Fry et al., 1978; Stolzenberg et al., 1980; Child et al., 1983; O'Brien, 1983). After irradiation of pre-implantation embryos by continuous wave ultrasound of therapeutic intensities the degradation or retarded growth of embryos was noted (Akamatsu, 1981). In these experiments the embryotoxic effects are undoubtedly the consequence of a thermal influence (Lele, 1985).

The diagnostic treatment of mammalian embryos at different stages of organogenesis does not result in any deleterious effect due to ultrasound. On the other hand, the diagnostic irradiation of pre-implantation embryos at the beginning of or during blastogenesis resulted, after transfer to the uteri, in a significantly higher proportion of resorbed embryos (Puissant et al., 1981; Demoulin et al., 1985).

In our recent electron microscopic study two series of experiments have been carried out. In the first series two-cell mouse embryos were irradiated in vitro by continuous wave ultrasound at SA intensity of $0.1\,W\,cm^{-2}$ for 5 min. The development of the irradiated embryos corresponded to that of embryos without any experimental intervention. The only unusual morphological finding was the occurrence of mitochrondria with tubular cristae in blastocysts, both in the trophoblast and the embryoblast cell (Fig. 1). This type of mitochondrion is considered to be a sign of steroidogenic activity of the cell. The occurrence of this type of mitochondrion

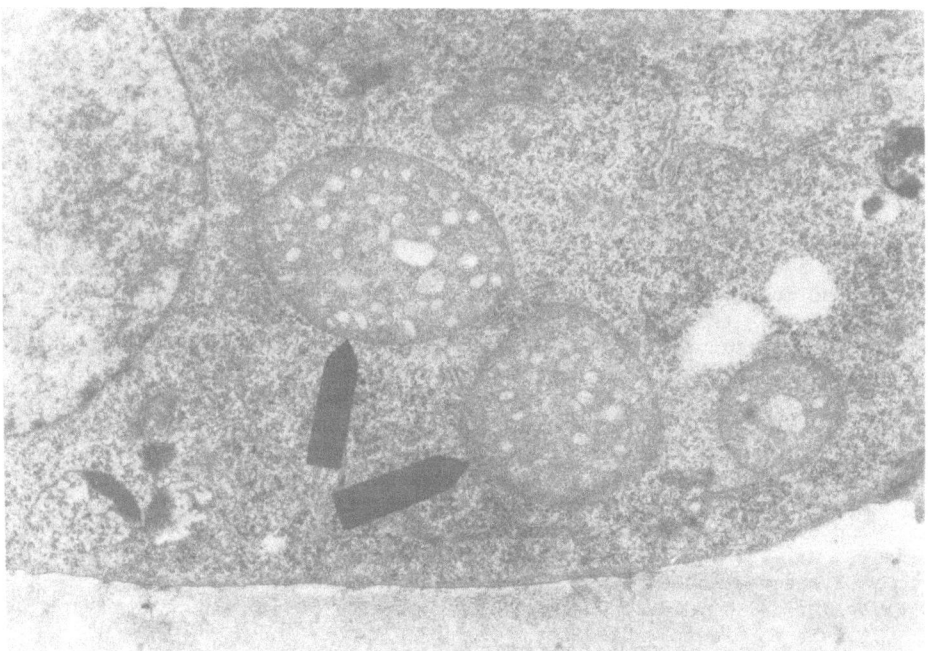

Fig. 1. Ultrathin section of mouse blastocyst. Cytoplasm of embryoblast cell containing mitochondria with tubulous cristae (arrows). Magn. x 16,000.

has not yet been described in blastocysts cultured in vitro. It might be considered a consequence of ultrasonically induced functional stimulation of the cells (Dvořák et al., 1987).

In the second series, adult female mice were irradiated in vivo immediately after mating under the same exposure parameters. After 72 hours regularly differentiated blastocysts were developed in both control and irradiated animals. No ultrastructural differences between the blastocysts of control and treated mice were demonstrated. Neither could mitochondria with tubular cristae, described after irradiation of two-cell embryos in vitro, be found in these experiments. The number of developed embryos did not differ in the experimental and control groups. However, the number of animals (5 in each group) was insufficient for a serious statistical evaluation.

CONCLUSIONS

From our experimental results the following conclusions can be drawn:

(i)   The demonstrated embryolethality and teratogenicity at higher intensities of ultrasound are most probably due to heating effects of ultrasound.

(ii) Pulsed ultrasound under diagnostic conditions, as well as continuous wave ultrasound at low intensities applied during organogenesis or later stages of embryonic development, has no deleterious effect on the embryo.

(iii) From the point of view of possible embryotoxic effects of ultrasound, the most critical stages are fertilisation and blastogenesis. These periods need further detailed investigation.

REFERENCES

Akamatsu, N., 1981, Ultrasound irradiation effects on pre-implantation embryos, Acta Obstet. Gynaec. Jpn., 33 : 969.
Barnett, S. B., 1983, The influence of ultrasound on embryonic development, Ultrasound Med. Biol., 9 : 19.
Child, S. Z., Carstensen, E. L., and Davis, H., 1983, A test for the effects of low-temporal average-intensity pulsed ultrasound on the rat fetus, Exp. Cell Biol., 52 : 207.
Demoulin, A., Bologne, R., Hustin, J., and Lambotte, R., 1985, Is ultrasound monitoring of follicular growth harmless?, Ann. New York Acad. Sci., 442 : 146.
Dvořák, M., Hrazdira, I., Šťastná, J., and Škorpíková, J., 1987, Influence of c.w. ultrasound on ultrastructure of pre-implantation mouse embryos cultured in vitro, Scripta Medica, 60 : 341.
Fry, F. J., Erdmann, W. A., Johnson, L. K., and Baird, A. I., 1978, Ultrasonic toxicity study, Ultrasound Med. Biol., 3 : 315.
Hrazdira, I., and Doskočil, M., 1986, Pulsed ultrasound and embryonic development, in: "Abstracts of UBIOMED VII", R. Millner, L. Pahl, and H.-J. Hein, eds., M. Luther Universität, Halle-Wittenberg : 70.
Lele, P. P., 1985, Ultrasound bioeffects and human reproduction, in: "Ultrasound Annual 1985", Raven Press, New York : 335.
O'Brien, Jr., W. D., 1983, Dose-dependent effects of ultrasound on fetal weight in mice, J. Ultrasound Med., 2 : 1.
Puissant, F., Lejeune, B., and Leroy, F., 1984, Effects on mature mouse oocytes of ultrasound treatment applied before in vitro fertilization, IRCS Med. Sci., 12 : 421.

Shoji, R., Murakami, U., and Shimizu, T., 1975, Influence of low-intensity ultrasonic irradiation on prenatal development of two in-bred mouse strains, Teratology, 12 : 227.

Stolzenberg, S. J., Torbit, C. A., Edmonds, P. D., and Taenzer, J. C., 1980, Effects of ultrasound on the mouse exposed at different states of gestation, Rad. Environ. Biophys., 17 : 245.

Taylor, K. J. W., and Dyson, M., 1973, Toxicity studies on the interaction of ultrasound on embryonic and adult tissues, in: "Proc. 2nd World Congress on Ultrasonic in Medicine", Excerpta Medica, Amsterdam : 353.

INTERACTION OF BLOOD WITH BIOMEDICAL POLYMERS - SOME BASIC ASPECTS

H. Wolf, R. Karwath and T. Groth

Biomaterials Research Unit
Humboldt University, School of Medicine (Charité)
Berlin, GDR

INTRODUCTION

During the last decade tremendous efforts have been made in the development and improvement of medical devices, artificial organs and other life support systems. One of the main problems to be overcome if further progress is to be achieved consists of the need to improve the biocompatibility of materials used for the above-mentioned devices and systems. Much basic research has been carried out to understand the complex interface reactions between living systems and biomedical materials to overcome the "trial and error principle" in the development of new biocompatible materials. This chapter summarises some basic aspects of blood-polymer interaction regarding the theoretical understanding of haemocompatibility.

BIOCOMPATIBILITY AND THE FACTORS WHICH DETERMINE IT

The term "biocompatibility" has not been defined precisely and it is usually considered to imply that the clinical application of biomedical materials will not create effects which endanger the life of patients. Considering biocompatibility as the interrelationship between non-biological surfaces and a living system, the following description is possible -

Biocompatibility includes:

 (i)   the energetics and kinetics of all physical, chemical and biochemical processes at the interface between the biomaterial and the biological system during the time of contact and the direct reactions of the biological system induced by these processes at the interface;

 (ii) the sum of the changes of physical and chemical properties of material, that is surface composition, surface energy, corrosion, biodegradation, and so on;

 (iii) the sum of the changes of the biological system outside the interface induced by the interaction (systemic effects such as immune reactions, carcenogenic or mutagenic effects).

If we want to understand the basic mechanisms which are responsible for the biocompatibility, and if we want to characterise the biocompatibility of a biomaterial, we have to focus our interest upon the following factors and processes, which are summarised schematically in Fig. 1.

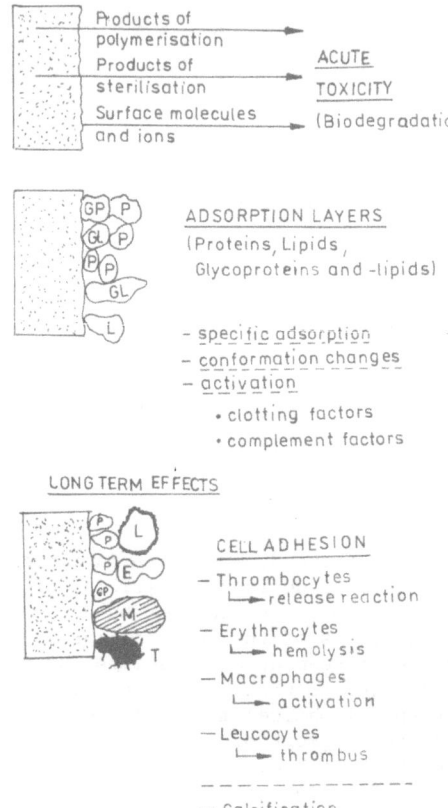

SHORT TERM EFFECTS

Products of
polymerisation

Products of
sterilisation

Surface molecules
and ions

ACUTE

TOXICITY

(Biodegradation)

ADSORPTION LAYERS

(Proteins, Lipids,
Glycoproteins and -lipids)

- specific adsorption
- conformation changes
- activation
  · clotting factors
  · complement factors

LONG TERM EFFECTS

CELL ADHESION

— Thrombocytes
     └─ release reaction

— Erythrocytes
     └─ hemolysis

— Macrophages
     └─ activation

— Leucocytes
     └─ thrombus

_ _ _ _ _ _ _ _ _ _ _ _

— Calcification

Fig. 1.   Factors and processes influencing biocompatibility of biomedical polymers.

The main problem in characterising biocompatibility and its evaluation
consists of proving unequivocal correlations between physical and chemical
properties of materials and the possible complex reactions of the biological system during interaction, as shown in Fig. 1.   In this state of affairs
not all the reactions mentioned are understood in detail.   Nevertheless,
there are some basic concepts in terms of which attempts were made to correlate directly physical surface parameters or chemical parameters such as
surface structure or surface composition with the biocompatible behaviour
of biomaterials.   Because most of these concepts were developed in connection with blood-polymer interaction, the remarks which follow will be restricted to this problem.

MAIN HYPOTHESES OF HAEMOCOMPATIBILITY

If blood comes into contact with a non-biological surface, several
initial events take place leading finally to thrombus formation as a function of time and rheological conditions (Bruck, 1977;   Wolf, 1981).   Under
constant rheological conditions the process of thrombus formation includes
three major steps occurring in a distinct time hierarchy (Bruck, 1977), as
shown in Fig. 2.

134

THROMBUS FORMATION   TIME HIERARCHY

```
┌─────────────────────────────┐      ┌─────────────────────────────┐
│ ADSORPTION OF PLASMA PROTEINS│      │ microseconds - milliseconds │
│ AND COAGULATION FACTORS      │      └─────────────────────────────┘
└─────────────────────────────┘

┌─────────────────────────────┐      ┌─────────────────────────────┐
│ CONFORMATIONAL CHANGES OF    │      │     seconds - minutes       │
│ ADSORBED PROTEINS            │      └─────────────────────────────┘
│- - - - - - - - - - - - - - - │
│ COAGULATION OF PLASMA PROTEINS│
└─────────────────────────────┘

┌─────────────────────────────┐      ┌─────────────────────────────┐
│ ⊢ ACTIVATION AND ADHESION OF │      │          minutes            │
│   PLATELETS AND WHITE CELLS, │      └─────────────────────────────┘
│ ⊢ ADHESION OF RED CELLS,     │
│ ⊢ RELEASE AND AGGREGATION OF │
│   PLATELETS                  │
└─────────────────────────────┘

        shear │ stress

   low    ╱        ╲  high

┌──────────┐   ┌──────────┐        ┌─────────────────────────────┐
│  RED     │   │  WHITE   │        │     minutes - hours         │
│ THROMBUS │   │ THROMBUS │        └─────────────────────────────┘
└──────────┘   └──────────┘
```

Fig. 2. Major steps of thrombus formation and its time hierarchy
at polymer surfaces.

During long-term interaction (more than two weeks) between blood and polymers, a second important event, the so-called "calcification", can take place which reduces the haemocompatibility of polymers (Harasaki et al., 1981; Herzlinger et al., 1981). This phenomenon consists of the formation of a surface layer that contains calcium and phosphorus compounds as well as organelles of platelets. The change of amorphous calcium/phosphorus deposits into crystallites is also possible. The reason for this calcification of polymer surfaces, which was also found to depend on pH and the calcium/phosphate ratio in blood serum, could not be clarified. There seem to exist some similarities with the process of arteriosclerosis observed on natural blood vessels which, however, have not to date been proved.

For long-term application of medical devices or artificial organs such as the total artificial heart, the calcification of polymer surfaces has to be considered as a serious unsolved problem. Thrombus formation was found to be strongly influenced by the four factors concerning the foreign surface listed in Table 1.

On the basis of much experimental work, several hypotheses were postulated concerning the inter-relationship between the factors listed in Table 1 and the optimum haemocompatibility of biomedical polymers.

The prevailing hypothesis for the past two decades has been that of Sawyer and co-workers (Ramasamy and Sawyer, 1977; Sawyer, 1964), by which it was postulated that the electrostatic repulsion between the negative charges of artificial surfaces and blood cells prevents cell adhesion and thrombus formation. Looking at this "negative fixed charge hypothesis" in

Table 1. Surface-related Factors Influencing
Haemocompatibility of Polymers.

---

1. Electrical surface charge
2. Surface free energy
3. Surface roughness
4. Chemical surface composition and molecular surface structure

---

detail, it was found that the adhesion mechanism of blood cells on to artificial surfaces could not be explained only in terms of electrostatic interaction (Wolf and Gingell, 1983). This stemmed from the spatial distribution of electrical charges at the cell surface and inside the protein adsorption layers on to solid surfaces (Voigt et al., 1983), as well as the
ability of biological membranes to make molecular contact with solid surfaces (Trommler et al., 1985; Wolf and Gingell, 1983).

Concerning the role of surface energy as a haemocompatibility-determining factor (Bayer, 1972), it was postulated that the so-called critical
surface tension of polymers determines the region of good biocompatibility
or good cell adhesiveness. The range of critical surface tension between
20 and 30 dynes per cm, where minimal cell spreading could be observed, was
concluded to be the region of good haemocompatibility.

This concept was extended and realised by the studies of Sharma (1981).
It was found that low dispersive - high polar surfaces provide surface energetics which appear to favour weak adsorption and retention of plasma proteins, that is a surface of poor haemocompatibility. Conversely, polymer
surfaces with high dispersive and low polar surface energy provide surface
energetics for achieving good haemocompatibility. Using this concept from
Andrade (1973) it was postulated that the sum of all the terms of interface
energy should approach a value of zero to obtain a surface of good blood
compatibility.

From Ratner et al. (1979) it was concluded that the ratio between the
hydrophilic and hydrophobic natures of a polymer surface must be well
balanced. From Coleman et al. (1982) both hypotheses were checked by means
of in vitro studies of platelet adhesion and activation of several clotting
parameters. The results led them to the conclusion that neither hypothesis
could explain the apparent conflict between the platelet adhesion data and
the data regarding coagulation time.

Several hypotheses of haemocompatibility are based on the chemical
composition and molecular structure of polymer surfaces. It was assumed
that the adsorption or covalent immobilisation of biologically active
molecules, which are able to inhibit the clotting system or the activation
of blood platelets, should increase the anti-thrombogenicity of polymer
surfaces. Therefore, compounds such as heparin, prostacyclin or heparin-
analogous compounds were used for the immobilisation on to polymer surfaces.
From Ebert and Kim (1982) it was shown that covalently immobilised heparin
improves the blood compatibility of polymer surfaces. The anticoagulant
activity was found to depend on the length of the diaminoalkane spacer
molecules between the polymer surface and the immobilised molecules. The
presence of the highly negatively charged heparin on the polymer surface
was shown to influence significantly the adsorption of plasma proteins and
the specific binding of clotting factors. Immobilised heparin, however,
was not able to penetrate the adsorbed plasma protein layer, indicating
that immobilised heparin does not directly interact with blood platelets.

From Josefowisz and his co-workers (Fougnot et al., 1979; 1984) it was found that heparin-like compounds immobilised at polymer surfaces are able to increase the thromboresistance of polymers. Regarding the underlying mechanism, it could be shown that these materials adsorb thrombin and its inhibitor, antithrombin, with a higher affinity for the protease than for the antiprotease. These complexes are responsible for the generation of thrombin-antithrombin complexes. The thrombin-antithrombin complex generated on the polymer surface with heparin-like compounds is desorbed by thrombin producing a catalytic anticoagulant effect.

A further approach to improve the haemocompatibility of biomedical polymers was made by Hennink et al. (1984) by means of pre-adsorption of albumin-heparin conjugates. The adsorbed albumin-heparin conjugates were shown to reduce the platelet adhesion on to the polymer surfaces and to prolong the clotting time of blood plasma. It could be demonstrated that antithrombin III interacting with albumin-heparin conjugates is involved in the inhibition of the intrinsic coagulation at the plasma-polymer interface. From these results it was concluded that the mechanism of action of heparin present in the surface-bound albumin-heparin conjugate is similar to that of bulk heparin.

Ikada (1984) attempted to unify the diverse hypotheses of haemocompatibility of polymers described above as far as possible. His basic assumption was that a polymer surface which does not interact with blood at all does not induce thrombus formation. In other words the polymer surface, which does not adsorb any plasma protein, must be blood-compatible. By means of theoretical calculations on the work of adhesion in water, it could be shown, qualitatively, that surfaces of high blood compatibility should possess either an extremely high hydrophobicity or an extremely high hydrophilicity. The concept of extremely high hydrophilicity of a polymer surface was put into practice by Ikada, polymerising uncharged water-soluble polymer chains on to the surface of a bulk polymer. These freely moving polymer chains were capable of keeping a large quantity of water inside and around them. Therefore, a mobile water layer is formed spontaneously on the surface if it comes into contact with water containing fluids such as blood. This "superhydrophilic diffuse" surface was shown to provide excellent in vitro and in vivo haemocompatibility in the first studies with blood or using test animals.

As was pointed out by Ikada, further experimental work is necessary to investigate the possibilities of medical application of these materials in blood pumps, blood vessels or the total artificial heart. Questions like sterilisation, mechanical and chemical stability of the superhydrophilic diffuse layer and its stability against calcification have not been clarified by experimental studies.

From the main hypotheses concerning blood-polymer interaction which have been described, it becomes clear that the formation and dynamic behaviour of plasma protein adsorption on to polymer surfaces play a key role in the mechanisms which determine haemocompatibility. The blood cells forming the thrombus do not adhere directly on to the polymer surface, but in most cases they should interact with adsorbed plasma proteins of native or changed conformation. Therefore, knowledge about the adsorption/desorption processes of plasma proteins and their possible surface-induced conformational changes is of great importance in obtaining a deeper understanding of the underlying molecular and cellular processes that determine the haemocompatibility of polymers.

Table 2. General Results and Principles Found for the Interaction
         of Proteins with Solid Surfaces.

---

1. In plasma protein mixtures preferential adsorption occurs at
   solid surfaces.
2. The structure of adsorption layer is influenced by physical
   and chemical surface properties of solid substrates and by
   physical and chemical properties of protein molecules.
3. Structural perturbations of proteins may occur as a result of
   adsorption.
4. Under static and dynamic conditions the replacement of adsorbed
   protein molecules is possible.

PROTEIN ADSORPTION AND HAEMOCOMPATIBILITY

The key role of protein adsorption layers on polymer surfaces as a
haemocompatibility-determining factor has been stressed in many investigations, which cannot be reviewed here in detail (Bohnert and Horbett, 1986;
Bornzin and Miller, 1982; Breemhaar et al., 1984; Brynda et al., 1984;
Chuang et al., 1978; Morrisey, 1977; Norde, 1980. Surveying this
literature it appears that the interface behaviour of proteins is a controversial problem. The reason seems to be that the adsorption of macro-
molecules, especially biopolymers, is different from that of low molecular
weight molecules. Consequently the application of thermodynamic laws that
are based on equilibrium is questionable. A comprehensive theory describing protein adsorption seems far away (Norde, 1980). Nevertheless, there
are some general results and some principles may be indicated, which are
summarised in Table 2.

Keeping these principles in mind it becomes clear that during the
interaction of blood with polymers, an adsorption layer of high complexity
results because of the complex composition of blood with more than 140
different proteins, lipids, glycoproteins and glycolipids (Breemhaar et al.,
1984; Chuang et al., 1978). This seems to be the reason that in the
literature many of the results are controversial as to the mechanisms by
which the adsorbed layer of plasma proteins can influence thrombus formation, complement activation or calcification on to synthetic polymers. In
connection with the main hypotheses of haemocompatibility described above,
which are based on physical or chemical surface properties of polymers such
as surface charge, surface energy or distribution of hydrophobic and hydrophilic regions at the surface, the question arises as to how the formed
blood elements recognise these surface properties through the protein adsorption layer.

For example, it was found that the electrical surface properties of
polymers are covered more or less after protein adsorption (Voigt et al.,
1983). Therefore, the surface properties of a bare polymer cannot be correlated directly with the adhesion behaviour of blood cells. On the other
hand, it was found that small differences in the surface properties of one
and the same polymer which had been coated with plasma proteins resulted
in a significant difference in platelet adhesiveness (Karwath and Wolf,
1986). From this result two conclusions are possible: (i) that the surface properties of polymers are not completely covered by the protein
adsorption layer, or (ii) that the polymer surface is completely covered
with plasma proteins, but the surface properties of polymers are transformed through the protein adsorption layer by means of an unknown mechanism.

Further important results that cannot be explained unequivocally are the following: the <u>activation of the complement system</u> during contact of blood with polymers of the cellulose type (Herzlinger et al., 1981), which, for instance, reduces the haemocompatibility of dialysers strongly, could be related to distinct hydroxyl groups of cellulose macromolecule. From this result it follows that the chemical surface structure or cellulose cannot be completely covered by the plasma protein adsorption layer because it was found that some proteins of mean molecular weight like that of human serum albumin are adsorbed after being in contact with blood for only microseconds.

On the other hand, it was concluded by Ebert and Kim (1982) that covalently immobilised heparin molecules cannot penetrate the protein adsorption layer despite the fact that the activation of clotting factors and platelet adhesion were prevented after heparinisation of the polymer surface.

Recent results (Bohnert and Horbett, 1986) emphasise that the protein adsorption layer on to polymers should be considered more in terms of dynamic changes inside this layer concerning the binding state of molecules and the composition of the layer. Extended theoretical concepts of haemocompatibility should, therefore, be based not only on the physical and chemical surface properties of polymers, but also include the dynamic behaviour of protein adsorption layers, which are generated at the surfaces of most types of polymer used at present for blood-contacting devices.

A further important aspect of haemocompatibility concerns the mechanism of blood cell interaction with polymer surfaces. Because in all cases of medical application of polymeric biomaterials there will be a protein adsorption layer on to the polymer surface, we need information about the adhesion mechanism between the adsorbed protein molecules and the formed blood elements. This knowledge, however, is very limited at the present time. Concerning the adhesion mechanism of red blood cells on to foreign surfaces, it could be demonstrated that the initial steps of contact formation are determined by physical forces (Bongrand et al., 1982; Trommler et al., 1985; Wolf and Gingell, 1983), including the spatial structure of the glycoprotein layer at the cell surface. Subsequent events, however, were shown to enable the red blood cells to make molecular contacts with the substratum. Such molecular contacts with foreign surfaces cannot be solved by mechanical forces without destroying the red cells (Trommler et al., 1985; Wolf and Gingell, 1983). Adsorption layers of human serum albumin were shown to increase the cell-substratum distance to about 25 nm. The well-known "adhesion reducing effect" of human serum albumin adsorption layers could, therefore, be explained in terms of this phenomenon (Wolf and Gingell, 1983).

The adhesion mechanism of platelets and of white blood cells on to protein-coated polymers, which is considered as a measure of the haemocompatibility of a polymer, is not understood in great detail. Generally, it is assumed that receptors at the cell surfaces and on the adsorbed protein molecules are involved in the interaction process (Kataoka et al., 1980). In the case of platelets, glycosyl transferases were found to be included in the adhesion mechanism as receptors. There is no doubt that the fluidity of cell membranes and their mechanical properties are also included in the adhesion mechanism. The activation of platelets and leukocytes in the adherent state seems also to be initiated and controlled via the cell membrane. The transducer mechanism at the molecular level is as yet unknown. The understanding of haemocompatibility determining mechanisms therefore requires an understanding of cell adhesion mechanisms on to protein layers adsorbed on to polymers.

CONCLUDING REMARKS AND OUTLOOK FOR THE FUTURE

It becomes clear that further development of new synthetic blood-compatible polymers will be more and more based on the extending knowledge of the relationships between the functional and structural surface properties of polymer surfaces and the complex interface reactions with blood. Because of the lack of biochemical signalling and feedback mechanisms between blood and non-biological surfaces as they exist between blood and biological surfaces (the vessel wall), the haemocompatibility of synthetic polymers will be both limited and time-dependent. Therefore, efforts to improve the haemocompatibility of polymers are directed towards this time limit, which is necessary for the application of every distinct device.

The main directions in <u>further investigations of haemocompatibility - determining factors and processes</u> will be the following:

(i)   investigation of the structure and dynamic behaviour of the adsorption layers of plasma proteins and plasma lipids in terms of their dependence on the structural parameters of polymer surfaces;

(ii) investigation of the structural and functional properties of adsorbed plasma proteins and plasma lipids;

(iii) investigation of the adhesion mechanism of blood cells on to polymer surfaces with and without protein and lipid adsorption layers;   and

(iv) development and standardisation of quantitative methods for the evaluation of haemocompatibility including the introduction of reference test surfaces.

The main directions in the <u>development of new polymers with improved haemocompatibility</u> will be based more and more on the current hypotheses of haemocompatibility described above. This will involve trying to put into practice the following concepts:

(i)   creation of energetically heterogeneous polymer surface structures at the molecular level to achieve protein adsorption layers that prevent the activation of the clotting and complement system as well as blood cell adhesion;

(ii) creation of polymer surfaces with minimum protein adsorption;

(iii) creation of polymer surfaces with adsorbed or covalently immobilised biomolecules or bioactive molecules that inhibit the activation of the clotting system and blood cells (platelets and white blood cells) as well as blood cell adhesion;   and

(iv) creation of polymers coated with living cell layers that simulate the function and structure of the natural surface of blood vessels.

These "bio-artificial" surfaces should be suited to overcoming many disadvantages of current materials which cannot be resolved for fundamental reasons. The development of this new generation of biomaterials will, however, need tremendous efforts in future basic research.

ACKNOWLEDGEMENT

This work was supported by HFR "Artificial Organs and Biomaterials" of the Ministry of Health of the GDR.

REFERENCES

Andrade, J. D., 1973, Interfacial phenomena and biomaterials, <u>Med. Instrum.</u>, 7 : 119.

Bayer, R. E., 1972, The role of surface energy in thrombosis, <u>Bull. N.Y. Acad. Med.</u>, 48 : 257.

Bohnert, J. L., and Horbett, T. A., 1986, Changes in adsorbed fibrinogen and albumin interactions with polymers indicated by decrease in detergent elutability, <u>J. Colloid Interface Sci.</u>, 111 : 363.

Bongrand, P., Capo, C., and Depieds, R., 1982, Physics of cell adhesion, <u>Progr. Surface Sci.</u>, 12 : 217.

Bornzin, G. A., and Miller, I. F., 1982, The kinetics of protein adsorption on synthetic and natural surfaces, <u>J. Colloid Interface Sci.</u>, 86 : 539.

Breemhaar, W., Brinkmann, E., Ellens, D. J., Beugeling, T., and Bantjes, A., 1984, Preferential adsorption of high density lipoprotein from blood plasma on to polymer surfaces, <u>Biomaterials</u>, 5 : 269.

Bruck, S. D., 1977, Current activities and future directions in biomaterials research, <u>Ann. N.Y. Acad. Sci.</u>, 283 : 332.

Brynda, E., Cepalova, N. N., and Stol, M., 1984, Equilibrium adsorption of human serum albumin and human fibrinogen on hydrophilic and hydrophobic surfaces, <u>J. Biomed. Mater. Res.</u>, 18 : 685.

Chuang, H. Y. A., King, W. F., and Mason, R. G., 1978, Interactions of plasma proteins with artificial surfaces: protein adsorption isotherms, <u>J. Lab. Clin. Med.</u>, 92 : 483.

Coleman, D. L., Gregonis, D. E., and Andrade, J. D., 1982, Blood-materials interactions: the minimum interfacial free energy and the optimum polar/apolar hypotheses, <u>J. Biomed. Mater. Res.</u>, 16 : 381.

Ebert, C. D., and Kim, S. W., 1982, Immobilized heparin-spacer arm effects on biological interactions, <u>Thromb. Res.</u>, 26 : 43.

Fougnot, C., Josefowicz, M., Samama, M., and Bara, L., 1979, New heparin-like insoluble materials, <u>Ann. Biomed. Eng.</u>, 7 : 429.

Fougnot, C., Josefowicz, M., and Rosenberg, R. D., 1984, Catalysis of the generation of thrombin-antithrombin complex by insoluble anticoagulant polystyrene derivates, <u>Biomaterials</u>, 5 : 94.

Harasaki, H., Murray, J. D., McMahon, J., Kiraly, R. J., and Nose, J., 1981, Calcification of left ventricular devices, <u>Artif. Organs</u>, suppl. 5 : 384.

Hennink, W. E., Kim, S. W., and Feijen, J., 1984, Inhibition of surface induced coagulation by preadsorption of albumin-heparin conjugates, <u>J. Biomed. Mater. Res.</u>, 18 : 911.

Herzlinger, G. A., Bing, D. H., Stein, R., and Cuming, R. D., 1981, Quantitative measurement of C3 activation at polymer surfaces, <u>Blood</u>, 57 : 764.

Ikada, J., 1984, Blood-compatible polymers, <u>Adv. Polymer Sci.</u>, 57 : 103.

Karwath, R., and Wolf, H., 1986, A modification of an in vitro method for assessment of blood compatibility, <u>Adv. Biomaterials</u>, 6 : 275.

Kataoka, K., Maeda, M., Nishimura, T., and Nitadori, Y., 1980, Estimation of cell adhesion on polymer surfaces with the use of "column bead method", <u>J. Biomed. Mater. Res.</u>, 14 : 817.

Morrisey, B. W., 1977, The adsorption and conformation of plasma proteins - a physical approach, <u>Ann. N.Y. Acad. Sci.</u>, 283 : 50.

Norde, W., 1980, Adsorption of proteins at solid surfaces, <u>in</u>: "Adhesion and Adsorption of Polymers, Part B", H. L. Lee, ed., Plenum Publishing, New York : 801.

Ramasamy, N., and Sawyer, P. N., 1977, Interfacial reactions and thrombosis, <u>Bioelectrochem. Bioenerget.</u>, 4 : 137.

Ratner, B. D., Hofmann, A. S., Hanson, S. R., Harker, L. A., and Whiffen, J. D., 1979, Blood compatibility - water content relationships for radiation grafted hydrogels, <u>J. Polym. Sci. Polym. Symp.</u>, 66 : 363.

Sawyer, P. N., 1964, Bioelectric phenomena and intravascular thrombosis, <u>Surgery</u>, 56 : 1020.

Sharma, C. P., 1981, Possible contributions of surface energy and interfacial parameters of synthetic polymers to blood compatibility, <u>Biomaterials</u>, 2 : 57.

Trommler, A., Gingell, D., and Wolf, H., 1985, Red blood cells experience electrostatic repulsion but make molecular adhesions with glass, <u>Biophys. J.</u>, 48 : 835.

Voigt, A., Wolf, H., Lauckner, S., Neumann, G., Becker, R., and Richter, L., 1983, Electrokinetic properties of polymer and glass surfaces in aqueous solutions: experimental evidence for swollen surface layers, Biomaterials, 4 : 299.

Wolf, H., 1981, Zur komplexen Wechselwirkung von Blut mit abiologischen Oberflächen - Ein Überblick aus biophysikalischer Sicht, Dt. Gesund.-wesen, 36 : 626 and 717.

Wolf, H., and Gingell, D., 1983, Conformational response of the glycocalix to ionic strength and interaction with modified glass surfaces: study of live red cells by interferometry, J. Cell. Sci., 63 : 101.

CHANGES IN VASCULAR REGULATION AFTER IMPLANTATION OF A TOTAL ARTIFICIAL

HEART

Jaromir Vašků, M. Dostál, Jan Vašků and S. Doležel

Institute of Pathological Physiology, Faculty of Medicine,
University of J. E. Purkyne, and
Research Centre for Heart Support and Total Heart Substitution,
Regional Institute for National Health,
Brno, Czechoslovakia

INTRODUCTION

After resection of biological ventricles, and after implantation of
artificial ventricles, the mechanisms of vascular regulation are markedly
changed. This situation has been observed in 42 long-term experiments on
calves and 1 experiment on a goat. In our research centre we have elaborated a functional concept of the causes of changes in the vascular capacitance area in long surviving animals and we have started to influence the
state of increase of central venous pressure (CVP) in these animals according to this concept. We have attained positive results, and we hope that
our contribution to the explanation of the CVP increase can help to keep
the CVP within acceptable limits in long-term survivors with a total artificial heart.

MATERIAL AND METHODS

The experiments were performed on 42 calves and 1 male goat. The
calves were mainly males of the hybrid Bohemian breed, with a substantial
component of the Holstein breed. In 23 calves the total artificial heart
TNS-BRNO-II was implanted, in 16 experiments the TNS-BRNO-VII, and in 4
experiments the Rostock heart was used. The surgical procedures, the post-
operative care and the system of clinical and laboratory measurements during
the survival time have been described elsewhere (Vašků, 1982; 1984).

Central venous pressure was measured twice a week by the direct method,
using aseptic puncture of the jugular vein. Also, the activity of the
atrial P-wave frequency was evaluated by means of an electrode implanted
into the atrial wall, or else by means of extremity leads. The amplitudes
of the P-wave recorded by an implanted electrode reached $100 - 200\,\mu V$, while
those taken from the extremity leads were $20 - 40\,\mu V$ (Vašků et al., 1987).

In some calves the aldosterone and thyroxine levels were measured and
the relevant parameters were calculated with the serum and urine levels of
sodium and potassium. The aldosterone level was estimated by means of the
RIA method, based on the competition between the labelled (aldosterone ^{125}I)
and non-labelled aldosterone in its binding on the antibody (Dostál et al.,
1987).

143

The main characteristics of our total artificial hearts and of our control and driving units have already been published (Vašků, 1982; 1986; Vašků et al., 1986a). The average survival is 107.4 days, the span of survival time being from 31 to 226 days. The mean body weight in these calves at the time of implantation was 86.2 kg; the male goat weight was 60 kg. Every animal was carefully autopsied, and detailed histological and histochemical evaluation was performed at the termination of every experiment. Also, organ evaluation using electron microscopy (EM) of the organs and further EM and SEM of the driving diaphragms were performed. Great attention was given to the evaluation of the neural elements in the atrial wall.

In 8 calves the following anti-hypertensive drugs were administered orally: Prazosin®, Capoten® and Dopegyt®. The dosage was modified according to the individual course of experiment in every calf.

RESULTS

In all the bearers of a total artificial heart (TAH), after some time, typically about 2 months after implantation, central venous pressure started to increase. This peculiar change of the vascular capacitance area has been observed by several authors (Fujimasa et al., 1986; Murakami et al., 1979). In only one calf, No. 85 "Cézar", during the whole course of the experiment was marked central venous hypotension observed (Vašků, 1986). Simultaneously, we observed in this calf a total absence of regeneration of the mono-aminergic neural terminals in the atrial wall, especially in the mouth of the venae cavae into the right atrium. This was probably caused by an inborn defect of the sympathetic neural system of this animal. Whereas in all the calves with CVP increase a marked hepatomegaly was a frequent finding with functional and morphological liver damage, due to extensive venostasis caused by the CVP increase, no hepatomegaly with liver damage was observed in the calf "Cézar". Also, the liver index (that is, the percentage relationship between the liver and body weight), which is an important indicator of the hepatic sequelae of increased CVP, and which is regularly markedly increased in the TAH bearers, was in this particular calf below the normal limit.

In comparison with the selected group of calves (8 animals) which were not treated by anti-hypertensive drugs, in all the 8 treated calves the liver index was significantly decreased, which was, aside from the decrease in CVP values, a confirmation of the positive effect of treatment (see Fig. 1).

Concerning atrial innervation, we could always observe in all the calves a marked regeneration of the mono-aminergic nerve terminals. Immediately after TAH implantation the atrial neural receptors completely disappear due to impaired vascular supply to atria. They are regenerated typically during the second month of pumping and this time coincides with the start of the CVP increase (Vašků, 1986; Vašků et al., 1986b). The P-wave, recorded by means of classical extremity leads, appeared to be too small (20 – 40 μV). The signal used was also sensitive to movement by the animals and was therefore hard to register. That was why implanted electrodes with a considerably higher signal level were used. The values of the P-wave frequencies varied over the range of 70 – 150 beats min^{-1}. Acoustic and mechanical excitation of the animal did not lead to frequency changes. However, a considerable difference (5 – 25 beats min^{-1}) appeared with changes of position (measurement in the standing and supine animal). The P-wave activity is real during the whole operation, often attended by mechanical contractions of the atria which can be registered by means of an inserted catheter. The general course of the P-wave frequency as a function of the length of survival is described in Fig. 2.

144

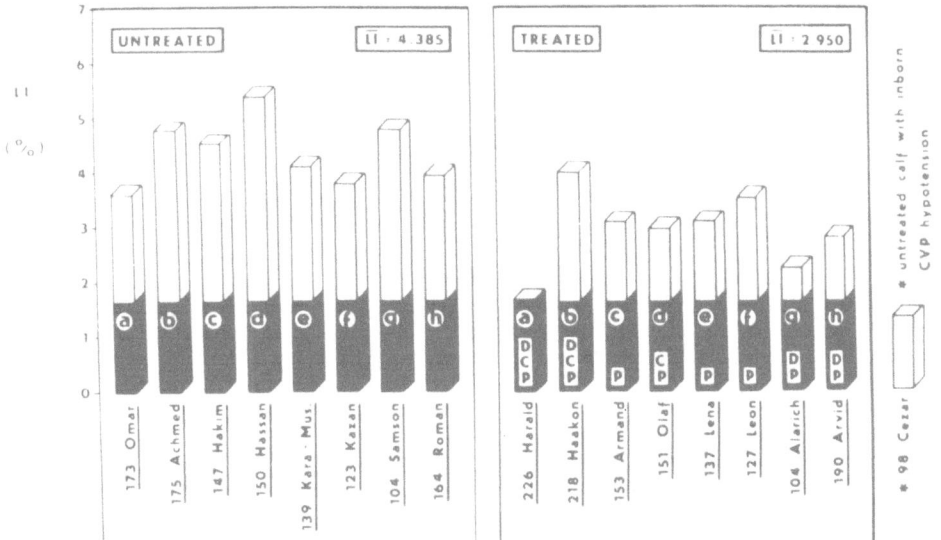

Fig. 1.• Comparison of liver indices (liver weight expressed as a percentage of body weight) in the group of 8 untreated and 8 treated calves with TAH. The treatments are the α₁ blocking agent Prazosin® (Pratsiol®, Deprazolin®) [P], Capoten® [C] and Dopegyt® [D], respectively. The decrease in values of the LI in the treated calves is significant.

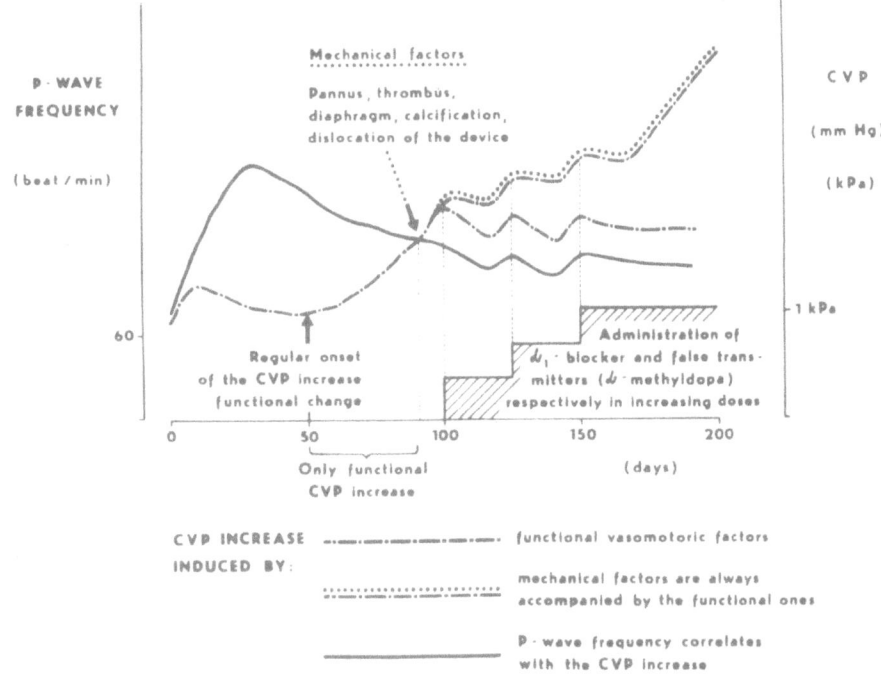

Fig. 2. The CVP increase caused by functional and mechanical factors, influenced by the drug therapy, and correlated with the atrial P-wave frequency.

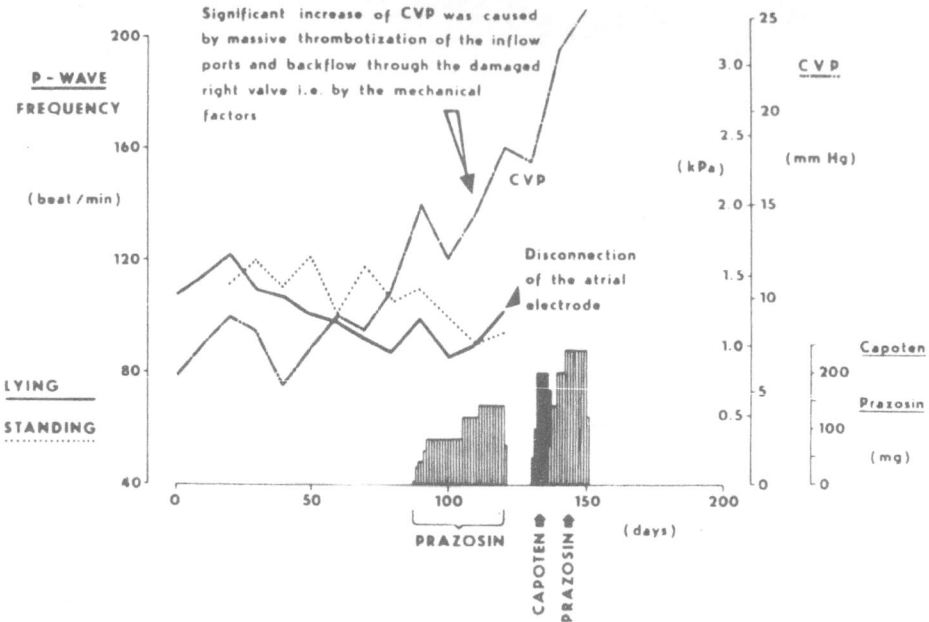

Fig. 3. Difference between P-wave frequency in supine and standing animal
(Experiment no. 112, "Olaf").

After the post-operative shock has decayed away, both amplitude and
frequency increase and reach maxima. During approximately 15 days' survival
there comes a fall, which is at first fast but later becomes slower (Fukumasu, 1986). During the experiments antihypertensive agents are applied in
a routine manner to compensate for the increased CVP. In the period of
administration of these antihypertensive agents, considerable correlations
were observed between P-wave frequency and the change of CVP level caused
by both vasomotor mechanisms, and mechanical factors (pannus, thrombus and
pump dislocation) were observed (see Figs. 3 and 4).

Although the number of the investigated animals with the TAH is for the
moment small, there is clear evidence of correlation between the P-wave frequency and CVP.

The estimation of the aldosterone and thyroxine levels gave us the following results. Serum aldosterone before surgery in the calves prepared for
TAH surgery was 325 ± 64 pmol l^{-1} ($\bar{x} \pm$ S.E.M.). In the control animals, kept
on the farm, the average value was 33 ± 6 pmol l^{-1}. On the 10th day after
TAH implantation it was 247 ± 62 pmol l^{-1}. During the life span with the TAH
a slight elevation was observed (see Fig. 5).

The level of serum thyroxine (T_4) before surgery was 47.2 ± 6.8 nmol l^{-1}.
On the 10th day after TAH implantation it was 40.5 ± 6.7 nmol l^{-1}. Statistically significant elevation was observed only from the 80th to the 100th
day.

DISCUSSION

Two main findings concerning the CVP increase are very important. The
first is the fact that CVP increases not immediately after TAH implantation,
but regularly after about 50 days of pumping, and that directly at that time

146

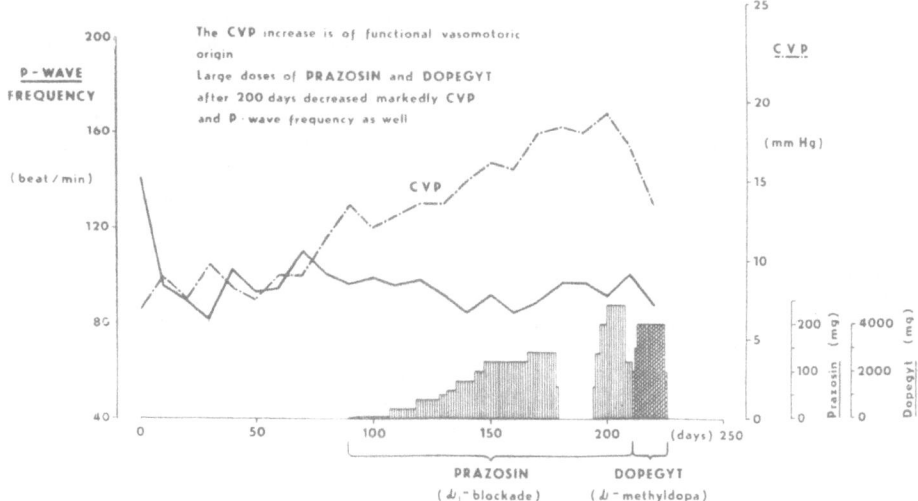

Fig. 4.   Correlation between P-wave and CVP during administration of anti-
          hypertensive drugs.

the regeneration of the atrial neural elements is completed, as can be seen
from mono-aminergic neural terminals stained by formaline-induced fluores-
cence according to Falck. In one calf with profound venous hypotension
reaching nearly pathological values, no mono-aminergic regeneration was
observed (Vašků, 1986).

     We have therefore concentrated on this important coincidence and our
concept tries to explain the increase in central venous pressure by mechanisms which originate in the atrial receptors. After the resection of the
biological ventricles, an extensive area, in which predominantly vasodepressive receptors are situated, is cut off. Due to this impairment of vascular
supply into the atria, all the atrial neural elements disappeared immediately

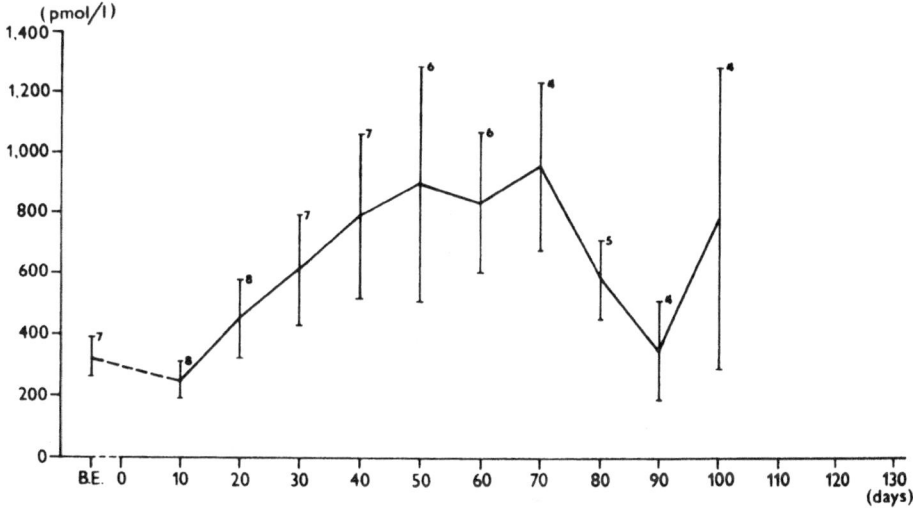

Fig. 5.   Changes in the serum aldosterone levels (x̄ ± S.E.M.) in the
          calves with a total artificial heart during long-term survival
          (TAH Nos. 100, 102, 103, 106, 107, 110 - 112.

147

after TAH implantation with regeneration occurring after about 50 days. Because in this time interval of 50 days both ventricular and atrial vaso-depressive receptors disappear, vasodepressive regulation is maintained by the baroreceptors in the carotid sinus, aortic arch and in some areas of the pulmonary artery bed, respectively. However, the atrial vasodepressive receptor area is gradually regenerated in the first two months, whereas the vasodepressive ventricular area is irreversibly removed by the resection of the cardiac ventricles and is substituted by mechanical ventricles (Vašků, 1986; 1988; Vašků et al., 1982; 1986).

Thus, the steady increase in the CVP after 50 days of pumping can be attributed to the steady increase of the constrictive impulses, originating in the atrial field and the gradual decrease of the dilating influences from the arterial baroreceptor fields due to "resetting" the arterial baro-depressive sensors, following the loss of the vast ventricular field of depressive mechanoreceptors. The increased atrial afferentation to the sympathetic bulbar area of brain stem reticular formation increases the stimulation of the vasoconstricting efferents into vascular periphery, with the increased activity of the alpha-1-receptors. Simultaneously, the afferentation to the bulbar depressive centre by way of the vagal and glosso-pharyngeal nerves is decreased after elimination of the vagal ventricular receptor area.

The splanchnic organs, particularly the liver, are greatly affected by the increased venous vasoconstriction. Because the splanchnic area seems to be of the utmost importance for cardiovascular stability, some splanchnic organs, such as the liver, can be very rapidly, and to a great extent, affected by the increased instability in the regulation of venous vascular tone in the splanchnic area and by the CVP increase, respectively. The liver is changed morphologically. There is extensive haemostasis, haemorrhage and finally fibrosis, a frequent finding in the liver. Marked changes are observed in liver enzymes, for example increases in transaminase and lactate-dehydrogenase (LDH), and signs of damage are observed very early in electron microscopic findings. The percentage relationship of liver weight to total body weight, the so-called liver index, shows the presence of hepatomegaly as a direct consequence of the CVP increase.

According to our concept of pathogenesis of CVP in the TAH bearers, we have concluded that several interventions can be useful in the control of CVP. The first is pharmacological, that is the influence of vasoconstricting efferentation, either by alpha-1-receptor blockade or by intervention into catecholamine metabolism by means of false transmitters (alpha-methyldopa). The others are chemical and pharmacological intervention into atrial afferents during implantation, and surgical interruption of afferent chan-nels which interrupt fibres in the cardiac plexus or cervical sympathetic afferents, respectively.

To date we have tried to influence vasoconstricting efferentation using alpha-1-blockers (Prazosin ®), angiotensin-convertase inhibitor (Capoten ®) and false transmitter alpha-methyldopa (Dopegyt ®).

We have observed that if we have started to administer the anti-hypertensive drugs between the 50th and 70th day following implantation, there is a marked retardation of the CVP increase, reduced liver damage, and a statistically significant decrease of the liver index in all 8 treated animals in comparison with the group of 8 controls.

The purely functional CVP increase can sometimes be complicated in the later stages of the experiment by mechanical factors, that is, pannus growth in the inflow tracts, especially on the right side, further TAH dislocation, and so on. Out of 8 calves we observed only 1 CVP increase that was purely

functional in origin. In this calf the vasodepressive therapy could maintain the liver index within completely normal limits. In the other animals,
where the purely functional cause of CVP increase was later complicated by
some mechanical factor, pharmacological therapy also decreased the values
of hepatic index. The difference between the group of treated animals and
the untreated group was highly significant.

We can say that our concept of the functional increase of CVP was
greatly confirmed by the pharmacological therapy. It remains to be seen
how effective other forms of therapy will be.

Just as in the pathogenesis of arterial hypertension, we must also
consider the changed pattern of humoral regulation. One of the humoral
indicators is aldosterone. We have shown that serum aldosterone levels in
TAH bearers were about 10 times higher than in the controls and that they
showed persistence of secondary aldosteronism. We have concluded that the
combination of transporting stress, renal hypoperfusion and possibly decreased formation of the atrial natriuretic factor play their role in the
development of aldosteronism in the animals with TAH.

Another hormone, thyroxine, is also changed. Its increased level can
possibly be explained by decreased mechanisms of de-iodination of the
thyroxine molecule in the peripheral tissues of TAH bearers. The exact
cause of this process remains to be elucidated.

A direct relationship between the CVP value and the atrial P-wave
frequency has been postulated. This fact deserves further experimental
elucidation.

CONCLUSION

Implantation of a total artificial heart represents a profound intervention into the vascular regulation of the organism. The ventricular,
predominantly vasodepressive, field is irreversibly removed and the atrial
vasoconstrictive receptor field is transiently damaged, but is regenerated
within two months. In the following period the vasoconstrictive afferents
from the atrial walls lead to increased stimulation of vasoconstrictor
efferents, which is mediated through central brain stem reticular formation.

The increased stimulation of alpha-1-receptors in the veins is the
functional cause of the CVP increase. This change is very dangerous
especially for the splanchnic area, where the most endangered organ is the
liver. There are pathological changes in liver tissue and the organ is
enlarged, which can be documented in TAH bearers by an increased liver
index. The liver damage greatly limits the life-span of the animals with
TAH. From several possible interventions aimed at preventing this state,
we have used the pharmacological one, employing antihypertensive drugs of
different modes of action. Thus we could positively influence CVP increase
and keep the liver index within its normal limits. The positive influence
of pharmacological treatment is evident also in those cases where the functional cause of CVP increase is complicated by mechanical ones.

The P-wave frequency correlates directly with CVP values (a higher
frequency is accompanied by a higher CVP value). Pharmacological treatment
leading to CVP decrease also diminishes the P-wave frequency.

It remains to be elucidated to what extent hormonal shifts contribute
directly to these vascular changes. The relationships between aldosterone
and the natriuretic atrial hormone may be of particular importance in this
context.

REFERENCES

Dostál, M., Bílková, B., Vašků, J., Černý, J., Gregor, Z., Guba, P., and
Šotolová, O., 1987, Hladiny aldosteronu v krevním séru telat s umělým
srdcem (předběžné sděleni), Bratislavské Lékarské Listy, 87 : 530 (in
Czech).

Fukumasu, H., 1986, Research of the total artificial heart, in: "Artificial
Heart 1", T. Akutsu, ed., Springer Verlag, Tokyo : 187.

Fujimasa, J., Imachi, K., Nakajima, M., Mabuchi, K., Tsukagoshi, S.,
Kouno, A., Ono, T., Takido, N., Motomura, K., Chinzei, T., Abe, Y.,
and Atsumi, K., 1986, Pathophysiological study of a total artificial
heart in a goat that survived for 344 days, in: "Progress in Artificial Organs - 1985", Y. Nosé, C. Kjellstrand, and P. Ivanovich, eds.,
ISAO Press, Cleveland : 345.

Murakami, T., Ozawa, K., Harasaki, H., Jacobs, G., Kiraly, R., and Nosé, Y.,
1979, Transient and permanent problems associated with the total
artificial heart implantation, Trans. Am. Soc. Artif. Int. Organs,
25 : 239.

Vašků, Jar., 1982, "Artificial Heart. Pathophysiology of the Total Artificial Heart and of Cardiac Assist Devices", University J. E. Purkyně
Press, Brno.

Vašků, Jar., 1984, Total artificial heart research in Czechoslovakia, in:
"Assisted Circulation", F. Unger, ed., Springer Verlag, Berlin.

Vašků, Jar., 1986, Total artificial heart research in Czechoslovakia:
pathophysiological evaluation of long-term experiments performed from
1979 to 1985, in: "Artificial Heart 1", T. Akutsu, ed., Springer
Verlag, Tokyo : 161.

Vašků, Jar., 1988, A contribution to further development of pathophysiology
of total artificial heart. Mutual relationships between total artificial heart and its living recipient in long-term survival, in:
"Assisted Circulation III", F. Unger, ed., Springer Verlag, Berlin
(in press).

Vašků, Jar., Černý, J., Dostál, M., Guba, P., Vašků, Jan, Urbánek, P.,
Vašků, A., and Doležel, S., 1982, Neurohumoral aspects of total artificial heart, in: "Int. Meeting on Heart Transplant. Total Artificial Heart and Assist Devices", Brussels, Abstracts : 43.

Vašků, Jan, Urbánek, P., Vašků, Jar., Černý, J., Smutný, M., Urbánek, E.,
Suchánek, J., Gregor, Z., Dostál, M., Guba, P., Sládek, T., Wendsche, P.,
Pavlíček, V., Trbůšek, V., Šotolová, O., Ůlehla, T., and Fiala, V.,
1986, Control and driving of pneumatic total artificial heart TNS -
Brno - II and III in long-term experiments, Artif. Organs, 10 : 145.

Vašků, Jar., Urbánek, P., Dostál, M., Doležel, S., Guba, P., Vašků, Jan,
Smutný, M., Sládek, T., Filkuka, J., Pavliček, V., Trbušek, V., and
Bednařik, B., 1986, Central venous pressure in calves surviving
several months with a total artificial heart, in: "Progress in Artificial Organs - 1985", V. Nosé, C. Kjellstrand, and P. Ivanovich,
eds., ISAO Press, Cleveland : 386.

Vašků, Jar., Vašků, Jan, Dostál, M., Černý, J., Guba, P., Doležel, S.,
Vašků, A., Mašek, J., Urbánek, P., and Krejčí, V., 1987, Electric
activity of biological atria in total artificial heart (TAH), in:
"4th IMEKO Conference on Advances in Biological Measurement", Bratislava, Abstracts : K-11.

TOPOGRAPHICAL METHODS IN ELECTROCARDIOLOGICAL DIAGNOSTICS

P. Kneppo and L. I. Titomir

Institute of Measurement and Measuring Technique,
Electro-physical Research Centre, Slovak Academy of Sciences,
Bratislava, Czechoslovakia; and
Institute of Problems of Information Transmission,
USSR Academy of Sciences, Moscow, USSR

INTRODUCTION

Electrocardiography (ECG), a method for the objective evaluation of
the heart state, is a standard diagnostic procedure possibly more frequently
used than any other instrumentation method for investigating the heart-
vessel system. Standardisation of ECG diagnostics has contributed to the
wide spread of the method and to the gathering of enormous clinical experience. However, the accepted standardisation of the method has also brought
about negative consequences. These include the impossibility of collecting
all the accessible information on the cardiac electrical field when using
only a limited number of standard leads, insufficient physical and physiological interpretation of the data, and an orientation towards a formal
"electrocardiographic" language relying on the particularities of the registered curves and their empirical and statistical comparison.

Works aimed at improvement of the ECG diagnostic method have been
carried out in various directions, over the last few decades, making use
of current developments in biophysics, and in measuring and computing techniques.

TOPOGRAPHICAL METHODS

Essential quality improvement of ECG diagnostics is possible, first
of all, at the cost of a more complex acquisition of the initial information on the cardiac electrical field using multiple leads distributed all
over the body (chest) surface instead of a limited number of standard
leads. Diagnostic techniques based on such potential measurements are
called topographical. However, the term "topographical methods" has to
be understood in a wider sense, also implying methods that use graphical
or visual representation of data in the form of maps with anatomical land-
marks (although the initial measurements are not necessarily performed
using multiple leads).

Having in mind only the aspect of measurement, there are several different types of topographic methods:

(i)   body surface potential mapping - that is, measurement of the elec-
      trical potentials of the heart all over the body or chest surface;
(ii) precordial mapping - that is, measuring the electrical potentials
      on the front side of the thorax; and
(iii) epicardial mapping - that is, measuring the electrical potentials
      directly on the heart surface.

    Sufficiently detailed information on developments in topographical
methods in electrocardiology is available from special monographs, reviews,
proceedings from articles in scientific journals and from conference proceedings (particularly from international electrocardiological congresses)
(Amirov, 1965; 1973; Japanese Circulation Society, 1981; Muromceva,
1983; Nelsona and Gezelovic, 1979; Okajima, 1981; Rush and Lepeschkin,
1974; Yamada, 1981; Yamada et al., 1983).

    In this chapter we consider only the most typical type of topographi-
cal measurement, that is, body surface potential mapping when potentials
are measured in a sufficient number of points equally distributed on the
chest surface, so that one can assume that the measured values are known
at every point of the body surface. As has been shown by a number of investigators, in this case measurements should be made with about one hundred
leads, that is, an order of magnitude greater than in the case of standard
electrocardiography.

    The widespread practical use of diagnostic methods based on topogra-
phical measurement of the cardiac electrical field is complicated by this
great quantity of initial information. Purely technical difficulties with
the measurements and digital signal processing are currently being overcome as a result of progress in the field of measuring instrumentation and
computers, particularly microprocessor-based measuring and computing
systems. Thus the problem of analysis and diagnostic interpretation of
topographical data becomes of prime importance. The difficulties in its
solution lie in the large quantity of initial data, in the absence of
optimal methods for data parameterisation and in the inadequate experience
in diagnostic interpretation. Under these conditions methods of automated
diagnostics, using purely statistical criteria, are less effective. It seems
that the most reasonable approach to analysis of topographical data in
electrocardiography consists of the combined use of modern computerised
methods together with the creative capabilities of the cardiologist, who
actively participates in the diagnostic procedure. He perceives the
acquired information including its analogue component, and analyses it
using his knowledge and internal intuitive model of the phenomenon under
study.

DATA PROCESSING

    A generalised automated system for electrocardiological diagnostics
using topographical measurements and assuming the active participation of
the physician-cardiologist in diagnostic decision-making, performs the
data processing in two basic parts: external and internal with reference
to the subjective procedure of decision-making by the cardiologist.

    The external processing includes automated data acquisition from the
subject under investigation using measuring equipment, signal pre-processing and basic digital data processing. The internal data processing consists of the interpretation of the information received by the cardiologist
on the basis of his own intuitive model and using his professional knowledge and collected medical experience for final diagnostic decision-
making. The activity of internal processing is executed jointly involving
the cardiologist, the external processing equipment and the subject being

investigated (if necessary it is permissible to change the parameters of the digital data processing or sometimes also the state of the subject).

Signal pre-processing in the measuring and computing system usually consists of the determination of certain time intervals of the measured signals, synchronisation of signals from various leads, calibration, filtering, analogue to digital conversion and other operations aimed at signal conditioning, elimination of random errors and storing the data in computers.

Basic digital data processing can be sub-divided into several steps. The most important of them is the primary mathematical description (parameterisation) of the data. After this step the data are usually represented as a set of functions of one or two variables; these functions serve as characteristics (parameters) of the accepted empirical method for data description, or of the biophysical model, and are preferably represented in an informative graphical form. The next step of the basic digital processing is the automated data classification according to diagnostic categories. In order to apply automated classification methods it is usual first to determine a reduced set of parameters chosen from the initial data or basic mathematical description (that is, to perform a secondary parameterisation) in correspondence with the particular diagnostic problem. As a result, after this step on the output of the measuring and computing system we obtain an "automated" decision about the diagnosis and graphical presentation of information which, together with supplementary clinical data, are perceived by the cardiologist and processed using his internal intuitive model.

In this chapter the main focus will be the method of receiving a basic mathematical data description as it plays an important role in the analysis and interpretation of topographical measurements. The characteristic building up of the basic mathematical description (now simply referred to as the mathematical description) holds the maximum possible continuous information on the subject under investigation. These characteristics, in an adequate graphical presentation, are evaluated visually as a consistent pattern characterising the investigated phenomena as a whole. At the same time, from this common description it is possible to select a smaller set of characteristic parameters convenient for further mathematical (and particularly statistical) analysis. Interpreting the characteristics of the mathematical description and using his intuitive model, the cardiologist at first makes use of his own experience with graphical representation of the particular data type and secondly makes use of his own knowledge on the electrophysiology of heart excitation. Analysing topographical data in electrocardiography, electrophysiological knowledge has considerable significance because the cumulation of purely formal experience from the interpretation of the graphical data presentations from all possible diagnostic categories is an extraordinarily difficult and long-term process. On the other hand, the demand for an electrophysiological evaluation leads to special requirements with regard to mathematical description of the data (stability, pattern representation, biophysical and electrophysiological content, and so on, see below).

It is obvious that the relative degree of usage of any data perceived by the cardiologist will depend on the level of his qualification and experience. The higher this level is, the more important is the role played by the pattern data from the cardiac electrical process together with additional clinical data. On the other hand, a cardiologist who is less well qualified basically relies on the conclusions made by the computerised classification.

MATHEMATICAL DESCRIPTIONS OF TOPOGRAPHICAL DATA

The methods for obtaining mathematical descriptions of topographical data in electrocardiography can be divided into two main categories: empirical and biophysical.

Empirical Methods

Empirical methods do not require any calculations based on electrodynamic models of the heart electric generator (cardiogenerator) and the body as a volume conductor implying an electric field of this generator. The mathematical description is characterised by the directly measured potential as a function of time and body surface point co-ordinates or by any derived characteristics of this potential (of amplitude, time or integral type).

In principle, a mathematical description of topographical data in the form of multiple electrocardiograms can be regarded as being empirical, that is, the set of time-varying basic functions which can be visually evaluated in an analogue manner as standard electrocardiographical curves. However, this approach has not found practical application, as it does not give the required general representation of the electrical process under investigation.

In the following section a number of well-known basic empirical methods for obtaining the mathematical description of topographical data, specifying them by corresponding characteristics of parameters, are introduced.

Momentary maps of surface potential. This widely used method assumes the evaluation and graphical representation of body surface potential distribution patterns at consecutive time instants of a given interval of the whole cardiac cycle or of its defined parts. It is necessary to accomplish a synchronous measurement of the potential in the topographical leads. The potential distribution is usually represented in the form of electrocardiotopograms - maps with isopotential lines with marked extrema on the unfolded chest surface. Sometimes they are represented as 3-dimensional axonometric images on which the intersection lines between the potential functions and the planes parallel to co-ordinate planes are plotted for greater insight into the spatial relief of the potential. For more convenient interpretation "differential maps" or "deviation maps" are sometimes used. They are obtained either as the difference between the specific map under consideration and the reference map obtained by averaging the maps from some typical group of subjects being investigated (for example, for a group representing normal patients), or as a part of the map under consideration, which deviates from the corresponding averaged reference map by more than a given value (which is proportional at each point to the standard deviation in the averaged reference map). For the selection of characteristic regions in the map, colouring, shading and other graphical representations are being used. During visual analysis, the general form of the maps is evaluated with particular attention given to potential extrema and determination of their number, polarity, value, times of existence, layout and movement during the cardiac cycle. Also, the layout and the area of positive and negative potential regions, the form of the border between them (the zero isopotential line and some typical peculiarities of the relief, such as, for example, saddles and sharp curvatures of isopotential lines) are taken into account.

Integral maps of the surface potential. This method is based on computation and graphical representation of functions giving the body surface distribution of some quantities characterising the measured cardioelectric signals described by such parameters as amplitudes and durations of their characteristic segments (for example, R-wave or ECG) recorded in topographical leads. Such quantities are the integral characteristics (in time) of the signal, in particular, integrals of the whole recorded cardiac cycle or of its individual parts (waves, complexes). Characteristics are determined for each surface point individually and independently of the meaning of the common time co-ordinate, thus the need for synchronous potential measurement is not obligatory. Also, difference maps can be computed for these quantities according to the same principle as for momentary maps. In general, the basis of the interpretation of integral electrocardiotopograms emerges from the same approach (evaluation of the general form, determination of the extrema magnitude and layout, and so on). In this case, however, the detailed change of the electrical field over time is not investigated, because the distribution of each parameter characterises the whole cardiac cycle, and not its internal dynamics.

Parameters for functional approximation of the surface potential. In order to obtain a mathematical description of the data, this method makes use of an approximation of primary measurements using a set of known (for instance trigonometric) functions aiming at compression or economic representation of the data. If necessary, recovery of information which might be lost during the measurement is provided. The illustrativeness of the graphical representation of the characteristics has only secondary importance as they are predominantly designated for subsequent automated processing.

Biophysical Methods. As well as empirical methods for the mathematical description of topographical data in electrocardiology, biophysical methods are being developed more and more intensively. They are based on mathematical modelling of the electrodynamic system created by the cardiogenerator and the body. The basic aim of the biophysical approach is to achieve a transformation of the measured data, which results in representing the heart state directly in terms of electropathology and heart anatomy. At the same time, this approach results in the elimination of interfering extra-cardiac factors, revealing of information hidden by a purely empirical approach, and an economic representation of the data is achieved. Topographical measurements of potential (usually along with data characterising the geometrical structure and passive electrical characteristics of the body) are used as input data for the solution of the inverse electrodynamic problem, and the parameters found are used directly for evaluation of the heart state. Biophysical methods for the interpretation of topographical data are described in more detail in Baum (1977), Nelsona and Gezelovic (1979) and Titomir (1980 and 1984). Let us illustrate the basic biophysical methods for obtaining a mathematical description of topographical data, specifying them by the names of corresponding models or their parameters.

Epicardial potential maps. In contrast to common surface maps, epicardial cardiotopograms represent the potential distribution on the heart surface itself, that is, in close proximity to the heart generator, and thus having a more complex pattern. To analyse them, one can in principle use the same methods as for the analysis of the potential distribution on the chest surface.

Fixed multiple dipoles. This model comprises a relatively small number (of 1st order) of elementary dipoles changing in time and forming the equivalent cardiac generator. The dipole moment magnitude of each dipole source is considered to represent immediately the intensity of the electrical process in the corresponding part of the heart muscle.

Multipoles. This model comprises a relatively small number (usually not greater than 3) of multipoles of low orders, situated at the specified origin of the co-ordinates. Their components changing in time are investigated as integral characteristics of bioelectric sources distributed in the volume of the heart muscle. Any method for multidimensional signal analysis can be used for these characteristics.

Fixed generator distributed on a surface. This model comprises a simple or double layer of current generators on a static surface closely surrounding the heart or identical with the surface of the excitable myocardium. The intensity characteristic (for example, the density of the dipole moment) is changing around the surface and in time, thus for its analysis analogue methods can be applied as for the analysis of surface and epicardial electrocardiotopograms. However, this characteristic is closely related to primary bioelectric processes in the myocardium, as it follows only from the configuration of the heart generator itself, in contrast to the potential distribution which is also influenced by the environment (extracardiac factors).

One or two moving dipoles. This model comprises one dipole or a set of 2 dipoles, with their dipole moments changing in time and both the dipoles changing their position in space. It is assumed that the changes of dipole moments and dipole position characterise the spatial evolution and changes of intensity of the heart generator (or of individual excitation waves) during the cardiac cycle. Both qualitative and quantitative methods can be used for model parameter analysis.

Moving uniform double layer. The model has the form of a uniform double layer of current sources describing a monolithic depolarisation wave, or, as the case may be, the border of an injured myocardial region. Its parameters generally characterise the size, spatial layout and peculiarities of the geometric form of the excitation wave or of the injured region respectively. They can be analysed using both qualitative and quantitative methods.

Heart excitation maps (chronotopocardiograms). The mathematical description of the cardiogenerator in the form of chronotopocardiograms represents a set of distribution functions on the heart surface describing the most important electrophysiological parameters - excitation delay with respect to the beginning of the heart depolarisation and the duration of the passage of the excitation through the heart wall. The method for obtaining chronotopocardiograms is described in detail in Titomir (1984) and Titomir et al. (1984). For chronotopocardiogram analysis, both qualitative and quantitative methods can be used which are analogous to the methods for integral electrocardiotopogram analysis.

DISCUSSION

In this way, biophysical methods make the transformation of topographical data according to an algorithm corresponding to an accepted biophysical model. Computerised characteristics or model parameters are used for immediate diagnostic interpretation. These characteristics represent either a relatively small set of time variable parameters which are given in a more or less specific electrophysiological sense, or surface distributed parameters which might be represented in an illustrative cartographical form. That is why the understanding of topographical methods in electrocardiography has a deeper dimension: not only are the methods for primary information acquisition taken into account, but also the methods for cartographical representation of any parameters for their direct visual diagnostic evaluation. Topographical methods for the graphical representation of analysed data are successfully implemented even in cases when the initial information acquisition is not carried out using a topographical method, but, for example, using orthogonal vectorcardiographical leads (Titomir, 1985). The efficiency of the methods of analysis and of the diagnostic interpretation of topographical measurements in electrocardiology

Table 1. Comparison of Methods for Analysing Topographical Measurements in Electrocardiology (+ method basically meets; – does not meet the requirement).

Requirements to the methods	Empirical					Biophysical (mathematical description of the data or a model of the cardiogenerator)					
	Multiple ECG	Instantaneous body surface potential maps	Integral body surface potential maps	Parameters of body surface potential approximation	Epicardial potential maps	Static multiple dipoles	Multipoles	Equivalent generator distributed on a static surface	One or two moving dipoles	Moving uniform double layer	Summary maps of heart excitation (chronoto-cardiograms)
1. Stability of the parameters in the presence of random input errors	+	+	+	+	–	–	+	–	+	+	+
2. Elimination of the extra-cardiac effects	–	–	–	–	–	+	+	+	+	+	+
3. Compressed representation of the data	–	–	+	+	–	+	+	–	+	+	+
4. Elucidation of electrophysiological configuration of the cardiogenerator	–	–	–	–	–	+	–	–	+	+	+
5. Pattern display of the parameters with anatomical landmarks	–	–	–	–	+	–	–	+	–	–	+

157

and the perspectives of their widespread practical application are dependent
on the degree to which they help to achieve the final aim of electrocardiological investigations - the most precise identification of the pathological
states of the heart along with objective and subjective conditions for the
solution of diagnostic tasks. The most serious step in the analysis of
topographical data (as mentioned above) is to obtain the basic mathematical
description of these data.

Here are the most important requirements for such a description:

   (i)   stability of parameters in the presence of random input errors;
  (ii)   elimination of extracardiac factors (particularly inhomogeneities
         of the body);
 (iii)   effective graphical representation of the data;
  (iv)   description of the data in terms of electrophysiology; and
   (v)   illustrative representation of the heart excitation process in
         relation to its anatomy.

At present, known methods of primary parameterisation meet the given
requirements at various levels, as shown in Table 1. The requirements
mentioned above correspond to the rows of the table and its columns correspond to the methods analysed.

It has been mentioned that the given classification of methods, and
correspondingly their comparative evaluation, is to some extent conditional
because the above methods are not often applied in a "pure form", but with
modifications, in combination with other methods, with the use of various
non-electrocardiological data, and so on. Nevertheless, such an evaluation
seems to be useful for the choice of optimal methodological approaches for
building up automated diagnostic systems. It follows from Table 1 that
biophysical methods satisfy these requirements to a greater extent than do
empirical ones and therefore they are preferable. Of particular value is
the method of chronotopocardiography.

REFERENCES

Amirov, R. Z., 1965, "Elektrokardiotopografija", Medicina, Moscow (in
    Russian).
Amirov, R. Z., 1973, "Integralnyje Topogrammy Potencialov Serdca", Nauka,
    Moscow (in Russian).
Baum, O. V., 1977, Modelirovanie elektitricheskoj aktivnosti serdca, in:
    "Biofizika Slozhnych Sistem i Radiacionnych Narushenij", Nauka, Moscow
    (in Russian) : 119.
Japanese Circulation Society, 1981, Clinical application of body surface
    mapping (Satellite Symposium of the 44th Annual Scientific Meeting
    of the Japanese Circulation Society), Jpn. Circ. J., 45 : 1171.
Muromceva, G. A., 1983, Nekotoryje metody analiza poverchnostnogo ras-
    predelenija kardioelektricheskogo potenciala i ich prilozhenie k
    diagnostike, in: "Elektricheskoe Pole Serdca", Izd. AN SSSR, Moscow
    (in Russian) : 58.
Nelsona, K. V., and Gezelovic, D. B., eds., 1979, "Teoreticheskije Osnovy
    Elektrokardiologii", Medicina, Moscow (in Russian).
Okajima, M., 1981, On current topics and future prospects in electrocardio-
    graphy, Jpn. Circ. J., 45 : 347.
Rush, S., and Lepeschkin, E., eds., 1974, "Body Surface Mapping of Cardiac
    Fields", S. Karger, Basel.
Titomir, L. I., 1980, "Elektricheskij Generator Serdca", Nauka, Moscow
    (in Russian).
Titomir, L. I., 1984, "Avtomaticheskij Analiz Elektromagnitnogo Polja
    Serdca", Nauka, Moscow (in Russian).

Titomir, L. I., 1985, "Obraznoe Predstavlenije Vektorkardiograficheskich
    Dannych", Preprint Instituta Problem Peredachi Informacii AN SSSR,
    Dubna (in Russian).
Titomir, L. I., Kneppo, P., Kricfalusi, M., and Cagán, S., 1984, Represen-
    tation of ventricular excitationchronotopography by means of multiple
    expansion, in: "Electrocardiology '83", I. Ruttkay-Nedecký and
    P. Macfarlane, eds., Excerpta Medica, Amsterdam : 115.
Yamada, K., 1981, Body surface isopotential map. Past, present and future,
    Jpn. Circ. J., 45 : 1.
Yamada, K., Harumi, K., and Musha, T., eds., 1983, "Advances in Body
    Surface Potential Mapping", Univ. Nagoya Press, Nagoya, Japan.

BIOMAGNETIC MEASUREMENTS IN UNSHIELDED ENVIRONMENT

H. Nowak, G. Albrecht, K.-H. Berthel, M. Burghoff,
W. Haberkorn and H.-G. Zach

PTI Jena, Academy of Sciences of the GDR, Jena, GDR,
FSU Jena, Department of Physics, Jena, GDR, and
ZWG Berlin, Academy of Sciences of the GDR, Berlin, GDR

INTRODUCTION

Measurements of biomagnetic fields are becoming more and more important in medical research. It is about 25 years since Baule and McFee (1963)
first used induction coils to detect the magnetic field of the human heart.
The weak magnetic fields of many organs of the body could not be studied
until Cohen et al. (1970) first used the SQUID (Superconducting Quantum
Interference Device) and showed that it could be effective in the study of
biomagnetic fields. SQUID systems are many orders of magnitude more sensitive to the low frequency fields arising within the organs of the body than
induction coils (see Fig. 1).

Performing biomagnetic measurements, scientists are faced with a twofold problem: very weak magnetic signals must be measured in an experimental area where there are disturbing fields which are orders of magnitude
stronger than those to be detected. Hence it is necessary not only to use
a sufficiently sensitive detector such as the SQUIDs mentioned above, but
also to reduce the detected ambient noise below the biomagnetic field to
be measured.

METHOD

Two approaches are commonly used to reduce the ambient external
noise. The first is large scale magnetic and electric shielding with the
advantage that a magnetometer or unbalanced gradiometers can be used (for
example, shielded chambers at MIT (Cohen, 1970), Helsinki University of
Technology (Kelhae, 1981), or PTB West Berlin (Mager, 1981)). The second
is a balanced gradiometer system, with the advantage that the instrumentation is not so expensive and the system can be moved easily to another
location. The rejection of disturbing fields by a gradiometer system is
quite satisfactory for many applications. This approach provides what can
be called "spatial discrimination", since it relies on the gradiometer
being sensitive to the field from a nearly inhomogeneous biomagnetic
source (ΔB) and insensitive to the field from distant homogeneous noise
sources (B) shown in Fig. 2.

The mechanical accuracy achievable in the construction of a

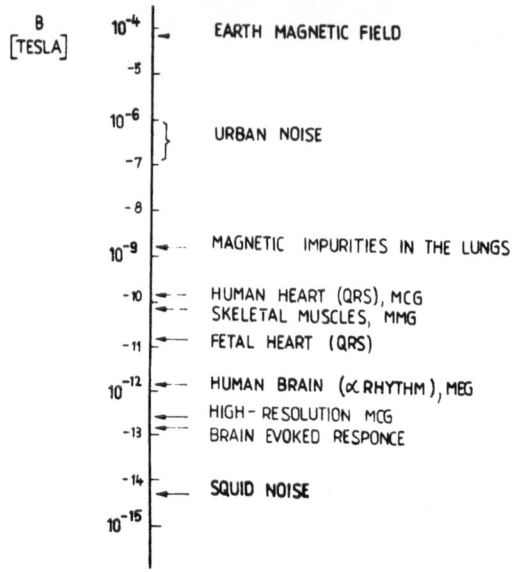

Fig. 1.   Average amplitude of the most
          studied biomagmetic signals. For
          the purpose of comparison the amp-
          litude of the earth's magnetic
          field and typical values of urban
          ambient noise are also reported.
          Adapted from Romani et al. (1982).

gradiometer is necessarily limited. This implies that the areas of the
coils are not exactly parallel or, for a second-order gradiometer, the base-
lines are slightly different. The degree of imbalance for a uniform field
can be quantified by expressing the fraction of the field that is still
detected by the gradiometer when the field lies parallel to the axis of the
sensor or along either of two orthogonal directions perpendicular to this
axis. The best achievable balancing when fabricating a second-order gradiometer is not more than a few parts per thousand with respect to a homogeneous field and even worse with respect to the gradient. These limitations
are due to imperfection in winding the wire on to a former to make the coils.

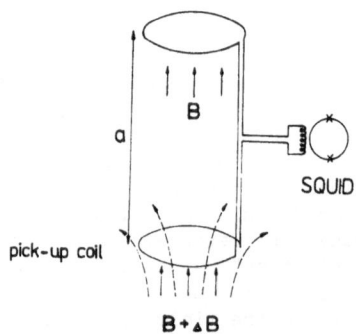

Fig. 2.   Scheme of a SQUID
          gradiometer.

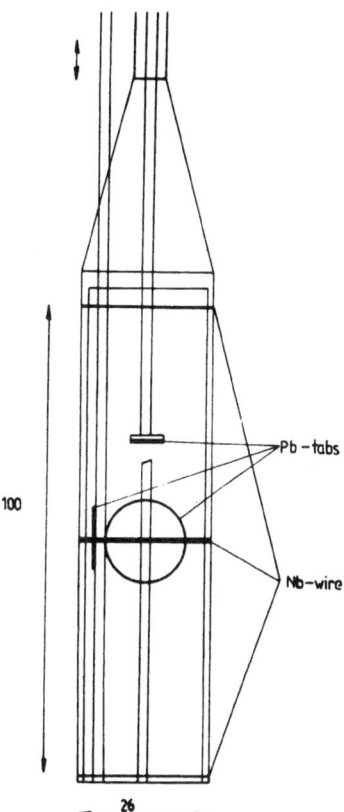

100

26

Fig. 3. Second-order gradiometer.
The equipment for balan-
cing consists of three
orthogonal movable super-
conducting tabs.

Several procedures for improving the field balance have been studied.
A simple method provides quite satisfactory results. Three small superconducting tabs with planes perpendicular to the directions x, y and z can be
accurately positioned with respect to the planes of the coils by use of
adjustment rods controlled from outside the dewar (see Fig. 3). The x- and
y-tabs are placed near the edge of a coil and can be moved to deflect the
respective components of the transverse field so that they pass through
the coil. This compensates for misalignment of the plane of the coil when
it is not truly transverse. To compensate for unequal areas of the coils,
the z-tab can be moved to an appropriate place between the pick-up coil and
the middle coil. The property of a superconductor to exclude magnetic
flux from its interior is used in such a way for reducing the sensing area
of one of the coils. In this way the product of area and number of turns
is equalised to render the detection coil insensitive to a uniform axial
field.

The biomagnetic measurement system in Jena consists of a d.c.-thin-
film-SQUID sensor (with an input coil inductance of about $1 \mu H$, flux resolu-
tion of about $1 \times 10^{-5} \Phi_o Hz^{-\frac{1}{2}}$, input current sensitivity $0.38 \mu A \Phi_o^{-1}$)
(Berthel et al., 1986) and a symmetrical second order gradiometer (diameter
26 mm, baselength 50 mm). Three superconducting adjustable discs are used

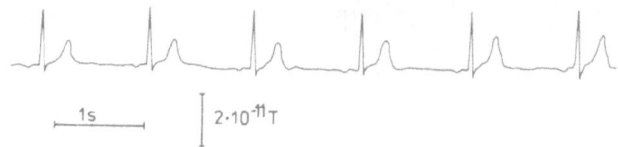

Fig. 4. Real time magnetocardiogram.

to compensate for the imbalance. A balance of about 100 000 was achieved in the unshielded laboratory. Provided that only the pick-up coil receives field energy, which is always possible with an adequate baselength, the magnetic field sensitivity of our equipment is about 20 fT Hz$^{-\frac{1}{2}}$. This sensitivity is sufficient for biomagnetic measurements such as magnetocardiography, inclusive of high-resolution magnetocardiography (Nowak et al., 1987) and magnetoencephalography.

EXPERIMENTAL RESULTS

The main advantages of magnetic measurements in relation to the human body are:

   (i)   the non-invasive and non-touching character;
  (ii)   the local resolution power; and
 (iii)   the measurability of frequency components down to 0 Hz.

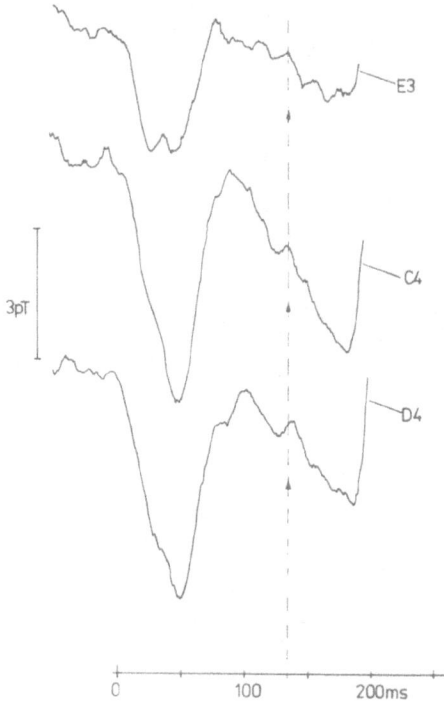

Fig. 5. High-resolution MCG of the PR segment at grid positions C4, D4, E3. Bandwidth 0.1 to 150 Hz.

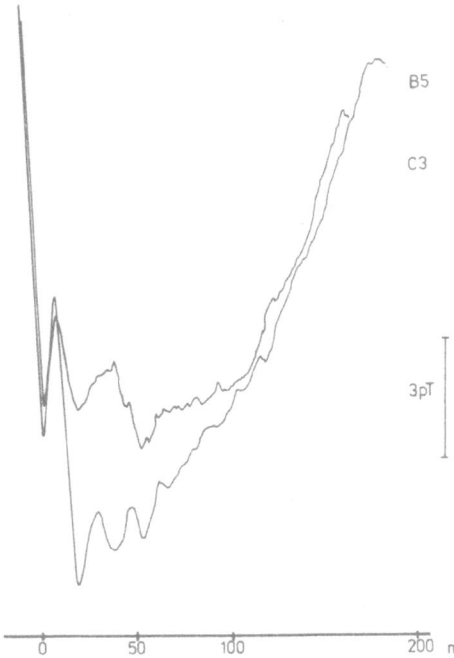

Fig. 6. High-resolution MCG of the ST segment according to the grid mentioned above (B5, C3) in a patient with electrical late potentials.

A typical real time magnetocardiogram (MCG) with P, Q, R, S and T waves is shown in Fig. 4. A map of MCGs measurements over the chest in conformity with a normalised rectangular grid on the frontal plane touching the chest with dimensions of 3.5 x 3.5 cm^2 is used for signal accumulation for high-resolution MCGs (Figs. 5 and 6) or isomagnetic field lines.

CONCLUSIONS

Finally, it should be noted that from the technical point of view high-resolution measurements without large-scale shielding and with averaging methods using one channel systems are very time-consuming. Therefore, biomagnetic research can be improved substantially by using multichannel systems. Moreover, it will be possible to look at simultaneous magnetic events at different points, procedures likely to reveal some interesting information. In line with other biomagnetic groups, we think that this new field of research and the application of superconducting devices will become important for both cardiomagnetic and neuromagnetic studies.

REFERENCES

Baule, G. M., and McFee, R., 1963, Detection of the magnetic field of the heart, Am. Heart J., 66 : 95.
Berthel, K. H., Albrecht, G., Kirsch, G., Nowak, H., Vodel, W., and Zach, H. G., 1986, Application of dc-SQUIDs for biomagnetic measurements, in: "Proc. 4th Czechoslovakian Symp. on Weak Superconductivity, Smolenice : 53.

Cohen, D., 1970, Large-volume conventional magnetic shields, <u>Rev. de Phys.</u>
     <u>Appl.</u>, 5 : 53.
Cohen, D., Edelsack, E., and Zimmerman, J. E., 1970, Magnetocardiograms
     taken inside a shielded room with a superconducting point-contact mag-
     netometer, <u>Appl. Phys. Lett.</u>, 16 : 278.
Kelhae, V. O., 1981, Construction and performance of the Otaniemi magneti-
     cally shielded room, <u>in</u>: "Biomagnetism", S. E. Erne, H. D. Hahlbohm
     and H. Leubbig, eds., Walter de Gruyter, Berlin : 33.
Mager, A., 1981, The Berlin magnetically shielded room (BMSR), <u>in</u>: "Biomag-
     netism", S. E. Erne, H. D. Hahlbohm and H. Leubbig, eds., Walter de
     Gruyter, Berlin : 51.
Nowak, H., Albrecht, G., Burghoff, M., Haberkorn, W., and Zach, H. G.,
     1987, Heart conduction system measurements and the identification of
     His-activity, <u>in</u>: "Functional Localization: A Challenge for Biomag-
     netism" (in press).
Romani, G. L., Modena, I., and Leoni, R., 1982, Biomagnetism: recent pro-
     gress in Italy, <u>in</u>: "Proc. of the Int. Conf. on Applications of
     Physics to Medicine and Biology", Trieste : 187.

METHOD FOR DETERMINING SOME ELEMENTS OF IMPEDANCE PLETHYSMOGRAPHY (IPG)

SIGNALS TO BE USED IN AUTOMATIC ANALYSIS

I. Stamboliev, H. Penev and A. Mushinski

Higher Institute of Mechanical and Electrical Engineering
Sofia
Bulgaria

INTRODUCTION

The detection of local distortion of the blood circulation by a non-invasive technique can be made using impedance plethysmography (IPG). This method is not widely used because of the complexity of the analysis of two curves, the $\Delta z(t)$ and $d(\Delta z)/dt$. The method proposed is suitable for automated calculation. The main problem of the automated analysis is the precise detection of the different components of the IPG signals.

ANALYSIS

The method is based on distinguishing the rheogram in two parts - arterial and venous. This enables the diagnostic possibilities to be extended. The method is realised by synchronised recording of the main wave $\Delta z(t)$ and its first derivative $d(\Delta z)/dt$.

We can determine many diagnostic parameters by identification and localisation in time of some of the significant points of $\Delta z(t)$ - A, B, C, D, E as well as by measuring the amplitude V1 pointed out in Fig. 1.

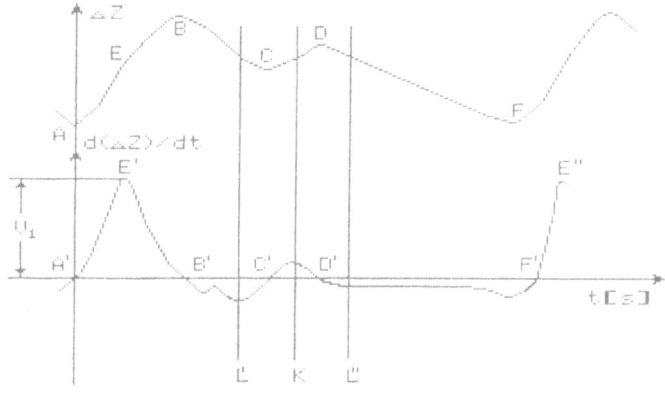

Fig. 1. IPG signal $\Delta z(t)$ and its derivative.

Firstly, an analog to digital conversion of both signals $\Delta z(t)$ and $d(\Delta z)/dt$ is made. Then the results are stored for a period of 20s. If the average frequency is out of the limits 30 – 240 strikes per minute the analysis is made. Secondly, a choice and calculations of the main parameters of five waves are made.

The zero-line level is determined as the minimum of the amplitude in the interval with magnitude of 15% of the whole period lying to the left side of the point E. Drift compensation for each separate wave is made (Stamboliev, 1985). The point A' defines the level of the isoelectric line for $d(\Delta z)/dt$ and corresponds to the point A.

The differentiated IPG curve has two local maxima in the isoelectric line and their position depends on the pathology of the person. The position of the maximum is determined by analysing the direction of $\Delta z(t)$ in the interval B-D. If the curve changes only its slope but not the direction, the local maximum will lie on the isoelectric line or under it. If the curve changes its direction the maximum will lie over the isoelectric line. On the basis of statistical data it was established that two of the local maxima are lying in the first part of the interval E-E' (Frolkis and Borisova, 1974; Lifshitz, 1970). This rule can therefore be used as a criterion for their detection. We need such a criterion because of the appearance of the other local maxima from point D' to the beginning of the next wave during the digital record. This problem does not exist if an analog recording is made. The middle of the interval E'-E" is denoted by L" and the middle of the interval L'-L" is denoted by K.

When the level of the curve between the points C and D is ascending the local maximum of the differentiated IPG in the interval C'-D' lies over the isoelectric line. The cross-points between the isoelectric line and $d(\Delta z)/dt$ detect the position of the incisure C and dicrotic peak D.

After that the minimum values of the function are to be found in the intervals L'-K and K-L". These two minima define the area in which we can search for the local maximum. Then we can easily determine the points C' and D', as well as C and D. These points are the cross-points between the curve and the isoelectric line.

The algorithm for determination of the points B', C' and D', as well as B, C and D, is shown in Fig. 2. If the local maximum is on or under the isoelectric line we can speak about an overlapping of the points C and

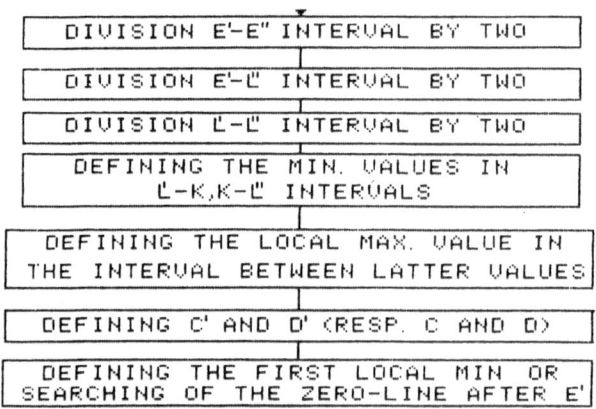

Fig. 2.   Algorithm for determination of B, B', C, C', D and D'.

D. The point B' is determined as the first cross-point between d(Δz)/dt and the zero-level line on the right side of the point E'. In abnormal cases the cross-point of the zero-level line occurs after the point L'. In this case the search for the local minimum is started after the point E'. This minimum defines the place of the point B' (that is B). The zone L'-L" can be appropriately defined by moving this interval to the left or to the right if necessary.

CONCLUSION

The method proposed constitutes a basis for developing computer-aided electronic devices for electroplethysmography.

REFERENCES

Stamboliev, I. B., Mushinski, A., and Penev, H., 1985, Determination of the isoelectric line in rheograph signals, in: "2nd Conference on Biomedical Techniques", Varna, Bulgaria (in Bulgarian) : 16.
Frolkis, A. V., Borisova, G. V., 1974, The use of harmonic analysis for quantity estimation of the rheograph's curve, Cardiology, no. 4 (in Russian) : 130.
Lifshitz, K., 1970, "Electrical Impedance Encephalography in Biomedical Engineering Systems", McGraw Hill, New York : chapter 2.

NEUROSTIMULATION AND NEUROPROSTHESES

M. Šramka and M. Mikuláš

Research Institute for Medical Bionics, and
Electrotechnical Faculty of the Slovak Technical University
Bratislava, Czechoslovakia

INTRODUCTION

Developments in medical bionics and microelectronics have enabled
specialists in neurology to work together with technologists to develop
means by which a man-made connection can be effected between technological
systems and the nervous system in order to replace damaged or lost functions.
It has been shown by experiments that stimulation of the cerebellum inten-
sifies the activity of the Purkyne cells. The use of stimulation in the
treatment of epilepsy is therefore a form of replacement for the Purkyne
cells and the technology needed for achieving this effect is termed a stimulator. Stimulators used for the treatment of epilepsy were later used also
for the stimulation of other brain structures, especially the nucleus caudatus (La Gruta, 1971; Šramka et al., 1975) and the nucleus ruber (Delgado,
1965). More recently stimulation has been performed directly in the epileptic focus (Šramka, 1985). A new area of interest in the treatment of
epilepsy originated in this way, which soon merged with a similar development in the treatment of pain (Mundinger and Neumuller, 1981), motility
disturbances (Galanda et al., 1980) and later also in the treatment of
mental disturbances.

This new direction in treatment is called therapeutic stimulation.
The stimulation is aimed at those brain structures which exert a suppressive
effect on such brain structures, and which participate in the motor function
of brain structures which themselves participate in motor function. Alternatively it is applied with the aim of influencing pathological activity,
to suppress or support the missing function and to enable the compensation
mechanisms of the nervous system to have their effect.

The problem in this case is one of prosthetics. Technological systems
mostly generate electric pulses having controllable parameters which are
directed on to the brain cortex via chronic electrodes, or introduced into
the deep brain structures. Apart from artificial electric parameters
generated by stimulators, research is also being carried out into the possibilities of transferring electrical activity sensed in the healthy structures
of the same individual to diseased structures with the aid of chronic electrodes (autotransplantation) or else transferring the normal electrical
activities from one individual to the diseased structures of the sick person
(homotransplantation). In the case of transport between different species
(heterotransplantation) the generated electrical pulse becomes a limiting

situation. Mechanisms of memory and also the mechanisms of interference depending on the pulses introduced are implemented in this case. The equipment used for the stimulation of nervous structures is termed a neurostimulator. This is a technological system situated in a biological environment with the task of replacing the missing natural stimuli by artificial or natural ones.

SYSTEM DESIGN

An implantable neurostimulator was developed in 1978 at the Research Laboratory for Clinical Stereotaxis in co-operation with the Neurosurgical Department of the Medical School of the Comenius University and the Tesla organisation. It is a partially implantable neurostimulating device, which should in the future be programmable and also fully implantable. The TESLA LSN 340 has been designed as a double convex disc of a diameter of a few millimetres, with an opening for the sealing and fixing plug and a space for the storage of the surplus length of the stimulating electrode. This shape makes implantation possible directly over the trepanating opening and also aids the fixing of the device with the help of a silicone cone. The stimulating electrode originates directly from the disc of the receiver, is introduced into the brain through the central opening of the disc, and is fixed with the plug.

The new design of the receiver facilitates implantation and does not need an operation on the neck and chest and the complicated connections on the skull, neck and chest. The decreased strain of the electrode enables a significantly slimmer stimulating electrode to be used and preserves brain tissue where the electrode is located. The implant is also more reliable because no contact system is needed between the electrode and the receiver, since the electrode is a part of the receiver. Important also is the cosmetic effect achieved by the shape of the receiver, which is hidden in the hairy part of the head.

After completing the development of the stimulator for the central nervous system, the development of an implantable device for the spinal cord was started. The implantable part of the neurostimulator proper is a round disc inside which two unipolar electrodes, or one unipolar and one reference electrode, may be connected. The electrodes are highly elastic. The ends of the electrodes have a wavy shape to ensure stability while located in the epidural space of the spinal channel. The electrodes are introduced with the aid of a rigid mandrel which is removed after the insertion of the electrode into the epidural space. The electrodes have a length of 50 - 60 cm (according to the desired location) and a width of 1.4 cm. The connectors are removable to facilitate introduction through an epidural needle, thus eliminating the necessity of surgical intervention. The electrodes are then placed under the skin and above the ribs with the aid of a special instrument to the contralateral side to provide for the simple location of the aerial over the receiver implanted in the subcostal space.

CLINICAL TESTING

The new type of brain neurostimulator was tested in 86 patients suffering from different diseases of the nervous system. Thirty-six neurostimulators were applied in epileptic patients for the stimulation of the nucleus caudatus and the gyrus hippocampi. Forty stimulators were used in the treatment of motility disturbances, especially in DMO, 5 stimulators in cases of central pain and 5 in mental diseases. The spinal neurostimulator was tested as a preclinical device in 10 patients. After externalisation of the implanted electrodes and testing the optimal stimulation programme, checking the parameters of stimulation and characteristics of stimulating electrodes, the electrodes were internalised again. Spinal cord neurostimulators were used in the treatment of pain and spasticity.

THE WAY FORWARD

Stimulation therapy with the aid of neurostimulators is considered to be a feasible physiological approach to the treatment of functional disturbances of the brain. Neurostimulators were called neuroprostheses because of the original idea according to which stimulation replaces and supports the suppressive action of some brain structures on motor fits in epilepsy. It has been found, however, that the use of neurostimulators in different diseases of the nervous system may also support or replace other damaged or missing functions of the nervous system. Stimulation therapy opens new and wide possibilities in influencing brain activity. This is not something which should be restricted according to current ideas, but should be used for the further development in combining technological systems with biological ones. At a time when transplantation of brain tissues with renewal of original functions is a matter of fiction, neuroprostheses show a way of replacing or changing the function of a damaged part of a nervous structure with the aid of therapeutic stimulation; something which, under certain conditions, may be considered to be a transplantation of function into the nervous system.

REFERENCES

Delgado, J. M. R., 1965, Sequential behaviour repeatedly induced by red nucleus stimulation in free monkeys, Science, 148:1361.
Galanda, M., Fodor, S., Šramka, M., and Nádvorník, P., 1980, Možnosti stereotaktického neurostimulačného liečenia v zadnej jame, in: "Czechoslovak Conference on Neurosurgery", SLS, Bratislava : 6 (abstract).
La Gruta, V., 1971, The control of amygdaloid and temporal paraxysmal activity by the caudate nucleus, Experientia, 27:278.
Mundinger, F., and Neumuller, H., 1981, Programmed transcutaneous (TNS) and central (DBS) stimulation for control of phantom limb pain and causalgia: a new method for treatment, in: "Phantom and Stump Pain", Springer Verlag, Berlin.
Šramka, M., Fritz, G., Galanda, M., and Nádvorník, P., 1975, Some observations on therapeutic stimulation of epilepsy, in: "Symposium on Stereotactic Treatment of Epilepsy", SLS, Bratislava : 78 (abstract).
Šramka, M., 1985, "Stereotaktické liečenie temporálnej epilepsie", Veda, Bratislava : 204.

PART 2

SIGNAL AND IMAGE PROCESSING

DETECTION OF LATE POTENTIALS OF THE HEART

A. Voss, B. Reime, U. Buhss, H. Fiehring and H. J. Kleiner

Academy of Sciences of the GDR
Central Institute of Cardiovascular Research
Berlin, GDR

INTRODUCTION

In modern industrialised countries chronic ischaemic heart disease is one of the major causes of death. In the course of its progression it frequently leads to acute myocardial infarction or, via different forms of heart-rhythm disturbance, to sudden heart death.

Basically, there are two approaches to identifying patients who have a high risk of sudden heart death:

(i) diagnosing coronary heart disease (mainly using conventional ECG registration, exercise-ECG, Holter-ECG, nuclear-cardiological methods and invasive investigations); and

(ii) recognition of the electric instability or inhomogeneity of the excitatory conduction on the heart (mainly using conventional ECG registration, Holter-ECG, programmed ventricular stimulation and recording of ventricular late potentials).

Thus, the detection of ventricular late potentials allows recognition of electrical instability or inhomogeneity of excitatory conduction (El-Sherif et al., 1985; Fiehring et al., 1986; Werner et al., 1986). In this way, and in combination with the methods mentioned above, an attempt can be made to single out patients who are under a high risk of sudden heart death and, furthermore, to develop a prophylactic therapy for preventing the catastrophe of sudden heart death (von Leitner, 1984).

PHYSIOLOGICAL BASIS

Regular micropotentials of the heart are understood to be potentials from the cardiac conduction system that are required physiologically, and are therefore necessary for the proper functioning of the myocardium. Such micropotentials are in particular the sinus-node potential, the potential of the AV-node and the His-bundle potential (Fig. 1).

Irregular micropotentials are an expression of pathological changes of excitatory conduction, and represent ventricular delayed depolarisation potentials (late potentials, Fig. 2). Therefore, they may be an indicator of serious electrical instability or inhomogeneity of the heart.

177

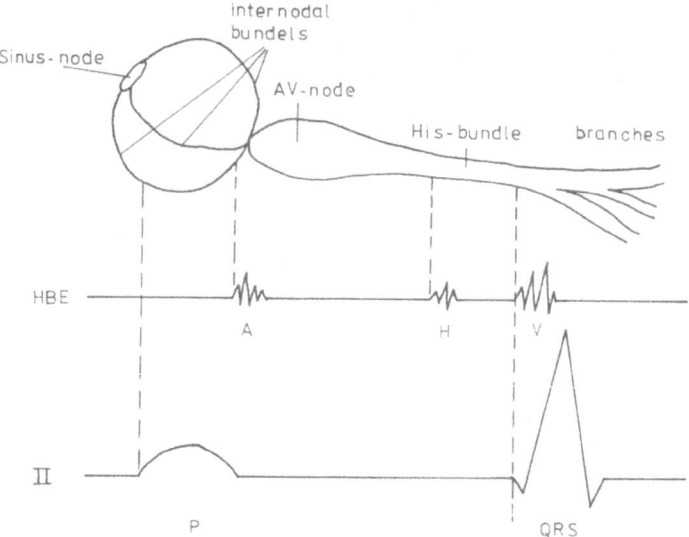

Fig. 1. His-bundle potential.

In a more general sense, the occurrence of such pathological signals could be regarded as a massive disturbance in the cell - conduction pathway - myocardium complex. Considering the complicated processes associated with the release of action potentials, one can well imagine the sensibility of this system to milieu alterations (for example, impaired substrate supply, impaired energy conversion, electrolyte shifts).

Therefore, it may be taken for granted that a functionally impaired cell is, in the worst case, electrically completely dead, or else becomes a pacemaker cell itself through an inconstant resting potential, or produces by variation of the ion-exchange process an action potential that is essentially prolonged and altered in its shape (Fig. 3). Furthermore, delays in

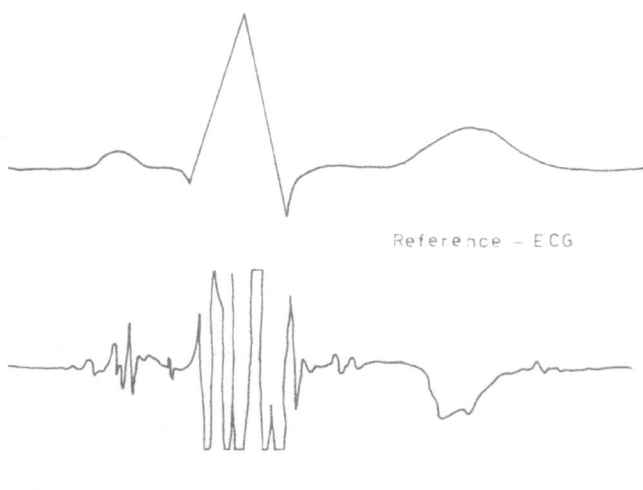

Fig. 2. High amplified averaged ECG.

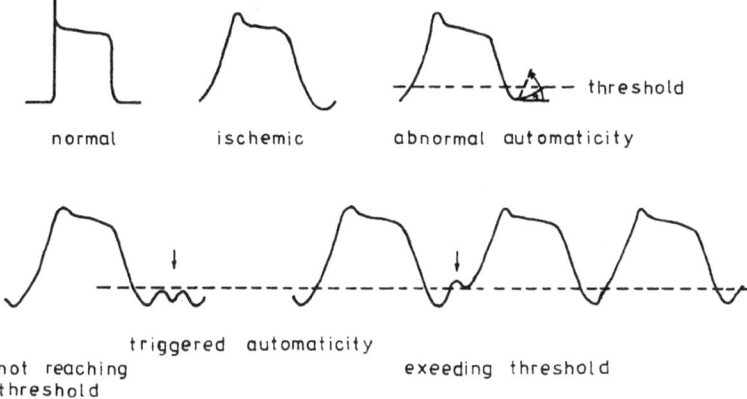

Fig. 3. Action potentials of normal and damaged cells.

pathways or conduction blockades as a result of structural changes occurring in the myocardium are conceivable. Additionally, conduction of the stimulus in the working muscle itself may be considerably delayed because of "diversions" around ischaemic and electrically mute areas, or by damaged but electrically still active areas, or may be retarded through isolated intact "islets of the myocardium" within a necrotic or infarction borderline zone. Thus, common to all these pathological processes is one property (with the exception of the focal pacemaker) - delayed depolarisation.

Under certain conditions (Bethge et al., 1985; Breithardt and Borggrefe, 1986; Kertes et al., 1984; Schwartz and Scheinmann, 1983; von Leitner, 1984) these late potentials may elicit single premature actions of the heart (ventricular ectopic beats) or, in the less favourable case, provoke life-threatening ventricular tachycardias. This mechanism is characterised as re-excitation of healthy myocardial areas by the same process of excitation, and pre-supposes a unidirectional block in part of the working muscle and a delayed conduction velocity in another area. It can be assumed that the delayed depolarisation potentials may directly elicit ventricular arrhythmias, whose coupling interval to the preceding QRS complex is greater than 200 to 220 ms (Bethge et al., 1985; von Leitner, 1984).

Finally, it has to be pointed out that the registration of putatively delayed depolarisation potentials may also be the result of methodological shortcomings or artefacts (Ideker et al., 1985). So, for instance, motion of the catheters used to derive the signals could mimic periodic late potentials as a result of potential shifts by variation of the ion distribution at the interface. Signals may be simulated by "ringing" when inappropriate filters are used.

The inadequate selection of sites for the derivation of signals may lead to non-existing frequencies through corresponding projection, which may feign fractionations. These are just a few examples of problems that have to be considered in the registration of "genuine" micropotentials.

TECHNICAL BASIS

The detection of micropotentials of the heart from the body surface is a process of obtaining a useful signal $S(t)$ (of the order of $1 - 20\mu v$) from the masking noise $N(t)$ (about $3 - 50\mu v$) and additional interfering

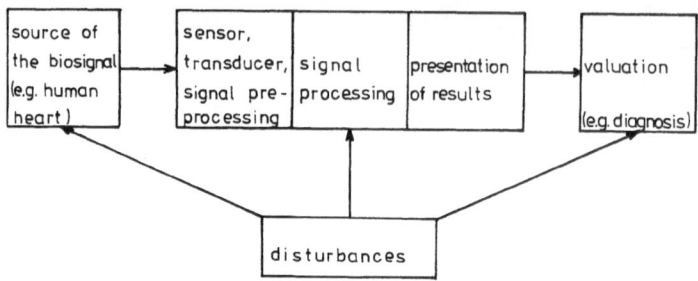

Fig. 4. Biological measurement chain.

signals. These undesirable disturbances appear within the whole biological
measurement chain (Fig. 4), thus preventing in the majority of cases the
detection of the micropotentials which are being sought, by simple amplification and filtering.

Consequently, the task is to improve the signal-to-noise ratio under
the condition that the useful signal, as a rule, possesses a lower effective
value than the interfering signal. The resultant measurable signal mixture
X(t) at the time t can therefore be defined as

$$X_i(t) = S(t) + N_i(t).$$ \hfill (1)

Now, whereas S(t) can be understood to be an undisturbed bioelectric
generator potential of the heart, the interfering signals consist of a
variety of undesirable phenomena (see also Ideker et al., 1985 and Zywietz,
1985):

   (i)   biological noise (for example due to the activity of the respiratory
         system, the oesophagus and of nerves close to the site of recording),
  (ii)   electrode noise (ionic currents, transition processes and electric
         polarisation currents),
 (iii)   motions of the electrodes and cables,
  (iv)   contact disturbances at the connector assemblies and with partly
         defective wires,
   (v)   capacitive and inductive intrusions by external fields such as
         arise from 50 Hz power-line interference,
  (vi)   amplifier noise,
 (vii)   errors in digitising the signal (sampling frequency, phase of
         sampling points, aperture time, number of quantisation steps, inter-
         polation methods for signal reconstruction),
(viii)   errors in signal processing (choice of interference voltage reduc-
         tion procedures - filter problems, determination of the start and
         finish of potentials), and
  (ix)   errors in the evaluation process (determination of reference values
         for amplitudes and time intervals).

Some of these disturbing factors can be reduced, or even completely
suppressed, by appropriate measures of optimal planning and performance of
the experiment:

   (i)   optimal site of recording,
  (ii)   favourable electrode material of low degree of polarisation and
         fitted equal electrode surfaces,
 (iii)   preparation of the skin prior to the application of the electrodes
         (low electrode impedance),

(iv)   fixation of the electrodes,
 (v)   relaxation of the patient, if possible,
(vi)   common mode signal feedback to the patient (active feedback),
(vii) impedance amplifiers close to the recording electrode (active electrodes),
(viii) shielded electrode connection,
(ix)   low-noise pre-amplifiers,
 (x)   opto-separation (floating input),
(xi)   optimally localised frequency band of the input stage,
(xii) shielded power lines, and
(xiii) recording in a shielded room (Faraday cage);

and by favourable selection of the corresponding signal processing procedures:

(i)   filters without ringing and phase shifts (digital filters),
(ii)   increase in sampling rate to diminish quantisation noise (Tiefenthaler, 1987), and
(iii) use of adequate methods for improving the signal-to-noise distance (averaging techniques and subsidiary procedures).

INTRACARDIAL RECORDING OF MICROPOTENTIALS

The endocardially recorded ECG has so far served mainly for investigating stimulus formation and conduction. Furthermore, it allowed us to register the behaviour of the excitatory-conduction system and the myocardium using electric stimulation (Kappenberger, 1986). For the detection of cardiac micropotentials, the catheter recording technique has now to be considered the only informative reference method because of the favourable site of deduction and the amplitude conditions prevailing there. This holds, in particular, for the multipolar recording technique (endocardial mapping). Disadvantages of the method include above all the invasive nature of the investigation technique itself, and the necessary indication which is often not possible only for the recording of micropotentials. Therefore, the epicardial and transmural recording techniques, used only in special cases (heart surgery), will not be dealt with in this paper.

TEMPORAL (SERIAL) AVERAGING TECHNIQUE

The principle of the serial signal averaging technique consists of a time-synchronous k-fold summation of the periodically repeating signal of interest with the corresponding superimposed stochastic interfering signal. In this way, amplitude summation to the kth value is achieved for the signal of interest, whereas the effective value of the disturbing signal adds up to a lower level obeying the laws of mathematical statistics. In this manner, the signal-to-noise ratio is improved (Fig. 5).

For the method to be used, the following pre-requisites should be fulfilled (see also Ros et al., 1981 and Tietze, 1975):

(i)   the noise must not be correlated with the signal of interest,
(ii)   the noise should be superimposed on the signal of interest in a stationary and additive manner,
(iii) the signal of interest must not be subject to temporal fluctuations (the problem of biological variance), and
(iv)   the threshold level must be stable and temporally constant with respect to the signal of interest.

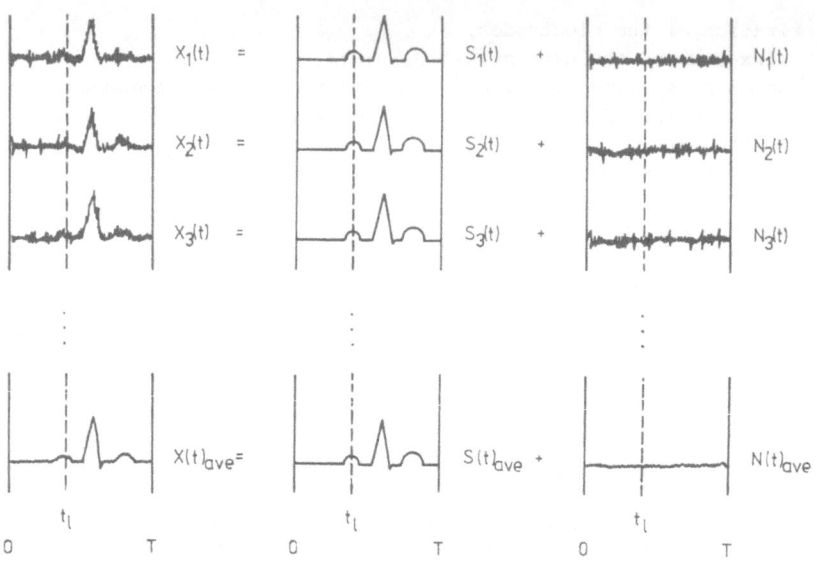

Fig. 5. Principle of temporal averaging.

Proceeding from the general formula

$$X(t) = S(t) + N(t) \qquad (2)$$

the k-fold addition of the k implementations of the stochastic process $N_i(t_i)$, i = 1, 2, ..., k gives

$$N(t_1) = \sum_{i=1}^{k} N_i(t_1) \qquad (3)$$

where t_1 is a fixed time in the averaging window.

Thus, the mean value and scatter become:

$$\mu_N = E\left[\sum_{i=1}^{k} N_i(t_1)\right] = \sum_{i=1}^{k} E[N_i(t_1)] = \sum_{i=1}^{k} \mu_i ; \qquad (4)$$

$$\upsilon_N - E[(N(t_1) - \mu_N)^2] = E\left[\sum_{i=1}^{k} (N_i(t_1) - \mu_i)\right]^2 \qquad (5)$$

If $N_i(t_1)$ does not correlate with $N_j(t_1)$ (i ≠ j) and, hence,

$$E\left[(N_i(t_1) - \mu_i)(N_j(t_1) - \mu_j)\right] = 0, \text{ then} \qquad (6)$$

$$\sigma_N = \sum_{i=1}^{k} E\left[(N_i(t_1) - \mu_i)^2\right] = \sum_{i=1}^{k} \sigma_i^2 \qquad (7)$$

Under the condition of stationarity of the stochastic process

$$\mu_1 = \mu_2 = \ldots = \mu_k \qquad (8)$$

and

$$\sigma_1^2 = \sigma_2^2 = \ldots = \sigma_k^2 \qquad (9)$$

From this it follows that

$$\mu_N = k\mu_i \qquad (10)$$

and

$$\sigma_N^2 = k\sigma_i^2 \qquad (11)$$

This yields the standard deviation

$$\sigma_N = \sqrt{k\sigma_i^2} \qquad (12)$$

The average values are

$$\mu_{Nave} = \frac{k\mu_i}{k} = \mu_i \quad ; \qquad (13)$$

$$\sigma_{Nave} = \frac{\sigma_N}{k} = \frac{\sigma_i\sqrt{k}}{k} = \frac{\sigma_i}{\sqrt{k}} \qquad (14)$$

Thus it has become obvious that the averaging procedure reduces the original standard deviation of the noise by the factor $\sqrt{k}$, thereby impro-ving the signal-to-noise distance by the factor $\sqrt{k}$. Fig. 6 illustrates the principle of our system working with time averaging. In addition to linear averaging, exponential averaging (Klötzner and Otto, 1985; Rakshit et al., 1985) is used predominantly in on-line representation of the effi-ciency of the averaging process. The algorithm has the following equation:

$$\hat{x}_i = \hat{x}_{i-1} + \frac{x_i - \hat{x}_{i-1}}{i} \qquad i = 1, 2, \ldots, k \qquad (15)$$

or

$$\hat{x}_i = ax_i + (1 - a)\hat{x}_{i-1} \qquad (16)$$

$\left(a' \text{ being the averaging constant, } a = \frac{1}{2^p}\right)$.

For large k, the following simplified expression results:

$$\hat{x}_k = s_k + \sqrt{\frac{a}{2 - a}} \ \sigma_N \qquad (17)$$

The remaining noise proportion is essentially determined by the averaging factor a.

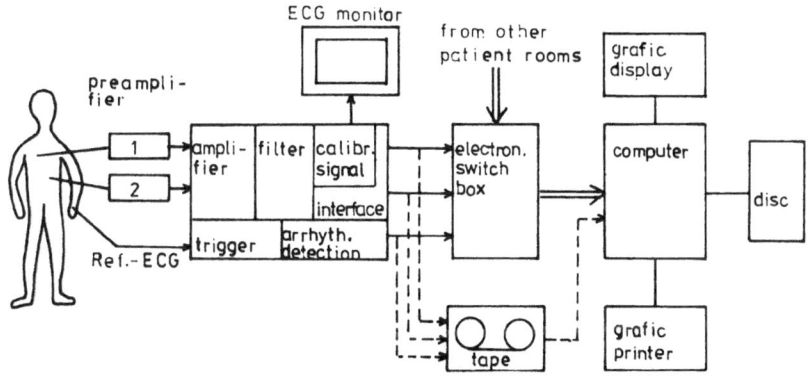

Fig. 6. Serial (temporal) averaging system.

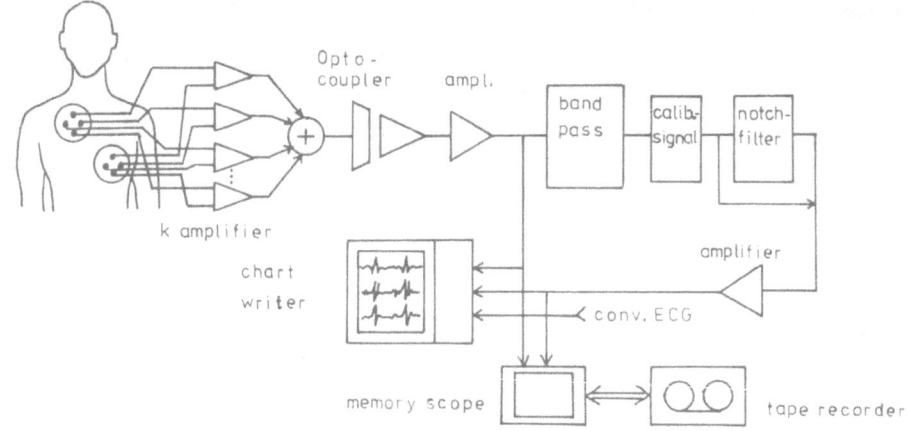

Fig. 7. Spatial averaging.

A disadvantage of exponential averaging is the greater inaccuracy caused by the abundance of mathematical operations due to truncation errors. In particular, negative effects on the result of averaging procedures have trigger jitters; they may either falsify the shape of the micropotentials being sought or suppress them completely. The fundamental condition for the use of the signal averaging technique - the configuration of bioelectric signals should be constant from beat to beat - has already been discussed by Durrer et al. (1961).

The non-invasive detection of His signals by means of the temporal signal averaging technique was first reported by Berbari et al. (1973), Flowers and Horan (1973) and Stopczyk et al. (1973). In 1978, Berbari and Fontaine (Berbari et al., 1978; Fontaine et al., 1978) were the first to describe a non-invasive registration of ventricular late potentials by means of the averaging technique. Historic accounts and subsequent studies are to be found in Breithardt et al. (1981), von Leitner (1984), Karbenn et al. (1985), Alperin and Sadeh (1986), Berbari et al. (1986), Craelius et al. (1986), Fiehring et al. (1986), Lindsay et al. (1986) and Oeff et al. (1986).

SPATIAL AVERAGING TECHNIQUE

With the spatial averaging technique, k bipolar ECG signals are ob-tained from 2k electrodes by means of k amplifiers, connected through the resistor network to the input of the amplifier of a summing network, which carries out the actual averaging procedure (Fig. 7).

Pre-requisites for using the method are (see also Peper, 1984 and Shvartsman et al., 1982):

  (i)   the noise should be distributed stochastically,
 (ii)   minimal influence of electric and magnetic disturbing fields,
(iii)   extremely low-noise pre-amplifiers,
 (iv)   special electrodes of lowest possible size but also with low impedances.

Proceeding from the general relation (1) (Shvartsman et al., 1982), with i = 1, 2, ..., k, the proportion of noise is:

$$N_k = \frac{\sqrt{\sum_{i=1}^{k} (G_i N_i)^2}}{k} \tag{18}$$

184

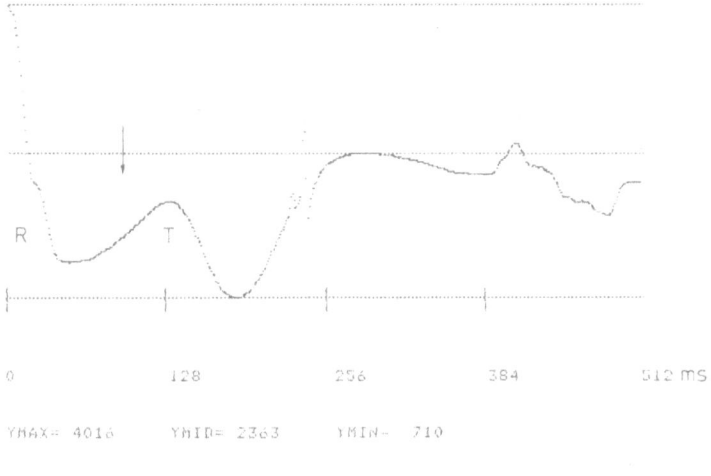

YMAX= 4016   YMID= 2363   YMIN= 710

PATID : DL1:220740.01D   LEAD : 1

Fig. 8.   Averaged ECG with artificial activity (non-detec-
          ted disturbance)

where G_i is the amplification in the i-channels and the signal proportion
is:

$$S_k = \frac{\sum\limits_{i=1}^{k} (G_i S_i)}{k} \tag{19}$$

With the simplifications:

$$G_1 \approx G_2 \approx G_3 \quad \ldots \quad \approx G \tag{20}$$

$$N_1 \approx N_2 \approx N_3 \quad \ldots \quad \approx N \tag{21}$$

$$S_1 \approx S_2 \approx S_3 \quad \ldots \quad \approx S \tag{22}$$

we obtain the relationships

$$N_k = \frac{\sqrt{k}GN}{k} = \frac{GN}{\sqrt{k}} \tag{23}$$

and

$$S_k = \frac{kGS}{k} = GS \tag{24}$$

Thus, under the above conditions, reduction of noise is possible by the
factor $\sqrt{k}$. The principle of a system using this spatial averaging technique is shown in Fig. 8.

The advantages of the procedure are that physiological signals can
be detected at any time and in each heart phase irrespective of biological
variances; the signals of interest may occur sporadically; the results
are available immediately; and no cumbersome triggering and arrhythmia
detection are required. On the other hand, there are a number of serious
disadvantages. So reduction of noise by the factor $\sqrt{k}$ cannot, in practice,
be achieved. Due to this compromise, electrode gap - electrode surface,
the biological noise at the electrodes with small gaps is in-phase (minimal

noise suppression) or, with large gaps, the amplitudes of the signals of interest vary in shape and size (lower amplification of the signal of interest, alterations of shape). Also, more equipment (k pre-amplifiers) is needed, together with the availability of special electrodes. The use of the method in shielded rooms and the relaxation of the patients at present restrict its application to cardiological centres.

The procedure of spatial averaging was presented by the working groups of Flowers et al. (1981a), Gebhardt-Seehausen et al. (1981), Stopczyk et al. (1981) and Wajszczuk et al. (1981) for detection of Pre-P or His-bundle potentials. Hombach et al. (1982) reported on the use of the method for the detection of ventricular late potentials. Subsequent and completing studies are to be found in Flowers et al. (1981b), Chen et al. (1986) and Werner et al. (1986).

TECHNICAL IMPLEMENTATION AND CLINICAL APPLICATION

Ventricular late potentials have been registered in our Institute on the basis of the serial signal averaging technique since 1981. The first system consisted of a one-channel analogue ECG pre-processing unit with a trigger and arrhythmia detector, with a tape unit for data intermediate storage, and a hybrid averaging computer. Since 1984 we have been using a more advanced equipment (see Fig. 6).

The signals from two ECG leads are pre-amplified by means of low-noise pre-amplifiers (input noise 1μv, common mode rejection 90 db), galvanically divided from the network via the opto-coupler (important for intracardial leads), submitted to band-pass filtering (upper limiting frequency 300 Hz, lower limiting frequency variable between 10 and 150 Hz) and, in a further amplification stage, raised to the level necessary for the A/D conversion (resolution 12-bit, sampling frequency 1 kHz). The size of the filters was selected such that the filter ringing was not recorded (test by loading the system with a unit step). To enable estimation of the quality of the signal already at registration, two calibration signals (10 and 50μv) are added to the highly amplified ECG signals, and thus the level of the disturbing signals can be checked on the monitor. The reference ECG is also obtained in a bipolar manner from a third lead, band-limited (1 - 35 Hz) and fed to two trigger stages (R-peak, steepest rise). Here, an arrhythmia detection circuit is built in to prevent falsification of the averaging result by undesirable disturbing signals or arrhythmias. The trigger error in using a phantom-ECG is less than 0.5 ms and, when recorded from a patient, less than 2 ms. The pre-processing unit is mobile, so that registration of cardiac micropotentials from any patient image in the supervision ward is possible. Data transmission to the rigidly installed computer system is controlled by means of a control interface in the analogous pre-processing unit. It is from here (that is, from the patient's bed) that sequential transmission of the ECGs (as a rule, 4 highly amplified leads - quasi-orthogonal, the corresponding 4 original ECGs and the reference ECG) to the computer is elicited. A complete ECG registration is supposed to take 30 min as a rule (including preparation of the patient, placement of electrodes and adjustments). The computer system for signal processing includes a minicomputer PDP 11/23 with disk drives, graphic terminal and graphic printer. Clinical testing of the system showed that, under certain unfavourable circumstances, arrhythmias and trigger errors were not detected, despite the hardware logic. So we developed two programs for detection of disturbing signals, which have been tested on 3200 ECGs.

The first one was based on pattern matching. At 4 dominant points pattern minima and pattern maxima were formed with which every single complex was compared (automatic minimum-maximum search of defined sites). The

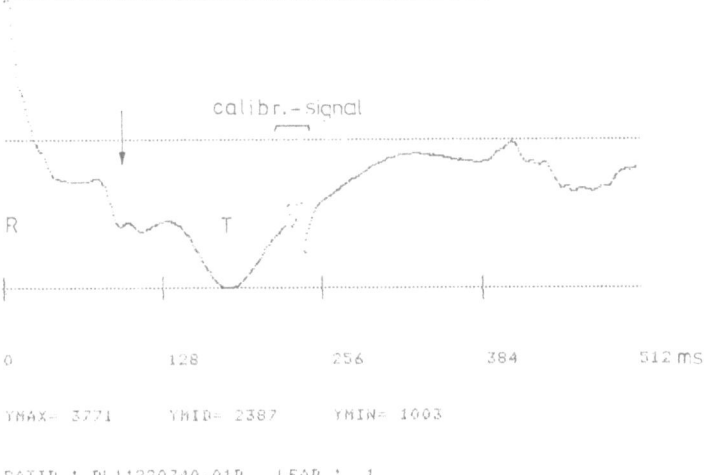

YMAX= 3771   YMID= 2387   YMIN= 1003

PATID : DL1:220740.01D   LEAD : 1

Fig. 9. Averaged ECG without detection errors.

detection rate of disturbing signals was 70%; the undetected 30% were disturbing signals lying between the points of observation. The second algorithm was based on the integral method, by which the area between the signal
curve and the zero line was determined. After pattern formation (averaging
over 4 cycles):

$$I_M = \left(\sum_{i=1}^{k} \left(\sum_{j=1}^{l} |x_j| \right) / l \right) / k \qquad (25)$$

where k is the number of averaging steps (4), l is the length of the
averaging window, and x_j are the discrete signal values, the individual
integrals are calculated:

$$I = \left(\sum_{j=1}^{l} |x_j| \right) / l \qquad (26)$$

and the respective difference

$$DIFF = |I - I_M| \qquad (27)$$

obtained.

Analysing the efficiency of the procedure by means of the ECG complexes
mentioned, the exclusive criteria were determined to be DIFF > l/16. It has
been demonstrated that this relatively simple algorithm is able to detect
100% of the disturbing signals and many of the strongly noisy sections of
the ECG and thus exclude them from averaging, which means an improvement
in the signal-to-noise distance. Figs. 8 and 9 underline the importance
of the 100% detection of the large disturbing signals.

The highly amplified ECGs were evaluated at the graphic terminal.
Having determined the QRS-end by the original ECG printouts and the stored
reference ECG (of great importance here is the problem of safe recognition
of ECG sections (Michaelis, 1985)), the averaged, highly amplified, ECG is
assessed visually and, if ventricular late potentials are present (we define
as late potentials those signals that occur after the end of the QRS and
show an amplitude of at least twice that of the residual noise level;
other definitions have been published, for example Breithardt and Borggrefe

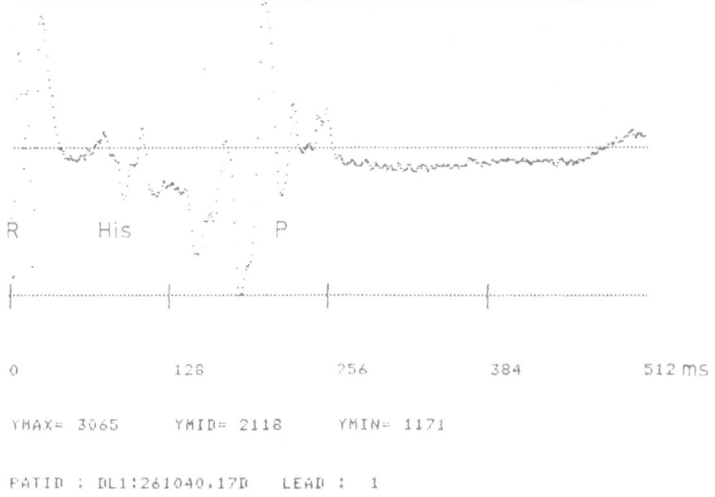

YMAX= 3065   YMID= 2118   YMIN= 1171

PATID : DL1:261040.17D   LEAD : 1

Fig. 10.   His potential.

(1986), Gang et al. (1986), Karbenn et al. (1985), Oeff et al. (1986) and
von Leitner (1984)), measure them by means of cursor control.   Thereafter
the measured curves are produced as graphic prints.   Figs. 10 and 11 show
examples for the registration of a His-signal and a ventricular late potential, respectively.

    Based on about 500 ECG recordings, taken partly in combination with
programmed stimulation and Holter-ECG, the following detection rates of
ventricular late potentials could be achieved:

   (i)   with patients suffering myocardial infarction in 40% of all cases,
         and
   (ii) in patients with ventricular tachycardias in 85% of all cases.

    The microcomputer system, which is now being clinically tested (third
generation equipment) for detecting micropotentials of the heart has the

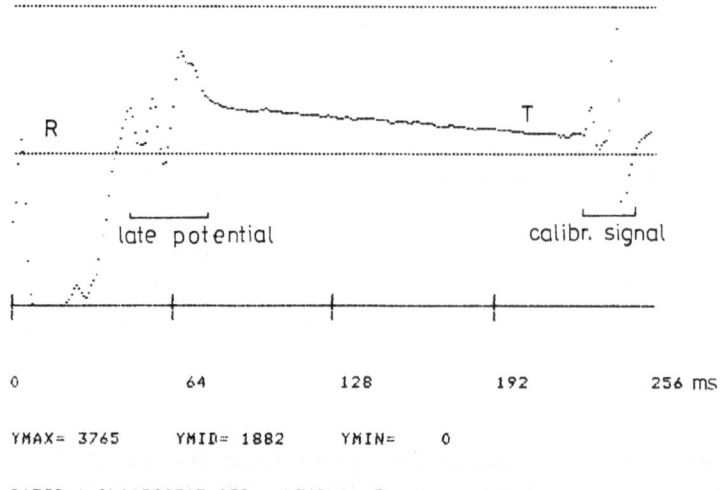

YMAX= 3765   YMID= 1882   YMIN=   0

PATID : DL1:200515.03D   LEAD : 2

Fig. 11.   Late potential.

following advantages:

(i)   real time processing of three ECGs simultaneously,
(ii) digital filtering in real time,
(iii) detection of the complete ECG cycle,
(iv) mobile system, and
(v)   reduction by 300% of the time required to investigate one patient.

Here is a brief description of the system. External bipolar input amplifiers mounted close to the site of measurement are connected to the 4-channel input. The signals enter a pre-amplifier unit allowing a programmable change of amplification in 4 steps and change-over of the high-pass filter (1 Hz/30 Hz). A fourth-order low pass filter ensures the band limiting necessary for the sampling (anti-aliasing). The sample and hold stages become effective in all channels simultaneously. The multiplexer selects one of the values, and passes it to the opto-coupler serving for potential separation. The non-linearity of the latter is compensated by an opto-coupler of the same kind in the negative feedback branch of an operational amplifier. The signal is converted into 8-bit data by an A/D converter (sampling frequency 1 kHz). Control of the input-side analogue unit and filtering of the samples are accomplished by the filter processor. The latter is connected with additional logic enabling rapid multiplication (8µs). This is a pre-requisite for realising a real-time digital filter. After being processed in the filter processor, the data are transmitted to the processing processor, which effects signal averaging, memorising of processed data, communication by means of a keyboard and passage of data to the D/A converter. The input values of the processing processor run into a pre-buffer shaped as a ring buffer in order to detect in time an area of 225 ms lying before the R-wave. The latter is a synchronisation or trigger point; the pre-buffer is arranged before, and the post-buffer after the triggering. The post-buffer can memorise a maximum of 576 samples, which is a time range of 576 ms (sampling rate 1 kHz). 871 ms are covered altogether. With a heart rate of at least 69 beats per minute, the whole ECG is processed. For the program testing stage, the processing processor was connected to test hardware and an EPROM programming device, as well as a coupling point with the software development tool, the minicomputer PDP 11/23. The output of the D/A converter leads to 4 sample and hold members functioning as a de-multiplexer for 4 analogue output channels. They follow reconstruction low-pass filters and form continuous signals from the discrete values. Transmission of parameters to the filter processor (amplification, high-pass selection, filter type) is accomplished by the processing processor via a serial connection. In order to improve the signal-to-noise ratio, the buffered samples are submitted to linear signal averaging. After every restart of the processing processor, the R-R gap of several heart cycles is measured and a pattern formed from it. The parameter "R-R difference" (RRF) input by means of a keyboard determines how large the part (RRDIF) of the mean value (RRMW) is by which the actual R-R gaps may deviate from the pattern value in order to admit the corresponding signal complex to averaging:

$$RRDIF = \frac{RRMW}{2 ** RRF} \tag{28}$$

We use additionally the integral procedure for reliable rejection of arrhythmias and strongly noisy parts of the ECG. In the beginning, a pattern is formed that has to be either confirmed or rejected by the investigator. To increase noise immunity further, the complexes immediately following the rejected ECG complexes are excluded from averaging. The results are displayed on a graphic printer, curve monitor and/or ECG chart recorder.

CONCLUSIONS

The identification of patients particularly at risk from dangerous arrhythmias represents a topical and important problem in the diagnosis and therapy of cardiovascular diseases.

Under certain unfavourable conditions, ventricular late potentials, induced by ischaemically damaged areas of the heart muscle, may elicit ventricular ectopic beats or even life-threatening ventricular tachycardias, which may then lead to the much feared sudden heart death. The detection of such ventricular late potentials in patients with severe heart disease is, in all likelihood, an indicator of the risk of occurrence of ventricular tachycardias and, in the worst case, of the occurrence of sudden heart death. The detection of these late potentials from the body surface is a process of obtaining a signal of interest from the superimposed noise and additional disturbing signals. These undesirable disturbances occur within the whole biological measuring chain, and their effects have to be suppressed by means of appropriate methods for improving the signal-to-noise ratio. In addition to temporal and spatial signal averaging techniques, a combination of these methods (Chen et al., 1986; Hombach et al., 1986; Jesus and Rix, 1985) is to be attempted and is, in part, already being used.

Furthermore, these techniques can still be improved by modern procedures of signal processing. We have implemented this improvement by a reliable and simple arrhythmia detection method. Special attention should also be paid to recent studies on frequency analysis of the highly amplified ECG (Lindsay et al., 1986) as a further method of improving the detection of electric instability and inhomogeneity of the heart.

REFERENCES

Alperin, N., and Sadeh, D., 1986, An improved method for on-line averaging and detection of ECG waveforms, Comput. Biomed. Res., 19 : 193.

Berbari, E. J., Lazarra, R., Samet, P., and Scherlag, B. J., 1973, Non-invasive technique for detection of electrical activity during the P-R-segment, Circulation, 48 : 1005.

Berbari, E. J., Scherlag, B. J., Hope, R. R., and Lazarra, R., 1978, Recording from the body surface of arrhythmogenic ventricular activity during the ST segment, Am. J. Cardiol., 41 : 697.

Berbari, E. J., Collins, S. M., and Arzbaecher, R., 1986, Evaluation of esophageal electrodes for recording His-Purkinje activity based upon signal variance, IEEE Trans. Biomed. Eng., BME-33 : 922.

Bethge, C., Gebhardt-Seehausen, U., and Recher, S., 1985, Zur Bedeutung endokardial abgeleiteter Spätpotentiale für ventrikuläre Extrasystolen beim chronischen Herzinfarkt, Herz-Kreislauf, 5 : 238.

Breithardt, G., and Borggrefe, M., 1986, Pathophysiological mechanisms and clinical significance of ventricular late potentials, Europ. Heart J., 7 : 364.

Breithardt, G., Becker, R., Seipel, L., and Abendroth, R.-R., 1981, Nicht-invasive Registrierung ventrikulärer Spätpotentiale-Methodik und erste Erfahrungen, Z. Kardiol., 70 : 1.

Chen, W. C., Zeng, Z. R., Chow, C., Xine, Q. Z., and Kou, L. C., 1986, Application of a new spatial signal averaging device for the beat-to-beat detection of cardiac late potentials, Clin. Cardiol., 9 : 263.

Craelius, W., Restivo, M., Assadi, M. A., and El-Sherif, N., 1986, Criteria for optimal averaging of cardiac signals, IEEE Trans. Biomed. Eng., BME-33 : 957.

Durrer, D., Formijne, P., van Dam, R. T. H., Büller, J., van Lier, A. A. W., and Meyler, F. Z., 1961, Electrocardiogram in normal and some abnormal conditions, Am. Heart J., 61 : 303.

El-Sherif, N., Gomes, J. A. C., Restivo, M., and Mehra, R., 1985, Late potentials and arrhythmogenesis, Pace (May – June) 8 : 440.

Fiehring, H., Kleiner, H. J., Voss, A., and Reime, B., 1986, "Entwickeltes System zur Erfassung und Verarbeitung von Mikropotentialen des Herzens", Research Report, ZIHK, Berlin.

Flowers, N. C., and Horan, L. G., 1973, His bundle and bundle-branch recordings from the body surface, Circulation, 47/48 (suppl. IV) : 102.

Flowers, N. C., Shvartsman, V., Sohi, G. S., and Horan, L. G., 1981a, Signal averaged versus beat-by-beat recordings of surface His-Purkinje potentials, in: "Signal Averaging Technique in Clinical Cardiology", V. Hombach and H. H. Hilger, eds., F. K. Schattauer Verlag, Stuttgart.

Flowers, N. C., Shvartsman, V., Kennelly, B. M., Sohi, G. S., and Horan, L. G., 1981b, Surface recording of His-Purkinje activity on an every-beat basis without digital averaging, Circulation, 63 : 948.

Fontaine, G., Frank, R., Gallais-Hamonno, F., Allali, J., Phan-Thuc, H., and Grosgogeat, Y., 1978, Electrocardiographie des potentiels tardifs du syndrome de post-excitation, Arch. Mal. Coeur., 71 : 854.

Gang, E. S., Peter, T., Rosenthal, M. E., Mandel, W. J., and Lass, Y., 1986, Detection of late potentials on the surface electrocardiogram in unexplained syncope, Am. J. Cardiol., 58 : 1014.

Gebhardt-Seehausen, U., Bethge, C., Bonke, F. I. M., and Merx, W., 1981, Continuous recordings of sinus nodal potentials, in: "Signal Averaging Technique in Clinical Cardiology", V. Hombach and H. H. Hilger, eds., F. K. Schattauer Verlag, Stuttgart.

Hombach, V., Kebbel, U., Höpp, H.-W., Winter, U. J., Braun, V., Deutsch, H., Hirche, H., and Hilger, H. H., 1982, Fortlaufende Registrierung von Mikropotentialen des menschlichen Herzens, Deutsche Med. Wochenschrift, 107 (51/52) : 1951.

Hombach, V., Höpp, H. W., Kebbel, U., Treis, I., Osterspey, A., Eggeling, T., Winter, U., Hirche, H., and Hilger, H. H., 1986, Recovery of ventricular late potentials from body surface using the signal averaging and high resolution ECG techniques, Clin. Cardiol., 9 : 361.

Ideker, R. E., Mirvis, D. M., and Smith, W. M., 1985, Late, fractionated potentials, Am. J. Cardiol., 55 : 1614.

Jesus, S., and Rix, H., 1985, ECG analysis by averaging and beat-to-beat approaches: Improvements and discussion, in: "XIV ICMBE and VII ICMP", Espoo, Finland : 1462.

Kappenberger, L., 1986, Das intrakardiale EKG, Schweiz. Rundschau Med., 75 : 1146.

Karbenn, U., Breithardt, G., Borggrefe, M., and Simpson, M. B., 1985, Automatic identification of late potentials, J. Electrocardiol., 18 : 123.

Kertes, P. J., Glabus, M., Murray, A., Julian, D. G., and Campbell, R. W. F., 1984, Delayed ventricular depolarization - correlation with ventricular activation, Europ. Heart J., 5 : 974.

Klötzner, J., and Otto, G., 1985, Digitales Messgerät zur Speicherung transienter Signale und zur Signalmittelung, Wiss. Beiträge der IH Zwickau, 11(3) : 53.

Lindsay, B. D., Ambos, H. D., Schechtman, K. B., and Cain, M. E., 1986, Improved selection of patients for programmed ventricular stimulation by frequency analysis of signal-averaged electrocardiograms, Circulation, 73 : 675.

Michaelis, J., 1985, Realisierung einer formalisierten diagnostischen Entscheidungsfindung am Beispiel der automatischen EKG-Analyse, in: "Verh. der Deutschen Gesell. für innere Medizin", K. Miehlke, ed., Bergmann Verlag, Munich : 389.

Oeff, M., Leitner, E.-R., Schwartz, W., and Schröder, R., 1986, Nichtinvasive Registrierung ventrikulärer Spätpotentiale bei Herzgesunden und bei Patienten mit intraventrikulären Reizleitungsstörungen, Z. Kardiol., 75 : 666.

Peper, A., 1984, Comments on "Multichannel signal processing based on logical averaging", IEEE Trans. Biomed. Eng., BME-31 : 483.

Rakshit, A., Bhattacharyya, S. N., and Choudhury, J. K., 1985, A microprocessor-based multipoint signal averager for repetitive biological
signals, Measurement, 3 : 169.

Ros, H. H., Koeleman, A. S. M., and v.d.Akker, T. J., 1981, The technique
of signal averaging and its practical applications in the separation
of atrial and His-Purkinje activity, in: "Signal Averaging Technique
in Clinical Cardiology", V. Hombach and H. H. Hilger, eds., F. K.
Schattauer Verlag, Stuttgart : 3.

Schwartz, A. B., and Scheinmann, M. M., 1983, Ventricular tachycardia:
newer insights, Clin. Cardiol., 6 : 307.

Shvartsman, V., Barnes, G. R., Shvartsman, L., and Flowers, N. C., 1982,
Multichannel signal processing based on logical averaging, IEEE Trans.
Biomed. Eng., BME-29 : 531.

Stopszyk, M. J., Kopec, J., Zochowski, R. J., and Pieniak, M., 1973, Surface
recording of electrical heart activity during the P-R-segment in man
by computer averaging, Int. Res. Comm. Syst., 83(11) : 21.

Stopczyk, M. J., Walczak, F., Kepski, R., Plucinski, Z., and Peczalski, K.,
1981, The history of noninvasive His-bundle recording, in: "Signal
Averaging Technique in Clinical Cardiology", V. Hombach and H. H.
Hilger, eds., F. K. Schattauer Verlag, Stuttgart.

Tiefenthaler, C., 1987, Weniger Rauschen durch Überabtastung, Elektronik,
7 : 121.

Tietze, G., 1975, Prinzip und Anwendung der Signal-Averaging-Technik,
Radio-Fernsehen-Elektronik, 24 : 485.

von Leitner, E. R., 1984, Prognostische Bedeutung nichtinvasiv registrier-
ter ventrikulärer Spätpotentiale, Herz , 9 : 26.

Wajszczuk, W. J., Palko, T., Przybylski, J., Bauld, T. J., and Rubenfire, M.,
1981, External recording of sinus node region activity in animals and
in man, in: "Signal Averaging Technique in Clinical Cardiology",
V. Hombach and H. H. Hilger, eds., F. K. Schattauer Verlag, Stuttgart.

Werner, G., Pries, W., and Nowack, R., 1986, "Registrierung kardialer Mikro-
potentiale von der Körperoberfläche", Research Report, MMA, Bad
Saarow.

Zywietz, C., 1985, Stabilität der Vermessung kleiner Biosignale in Abhängig-
keit von Abtastrate und Störpegel, BMT Ergänzungsband, 30 : 161.

ECG INTERPOLATION USING TAUT CUBIC SPLINES

E. Tkacz

Technical University of Silesia
Department of Medical Electronics
Gliwice, Poland

INTRODUCTION

Traditional interpolation methods for the analysis of sampled ECG
signals are inaccurate. Hence the application of interpolating cubic
splines is of great significance. However, this class of functions in
their ordinary form often does not give satisfactory results when used for
non-uniformly sampled ECG signals. In the case of the uniformly sampled
signal, its reconstruction using ordinary cubic splines is relatively simple.
However, the model of the ECG signal obtained in this way is far from
effective.

Let us restrict our consideration only to the non-uniformly sampled
signal. Before that, it is necessary to say a few words about this form
of sampling. Its main aim is the effective reduction of the number of
samples, that is, data compression such that the signal can be reconstructed
and the error of the procedure of this type satisfies the respective norms.

THE EXTRACTION OF SIGNIFICANT POINTS

In the first stage the following characteristic points are extracted
(detailed in Tkacz, 1987) from the uniformly sampled ECG signal:

- all local extrema of a signal were chosen as follows (see Fig. 1):
  - evaluation of the first difference

$$x(t_1) - x(t_0) = a \tag{1}$$

  - if $a > (<)$ 0 and $x(t_2) - x(t_1) <(>)$ 0, then the local maximum (minimum)
    occurs at t_1,
  - and the $a := x(t_2) - x(t_1)$, and so on.

With regard to the properties of the ECG signal, we can see that the
characteristic waves of an ECG alternate mathematically and we try to avoid
the choice of those points which are in the two succeeding samples of a
signal and in which lie extrema of the same type, such as, for example,
$x(t_2)$ and $x(t_3)$. Then, with regard to the comparison of the frequency of
an ECG signal with the sampling frequency, we reduced the extrema chosen
to include only important ones, that is, those which do not occur in

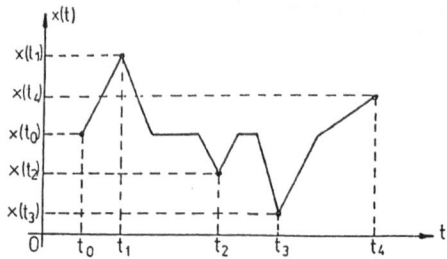

Fig. 1.   The choice of extrema.

two neighbouring samples.

- all local inflection points of a signal were chosen as follows (see
  Fig. 2):
  - points of inflection are found between each pair of extrema found
    earlier,
  - the inflection points are found using the method which is a modifica-
    tion of the Jensen inequality,
  - the point $(t_3, x(t_3))$ is a point of inflection if the following in-
    equalities are satisfied:

$$(x_s(t_4) \geqslant x(t_4)) \text{ and } (x_s(t_2) < x(t_2)) \tag{2}$$

or

$$(x_s(t_4) < x(t_4)) \text{ and } (x_s(t_2) \geqslant x(t_2)) \tag{3}$$

where $x(t)$ is the examined signal, and $x_s(t)$ is the chord which connects
the points $(t_1, x(t_1))$ and $(t_5, x(t_5))$

- the subscript is incremented, and so on.

Also, in this case the number of points of inflection is reduced to only
the important ones in the above-mentioned sense.

- all the points which begin and end the characteristic waves of the ECG.
  For this purpose we use the method described in Balda et al. (1977).
  This depends upon finding an auxiliary signal WBI (wave boundary indica-
  tor) as follows:

$$\text{WBI}(k) = C_1 |x'(k)| + C_2 |x''(k)| \tag{4}$$

with

$$x'(k) = x(k+1) - x(k-1) \tag{5}$$

$$x''(k) = x(k+2) - 2*x(k) + x(k-2) \tag{6}$$

This auxiliary signal has its greatest magnitude at the place where the
most dynamic change of an ECG signal is the fastest, so the interesting
points are chosen in those places where WBI changes from a small magnitude to a large one, and from a large magnitude to a small one.   This
method has the essential defect that it is sensitive to noise.

     In the second stage we try to reconstruct the non-uniformly sampled
ECG signal using ordinary cubic splines. Unfortunately, in cases of this
type two problems appear, namely the "lack of data" and the "extraneous"
points of inflection which result in the reconstructed signal differing
considerably from the initial ECG signal. Therefore, taut cubic splines
were used as the remedy.

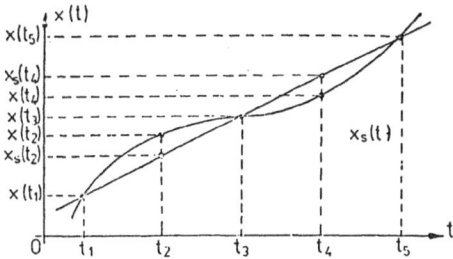

Fig. 2. The choice of the points of
inflection.

LACK OF DATA

This is an intuitive and imprecise term which is meant to describe the
following situation. A curve or function has been approximated by some
means, making use of the data provided. It fits the data, yet the user of
the approximation protests that it does not look right or that he would
have drawn it differently. In effect, the data provided were not sufficient
to yield the approximation procedure which is desired. Of course, such
user complaints are at times unreasonable and can be dealt with simply by
requesting more data in the area where the approximation was felt to be
wrong. However, in at least one situation, there is justification for the
complaint, and this is in the case of "extraneous" points of inflection.

"EXTRANEOUS" POINTS OF INFLECTION

We are given data $(\tau_i, g(\tau_i))_1^n$ to be interpolated. Let f be a smooth
interpolant to these data. We know that for each i there is a point η_i in
the open interval (τ_{i-1}, τ_{i+1}) such that

$$[\tau_{i-1}, \tau_i, \tau_{i+1}]g = f''(\eta_i)/2 \qquad (7)$$

therefore, corresponding to each sign change in the sequence $(\delta)_2^{n-1}$, with

$$\delta_i := [\tau_i, \tau_{i+1}]g - [\tau_i, \tau_{i-1}]g \qquad (8)$$

there must be a sign change in the second derivative of f, that is a point
of inflection of f. We call all other points of inflection extraneous.
To be more precise, we call a point of inflection of interpolant f in the
open interval (τ_i, τ_{i+1}) extraneous if $\delta_i \delta_{i+1} > 0$.

In effect, we insist that the interpolant preserve convexity/concavity
of the data in the following way: if the broken line interpolant to the
data is convex/concave in the interval (τ_{r-1}, τ_{s+1}), then a "good" interpolant should be convex/concave in the interval (τ_r, τ_s).

SPLINE IN TENSION

Schweikert (1966) was the first to deal with the problem of extraneous
points of inflection. He proposed the tensioned spline as a remedy. In
constructing this spline, he was guided by the physical notation of the
spline as an elastic band which goes through rings at the interpolation
points in the plane and which is pulled until all extraneous kinks have
been straightened out. In the limit, that is with a very strong pull, he
obtained the broken line interpolant. In mathematical terms, Schweikert's

195

tensioned interpolant has two continuous derivatives and satisfies, between interpolation points, the following differential equation

$$(D^2 - p^2)D^2 y = 0 \qquad (9)$$

with p as the tension parameter. For p = 0, this means that the interpolant is piecewise cubic, that is just the customary cubic spline interpolant. However, for p > 0, each piece of the interpolant is a linear combination of the four functions:

$$1, \ x, \ e^{px}, \ e^{-px}.$$

Spath (1974) has generalised this to allow the tension parameter p to vary from interval to interval, thus making the procedure more responsive to the local behaviour of the data. In fact, Spath deals with more general interpolants which, over the interval (τ_i, τ_{i+1}), have the following form:

$$f(x) = A_i u + B_i v + C_i \phi(u) + D_i \phi(v) \qquad (10)$$

with

$$u := u(x) := (x - \tau_i)/\Delta\tau_i, \quad v := 1 - u \qquad (11)$$

and ϕ_i a function on (0,1) with $\phi_i(0) = \phi_i(1) = 0$ and

$$\phi_i''(0)^2 \neq \phi_i''(1)^2. \qquad (12)$$

For $\phi_i(s) = s^3 - s$, with s being arbitrary, we obtain the cubic spline. The piecewise tensioned spline fits into this schema with the choice:

$$\phi_i(s) = \sinh(p_i s) - s*\sinh(p_i) \qquad (13)$$

with p_i the tension parameter for the ith interval.

A different scheme has been proposed by Soanes (1976), who uses an interpolant which, on the interval (τ_i, τ_{i+1}), has the following form:

$$f(x) = A_i u + B_i v + C_i u^{m_i} + D_i v^{n_i} \qquad (14)$$

with the possibly fractional exponents m_i, n_i to be chosen appropriately. The choice $m_i = n_i = 3$ reduces this scheme to a cubic spline interpolation. While the Schweikert-Spath tensioned spline has its parameters chosen from interval to interval, Soanes (1976) chooses the ratio m_{i-1}/n_i from data point to data point. The only real objection to these and other schemes is that they use exponential functions instead of polynomials, thereby making the evaluation of the interpolant more expensive and, on some microcomputers, even impracticable. The trick which makes it possible to adopt polynomials for constructing the interpolant is to use a cubic spline with additional knots placed so that the interpolant can make sharp bends where required without breaking out into oscillations as a consequence.

TAUT CUBIC SPLINE

There exist some ways for choosing the additional knots of approximation, but the most typical is to choose first the abscissa and then to experiment with the function value until the curve is satisfactory. Below we describe a more systematic way of introducing additional knots where required. We begin with the following special case. We are to interpolate the following data:

$$f(0) = 0, \quad f'(0) = s_0, \quad f(1) = 0, \quad f'(1) = s_1 \tag{15}$$

in the interval $(0,1)$ by a cubic polynomial.

The interpolant for this special case can be written in the following form:

$$f(x) = s_0 x - (2s_0 + s_1)x^2 + (s_0 + s_1)x^3 \tag{16}$$

By computing the divided difference of the function f we get:

$$[0,0,1]f = s_0 \quad , \quad [0,1,1]f = s_1 \tag{17}$$

hence, a point of inflection in $(0,1)$ would be extraneous for the case $s_0 s_1 < 0$.

On the other hand:

$$f''(0) = -2(2s_0 + s_1) \quad , \quad f''(1) = 2(s_0 + 2s_1) \tag{18}$$

Thus, f has no extraneous point of inflection in $(0,1)$ if, and only if,

$$(2s_0 + s_1)(s_0 + 2s_1) \leqslant 0 \tag{19}$$

We can write this condition also as:

$$(2(s_0 - s_1) + 3s_1)(s_0 - s_1 + 3s_1) \leqslant 0 \tag{20}$$

or, dividing through by $(s_0 - s_1)^2$ and using the substitution $z := s_1/(s_0 - s_1)$, as:

$$(2 - 3z)(1 - 3z) \leqslant 0. \tag{21}$$

It follows that the cubic interpolant in this case reproduces the convexity/concavity of the data only if

$$1/3 < z := s_1/(s_1 - s_0) \leqslant 2/3 \tag{22}$$

For values of z outside this range, we propose to change the interpolating function as follows. Consider first the case $z > 2/3$. Then the interpolant is required to bend near 1 more sharply than a monotonic cubic polynomial can. We therefore replace that basic function among the four, x, $1 - x$, x^3, $(1 - x)^3$, which bend more sharply near 1, by a cubic spline ϕ with one knot which bends more sharply near 1 than does x^3. For the sake of uniformity, we also insist that $\phi(0) = \phi'(0) = \phi''(0) = 0$, $\phi(1) = 1$, and $\phi'' > 0$ on $(0,1)$. This ensures, amongst other things, that the interpolant has a point of inflection in $(0,1)$ if, and only if, the coefficients of ϕ and $(1 - x)^3$ are of opposite sign, as we will see in a moment. The resulting function ϕ is given by

$$\phi(x) := \phi(x;z) := \alpha x^3 + (1 - \alpha)\left(\frac{x - \zeta}{1 - \zeta}\right)^3 \tag{23}$$

with $\zeta = \zeta(z)$ the additional knot and $\alpha = \alpha(z)$ a number in $(0,1)$. We take both ζ and α to be continuous monotonic functions of z, with $\alpha(2/3) = 1$ in order to connect continuously with the usual cubic polynomial when $z \leqslant 2/3$, and with $\zeta(1^-) = 1$. In terms of this new basis, the interpolant can be written as follows:

$$f(x) = Ax + B(1 - x) + C\phi(x;z) + D(1 - x)^3 \tag{24}$$

197

where

$$-A = C = (s_0 + 2s_1)/(3(2p - 1)) \tag{25}$$

$$-B = D = -((3p - 1)s_0 + s_1)/(3(2p - 1)) \tag{26}$$

and

$$p := \phi'(1;z)/3 = \alpha + (1 - \alpha)/(1 - \zeta). \tag{27}$$

It follows that

$$f''(x) = 6(C(\alpha x + (1 - \alpha)(x - \zeta)_+/(1 - \zeta)^3) + D(1 - x)) \tag{28}$$

This is of one sign in (0,1) if, and only if, C and D are of the same sign. The condition that f has no extraneous point of inflection in (0,1) (in case $s_0 s_1 < 0$) thus becomes:

$$((3p - 1)s_0 + s_1)(s_0 + 2s_1) \leqslant 0 \tag{29}$$

Slightly re-arranging, dividing through by $(s_0 - s_1)^2$, and using again $z := s_1/(s_1 - s_0)$, produces the equivalent condition that

$$((3p - 1) - 3pz)(1 - 3z) \leqslant 0 \tag{30}$$

Preservation of convexity/concavity therefore demands that

$$1/3 \leqslant z \leqslant (3p - 1)/(3p), \tag{31}$$

that is

$$3p \geqslant 1/(1 - z) \tag{32}$$

In terms of α, this condition can be written as follows:

$$\alpha \leqslant (1 - (1 - \zeta)/(3(1 - z)))/\zeta \tag{33}$$

Thus, choosing α to be as large as possible, we obtain the formula:

$$\alpha(z) = (1 - (1 - \zeta)/(3(1 - z)))/\zeta \tag{34}$$

Note that $\alpha(2/3) = 1$ regardless of ζ. This also shows that we need

$$3(1 - z) \geqslant 1 - \zeta \tag{35}$$

to make certain that $\alpha \geqslant 0$. This condition is satisfied for the choice $1 - \zeta(z) = \gamma(1 - z)$ as long as $\gamma \leqslant 3$.

The case $z < 1/3$ is handled in an analogous manner, with the basic function $(1 - x)^3$ replaced by $\phi(1 - x; 1 - z)$ and ϕ determined as in the case $z > 2/3$. We can combine these various cases conveniently into one by considering the approximation to be of the form:

$$f(x) = Ax + B(1 - x) + C\phi(x;z) + D\phi(1 - x; 1 - z) \tag{36}$$

with

$$\phi(x;z) := \alpha x^3 + (1 - \alpha) \left(\frac{x - \zeta}{1 - \zeta}\right)_+^3 \tag{37}$$

$$\alpha(z) := (1 - \gamma/3)/\zeta \tag{38}$$

$$\zeta(z) := 1 - \gamma \min\{1 - z, 1/3\} \tag{39}$$

for some $\gamma(0,3)$. As γ increases, both α and ζ decrease for fixed z.

198

The more general case of interpolation to data $f(0) = f_0$, $f'(0) = s_0$, $f(1) = f_1$, $f'(1) = s_1$ is reduced to the present one by substracting first the straight line $f_0 + (f_1 - f_0)x$. This subtraction does not affect questions of convexity and concavity, and changes the quantities s_0 and s_1 used above into $s_0 - (f_1 - f_0)$ and $s_1 - (f_1 - f_0)$. If, even more generally, interpolation to data $f(\tau_j) = f_j$, $f'(\tau_j) = s_j$, $j = i$, $i+1$ is considered, then, instead of s_0 and s_1, the relevant quantities are

$$s_i - [\tau_i, \tau_{i+1}]f \quad \text{and} \quad s_{i+1} - [\tau_i, \tau_{i+1}]f.$$

We make use of these considerations in the construction of a "taut" cubic spline interpolant to given data $(\tau_i, g(\tau_i))_1^N$ by choosing the basis in each interval (τ_i, τ_{i+1}) in the above manner, with the role of s_i played by slope $[\tau_{i-1}, \tau_i]f$ and that of s_{i+1} played by $[\tau_{i+1}, \tau_{i+2}]f$. This means that the interpolant has the following form:

$$A_i + B_i u + C_i \phi(u;z) + D_i \phi(i-u; 1-z) \tag{40}$$

in (τ_i, τ_{i+1}) with

$$u(x) := (x - \tau_i)/\Delta\tau_i \tag{41}$$

$$z := \begin{cases} \delta_{i+1}/(\delta_i + \delta_{i+1}) & \text{if } \delta_i \delta_{i+1} \geqslant 0 \quad \delta_i + \delta_{i+1} = 0 \\ 1/2 & \text{otherwise} \end{cases} \tag{42}$$

where

$$\delta_j := [\tau_j, \tau_{j+1}]f - [\tau_j, \tau_{j-1}]f \quad \text{for all } j \tag{43}$$

as before. We set $z = 1/2$ in case $i = 1$ or $i = N-1$.

The particular choice (42) of z makes the interpolant discontinuous as a function of the data. This means that a small change in the data which changes the sign of some $\delta_i \delta_{i+1}$ may change the interpolant drastically. For this reason we use a modified version of (42), namely:

$$z := \begin{cases} |\delta_{i+1}|/(|\delta_i| + |\delta_{i+1}|) & \text{if} \quad |\delta_i| + |\delta_{i+1}| > 0 \\ 1/2 & \text{otherwise} \end{cases} \tag{44}$$

which removes an obvious source of discontinuity.

In terms of the quantities f_j, f_j'', $j = i, i+1$ and the number $h := \Delta\tau_i$, the interpolant has the coefficients:

$$A_i = f_i - D_i \tag{45}$$

$$B_i = h*[\tau_i, \tau_{i+1}]f - (C_i - D_i) \tag{46}$$

$$C_i = h^2 * f_{i+1}''/\phi''(1;z) \tag{47}$$

$$D_i = h^2 * f_i''/\phi''(1;1-z) \tag{48}$$

Note that

$$\phi''(1;z) = 6(\alpha + (1 - \alpha)/(1 - \zeta)^2) \tag{49}$$

hence

$$1/\phi''(1;z) = (1/6) * (1 - \zeta)^2/(\alpha(1 - \zeta)^2 + (1 - \alpha)) \tag{50}$$

We determine the vector $(f_i'')_1^N$ so that the resulting interpolant has two continuous derivatives. This gives, at each interior data point τ_i, the equation:

$$f'(\tau_i^-) = f'(\tau_i^+) \tag{51}$$

or

$$(\Delta\tau_{i-1}/\phi''(1;1-z))f_{i-1}'' + [\Delta\tau_{i-1}(\phi'(1;z_{i-1}) - 1)/\phi''(1;z_{i-1}) +$$
$$+ ((\phi'(1;1-z_i) - 1)/\phi''(1;1-z_i))\Delta\tau_i] + \Delta\tau_i f_{i+1}''/\phi''(1;z_i) = \delta_i,$$

$$i = 2, \ldots, N-1 \tag{52}$$

Note that

$$(\phi'(1;z) - 1)/\phi''(1;z) = ((1-\zeta)/6)*((3\alpha-1)(1-\zeta)$$
$$+ 3(1-\alpha))/(\alpha(1-\zeta)^2 + (1-\alpha)) \tag{53}$$

The evaluation of the coefficients in (52) in the manner of (50) and (53) ensures accuracy as $\zeta \to 1$.

If no additional knots are introduced, that is if $\alpha = 1$ throughout, than (50) and (53) show that (52) reduces to

$$(\Delta\tau_{i-1}/6)f_{i-1}'' + (\Delta\tau_{i-1}/3 + \Delta\tau_i/3)f_i'' + (\Delta\tau_i/6)f_{i+1}'' = \delta_i \tag{54}$$

the equation which describes the ordinary cubic spline interpolation. We need two more equations to obtain N equations in the N unknowns $(f_i'')_1^N$. Of the several possible choices of boundary conditions we consider here only the "not-a-knot" condition. This means that we adjoin the two equations

$$jump_{\tau_2} f''' = 0 = jump_{\tau_{N-1}} f''' \tag{55}$$

More explicitly, from (40) and (45 - 48) the first of these can be written in the following form:

$$(f_2'' - f_1'')/\Delta\tau_1 = ((\phi'''(0;z_2)/\phi''(1;z_2))f_3''$$
$$- (\phi'''(1;1-z_2)/\phi''(1;1-z_2))f_2'')/\Delta\tau_2 \tag{56}$$

and the second looks similar. Note that

$$\phi''(1;z)/\phi'''(1;z) = (1-\zeta)*(\alpha(1-\zeta)^2 + (1-\alpha))/(\alpha(1-\zeta)^3 + (1-\alpha)) \tag{57}$$

so that the equation is in computational difficulty if $z = 0$. For this reason, we write the boundary equation in the form:

$$(\phi''(1;1-z_2)/\phi'''(1;1-z_2))\Delta\tau_2(f_2'' - f_1'')$$
$$= ((\phi''(1;1-z_2)/\phi'''(1;1-z_2))*(\phi'''(0;z_2)/\phi''(1;z_2))*f_3'' - f_2'')\Delta\tau_1 \tag{58}$$

with an analogous version for the equation $jump_{\tau_{N-1}} f''' = 0$.

Finally, a detailed analysis shows that these equations should be the second, and the last but one, in the total system in order to avoid the necessity of pivoting.

CONCLUSION

Results, which are obtained using taut cubic spline interpolation (see Fig. 3), make it possible to create the form of database of an ECG

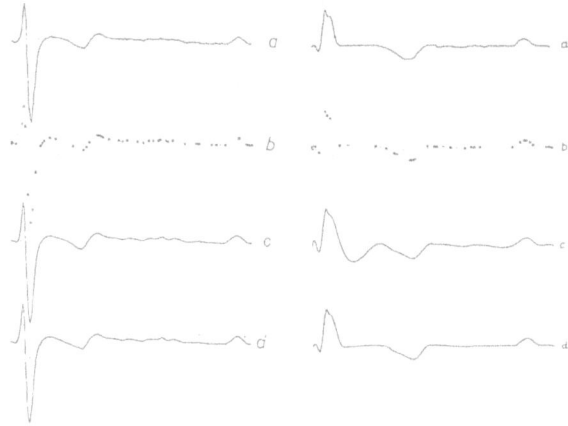

Fig. 3. (a) Initial morphology of ECG signal;
(b) Significant point extracted from
signal; (c) Ordinary cubic spline
interpolant; (d) Taut cubic spline
interpolant.

signal in which each signal is represented by a network of its characteristic points, that is in a reduced form. In this way, we also obtain an algorithm for effective data compression. The two examples considered above illustrate this fact. The compression ratio is in the worst case 5:2.

REFERENCES

Balda, R. A., Diller, G., Deardorff, E., and Doue, J., 1977, The HP ECG
    analysis program, in: "Trends in Computer Processed Electrocardio-
    grams", North Holland, Amsterdam.
Schweikert, D. G., 1966, An interpolation curve using a spline in tension,
    J. Math. Phys., 45 : 312.
Soanes, Jr., R. W., 1976, VP - splines, an extension of twice differen-
    tiable interpolation, in: "Numerical Analysis and Computer Conf.",
    Triangle Park, NC : 141.
Spath, H., 1974, "Spline Algorithms for Curves and Surfaces", Utilitas
    Mathematical Publishers, Winnipeg, Manitoba : 301.
Tkacz, E., 1987, "The Modelling of the ECG Signal Using Cubic Spline Inter-
    polation", PhD Dissertation, Technical University of Silesia.

A QUANTITATIVE ESTIMATION OF PROXIMITY BETWEEN ECG WAVEFORMS

C. Levkov

Institute for Biomedical Instrumentation
Medical Academy
Sofia, Bulgaria

INTRODUCTION

Two quantities are used most often to compare the proximity (or similarity) of digitised ECG waveforms: the correlation coefficient r and the
normalised mean-square distance D. This latter quantity will be referred
to as "distance". The formulae for r and D are shown in Fig. 1. The same
figure shows two ECG waveforms, A(i), B(i), and the values r and D calculated for them. Further on we shall assume that the waveforms we compare
are aligned so that there is no time shift between them. That means that
the corresponding points i from A(i) and B(i) belong to the same phase of
the cardiac cycle.

ANALYSIS

For identical waveforms r = 1 and D = 0. The quantities r and D are
sensitive to different features of ECG waveforms. The correlation coefficient r is most sensitive to differences in the shape of the curves. For
two waveforms with the same shape r is always 1 instead of their amplitude
differences. D is sensitive to the amplitude and to the shape differences.

An important note - the waveforms to be compared must be without offset related to some reference value, otherwise meaningless results will be
obtained. They will be an error if we assume the reference value to be the
mean value of all samples in the waveform (as r is usually calculated in
statistics). The offset must be calculated towards a point which has some

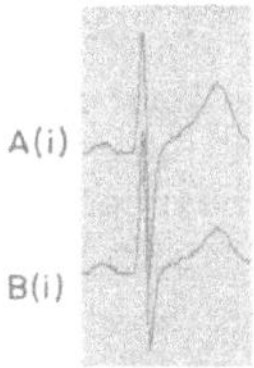

$$r = \frac{\sum_i A_i B_i}{(\sum A_i^2 \, \sum B_i^2)^{\frac{1}{2}}} \cdot 0{,}87$$

$$D_{AB} = \left(\frac{\sum (A_i - B_i)^2}{\sum A_i^2} \right)^{\frac{1}{2}} \cdot 0{,}44$$

Fig. 1.   Two ECG waveforms A(i) and
          B(i) together with their values
          of correlation coefficient, r,
          and normalised mean-square distance, D.

203

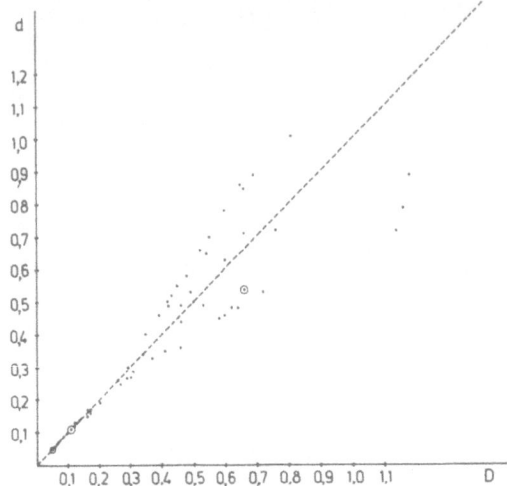

Fig. 2. Correlation between "invariant
distance", d, and normalised mean-
square distance, D.

physical meaning. In our case the reference value is the isoelectric level
of the ECG - the point just preceding the beginning of the QRS complex.

Actually, when using these quantities, some shortcomings are met and
we shall discuss them and try to find a way of obtaining some better solutions.

(1) The D quantity is not invariant to the order in which the waveforms are compared - $D(AB) \neq D(BA)$. Intuitively one feels that a distance
quantity must be invariant. Accordingly, a new quantity, denoted d, is
introduced which will be called "invariant distance":

$$
d = \left(2 \frac{\sum_i (A_i - B_i)^2}{\sum_i A_i^2 + \sum_i B_i^2} \right)^{1/2} \tag{1}
$$

The idea is obvious - the normalising factor is the sum of areas of
both the waveforms. Fig. 2 shows the correlation between d and D. Notice
the marked dispersion when the differences between waveforms increases.
The 45° line is drawn (also on the following figures) to show the trend.
All the data presented in the following figures are computed from a real
ECG database. There is no drift, and 50 Hz interference is removed digitally from the records. The proximity quantities are computed only in the
region of the QRS interval.

(2) To compute proximity quantities one has to perform multiplications. These multiplications are the slowest part of the program and for
real time applications another approach can be used. By analogy with D
and d new quantities DA and dA are suggested where absolute value calculations are used:

$$
DA = \frac{\sum_i |A_i - B_i|}{\sum_i |A_i|} \tag{2}
$$

$$dA = 2 \frac{\sum_i |A_i - B_i|}{\sum_i |A_i| + \sum_i |B_i|} \tag{3}$$

The correlation between mean-square distances and absolute value distances is shown in Fig. 3 and Fig. 4. A regression line can be drawn, but such fitting is not meaningful. The important thing is that a strong correlation exists and we can use absolute value quantities as a fast procedure for measuring the proximity between ECG waveforms.

Unfortunately for the correlation coefficient r an adequately simple computational formula for absolute values has not been found and, indeed, may not even exist.

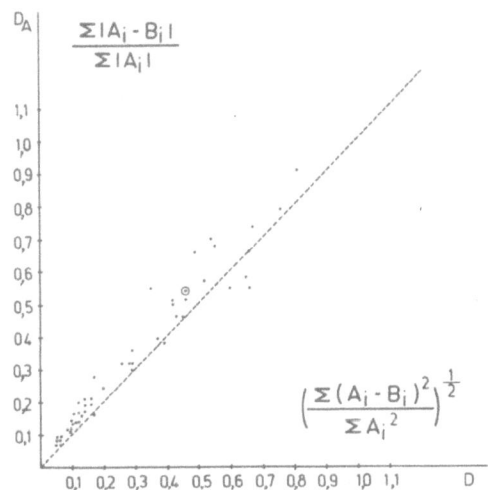

Fig. 3.   Correlation between DA and D.

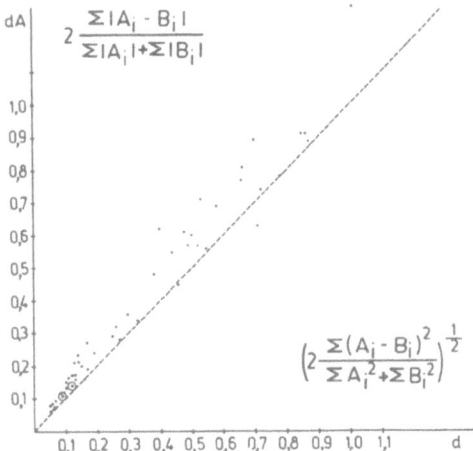

Fig. 4.   Correlation between dA and d.

The invariant distance d can be generalised for multidimensional space:

$$d = \left(2 \frac{\sum_i (\vec{A_i} - \vec{B_i})^2}{\sum_i (\vec{A_i})^2 + \sum_i (\vec{B_i})^2} \right)^{1/2} \qquad (4)$$

$\vec{A}(i)$ and $\vec{B}(i)$ are the vectors from the zero level point to the i-point.

CONCLUSIONS

The correlation coefficient r and the invariant mean-square distance d are simple and suitable quantities to measure the proximity between ECG waveforms, and for most cases of ECG waveforms pattern recognition in r-d space is possible.

The absolute value distance dA is closely related to the corresponding mean-square distance. This quantity can be used as a very fast criterion for ECG wave recognition using a template technique. The minimal shift between ECG waveform and the template can be obtained by the procedure that minimises dA. This minimal value of dA will show whether the particular waveform belongs to the template class of waveforms. The computation time for dA is 5 - 500 times shorter (depending on the MPU type) than the computation time of the correlation coefficient r which is most widely used in template recognition methods.

R@-SYSTEM - THE SOFTWARE SYSTEM FOR REAL BIOMEDICAL DATA ACQUISITION AND

PROCESSING WITH REGARD TO CLINIC AND RESEARCH

P. Heřman

Computing Centre
Faculty Hospital in Motol
Prague, Czechoslovakia

INTRODUCTION

A number of technical problems arise in the acquisition and processing of real biomedical data (biosignals - such as EEG, EMG, ECG, some evoked potentials, and so on). In general, physicians would rather buy standard equipment for their clinical examinations. However, by so doing they are totally dependent on the market supply, because when installed the equipment is usually hard-wired and not changeable by the user. Therefor problems may arise when they wish to update their equipment and experiment with some new non-standard method. In such cases scientists would rather construct their own equipment. In practice, they often buy some minicomputer or personal computer with a graphical display, equip it with a plotter and A/D converters and connect it to some EEG, EMG or ECG apparatus. Then they write software for their own methods of examination themselves. By operating in this manner, writing their particular routines takes a great deal of time. The work may be difficult and the results are non-standard, and usually non-compatible with other systems, such that they are hardly usable elsewhere.

These and other reasons have led to the development of the R@-SYSTEM which can be regarded as:

   (i)   a supplement to the host operating system providing control of A/D and D/A conversions, appropriate real-data file handling, displaying, special task execution, time sharing, and so on;

  (ii)   a utility package containing a set of standard and special algorithms for biosignal acquisition and processing;

 (iii)   an environment for the development and testing of new algorithms;

  (iv)   a set of standardised methods facilitating a comparison between the results of different laboratories;

   (v)   as a data format standardisation, providing easy transmission of real biomedical data which have been acquired;

  (vi)   as a program format standardisation, providing easy exchange and installation of developed programs;

 (vii)   a supplement for software interface between the biosignal acquisition and processing domain and a hospital information system (network); and

(viii)   in totality the R@ represents an attempt at developing a system which could bring research and clinical work together.

The original motivation for development of the R@-SYSTEM arose from the demand for some software for a computer used in neurophysiological examinations. A number of similar problems had been solved using various mini- and microcomputers. The computers were: first a LINK-8 (DEC, 12-bit mini, 6 KB core RWM), then JPR-12 (Tesla, 12-bit mini, 6 KB core RWM), HP-30, LSI-11 (Heath, like a PDP-11/03, 16-bit LSI, 48 KB RWM), PDP-11/34 (DEC, 16-bit mini, 128 KB RWM), VT-20A (Videoton, 8-bit micro, Z80 CPU, 64 KB RWM) and JPR-1 (Tesla, 8-bit micro, 8080 CPU, 48 KB RWM). During this work the idea developed to adopt a more general approach. Systems used for research as well as those used in the clinic can be based on the same principles. It is therefore better to produce one unified system with a part which is hardware-independent. This concept can also lead to the solution of two complementary problems: how to provide sufficient software support for clinical laboratories for routine examinations, as well as for scientific work, and how to provide sufficient data support for research laboratories.

The first version 1.0 of the R@-SYSTEM was developed to work on any 8-bit machine with the CP/M operating system. The particular configuration adopted at that time was an MZ-3541 (Sharp, 8-bit business microcomputer with Z80 CPU 4 MHz, 64 KB oper. RWM + 176 KB RAM-disk, two $5\frac{1}{4}$" and two 8" floppy disk units, 400 * 640 pixel graphical display, line printer, home-made 12-bit A/D converter with up to 16 multiplexed channels, EOS - with CP/M 2.2+ compatible operating system). The C-language was used for its good portability and only small parts of some programs were written in macroassembler.

Example

The following is an example of a problem that has been solved with the aid of R@-SYSTEM version 1.0. The human EEG is considered, using the approach of the theory of deterministic chaos, to be a chaotic process with some fractal dimension. The R@-SYSTEM also includes a program that allows estimation of its so-called "correlation dimension" by computing a number of "correlation integrals". (This problem is considered in detail by Dvořák and Šiška (1986) and Dvořák et al. (1986)). Only some important characteristics of the program are shown here with an example of the methods of data acquisition and processing.

The EEG signal is registered using normal 8-channel EEG apparatus (Bioscript BST 2100) during a standard EEG examination of a child and is stored on an FM recorder. Then all 8 (max. 16) channels are digitised at 204.8 Hz and with 8-bit (max. 12-bit) precision and by simultaneous moni-toring placed on RAM-disk. The maximum length of the file is 160 KB and it can hold up to 100 seconds of the 8-channel EEG record. There is an option for the file to be stored on the floppy disk. Any channel and any 80-second length section of the record can be selected for analysis and placed into 16 KB space in RWM. Then the delay can be chosen to construct multidimensional vectors, the maximal dimension of which will be computed (max. 50), together with the number of pairs of vectors for which the dis-tances will be computed. (Not all possible pairs are examined due to the availability of machine time. Usually about 100 vectors are randomly selected forming pairs with all the remaining ones.) The distances are sorted into 256 epsilon-classes for each dimension. The n   er of v   or pairs with their mutual distance less than some value of   re. sents the "correlation integral" C(eps). From the dependence   upon log(eps) the "correlation dimension" CD is estimated for   d dimension (dim). Finally, from the dependence of CD (dim) up(ie fractal dimension of the process being examined can be estim

The calculation of all correlation integrals takes less than 15 minutes (for max. dimension = 20 and for one million vector pairs). This program is relatively simple (about 200 source lines written in 'C' and less than 100 lines for the macroassembler) and its development took two weeks of work, thanks to support from the R@-SYSTEM.

A second version 2.0 of the R@-SYSTEM has now been developed on the same principles as the first one. This uses an IBM-AT personal computer (640 KB RWM + 2 MB Magic Card, one $5\frac{1}{4}$" high-capacity 1.2 MB drive, two 30 MB Winchester disks, EGA, one dot-matrix printer, one wheel-printer, a six-colour Hewlett-Packard plotter, a mouse, four cards with four 12-bit A/D and two D/A converters each, MS-DOS 3.2) connected on-line with the 16-channel ink jet Siemens-Elema EEG apparatus.

R@ PHILOSOPHY

When a personal computer is being used for text-processing, the most common and important format for a data file is a text file. The entry of numeric data via a keyboard and output via the display or printer are also text-files. The commands are text-strings too. Consequently, the operating system is primarily oriented to process these files: it can create them, modify (edit), display, print, join, search, sort, and so on.

Now, if we want to connect our personal computer with any real process, the situation will be more complicated. The operating system itself is not sufficient for reasonable handling of the typical quantities of data with some desirable format that result. In such cases the system must be supplemented. The R@-SYSTEM provides that supplement on a number of different levels:

(i) The BIOS (Basic Input/Output System) is supplemented to operate with the additional I/O cards, A/D - D/A converters and so on in any standard way. (In the 2.0 version of R@-SYSTEM this is accomplished naturally by means of "installable device drivers".)

(ii) The BDOS (Basic Disk Operating System) is supplemented to provide some more complicated operations with real data sequences.

(iii) For writing a program in any high-level language we need some library of procedures and functions to operate with those I/O devices and to handle the quantities of data obtained. So the 'C'-language library is first prepared for the R@-SYSTEM.

(iv) Extension of the console command processor: It normally performs so-called internal commands. Just as the TYPE command can display the contents of any text-file, the DISPLAY command can display the contents of any real-data file, for example EEG signals. It provides the option to display it on the screen or plot or print it with a dot-matrix printer or return it via D/A converters to be painted by the EEG-apparatus.

(v) The utilities are programs which enable relatively complicated operations to be performed with any amount of data in a simple manner - by external commands like convenient CP/M or MS-DOS commands. There are two kinds of these commands: (a) input/output commands, or (b) filters. For example:
(a) Just as some text-editor can be used for any text-data acquisition, the SAMPLE program can acquire the real data from A/D converters.
(b) On the one hand there can be relatively simple operations with real-data files such as to cut or join them or to select any

parts of them, whilst, on the other, there can be relatively complicated transformations - such as the Fourier transform.

(vi) Batch-processing commands: they are one of the most powerful and useful features of the operating system. With R@-batch-processing files the work of preparing some original method of electrophysiological examination comes to the top. Using reliable programs any experimental process can be arranged and saved into that file.

R@ Files

The R@-SYSTEM can handle various kinds of files, including:
text-files (for example, source-text files, batch-command files, text-data files),
compiled files (for example, programs in machine code or any intermediate product),
real-data files (such as saved time-series of sampled biosignals and/or their transformations).
The first two are well-known types of file which every operating system can handle. The real-data files are more interesting. For versatility there are three forms of such files:

(i) Binary form: The data are stored as twos complement digits. These files contain maximal volume of information with minimal storage requirement. However, such files perhaps cannot be passed through some communication channel or device which is designed for character transmission only, for example 7-bit USART.

(ii) Encoded form: A conversion is made so that every original sampled value is broken down into 6-bit digits and these are encoded using ASCII printable characters. However, some alien programs may not understand it.

(iii) Explicit form: The real data are represented as a sequence of decimal numbers formatted into rows and columns. This form is less economical but the most universal one.

By the use of special programs, one form can be converted into another, and vice-versa. The last two forms are used when communicating with any auxiliary device or other system.

File Headers

No matter whether it is a real-data file, source-text file or compiled binary file, every R@-file has to have a "label" to identify it. Even if the data are stored in binary form the R@-file contains an ASCII header which bears useful information concerning the data which follow. When some other program modifies the data contained in that file, it will also mark the change into the header. By this means every file carries notice of its whole history on itself.

For example (simplified): When a program SAMPLE samples and stores the EEG signals from any patient, it will also ask for his name, date of birth, diagnosis and so on, and it will store this information into the file-header. It will also place there its own name ("SAMPLE v.2.0"), the current time and date, data format used and other parameters of the measurement; and every further program that will modify the data stored in that file must also append its relevant message to the header.

This feature is very substantial and useful for the R@-SYSTEM, chiefly for the following reasons:

(i) identification of files (data contents description);
 (ii) proper processing of files (data format description);
(iii) documentation;
 (iv) data base creation;
  (v) data communications.

Time Sharing

Much time may be consumed in performing calculations with large amounts of data. In most cases these computations run automatically without the need of any human intervention. On the other hand, much machine time is wasted by waiting for a key depression or for some device which is not ready. For this reason it is very important for the R@-SYSTEM that it allows the running of two dependent or independent tasks simultaneously. For example, in the foreground a nurse can be inputting the personal data of a patient using a pad, while the processing of his electrophysiological data continues in the background.

Standard Commands

The most usable commands contained in the R@-SYSTEM are described briefly below:

Input/output:

  (i) SAMPLE provides sampling of real data from some number of A/D converters. It can also accept some personal data from the keyboard.
 (ii) FORM sets the parameters which control the sampling and file forming processes.
(iii) DISPLAY displays the sampled and also the transformed data. It can also plot them or send them to the D/A converters back to the EEG apparatus to be plotted on it or to be recorded, and it can make a hard copy on the dot-matrix printer too.
 (iv) MARK makes possible some interventions into real-data files - for example, remarks, changes, selections.

Filters:

  (v) NOTCH suppresses the mains frequency contained in the sampled data
 (vi) FTR Fourier transform
(vii) WTR Walsh transform
(viii) GTR Gabor transform (Gabor, 1946; Bastiaans, 1980)
 (ix) OTR general orthogonal transform
  (x) ACOR autocorrelation function
 (xi) POW power spectra

and others.

NON-LINEAR ANALYSIS OF EVOKED POTENTIALS

Evoked potentials (EP) are defined as the sequence of voltage changes generated in the brain and sense organs and in pathways leading to the brain, evoked by a transient stimulus of any sensory modality. The most widely known EP are (according to the kind of stimuli):

    VEP: Visual EP
          ERG: Electroretinography

    AEP: Acoustic EP
          LLP: Long Latency Potentials (AEP components occurring
                between 60 and 300 ms)

Table 1. Use of Evoked Potential Examination in Clinical Practice

	VEP	ERG	LLP	MLP	BAEP	ECoG	SEP	SSEP	ERP
Neurology	x	x		x	x		x	x	
Psychology and Psychiatry			x						x
Otolaryngology				x	x	x			
Ophthalmology	x	x							
Surgery	x				x		x	x	
Paediatrics	x	x			x				x

```
    MLP:  Middle Latency Potentials (8 ... 60 ms)
    BAEP: Brainstem AEP (under 8 ms)
    ECoG: Electrocochleography

  SEP:  Somatosensory EP
  SSEP: Spinal somatosensory EP

  ERP:  Event-Related Potentials (evoked by some mental activity)
```

Examination of these EP is widely adopted in current clinical practice.
Its most common use in different departments is as listed in Table 1.
(Only the most typical examples are mentioned here, but their applicability
in medicine is indeed wider.)

From a cybernetic point of view the problem of some EP examination is
equivalent to the problem of system identification which is one of finding
the functional relationship that determines the output of a system (some
control plant or "black-box") in response to any relevant input. A variety
of well-known techniques of the system identification have been used extensively both in industrial applications as well as in the study of some
biological systems.

To express the problem more rigorously, the identification of a system
S is the task of associating S with a member of a class of systems to which,
on the basis of a priori information, S is known to belong. The result is
the derivation of a model that allows the prediction of the system output
for a given input from a wide class of input functions. So the primary
question is how to determine the adequate class of systems to be considered.
In most cases the solution is more arbitrary and dependent on our technical
equipment rather than on our theoretical prepositions. For instance, most
of the methods of clinical EP examination are based on a linear model, even
though the substance of life itself is non-linear.

So the most widespread method of determining some EP is averaging.
From the theoretical viewpoint, the dependence of the output y(t) of any
linear dynamic system on its input x(t) is represented by the convolution
integral

$$y(t) = \int_0^\infty h(\tau)\, x(t - \tau)\, d\tau \tag{1}$$

where $h(\tau)$ is the so-called transient function which determines the behaviour of that system. Because this function represents the system's response to a Dirac impulse (the integral (1) reduces the identity $y(t) = h(t)$
if $x(t)$ is the Dirac function), it could be determined in this way. However, in practice there is some accompanying noise on the output of the
system, noise which can be substantially greater than the true output

signal itself. That noise can be reduced considerably by signal averaging which consists of collecting many repetitions of the transient signal and adding them together precisely in phase; out of phase components such as randomly occurring noise will thus be reduced. The vast majority of methods of clinical examination of EP are based on that principle. Using this linear method, however, it is, of course, not possible to discover many of the interesting and essential non-linearities.

During the last decade a variety of methods of non-linear system identification have become established, but usually only in a research context. Some of the non-linear models are based on the natural expansion of the linear functional (1), called the Volterra functional series:

$$y(t) = \int_0^\infty h_1(\tau) \, x(t - \tau) \, d\tau$$

$$+ \int_0^\infty \int_0^\infty h_2(\tau_1, \tau_2) \, x(t - \tau_1) \, x(t - \tau_2) d\tau_1 \, d\tau_2 + \ldots \tag{2}$$

Only a few terms are considered in practice rather than the infinite series given in (2). Many methods consider only the first two terms. However, even this approximation is substantially nearer to the non-linear reality than the linear one.

As in (1), $h_1(\tau)$, $h_2(\tau_1, \tau_2)$, ... are the functions characterising the non-linear system. They are called kernels. However, problems can arise trying to compute these kernels from experimental data because of non-orthogonality of the base system of terms in (2). So the higher terms can affect the estimation of the lower ones. For this reason Wiener (1958) showed that the relationship (2) can be transcribed as:

$$y(t) = \sum_{n=0}^\infty G_n[h_n, x(t)] \tag{3}$$

where $\{G_n\}$ is a complete set of orthogonal functionals with respect to Gaussian white-noise on the input. The functionals of the 1st and 2nd order are:

$$G_1[h_1, x(t)] = \int_0^\infty h_1(\tau) \, x(t - \tau) d\tau$$

$$G_2[h_2, x(t)] = \int_0^\infty \int_0^\infty h_2(\tau_1, \tau_2) \, x(t - \tau_1) \, x(t - \tau_2) d\tau_1 \, d\tau_2$$

$$- P \int_0^\infty h_2(\tau_1, \tau_1) d\tau_1 \tag{4}$$

where $\{h_i\}$ is the set of Wiener kernels that has to be identified, assuming that the power density spectrum of white noise x(t) is $\Phi_{xx}(f) = P$.

Lee and Shetzen (1965) used the orthogonality of (4) to show that the kernels can be estimated by input-output cross-correlations. This discovery allowed simple computation of Wiener kernels and so the era of practical adoption of non-linear methods of system identification in biology began (Marmarelis, 1972; Marmarelis and McCann, 1973; Marmarelis and Naka, 1972; 1973; and Seelen and Hoffmann, 1975). French and Butz (1973) employed the fast Fourier transform algorithm in the most efficient measurement of the kernels, and later (French and Butz, 1974) they increased its efficiency even more using Walsh functions with which the kernels were expanded. Kadri (1972) showed that the cross-correlation functions could be calculated using multidimensional Laplace transforms for any input (for example, a pseudo-random binary sequence). Krausz (1975) published a method analogous to that of Wiener but using a random impulse train on the inputs instead of white noise. This method is particularly suited to EP examination because of the discrete character of the stimuli used. Marmarelis and Naka (1974) extended the Wiener theory to multi-input systems. The special beauty of non-linear theory in this case arises from the fact that it allows the mutual interactions between signals going through different pathways to be studied, something that linear theory does not allow. Moore et al. (1977) and Herman (1978) showed that epileptical spikes could be evoked by a rat by simultaneous acoustic and visual stimulation at a certain rate. Further examples can be found in French (1976), Koblasz (1978), Koblasz et al. (1980), Larkin et al. (1979), Trimble and Phillips (1978), Wickesberg and Geisler (1984) and Yasui et al. (1979). The entire Wiener theory as applicable to biological experiments is fully explained in Marmarelis and Marmarelis (1978).

A somewhat different approach to non-linear system identification has been adopted by Rajbman (Rajbman and Terechin, 1965; Rajbman and Hanš, 1967; and Rajbman, 1981). In the Wiener theory any system is primarily assumed to be deterministic and its stochasticity is explained as some accompanying noise. Rajbman considers the problem from the opposite point of view and his method is primarily statistical. It is interesting to note, however, that regardless of their theoretical origins both approaches are very similar in practice (Herman, 1978).

It is a pity that such methods of non-linear analysis of systems have not yet found their place in routine clinical examination. The introduction of non-linear methods into clinical practice is one of the main reasons for the development of the R@-SYSTEM.

DISCUSSION

What is the merit of the system being described here? That is the question which can now be discussed. It is not totally true to say that the market supplies special apparatus for some electrophysiological examination with hard-wired program equipment only. In fact, many manufacturers put up very flexible and programmable devices for sale which are designed for a whole battery of such examinations. However, everything has its limitations. It is therefore appropriate to examine the need for the R@-SYSTEM by considering some of the examples of devices which have been used:

(i)   Link-8 (DEC, USA): This was the first machine used by the author. It was small but nice. It was developed in the 1960s when the technology was comparatively primitive. However, its beauty lay in its simplicity. It contained all that was necessary and is still in daily use.

(ii) Plurimat (France): This was a very good machine albeit expensive. It was designed for signal processing in real time, but it had only Basic as a high-level language.

(iii) System-1200 (DISA, now DANTEC, Denmark) + Datalab (DEC, USA): This was a tandem of 4-channel modular EMG apparatus connected on-line with the 16-bit microcomputer system. The EMG was equipped both for some EMG as well as EP examinations with A/D converters, display and 20-bit hard-wired arithmetic controlled with 8080 CPU. The computer was the well-known PDP-11/03 (LSI technology) with 48 KB RWM, graphic display, floppy disk drives and dot-matrix printer. The special interface ("Interpack") connecting EMG system and PDP provided bilateral communication which was sufficiently supported with available software.

(iv)   Evomatic 2000 (Dantec, Denmark): This was the next step in the development, representing a cheap and compact piece of equipment for a number of standard EMG and EP clinical examinations. The programs are stored in ROMs, but Evomatic incorporates a built-in IEEE interface for communication with external computers (as standard with the Apple). This enables the further compiling and processing of patient data.

(v)   Evomatic 4000/8000: The number of differential input channels was extended from two to four or eight and the control programs are loaded from $3\frac{1}{2}$" floppy disks on which patient data can also be stored. However, the computer-like keyboard incorporated cannot be used for any programming, but only as a pad for personal data and some comments. Only the manufacturer himself, therefore, can develop the program disks.

(vi)   Tesy (Tonnies): 1 to 8 channels modular EMG and EP apparatus. Some of the programs can be stored in EPROMs and hence they can be changed. However, some other computer must be connected and em-ployed for subsequent program development.

(vii) Nicolet CA2000 (USA): Enables 2 to 4 channels EMG and EP to be captured. 8-bit Z-80 (master) and 16-bit AMD 2903 (display/acquisition) CPUs, $5\frac{1}{4}$" floppy disk drive, user programmable in APEX (Nicolet's Automated Process Execution language) only.

(viii) Nicolet Viking (USA): An efficient machine for EMG examinations - 2 to 8 channels, 1 MB RWM, floppy disk drives, Winchester disk. Software development package (includes RMX op. system, Fortran 77, System utilities and interface) as option.

(ix)   Nicolet Pathfinder I/II: the top product of the range designed for a variety of EEG and EP studies including brainmapping. Very flexible. 16 channels, 20-bit special CPU with hard-wired arith-metic, floppy disk drives, Winchester disk, tape streamer, colour graphic display, plotter, dot-matrix ink jet. User programmable in Mecol (special language) or Fortran 77 only.

(x)   Brain Atlas (Bio-logic): This apparatus is based on a Zenith-100 (IBM-PC/XT compatible) computer. However, its software equipment only includes programs for some special examinations with no sup-port for development.

    This is only a brief survey. Some of the machines are user program-mable, but in one or two high-level languages only. They often contain some special CPU and assembler language coding is precluded as well as the utilising of any extraneous software. Most of the machines mentioned above may be connected with any general-purpose computer. However, any software support for new program development is missing. It is this absence of adequate software support for such a widespread computer as the IBM-PC/XT/AT that has led to the developments described in this chapter.

REFERENCES

Bastiaans, M. J., 1980, Gabor's expansion of a signal into Gaussian elemen-
    tary signals, Proc. IEEE, 68 : 538.
Dvořák, I., and Šiška, J., 1986, On some problem encountered in calcula-
    ting the correlation dimension of EEG, Phys. Letters A, 118(2) : 63.

Dvořák, I., Šiška, J., Wackerman, J., Hrudová, L., and Dostálek, C., 1986, Evidence for interpretation of EEG as a deterministic chaotic process with a low dimension, Activitas Neurosa Superior, 28(3) : 228.

French, A. S., and Butz, E. G., 1973, Measuring the Wiener kernels of a non-linear system using the Fast Fourier Transform algorithm, Int. J. Contr., 17 : 529.

French, A. S., and Butz, E. G., 1974, The use of Walsh Functions in the Wiener analysis of nonlinear systems, IEEE Trans. Comput., C-23(3) : 225.

French, A. S., 1976, Practical nonlinear system analysis by Wiener kernel estimation in the frequency domain, Biol. Cybern., 24 : 111.

Gabor, D., 1946, Theory of communication, J. IEE: 429.

Heřman, P., 1978, Nonlinear Analysis of Electroneurological Preparats, MMF UK, Prague (in Czech).

Kadri, F. L., 1972, A method of determining the crosscorrelation functions for a class of non-linear systems, Int. J. Contr., 15 : 779.

Koblasz, A. J., 1978, Nonlinearities of the human ERG reflected by Wiener kernels, Biol. Cybern., 31(4) : 187.

Koblasz, A., Rae, J. L., Correia, M. J., and Ni, M.-D., 1980, Wiener kernels and frequency response functions for the human retina, IEEE Trans. Biomed. Eng., BME-27(2) : 68.

Krausz, H., 1975, Identification of nonlinear systems using random impulse train inputs, Biol. Cybern., 19 : 217.

Larkin, R. M., Klein, S., Odgen, T. E., and Tender, D. H., 1979, Nonlinear kernels of the human ERG, Biol. Cybern., 35(3) : 145.

Lee, Y. W., and Schetzen, M., 1965, Measurement of the Wiener kernels of a non-linear system by cross-correlation, Int. J. Contr., 2 : 237.

Marmarelis, P. Z., 1972, Nonlinear identification of bioneural system through white-noise stimulation, in: "Proc. 13th Joint Automatic Control Conference", Stanford : 117.

Marmarelis, P. Z., and Naka, K. I., 1972, White noise analysis of a neuron chain: an application of the Wiener theory, Science, 175 : 1276.

Marmarelis, P. Z., and McCann, G. D., 1973, Development and application of white-noise modeling techniques for studies of insect visual nervous system, Kybernetik, 12 : 74.

Marmarelis, P. Z., and Naka, K. I., 1973, Non-linear analysis and synthesis of receptive field responses in the catfish retina, J. Neurophysiol., 36 : 605.

Marmarelis, P. Z., and Naka, K. I., 1974, Identification of multi-input biological systems, IEEE Trans. Biomed. Eng., BME-21(2) : 88.

Marmarelis, P. Z., and Marmarelis, V. Z., 1978, "Analysis of Physiological Systems", Plenum, New York.

Moore, G., Heřman, P., Krekule, I., and Bureš, J., 1977, Application of the nonlinear Wiener theory of the analysis of evoked activity of an epileptic focus in the motor cortex in rats, in: "3rd Conference of Biomedical Engineering", Brno.

Rajbman, N. S., and Terechin, A. T., 1965, Dispersion methods of random functions and their use for study of nonlinear control plants, Avtomatika i Telemechanika, 26 : 500 (in Russian).

Rajbman, N. S., and Hanš, O. F., 1967, Dispersion methods of identification of multidimensional nonlinear control plants, Avtomatika i Telemechanika, 28(5) : 5 (in Russian).

Rajbman, N. S., 1981, "Dispersion Analysis", Nauka, Moscow (in Russian).

Seelen, W.von, and Hoffmann, K.-P., 1976, Analysis of neuronal networks in the visual system of the cat using statistical signals, Biol. Cybern., 22 : 7.

Trimble, J., and Phillips, G., 1978, Nonlinear analysis of the human visual evoked response, Biol. Cybern., 30(1) : 55.

Wickesberg, R. E., and Geisler, C. G., 1984, Artifacts in Wiener kernels estimated using Gaussian white noise, IEEE Trans. Biomed. Eng., BME-31: 454.

Wiener, N., 1958, "Nonlinear Problems in Random Theory", Wiley, New York.

Yasui, S., Davis, W., and Naka, K. I., 1979, Spatio-temporal receptive field measurement of retinal neurons by random pattern stimulation and cross-correlation, IEEE Trans. Biomed. Eng., BME-26 : 263.

A FIRST EXPERIENCE WITH THE AUTOMATIC EXTRACTION OF COMPRESSED

INFORMATION FROM THE EEG IN CLINICAL PRACTICE

V. Krajča, I. Rottová and S. Petránek

Department of Neurology
Postgraduate Medical and Pharmaceutical Institute
Prague, Czechoslovakia

INTRODUCTION

Modern mathematical and cybernetic methods to an increasing extent find their way into the diverse disciplines of medicine and biomedical engineering. Hardly anywhere, however, is their influence so considerable as in the field of biological signal analysis.

One of the most important areas of penetration of these methods is in the analysis and evaluation of the electroencephalogram (EEG), a complex electrical biosignal which is known to reflect brain activity - stages of sleep, different states of wakefulness, epileptic seizures, or some other neurological disorders. It is used for monitoring the depth of anaesthesia and as a non-invasive tool for the diagnosis and research of the brain.

Because of its great complexity and non-stationary behaviour, with a large amount of noise, analysis of the EEG is a tedious and difficult task even for a skilled expert. In routine clinical practice, about 20 minutes of EEG is recorded (that is, 35 m of paper chart), so there is also a need for data reduction with regard to the considerable volume of assembled information, and for creating widely available standardised databases.

In order to support the physician in decision-making and for data reduction, we have suggested a new approach to the extraction of compressed information from the EEG based on fuzzy set theory (Krajča, 1985; Krajča et al., 1985). The first experience of applying this method in clinical practice is presented.

ADAPTIVE SEGMENTATION OF THE EEG

The first step in signal processing is the extraction of features from the EEG. For this, the signal is usually divided into smaller parts capable of being processed. If the length of these sections is kept constant (as, for example, during the application of the Fast Fourier Transform for power spectrum estimation), the fixed segment boundary has no relationship to the events in the EEG signal. Segments with a mixture of different waveforms can arise, with negative effect not only for spectrum estimation, but also for the subsequent classification. If the complex structure of the EEG is to be saved, it would be better to express the record in approximately stationary segments of varying duration, depending on the occurrence of non-stationarities in the signal and to describe them with a small number of parameters to improve data reduction.

For this purpose, the method of adaptive segmentation by means of the autocorrelation function, originally suggested by Bodenstein and Praetorius (1977) and lately improved by Michael and Houchin (1979), was used to split the signal into quasi-stationary (piece-wise stationary) segments of variable length and to extract the spectral features from them.

The signal is scanned through the moving test window and the deviation from the fixed reference window, placed at the beginning of the segment, is monitored. The departure from stationarity (difference measure) is estimated from the changes of autocorrelation function of the EEG in the test and reference regions. The segment boundary is indicated as soon as the deviation from stationarity reaches a certain threshold, set in advance (Bodenstein and Praetorius, 1977).

The example of segmentation of 2 minutes of the EEG (derivation Fpl-Al) including complex spikes is shown in Fig. 1. Vertical lines represent the segment boundaries; their correct position is estimated by a rather complicated adaptive procedure (Michael and Houchin, 1979). The nine coefficients of the autocorrelation function of the signal in each segment were used as features for subsequent clustering.

EXTRACTION OF COMPRESSED INFORMATION FROM THE EEG BY FUZZY CLUSTER ANALYSIS

The next phase of processing is the classification of the segments obtained into homogeneous classes, significant from the diagnostic viewpoint. For this reason, cluster analysis methods are commonly used. Their task is to find the "natural" grouping of the data under study into homogeneous classes (clusters), so that the objects in the same class are more "similar" to each other than the elements of different classes.

One of the drawbacks of the "classical" or "hard" clustering methods is the disjunctive classification. Thus every object belongs to just one class, a situation that does not match reality: hybrid EEG segments can have membership in miscellaneous classes at the same time. This disadvantage is overcome by the use of fuzzy sets theory for cluster analysis.

In conventional set theory, the element either belongs, or does not belong, to some set. Its characteristic function may possess only two values: 0 or 1. In contrast, fuzzy set theory (Bezdek, 1981) generalises this approach and admits the possibility of only partial membership of an element in a (fuzzy) set (in the whole range of the closed interval <0,1>).

A very useful tool for the interpretation of fuzzy clusters is their α-cores (α-level sets), that is, (classical) sets of those members of fuzzy clusters which have a greater value of grade of membership than some fixed threshold α. They may be used for the "purification" of clusters, that is for elimination of untypical (hybrid) segments, which have a grade of membership lower than the threshold α (Krajča et al., 1985).

We have tried cluster analysis of EEG segments by means of Bezdek's fuzzy c-means (FCM) algorithm (Bezdek, 1981). This approach, in contrast to classical "hard" clustering, has the advantage of the possibility of improving cluster homogeneity by eliminating the less typical cluster members with membership in the fuzzy cluster lower than the preset threshold α.

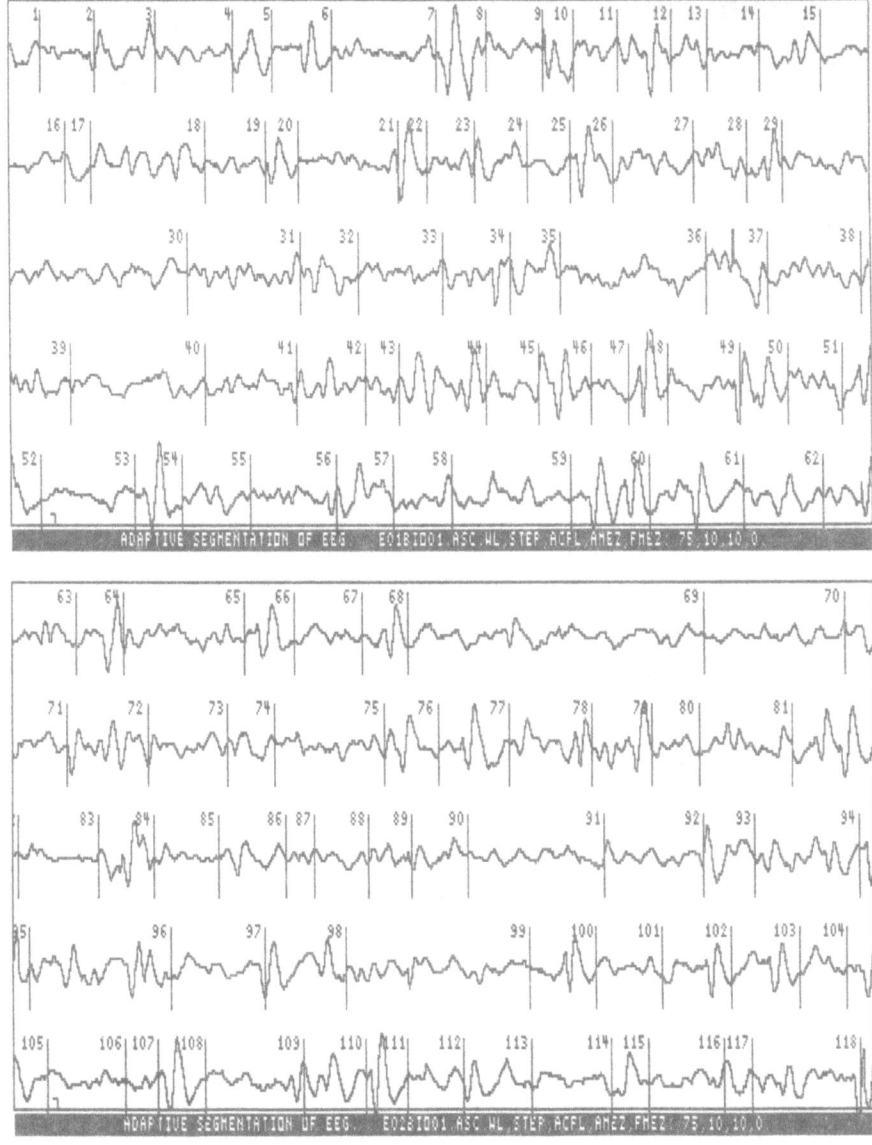

Fig. 1. Adaptive segmentation of EEG signal with complex spikes

The result of the fuzzy classification into 5 classes is shown in Fig. 2(a and b). The clusters of EEG segments are ordered according to the magnitude of the mean variance of the signal in every cluster, so that there are EEG waves with small amplitude in the first class (the largest one), and the last class exhibits the most dramatic changes in the EEG.

From the fuzzy clusters, the segments with maximum class membership were chosen as the most typical, representative elements of each fuzzy set. The result for the three prototypes of every set is shown in Fig. 3 in the form similar to that used by Creutzfeld et al. (1985), that is, including the average power spectrum of each class computed by means of an AR model, percentage occurrence of types of EEG segments in the record, and the time

221

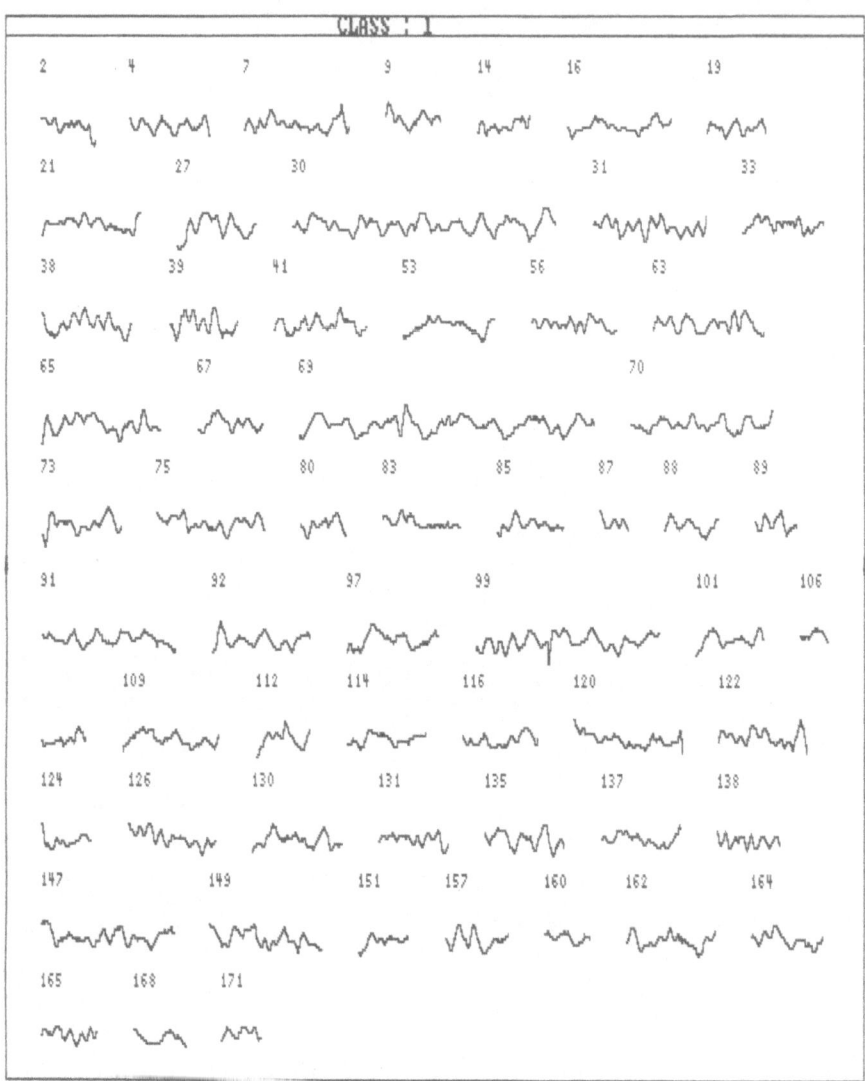

Fig. 2a and b. Fuzzy cluster analysis of EEG segments

profile of the EEG record, that is the diagram of occurrence of the EEG
segments in the relevant class during the whole recording time. The occurrence of segments not included in the classification because of low membership (possibly artefacts) is marked "NS" (left side of the time plot).
The computer programs for the methods described above were written in Turbo
Pascal and implemented on a Zenith Z-150 computer.

DISCUSSION

     The approach suggested for the extraction of information from the EEG
in compressed form, based on adaptive segmentation, subsequent fuzzy

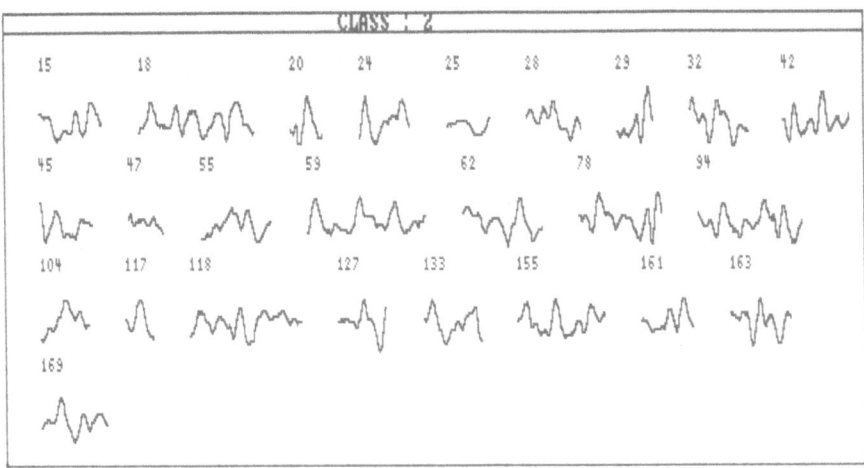

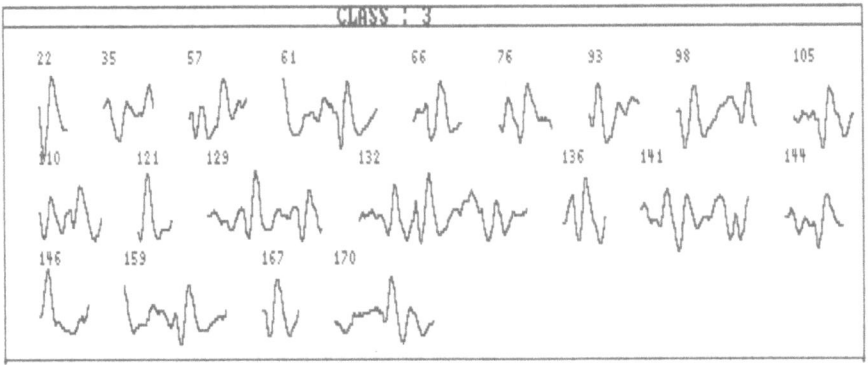

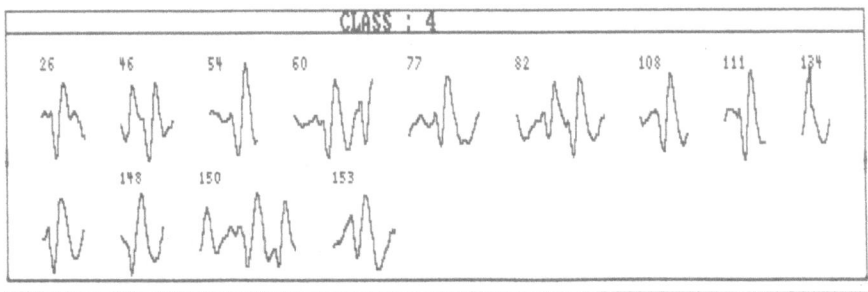

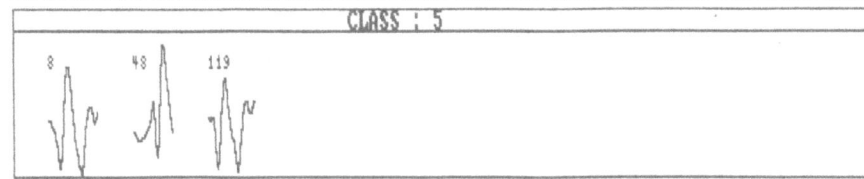

Fig. 2b. Fuzzy cluster analysis (continued)

clustering, and searching for representative segments may be useful for efficient quantitative characterisation of long or sleep EEG records, for data reduction and to help the physician in evaluation of routine clinical EEG records.

The first results of application of this method in clinical practice

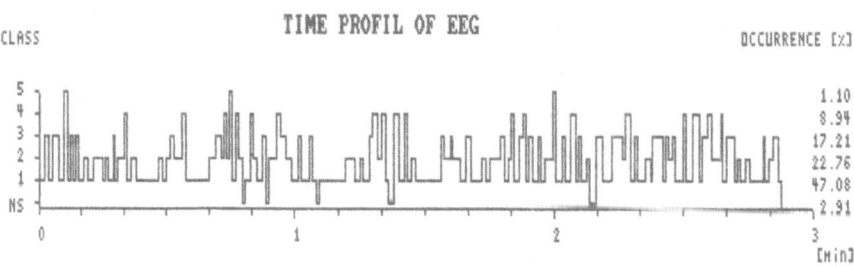

CLASS : 1

149   122   75

OCCURENCE :
47.08 [%]

[DB]   LP SPECTRUM

FMAX :
1.56 [Hz]

0   25 [Hz]

CLASS : 2

78   169   118

OCCURENCE :
22.76 [%]

[DB]

FMAX :
3.91 [Hz]

0   25 [Hz]

CLASS : 3

170   105   66

OCCURENCE :
17.21 [%]

[DB]

FMAX :
4.69 [Hz]

0   25 [Hz]

CLASS : 4

150   82   148

OCCURENCE :
8.94 [%]

[DB]

FMAX :
4.69 [Hz]

0   25 [Hz]

CLASS : 5

48   8   119

OCCURENCE :
1.10 [%]

[DB]

FMAX :
3.91 [Hz]

0   25 [Hz]

TIME PROFIL OF EEG

CLASS   OCCURRENCE [%]

5   1.10
4   8.94
3   17.21
2   22.76
1   47.08
NS   2.91

0   1   2   3
[Min]

Fig. 3.   Representative segments of the EEG

show that in future research it will be necessary to concentrate on the automatic setting of the segmentation threshold (its value varies for different patients). A very interesting problem is also the question of cluster validity, that is, how many clusters are present in EEG data.

Fuzzy cluster analysis improves the homogeneity of clusters of EEG segments and the extraction of representative segments from the EEG by means of α-cores seems to be much more efficient and flexible than the alternative method used by Krajča (1984). The fuzzy approach makes the interpretation of the results closer to the problems of the "real world".

REFERENCES

Bezdek, J., 1981, "Pattern Recognition with Fuzzy Objective Functions
    Algorithms", Plenum, New York.
Bodenstein, G., and Praetorius, H. M., 1977, Feature extraction from the
    EEG by adaptive segmentation, Proc. IEEE, 65 : 642.
Creutzfeld, O. D., Bodenstein, G., and Barlow, J. S., 1985, Computerized
    EEG pattern classification by adaptive segmentation and probability
    density function classification. Clinical evaluation, Electroenceph.
    Clin. Neurophysiol., 60 : 373.
Krajča, V., 1984, Automatic classification of EEG segments and the extrac-
    tion of representative ones by the dynamic clusters method, Activ.
    Nerv. Sup., 26(2) : 118.
Krajča, V., 1985, Teorie fuzzy mnozin jako prostredek pro extrakci
    zhustené informace z EEG signálu, Automatizace, 28 : 7 (in Czech).
Krajča, V., Formánek, J., and Petránek, S., 1985, Application of fuzzy
    cluster analysis in EEG computer assisted decision-making, in:
    "Medical Decision Making", J. H. van Bemmel, F. Grémy, and J. Zvárová,
    eds., North-Holland, Amsterdam : 169.
Michael, D., and Houchin, J., 1979, Automatic EEG analysis: a segmentation
    procedure based on autocorrelation function, Electroenceph. Clin.
    Neurophysiol., 46 : 232.

NON-LINEAR MODELS OF TIME SERIES AND APPLICATIONS TO ANALYSIS OF EEG
SIGNALS

J. Anděl, T. Cipra, J. Dvořák, J. Formánek, J. Kubát and
Z. Prášková

Department of Statistics, Charles University, and
Institute of Hygiene and Epidemiology
Prague, Czechoslovakia

INTRODUCTION

Over the last decade remarkable progress has been made in the non-
linear modelling of discrete time series data. Bi-linear models, exponential autoregressive models, threshold models and others have been developed
in order to gain a deeper understanding of the structure of real data.

It is known that many frequency-domain phenomena observed when
studying various kinds of biosignals cannot be properly explained if a
linear model is assumed. Threshold models first introduced by Tong in
1977, and systematically studied in the monograph by Tong (1983), seem to
be an appropriate tool for analysing such phenomena.

NON-LINEAR VIBRATIONS AND THRESHOLD MODELS

Tong proposed a class of piece-wise linear models that capture such
frequency-domain phenomena as limit cycle, jump resonance, amplitude-
frequency dependency and other features known in the theory of non-linear
vibrations. Let us consider each of them in turn.

The stable limit cycle represents the stationary state of sustained
oscillations which does not depend on the initial conditions but depends
exclusively on the parameters of the system. The existence of a limit
cycle in a time series X_t can be observed on the phase diagram of the
points $[X_t, X_{t-1}]$ (see Fig. 1).

A jump resonance phenomenon may appear as a jump of the output amplitude depending on whether the input frequency is monotonically increasing
or monotonically decreasing. Fig. 2 shows a decline of the output amplitude at the frequency f_2 if the input frequency is monotonically increasing,
and, conversely, a rise of the output amplitude at the frequency f_1 when
the input frequency is monotonically decreasing. Similarly, the output
amplitude may have a resonance jump when the input amplitude (of constant
frequency) is monotonically increasing or monotonically decreasing.

In a non-linear system the output signal may show different frequencies of oscillation for different amplitudes. This phenomenon is called

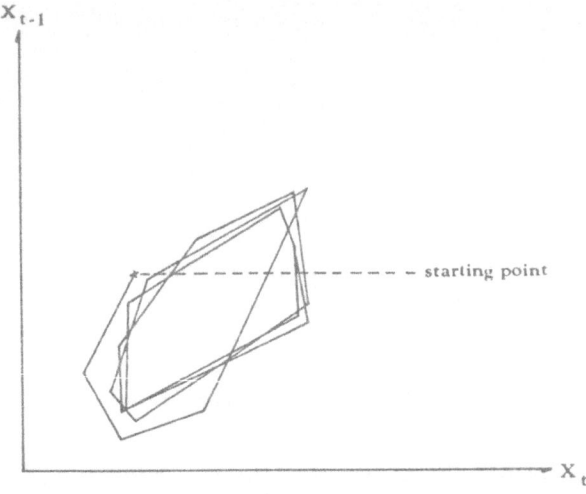

Fig. 1.   Stable limit cycle.

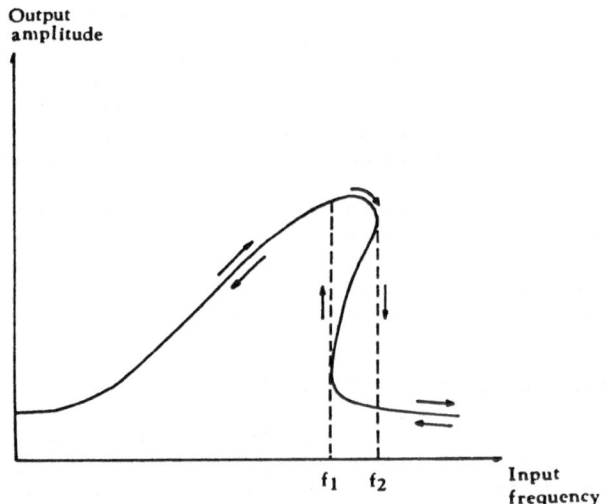

Fig. 2.   Jump resonance.   Dependency of output
          amplitude on input frequency (input
          amplitude is constant).

amplitude-frequency dependency.   Examples of such phenomena are demonstrated
in Figs. 3 and 4.

It has been shown by means of simulation experiments that threshold
models exhibit the above-mentioned properties of non-linear vibrations
(Tong, 1983).   Threshold models are defined by a set of certain values
(termed thresholds).   When the process exceeds such a value, then depending
upon the feedback control mechanism involved the behaviour of the system
changes.   As a simple model of a feedback system, an autoregressive model
can be used.   So we talk about threshold autoregressive (TAR) models.   Now
we can introduce the mathematical definition of a general TAR model, con-
fining ourselves to the simplest case.

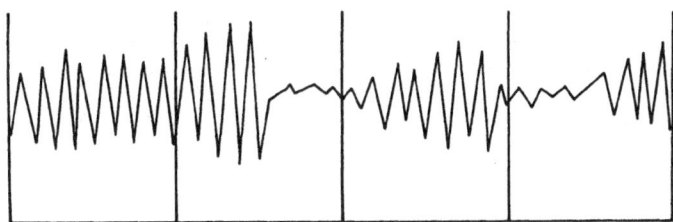

Fig. 3. Amplitude – frequency dependency. High [low]
amplitudes – high [low] frequencies.

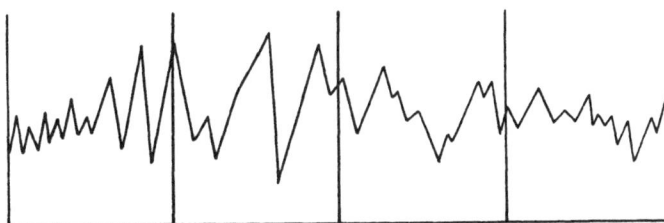

Fig. 4. Amplitude – frequency dependency. High [low]
amplitudes – low [high] frequencies.

Consider real numbers $r_1 < \ldots < r_{n-1}$ and the intervals

$$R_1 = (-\infty, r_1), \qquad R_2 = (r_1, r_2), \quad \ldots, \qquad R_n = (r_{n-1}, \infty). \qquad (1)$$

Define the process X_t by the formula

$$X_t = a_{o,j} + \sum_{i=1}^{k_j} a_{i,j} X_{t-i} + \varepsilon_{t,j} \text{ if } X_{t-d} \in R_j, \ j = 1, \ldots, n, \qquad (2)$$

where $a_{i,j}$ are some constants, $\varepsilon_{t,j}$ are independent strict white noise
sequences with zero means and finite variances, d is a positive number.
The process X_t is called the self-exciting threshold autoregressive (SETAR)
process of order (n; $k_1, \ldots, k_n$). The values $r_1, \ldots, r_{n-1}$ are the
thresholds and d is the time delay connected with a feedback control mechanism. The process is locally autoregressive. The behaviour of the process
at time t depends on the past value X_{t-d}. Very often we have d = 1. More
general threshold autoregressive models, or even threshold autoregressive-
moving average models, can be considered (see Tong (1983) for further information).

The problem of stationarity of the SETAR model has been studied by
several authors. Given the distribution of white noise, the problem of
finding the stationary distribution of the SETAR process reduces to that
of solving an integral equation for the density function, the explicit
solution of which is available only in some special cases (see, for example,
Anděl et al., 1984, or Anděl and Bartoň, 1986, where the explicit formulae
for the stationary density of simple SETAR processes were given). Having
established the stationary distribution, we can calculate characteristics
of the process such as moments, skewness, excess, correlation coefficients
and others.

IDENTIFICATION OF SETAR MODELS

Fitting methods originally developed for the identification of autoregressive processes can be modified and extended to SETAR processes in

view of the piece-wise linearity of the latter models. Tong and Lim (1980) proposed the recursive use of a procedure based on Akaike's Information Criterion (AIC). The procedure is described in some detail by Tong and Lim (1980) and the reader can also find it in Tong (1983). Alternatively, other procedures for fitting autoregressive models can be used. On the basis of our earlier Monte Carlo studies, we propose to use a procedure recursively, making use of the technique developed by Hannan and Quinn for fitting auto-regressive models (see Andĕl and Cipra, 1981; 1982).

Let us briefly recall both these methods of fitting autoregressive models (see, for example, Andĕl, 1982). Consider an autoregressive process $\{Y_t\}$ of order p, given by

$$Y_t = a_1 Y_{t-1} + \ldots + a_p Y_{t-p} + \varepsilon_t \tag{3}$$

where $a_1, \ldots, a_p$ are parameters, $a_p \neq 0$ and $\{\varepsilon_t\}$ is white noise. The problem is how to determine the order p when only $Y_1, \ldots, Y_t$ are observed. Let us suppose that $p \leqslant K$ where K is a prefixed number, $K < T$.

For $k = 0, 1, \ldots, K$ denote

$$\hat{\sigma}_k^2 = \frac{1}{T - k} S_k (\hat{a}) \tag{4}$$

where

$$S_k (\hat{a}) = \sum_{t=k+1}^{T} \left(Y_t - \hat{a}_1 Y_{t-1} - \ldots - \hat{a}_k Y_{t-k}\right)^2 \tag{5}$$

and $\hat{a}_1, \ldots, \hat{a}_k$ are the least square estimates of $a_1, \ldots, a_k$.

While the Akaike method determines the order p by the minimisation of the function

$$AIC(k) = T \ln (S_k(\hat{a})/T) + 2(k + 1), \tag{6}$$

the method developed by Hannan and Quinn minimises the function

$$H(k) = \ln \hat{\sigma}_k^2 + 2 kc(\ln \ln T)/T \tag{7}$$

for $k = 0, 1, \ldots, K$.

Here c is an arbitrary number greater than 1. Our simulation study recommends the use of $c = 3$.

Having performed this modification, we can proceed in a manner analogous to that of Tong and Lim (1980). For observations $X_1, \ldots X_T$, fixed values of n (the number of intervals) and K (an upper bound for the order of each of n local autoregressive models), and for a given set of potential candidates for the threshold values, the procedure determines the delay parameter d, the thresholds $r_1, \ldots, r_{n-1}$ and yields estimates of the order k_j as well as of parameters of the jth local autoregressive process, $j = 1, \ldots, n$. The choice of n, K and the set of potential thresholds depends upon the data. The procedure has been applied to simulated as well as to real data.

APPLICATION TO EEG SIGNALS

We have analysed an EEG signal record of an experimental animal from an experiment of CO inhalation. Ten series were available, each of them consisting of 250 items. The data are illustrated in Figs. 5 and 6 (data

230

A and B respectively). Before applying the identification procedure, all the series were centred by subtracting the corresponding arithmetic mean. The delay d remained fixed during the computations.

Data A were fitted to the following autoregressive model of order 4:

$$X_t = 0.626 \ X_{t-1} + 0.415 \ X_{t-2} - 0.440 \ X_{t-3} + 0.265 \ X_{t-4} + \varepsilon_t. \tag{8}$$

The same data can be fitted to SETAR (2; 1, 4) given by

$$X_t = 0.746 \ X_{t-1} + \varepsilon_{t,1} \qquad\qquad\qquad \text{if } X_{t-1} \leqslant -5 \tag{9}$$

$$= 0.568 \ X_{t-1} + 0.522 \ X_{t-2} - 0.456 \ X_{t-3} + 0.282 \ X_{t-4} + \varepsilon_{t,2}$$
$$\text{if } X_{t-1} > -5. \tag{10}$$

Similarly, data B were identified with SETAR (3; 3, 2, 2,),

$$X_t = 0.552 \ X_{t-1} + 0.618 \ X_{t-2} - 0.365 \ X_{t-3} + \varepsilon_{t,1}, \qquad \text{if } X_{t-1} \leqslant -10.7 \tag{11}$$

$$= 0.545 \ X_{t-1} + 0.407 \ X_{t-2} + \varepsilon_{t,2} \qquad \text{if } -10.7 < X_{t-1} \leqslant 19.3 \tag{12}$$

$$= 0.276 \ X_{t-1} + 0.603 \ X_{t-2} + \varepsilon_{t,3} \qquad \text{if } X_{t-1} > 19.3 \tag{13}$$

Fitting these data to the autoregressive model we obtain

$$X_t = 0.639 \ X_{t-1} + 0.378 \ X_{t-2} - 0.268 \ X_{t-3} + 0.317 \ X_{t-4} - 0.240 \ X_{t-5} + \varepsilon_t \tag{14}$$

It can be seen that local autoregressive models within the framework of SETAR models are usually simpler than the global autoregression.

EXPONENTIAL AUTOREGRESSIVE MODELS

In this section we briefly mention other non-linear models which could be applied to the analysis of EEG signals. These models have been created for the modelling of periodic time series with an amplitude-frequency

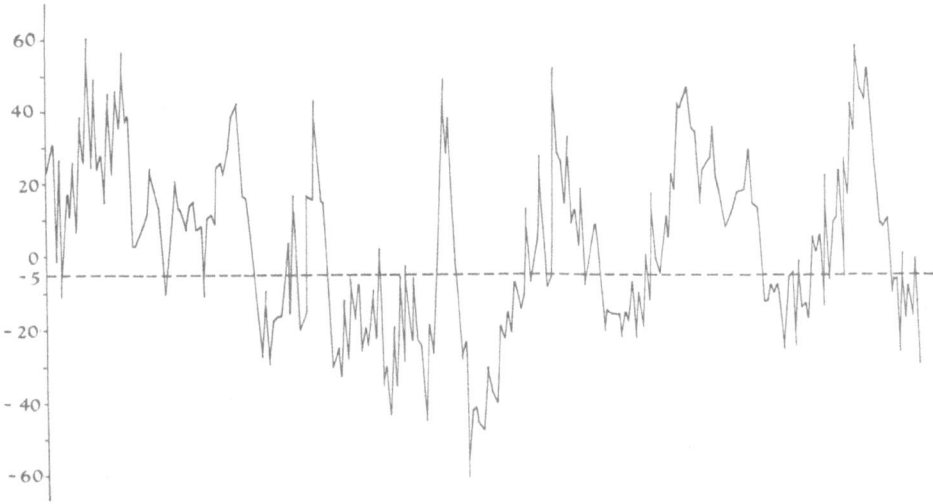

Fig. 5.   Data A - original data were centred by subtracting the arithmetic mean.

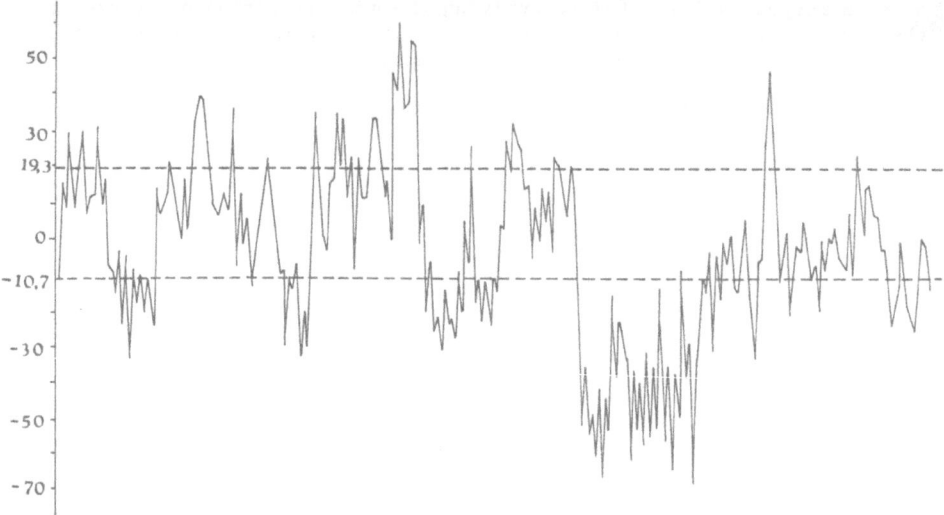

Fig. 6. Data B - original data were centred by subtracting the arithmetic mean.

dependency. We call them exponential autoregressive models (EAR). As an example, let us introduce the following EAR model $\{x_t\}$ of order p, given by

$$x_t = (\phi_1 + \pi_1 \exp\{-\gamma x_{t-1}^2\}) x_{t-1} + (\phi_2 + \pi_2 \exp\{-\gamma x_{t-1}^2\}) x_{t-2}$$

$$+ \ldots + (\phi_p + \pi_p \exp\{-\gamma x_{t-1}^2\}) x_{t-p} + \varepsilon_t \qquad (15)$$

where ϕ_i, π_i, $i = 1, \ldots, p$ and $\gamma > 0$ are the parameters and $\{\varepsilon_t\}$ is white noise.

Obviously, the coefficients of this model at time t are dependent upon the amplitude $|x_{t-1}|$ of time series at time t-1. The necessary conditions for the existence of the stable limit cycle are as follows:

(i)   at least one of the roots of the polynomial

$$\lambda^p - (\phi_1 + \pi_1) \lambda^{p-1} - \ldots - (\phi_p + \pi_p)$$

lies outside the unit circle;

(ii)   all roots of the polynomial

$$\lambda^p - \phi_1 \lambda^{p-1} - \ldots - \phi_p$$

lie inside the unit circle;

and the following expression holds

(iii) $\left(1 - \sum\limits_{i=1}^{p} \phi_i\right) \Big/ \sum\limits_{i=1}^{p} \pi_i > 1$ or $\left(1 - \sum\limits_{i=1}^{p} \phi_i\right) \Big/ \sum\limits_{i=1}^{p} \pi_i < 0.$

The parameters ϕ_1, π_1, $\ldots$, ϕ_p, π_p can be estimated by the regression; parameters γ and p can be estimated using the AIC criterion or in some other similar way (see, for example, Haggan and Ozaki, 1981).

232

APPENDIX: METHODS OF EEG SIGNAL MEASUREMENT

The EEG signal analysed in this chapter was measured using implanted electrodes and commercial EEG pre-amplifiers, and was recorded by an FM magnetic tape recorder. The off-line discretisation was performed with a 10-bit A/D converter and data stored using a minicomputer HP 2116C for further processing, including spectral analyses and other procedures (for details see Dvořák et al., 1981). The discretisation frequency was $100\,s^{-1}$.

REFERENCES

Anděl, J., 1982, Fitting models in time series analysis, Math. Operations-forsch. Statist., Ser. Statistics 13 : 121.

Anděl, J., and Bartoň, T., 1986, A note on the threshold AR (1) model with Cauchy innovations, J. Time Ser. Anal., 7 : 1.

Anděl, J., and Cipra, T., 1981, "Determination of the Order of an Autoregressive Model and its Application to Spectral Density Estimation", Research Report, Institute of Hygiene and Epidemiology, Prague (in Czech).

Anděl, J., and Cipra, T., 1982, "Threshold Models", Research Report, Institute of Hygiene and Epidemiology, Prague (in Czech).

Anděl, J., Netuka, I., and Zvára, K., 1984, On threshold autoregressive processes, Kybernetika, 20 : 89.

Dvořák, J., Formánek, J., Kubát, J., Plevová, J., Vaničková, M., Fireš, M., Anděl, J., Cipra, T., Tomášek, L., Prášková, Z., Holoubková, E., and Fabián, Z., 1981, Analysis of the time series of the EEG frequency spectra and of EEG spectral power densities, Activ. Nerv. Super. (Prague), 23 : 157.

Haggan, V., and Ozaki, T., 1981, Modelling nonlinear random vibrations using an amplitude-dependent autoregressive time series model, Biometrika, 68 : 189.

Tong, H., and Lim, K. S., 1980, Threshold autoregression, limit cycle and cyclical data, J. Roy. Statist. Soc. Series B, 42 : 245.

Tong, H., 1983, "Threshold Models in Non-linear Time Series Analysis", Springer-Verlag, New York.

AMPLITUDE AND CORRELATIONAL MAPPING OF EEG - A SOFTWARE SYSTEM

J. Wackermann, J. Hönig, L. Hrudová and C. Dostálek

Institute of Physiological Regulations
Czechoslovak Academy of Sciences
Prague, Czechoslovakia

INTRODUCTION

Since the times of Berger's discovery (Berger, 1969), electroencephalography has undergone three major changes. There was a transition from
single-channel to multi-channel recording; from analog to digital acquisition, storage and data-processing techniques; and from studies of
punctually-derived potential differences to those of global potential field
distributions, that is, from classical electroencephalography to topo-
electroencephalography (see Petsche, 1976, for a historical review). Nowadays, topo-electroencephalography (TEEG) is becoming important, both in
clinical practice as well as in the field of research. As a result, professional systems providing TEEG or, more popularly, "Brain Mapping" facilities
have become commercially available, such as Brain Atlas (Biologic Systems),
Pathfinder (Nicolet), or Brain Imager (Neuroscience). It would be preco-
cious, however, to consider "home-made" systems obsolete; on the contrary,
there is still an urgent need for methodical innovations, especially in
experimental research work.

In this chapter the structure and functioning of the software system
UFR-BEAM will be described. The system has been developed in the Institute
of Physiological Regulations since 1985; it should find its use mainly in
research on higher nervous functions, especially in the investigation of
EEG correlates of conditioned reflexes. In accordance with the aim of this
book, the explanation will focus upon the technical and programming concepts.

UFR-BEAM: GENERAL DESCRIPTION

The system is conceived as "pure software"; what is meant by this
expression is that it is virtually independent of both the recording apparatus and the computer. Primary data acquisition and whatever is related to
it are external to the system; it begins with a definition of a unified
data structure, which is taken for granted. The data-processing programs
are written in a high-level language (Pascal-2 by Oregon Software) and are
thus hardware independent, too.

Only a few words are needed about the hardware. For EEG recordings,
the MINGOGRAF EEG 21 (Siemens-Elema) apparatus is used; after amplification,

the signal is fed into the computer laboratory where it is digitised by means of a 12-bit A/D converter. Although the programs for data acquisition and file handling are not included within UFR-BEAM, an all-purpose software package has been designed and written by one of the authors (J.H.). The hardware configuration consists of a mini-computer PDP 11/34 (DEC) with two disk drives RL-01 (5 MB) and two floppy drives RX-02 (0.5 MB), connected to a terminal VT100 and to a line printer LA120. The graphics are performed by a graphic processor unit VS-11 (DEC) connected to the host computer and to a colour graphics monitor (TESLA). Transfer of the software to another configuration should not present any special problems, except regarding the programs for graphical output. In accordance with contemporary trends in biological data analysis, we believe that a major part of the system should be transferred to an IBM-PC compatible micro-computer within a short time.

Three principal requirements needed to be constantly kept in mind in the outline of the system: its versatility, flexibility, and modularity. Thus, the system has been designed as consisting of a number of relatively independent programs which perform rather circumscribed operations on data stored as disk files of a uniform structure. This philosophy is largely supported by the operating system RSX-11M under which the program system is run. It is then easy to create a problem-oriented package in the form of a command file submitted to the indirect command line interpreter (AT) controlling the subsequent execution of appropriate tasks. Alternatively, the tasks can be activated directly from the operator's terminal. Let us now introduce some fundamental concepts intrinsic to the UFR-BEAM system.

Data Representation

All the data, regardless of their nature (for example, amplitude measures, power spectra, correlation functions) are converted into 2-byte integers obtained by using the formula:

$$z = \text{round}(3.2E3 * x/x_{max}) \tag{1}$$

and stored as binary files. This provides better than 0.002% precision along with a certain reduction of the required disk space when compared to the usual floating-point representation.

Block-oriented Logical Structure (BLS)

This concept is of crucial importance in UFR-BEAM. An ordinary binary file consisting of N records, each of them being K words long, can be considered as a matrix of $N \times K$ values; this is a natural representation of an EEG recording consisting of N samples in K channels. The BLS is a generalisation of this simple structure (see Fig. 1). Any file is considered to consist of N blocks, each of them containing L records. Even the "record" is only a logical concept, for it can be a sample of original data and/or a vector of calculated values, in both cases K words long; but it can be an interpolated map as well. In other words, the "record" is an elementary unit of access to a file within its BLS.

In an analogy to the simplest file structure presented above, where samples are ordered by increasing values of time, there are similar notions in the BLS concept. Blocks in a file, and records in a block, are thought of as ordered by increasing values of a "File Ordering Variable" (FOV), and of a "Block Ordering Variable" (BOV), respectively. These variables can, for example, be time and frequency. Moreover, the records can be thought of as "single" or "dual"; in the latter case, there are two records corresponding to any value of BOV rather than one. Such a "dual" file may be considered a file of pairs of certain closely related values, such as mean and standard deviation of averaged EPs, or real and imaginary part of a

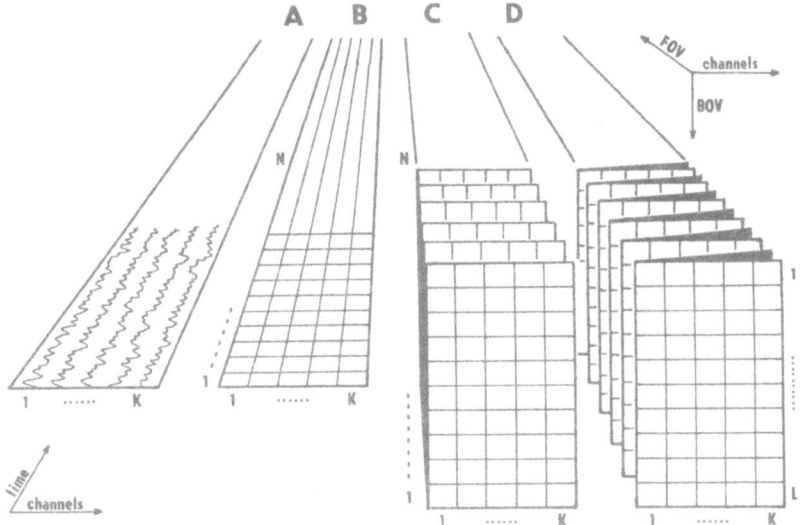

Fig. 1. Explanation of Block-oriented Logical Structure. A: an
ordinary analog recording; B: a digitised recording $N \times K$;
C: a data file of the structure $N \times L \times K \times 1$ (single);
D: a data file of the structure $N \times L \times K \times 2$ (dual).

spectral density function, or symmetric and antisymmetric components of a
scalp field distribution. Obviously, the concept of BLS provides the possibility of representation and handling of rather complicated data structures.

The BLS is, however, a purely logical concept; it has been introduced
in order to make the data transformations more transparent to the user. In
fact, all the files in the UFR-BEAM system have the same structure, namely
a binary file of a fixed (physical) record length equal to 512 bytes.
Access to the data is provided by means of several find/read/write procedures, so that even the programmer may completely ignore the relations between the BLS and the actual structure of the file. In the case of logical
"records" of variable length (such as compressed maps - see below), access
is supported by forward and backward pointers placed before and after each
"logical" record.

Header

The first 512 bytes of any file are reserved for some fundamental information about the file, such as subject identification, BLS descriptors,
names and physical dimensions of FOV and BOV.

File Attributes

There are 16 Boolean variables which specify some essential features
of the file, for example "single/dual" and "data/maps". Some attributes
can be defined by the user. The values of the attributes are stored in a
16-bit region of the header. When a file is submitted to a UFR-BEAM program, the state of attributes is checked for their admissibility.

Events

An "event" is an external or internal phenomenon, occurrence of which
can be marked in the data file, for example a stimulus onset or a motor

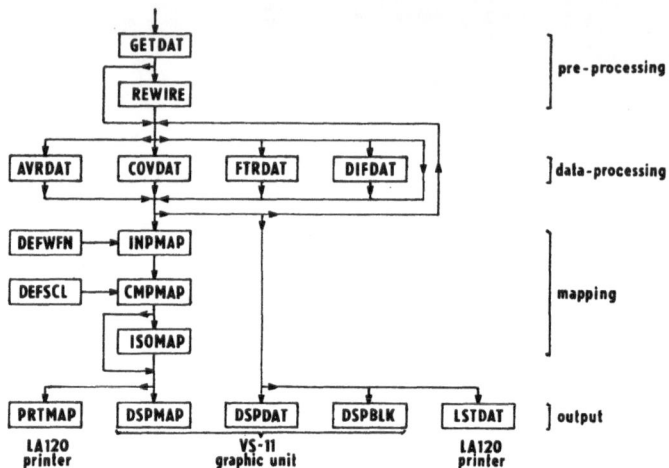

Fig. 2.   Flow-chart of the UFR-BEAM system.

reaction of the subject.   Up to 16 independent events can be defined;   they are stored in an auxiliary channel in the file as 2-byte event codes.   Their function is not only informative;   they can also serve as "tags" for those programs that operate on selected segments of data, for example in EP averaging and similar applications.

The programs can be divided into several classes according to their function:

(1)   pre-processing: GETDAT (calibration, baseline subtraction);   REWIRE (transforming of bipolarly recorded data to a common reference);

(2)   data-processing: AVRDAT (averaging of data segments for EP studies); COVDAT (computation of auto- or cross-covariance functions);   FTRDAT (Fourier transformation of original data or that of output data from COVDAT);   DIFDAT (computation of differences in time or space domain);

(3)   mapping: INPMAP (calculation of interpolated maps);   CMPMAP (scaling and compression of maps into a space-sparing format);   DEFWFN (preparation of weight coefficients for the interpolation); DEFSCL (preparation of a set of bound values for scaling);

(4)   output: LSTDAT (listing of data files on printer);   DSPDAT (graphic display of data);   DSPBLK (graphic display of the contents of a data file by logical blocks);   DSPMAP (dynamic graphic display of a sequence of interpolated maps);

(5)   auxiliary: ALTBLS (modification of the file structure);   INSHDR (putting info-records into the file header);   LSTHDR (listing of the header contents).

The flow of data-processing is summarised in Fig. 2.   In this form, UFR-BEAM is a versatile tool for most data-analytic tasks.   Its main purpose, however, is to provide facilities for studies in the spatio-temporal dynamics of EEG processes.   In the following section we shall discuss this topic in detail.

EEG MAPPING: SPECIFIC PROBLEMS

In principle, it is possible to calculate interpolated maps from any data file, irrespective of its contents, so that spatial distributions of

238

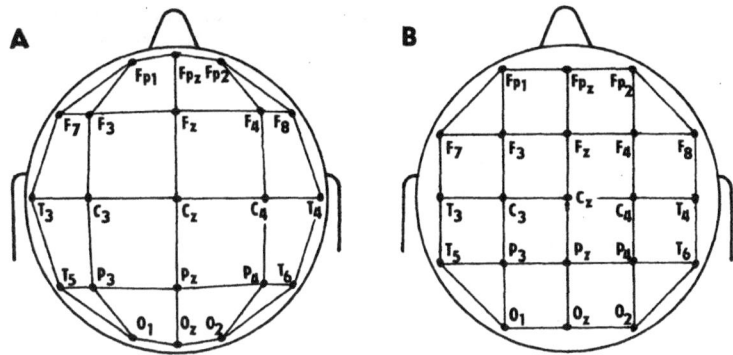

Fig. 3. Projections of head surface into image plane.
         A: a true geometrical projection; B: a conventional
         regularised projection.

any reasonable EEG parameter (amplitude, power, correlation or coherence
function) can be analysed. The specific issues of EEG mapping will now be
discussed without any special reference to the nature of the original data.

     A necessary pre-requisite for the mapping is to have a well-defined
common reference. We believe, together with other authors, that the
averaged common reference is "the least bad choice" (Walter et al, 1984),
if not the best solution. There is no need for an "inactive" electrode in
this approach; nor is it necessary to make recordings against a unique
electrode. The program REWIRE, as it is suggested by its name, transforms
the data obtained from any electrode arrangement into the equivalent values
with respect to a conventional "zero".

     Another problem to be solved is the projection of the head surface
into the image plane. Instead of a true geometrical projection, a simple
mapping is used in which the standard electrode locations, as defined by
the 10/20 system, are assigned to the nodes of a 5×5 rectangular grid,
with its four corners cut off (Fig. 3). A great advantage of this solution
lies in the uniformity of elementary areas defined by any four neighbour
nodes, so that the interpolation procedure becomes simple. On the other
hand, some distortions, especially near the frontal and occipital poles,
are inevitable. This, however, is only a minor disadvantage. The projection pattern is thus defined by 21 electrode locations; for any specific
arrangement, their assignment to the channel numbers is stored in the file
header. In most cases, there are some locations which remain unoccupied.
As the interpolation program requires all the nodes within the mapped area
to have defined values, the missing node values are to be supplemented by
means of the program PREINP.

     The interpolation program INPMAP is the heart of the mapping branch
of the UFR-BEAM system; it transforms data files into map files. The
elementary square areas defined by four neighbour nodes are divided into
32×32 pixels, the full format of maps being thus 129×129 pixels. During
the interpolation, the value assigned to any point is calculated as a
weighted average from the four known values assigned to the corners of the
elementary area containing the point. More specifically, the following
formula is used:

$$u^*(x, y) = \sum_{i=0}^{1} \sum_{j=0}^{1} w\left(i + (1 - 2i)\frac{x}{L},\ j + (1 - 2j)\frac{y}{L}\right) \cdot u_{ij} \qquad (2)$$

where $u_{00}, \ldots, u_{11}$ denote the known node values, $u^*(x, y)$ represents the estimated value in the point (x, y), L is the size of the elementary square side, and $w(\cdot, \cdot)$ is a weight function defined over the square $[0,1] \times [0,1]$. This function has to meet the following requirements:

(i) $w(0,0) = 1$;
(ii) $w(\xi, \eta) = w(\eta, \xi)$ for any ξ, η;
(iii) $w(1, \eta) = 0$ for any η; and   (3)
(iv) w is non-increasing in both its arguments.

The choice of the weighting function presents some rather delicate problems. After some experimenting it has been found that the function

$$w(\xi, \eta) = (1 - \xi)(1 - \eta) \qquad (4)$$

provides the most satisfying results. Nevertheless, there is the possibility of choosing other weighting functions according to the user's needs; the program DEFWFN calculates the weighting coefficients for a sufficiently large family of functions. It may be worth mentioning, however, that weighting functions proportional to the rth power ($r > 0$) of the Euclidean distance of the point from a node, as have been proposed by some authors (Buchsbaum et al., 1982; Walter et al., 1984), do not fulfil the conditions (3)-(iii), (iv) and so they do not appear very convenient.

It should be emphasised that the interpolation according to (1) has the property of linearity, that is

$$M\left(c_1 u^{(1)} + c_2 u^{(2)}\right) = c_1 M\left(u^{(1)}\right) + c_2 M\left(u^{(2)}\right), \qquad (5)$$

where $u = (u_1, \ldots, u_N)$ denotes the vector of the known node values, M represents the interpolation operator, and c_1, c_2 are any scalars. Due to this fact, the interpolation commutes with another linear operation (such as averaging, for example), so that such operations may be applied on the input data first rather than on the resulting maps, with the same result. This is an important reason for keeping to these procedures, even if higher-order methods (2-D splines) provide an undoubtedly better solution to the interpolation problem (Perrin et al., 1987).

After interpolation the resulting matrix is submitted to the packing program CMPMAP for scaling and subsequent compression of the maps. The maps are scanned line-by-line; to each pixel its scale value in the range 0 - 15 is assigned, and then contiguous segments of pixels of the same scale value are identified and represented within 2 bytes in the form [length, value]. By this procedure, a considerable reduction in the disk space required by the final map is achieved. As an option, the scaled map can be submitted to the program ISOMAP for detection of contour lines between the areas of the same scale value and thus represented by isolines rather than by colour or brightness levels.

The scaling is linear through the full range of values in most cases; alternatively, a non-linear scaling can be defined by the user (program DEFSCL). The contents of the map files can be either presented visually on a colour graphic display (program DSPMAP), or printed in a semi-graphical form on a line printer (program PRTMAP).

The program DSPMAP provides several display layouts: it is possible to display either one map in expanded format (257 x 257), or one map in standard format (129 x 129) together with the original data tracings, or two "dual" maps in standard format simultaneously. The scale values are converted into colour/brightness levels via a look-up table, which can be selected from several pre-defined "palettes" by the user. Together with the maps, some fundamental information about the file, extracted from the header, is displayed.

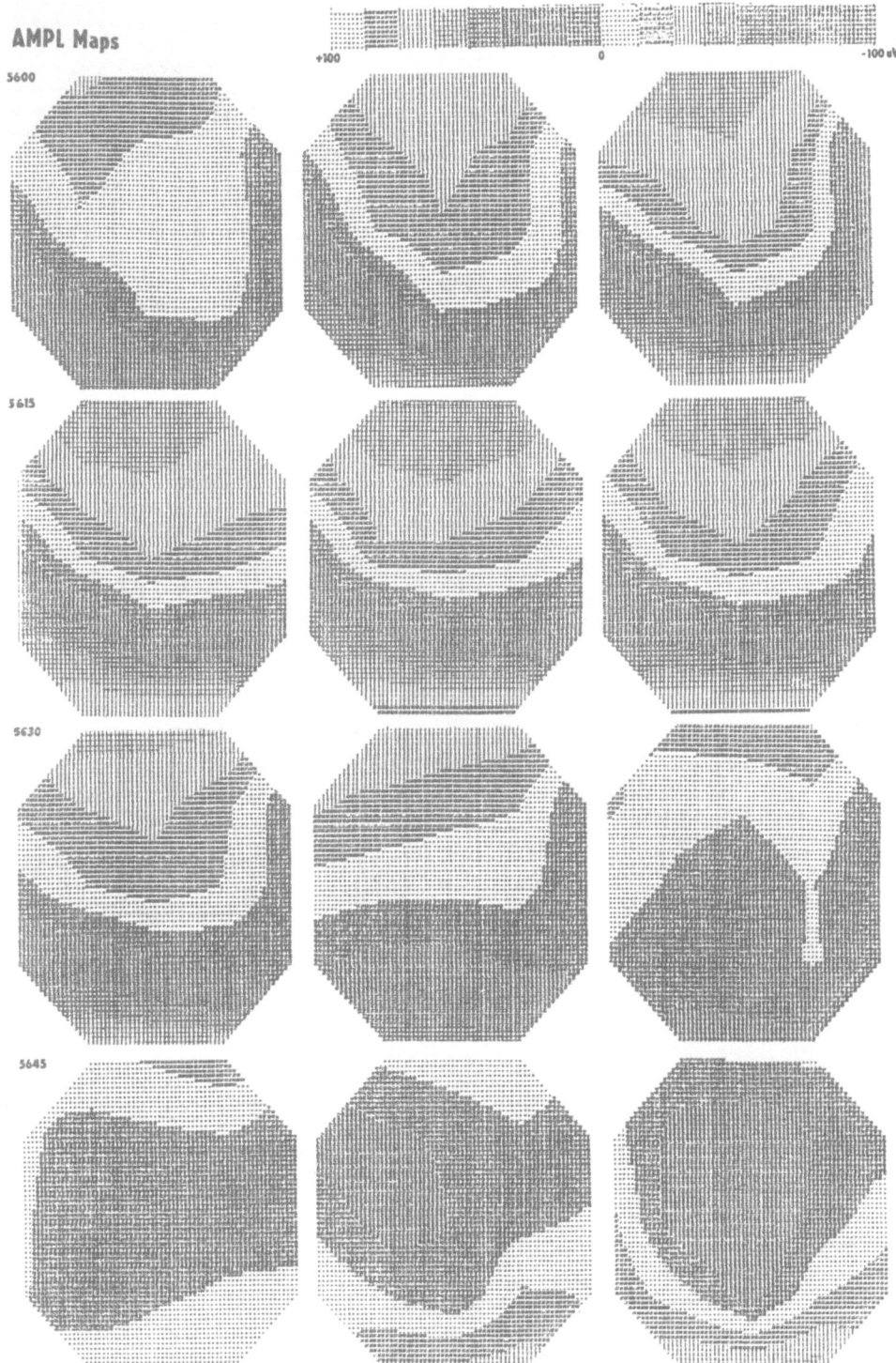

Fig. 4a and b. Example of amplitude mapping. The numbers indicate the
current time (ms) with respect to the start of recording.

241

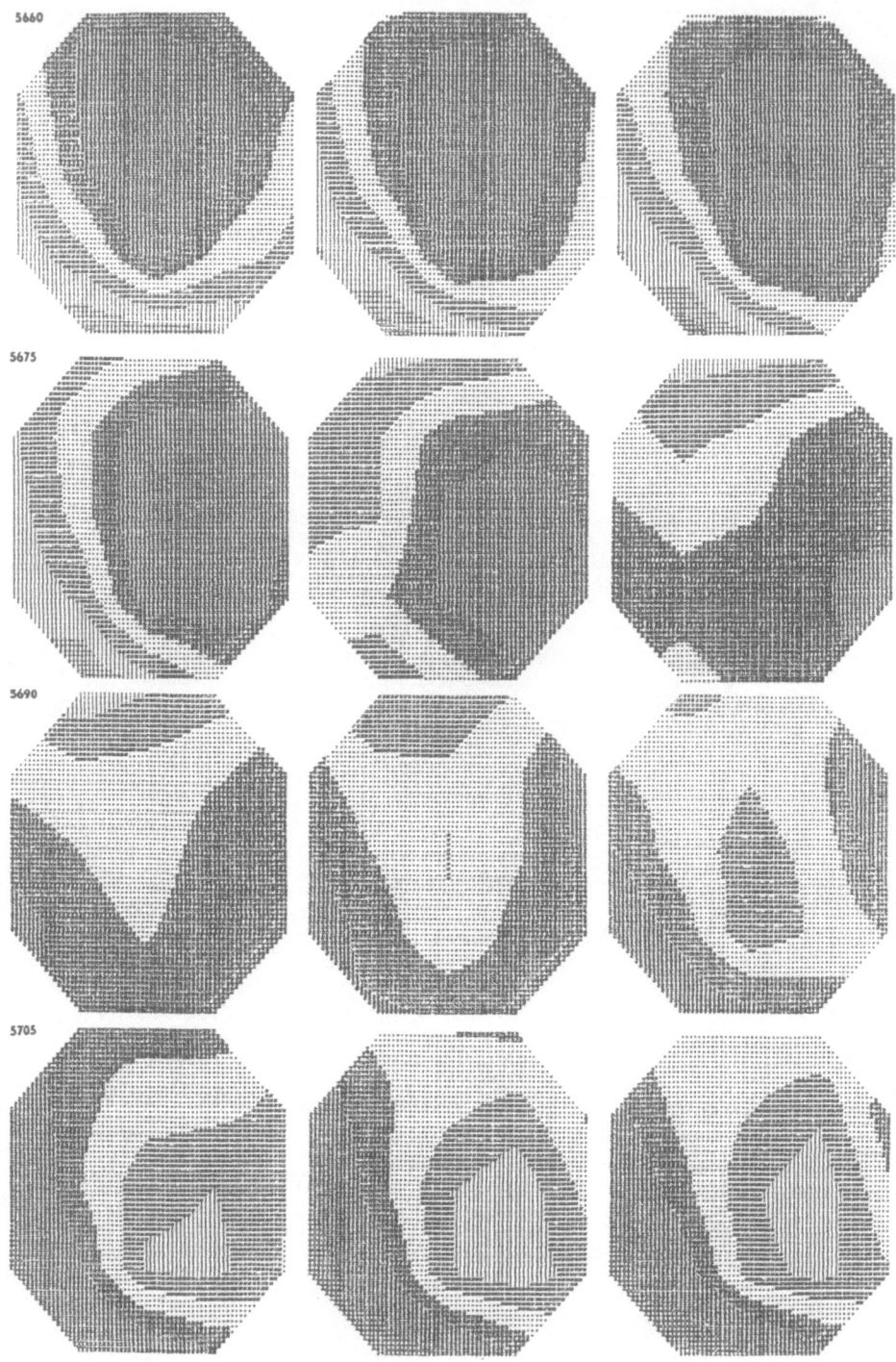

Fig. 4b. Example of amplitude mapping (continued)

The course of display can be controlled manually by the operator; it is therefore possible to alter colours, switch the display layouts, move forward or backward through the map file according to its BLS, and so on. The event-codes, if present, can be displayed in alpha-numerics or by means of pre-defined graphic symbols. For a summarising review, the map file can be displayed in an automatic mode like a movie, at the maximum rate of 4 snaps s^{-1}. This provides rather impressive insight into the dynamics of the EEG phenomena being studied.

The program PRTMAP transforms the contents of map files into text files which are supposed to be printed on an ordinary line printer. The resolution is reduced twice, so that a full-format map is printed in a grid 65 x 65 characters; a pair of "dual" maps can then be printed simultaneously on a 132-column form. The scale values are converted into characters of various optical density, as can be seen in Figs. 4 - 6.

APPLICATIONS

The system has been designed primarily to study spatio-temporal patterns of the electric activity of the functioning brain, with special emphasis on time/phase relationships between the activity of various cortical areas corresponding to the projections of stimuli used in conditioning experiments. Thus, the most frequently used techniques will be concentrated upon analysis in the time domain rather than in the frequency domain, unlike the ordinary power mapping used in clinical practice. Examples of amplitude and correlational mapping will be presented.

These examples are based on EEG data recorded from a healthy male subject, age 30, in a control session: the subject was sitting in a dark, soundproof room, relaxed and with closed eyes. The recordings were derived in a bipolar manner from 16 locations in relation to C_z (see Fig. 3; Fp_z, O_z, T_3, T_4 omitted). The signal was digitised at a rate of 200 s^{-1}. After calibration the data were transformed to a common reference as defined by the total sum equal to zero. The values for the four missing locations were computed as the average of the neighbouring locations, so that the full 21-channel data file was obtained.

In Fig. 4 there is an example of a sequence of maps interpolated from these original data. It illustrates the dynamics of the brain potential field with its characteristic changes of pattern, described by Livanov (Livanov and Ananiev, 1960) as "migration" (pereliv) and later extensively studied by Lehmann (1977; 1984). Observing the map sequence at greater length, it is possible to distinguish two main dynamic principles in the process, namely "fluttering" and "swinging" of field distribution; these changes occur at a rate equivalent to the dominant alpha-frequency (10.8 Hz, as revealed both by spectral analysis and by visual inspection).

For the study of spatio-temporal patterning of the brain field dynamics, correlational mapping appears to be a suitable tool. In this approach, cross-covariance or cross-correlation functions between some reference location and all the remaining locations are evaluated for a wide range of time lags and then these values are presented as a sequence of maps with the time lag as the BOV. In the case of an EEG dominated by a unique frequency, as in the example presented, the correlations correspond directly to the phase shifts; in other cases, a mapping of cross-spectral density or coherence functions would be more appropriate. In Fig. 5 there is an example of a series of such cross-correlation maps for some selected time lags. The correlation functions with respect to O_z were calculated for a 1-second epoch (5000 - 6000 ms), from which the example in Fig. 4 also originates.

The last example demonstrates the use of a "mirror" option of the program DIFDAT. By summation and subtraction of a map and its mirror image along the midline, a pair of "dual" maps is obtained, which represent symmetrical and anti-symmetrical components of the distribution; the algebraic sum of the two maps equals the original map. In Fig. 6 this is done for a strongly asymmetric (B) and a slightly asymmetric (A) map. An examination of such pairs of Sy/Asy maps provides a comprehensive look into interhemispheral differences in the distribution of the investigated EEG parameter.

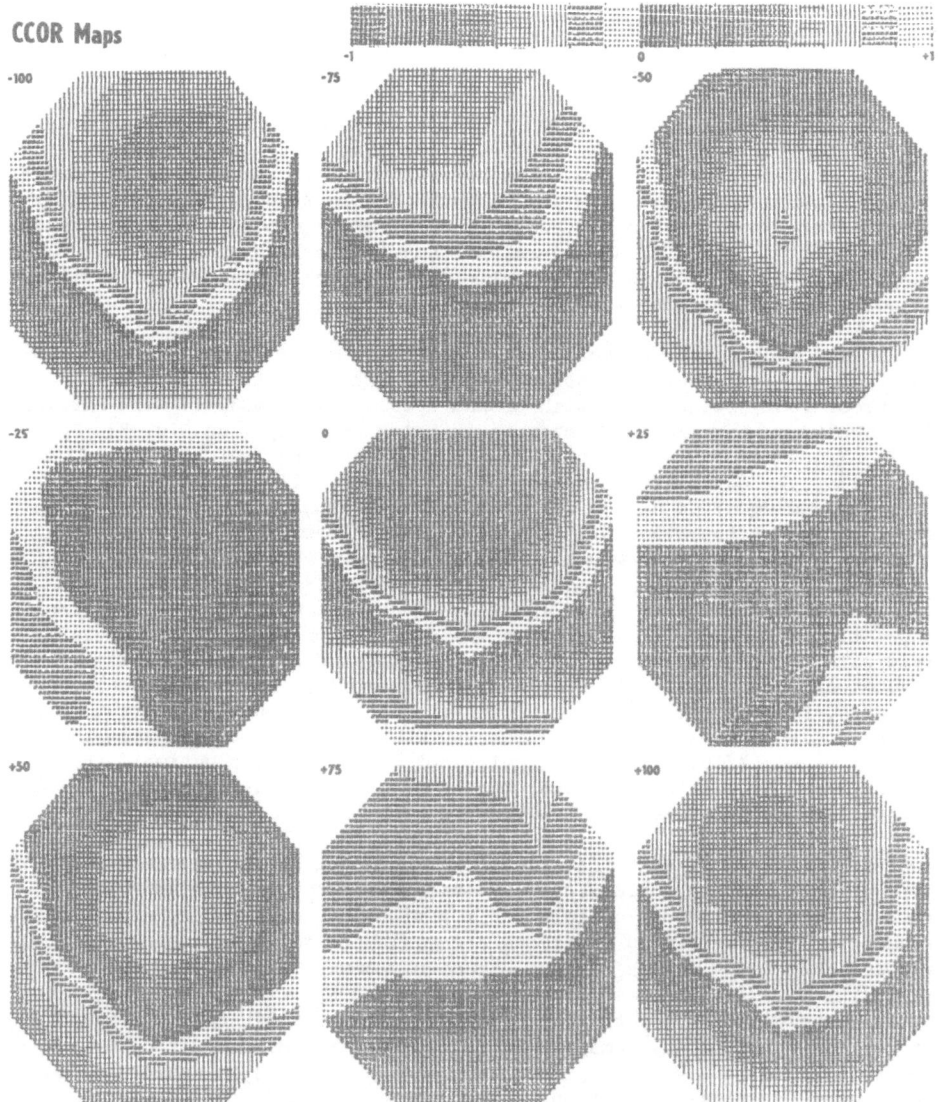

Fig. 5. Example of correlational mapping. The numbers indicate the time delay between the reference (O_z) and other locations.

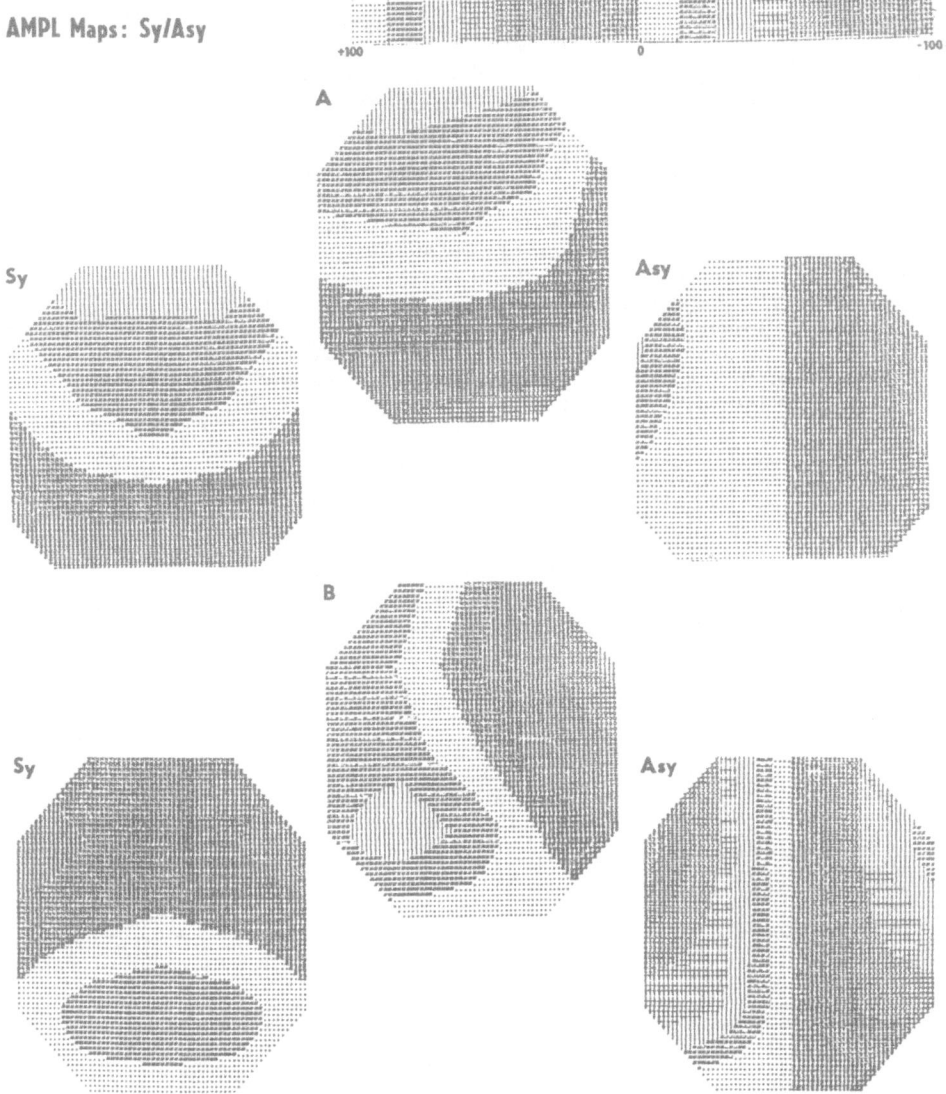

+100   0   -100 uV

Fig. 6. Evaluation of the field asymmetry for a slightly (A) and strongly
         (B) asymmetric map.

SUMMARY

        In this chapter a general description of the software system UFR-BEAM
has been given and its fundamental concepts explained. Topics related to
the topo-electroencephalography of brain processes have been discussed.
The use of the system has been demonstrated by some examples of amplitude
and correlational mapping of a spontaneous EEG activity.

REFERENCES

Berger, H., 1969, On the electroencephalogram of man (P. Gloor, transl.
     and ed.), EEG Clin. Neurophysiol. (suppl. 28).

Buchsbaum, M. S., Rigal, F., Coppola, R., Cappelleti, J., King, C., and
    Johnson, J., 1982, A new system for gray-level surface distribution
    maps of electrical activity, EEG Clin. Neurophysiol., 53 : 237.
Lehmann, D., 1977, The EEG as scalp field distribution, in: A. Remond, ed.,
    "EEG Informatics", Elsevier, Amsterdam : 365.
Lehmann, D., 1984, EEG assessment of brain activity: spatial aspects, seg-
    mentation and imaging, Int. J. Psychophysiol., 1 : 267.
Livanov, M. N., and Ananiev, V. M., 1960, "Electroencefaloskopiya", Moscow
    (in Russian).
Perrin, F., Pernier, J., Bertrand, O., Giard, M. H., and Echallier, J. F.,
    1987, Mapping of scalp potentials by surface spline interpolation,
    EEG Clin. Neurophysiol., 66 : 75.
Petsche, H., 1976, Topography of the EEG: survey and prospects, Clin.
    Neurol. Neurosurg., 79 : 15.
Walter, D. O., Etevenon, P., Pidoux, B., Tortrat, D., and Guillou, S.,
    1984, Computerized topo-EEG maps: difficulties and perspectives,
    Neuropsychobiology, 11 : 264.

FOETAL MONITORING: AUTOCORRELATION OF ULTRASOUND FOETAL SIGNAL IN THE

DEVICE AM-2

E. Novakov and M. Dimitrova

Institute of Biomedical Engineering (CLEMA)
Medical Academy
Sofia, Bulgaria

INTRODUCTION

Foetal heart rate (FHR) monitoring during pregnancy and delivery provides very important diagnostic information (Goodelin, 1979).

The Foetal Monitor AM-2 is a completely non-invasive instrument
designed for continuous monitoring simultaneously of the foetal heart rate
(FHR) and the uterine contractions of the mother (TOCO signal). Both signals are obtained from external transducers, attached to the mother's abdo-
men by special belts.

The signal from the foetal heart activity is obtained by the continuous
Doppler ultrasound method (Wells, 1977). Fig. 1 shows the block diagram of
the device, whilst Fig. 2 shows the real signals. The signal U2 is fed to
an analog to digital converter (ADC) and a microprocessor system. The
sampling rate is 125 Hz with 8-bit resolution. A 4s epoch (500 sampled
values) is stored in the microcomputer memory.

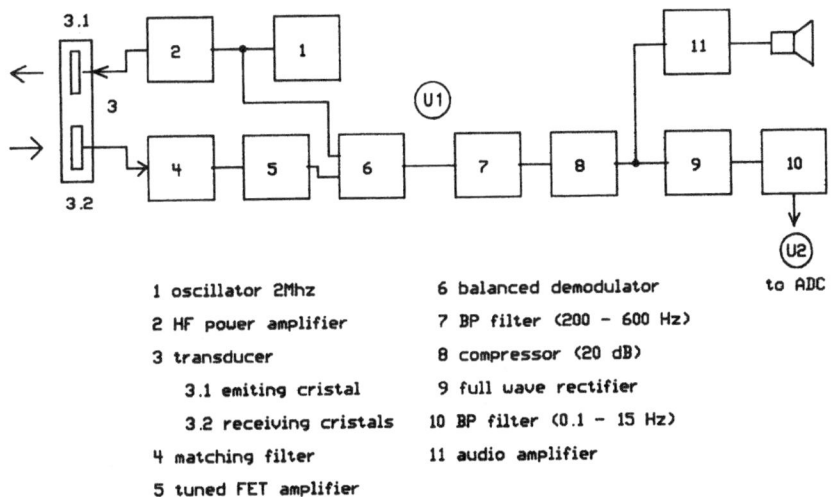

1 oscillator 2Mhz

2 HF power amplifier

3 transducer

   3.1 emiting cristal

   3.2 receiving cristals

4 matching filter

5 tuned FET amplifier

6 balanced demodulator

7 BP filter (200 - 600 Hz)

8 compressor (20 dB)

9 full wave rectifier

10 BP filter (0.1 - 15 Hz)

11 audio amplifier

to ADC

Fig. 1. Block diagram of the Foetal Monitor AM-2.

247

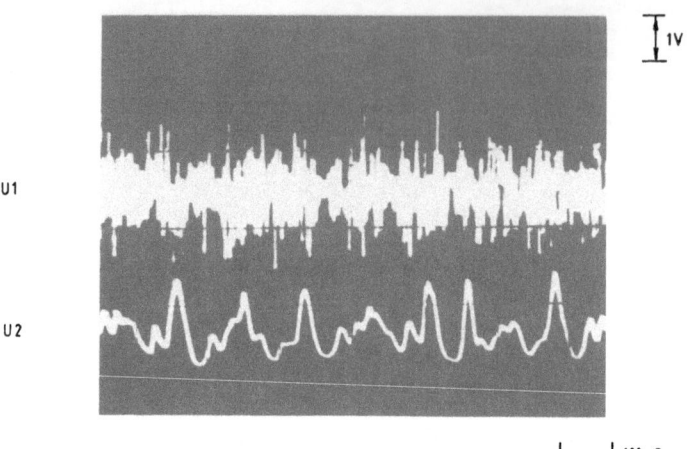

Fig. 2.   Typical ultrasound foetal signals.

THE AUTOCORRELATION METHOD

An autocorrelation method for FHR calculation is used due to the low signal-to-noise ratio and the irregularity of the heart signal of the foetus (Takeuchi and Hogaki, 1987; Tuck, 1981).

The signal stored in the microcomputer memory is filtered using a finite impulse response (FIR) digital filter with the following equation:

$$xf(i) = [1/(2n+1)] \sum_{i=-n}^{n} x(i) \; , \quad n = 2, \; i = 2 \text{ to } 498 \tag{1}$$

The autocorrelation function (ACF) is computed by the equation:

$$C_{xx}(p) = \sum_{i=0}^{N-p-1} xf(i)xf(i+p) \quad N = 500, \; p = 0 \text{ to } 250 \tag{2}$$

The normalised ACF or correlation coefficient is computed by:

$$\overline{C_{xx}}(p) = C_{xx}(p)/C_{xx}(0) \tag{3}$$

Fig. 3 shows the real signals on the left hand side and their ACF on the right hand side. The main problem in the autocorrelation analysis is the large number of multiplications to be performed leading to an increased computation time.

A reduction from 8-bit to 4-bit accuracy is made in order to reduce the computation time of ACF. The dynamic range of the signal is (Fig. 4):

$$D = A_{max} - A_{min} \tag{4}$$

The new resolution level is:

$$Q = D/15 \tag{5}$$

and the new sampled value is:

$$x4(i) = INT[xf(i)/Q] \tag{6}$$

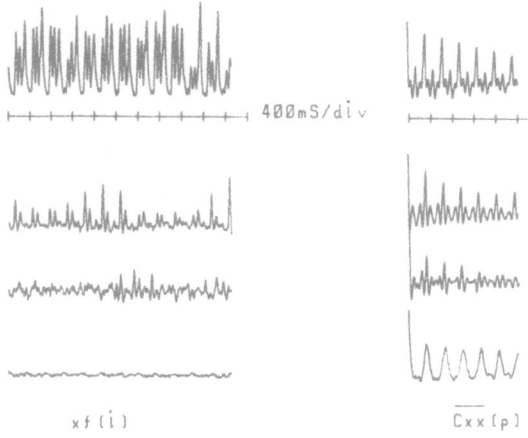

$$xf(i) \qquad \overline{Cxx}(p)$$

Fig. 3. Real signals (left) and their ACF
         (right).

where INT[ ] is the integer part of the division. Fig. 5 shows the original 8-bit sampled signal xf(i) and the same signal with a 4-bit accuracy
x4(i) with their respective ACF. It is obvious that the error ERR is not
correlated to the signal. The ACF computation time is reduced substantially
because a 4-bit multiplication is performed.

An amplitude-interval analysis is made on the ACF in order to obtain
the maxima needed for FHR determination. The time occurrences of the three
local maxima with highest amplitudes (T1, T2 and T3) are determined and the
following conditions are checked (Fig. 4):

$$\left| 1 - T1/T2 \right| < k \qquad\qquad\qquad (7)$$

$$\left| 1 - T2/T3 \right| < k, \quad k = 0.04 \qquad\qquad (8)$$

If the conditions are fulfilled the FHR is calculated as follows:

$$FHR = 60000/T1 \qquad\qquad\qquad (9)$$

where T1 is in ms and FHR in beats per minute.

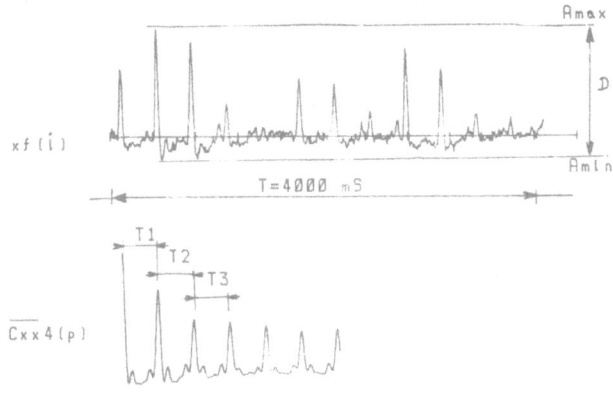

Fig. 4. Dynamic range of the signal.

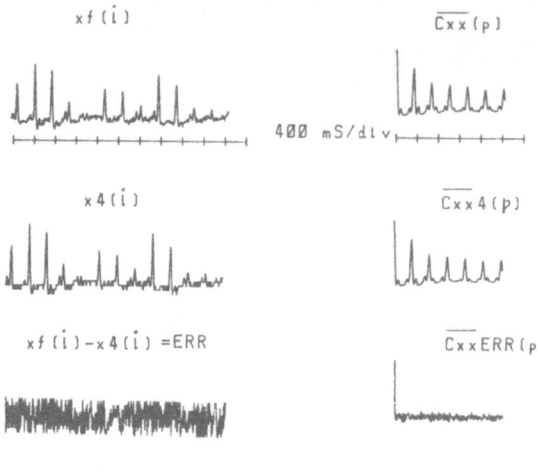

Fig. 5. Original 8-bit sampled signal xf(i)
and the same signal with a 4-bit
accuracy together with their ACF.

An MC6809 microprocessor is used in AM-2. This particular microprocessor has an unsigned 8 x 8-bits multiplication command giving the possibility of multiplying signed 4-bit numbers. The data acquisition and the ACF computation are performed in parallel mode (the so-called ping-pong mode) without any information loss (see Fig. 6). In the Ti time interval the signal acquisition is performed and the analysis is performed in Ti + 1. At the same time a new epoch is sampled from the input signal. All the algorithms are implemented in software.

SYSTEM FEATURES

A tensometric transducer produces the TOCO signal. The same microprocessor processes the TOCO signal (10 Hz sampling rate and 8-bit resolution). A silent microdot thermal printer plots both the FHR and TOCO signals - Fig. 7.

Additionally, the AM-2 (Fig. 8) has the following features:

  (i)   zero line clamping of the TOCO signal;
 (ii)   a monitor mode: when FHR loss of information occurs, an audible
        alarm is activated;
(iii)   an LED panel display showing the Doppler signal quality;
 (iv)   a test signal and a self-diagnostic firmware program;

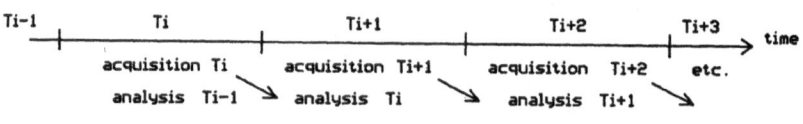

$(Ti-1) = (Ti) = (Ti+1) = (Ti+2) = 4$ S

Fig. 6. Data acquisition and ACF computation performed in parallel
mode.

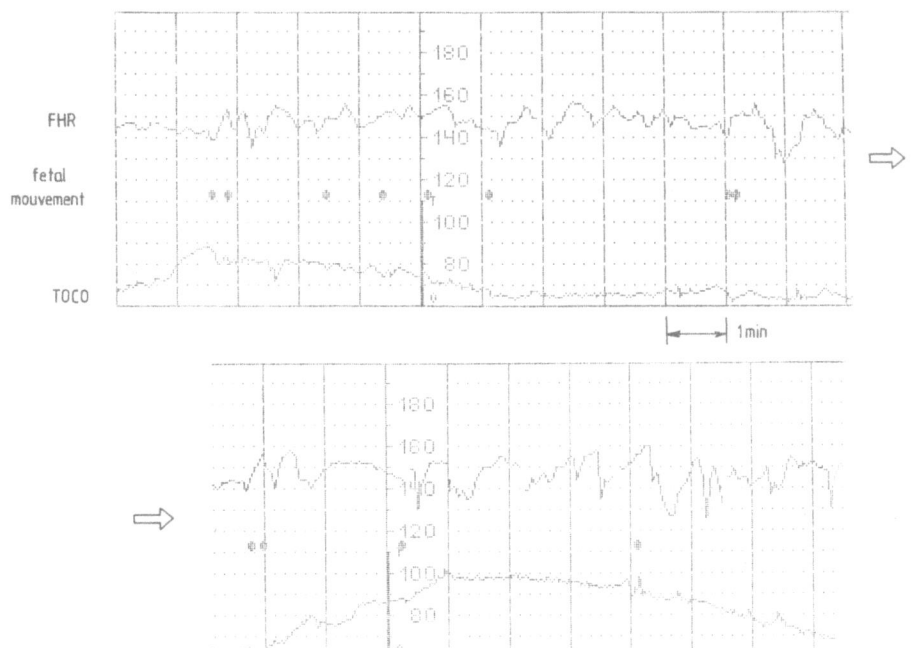

Fig. 7.   Plots of FHR and TOCO signals.

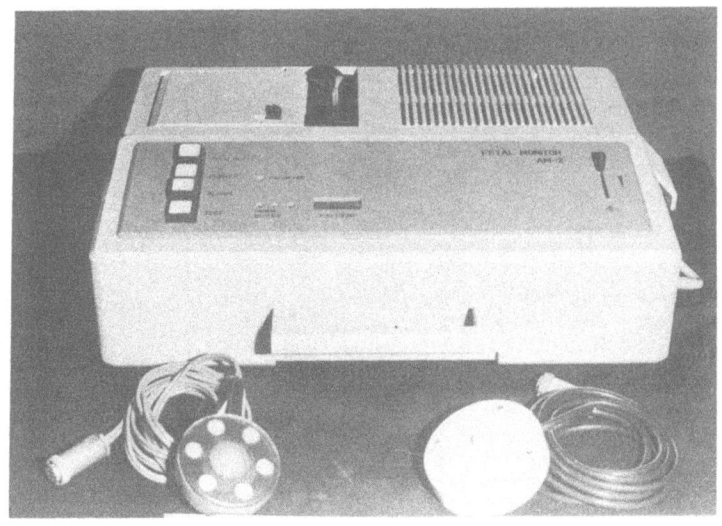

Fig. 8.   The Foetal Monitor AM-2.

(v)   an interface for the central monitoring station (optional);   and
(vi) a hand-held button for the mother, marking on the record movements
      of the foetus which are sensed.

REFERENCES

Goodelin, R., 1979, History of fetal monitoring, <u>Am. J. Obst. Gynecol.</u>,
    133 : 323.
Takeuchi, Y., and Hogaki, M., 1987, An adaptive correlation rate meter:
    a new method for Doppler fetal heart rate measurements, <u>Ultrasonics</u>,
    16(3) : 127.
Tuck, D., 1981, Improved Doppler ultrasonic monitoring of the foetal heart
    rate, <u>Med. Biol. Eng. Comput.</u>, 19 : 135.
Wells, P., 1977, "Biomedical Ultrasound", Academic Press, Oxford.

ANALYSIS OF FREQUENCY PROPERTIES OF IMAGES FROM ULTRASOUND SCANNERS

J. Jan

Department of Medical Electronics
Technical University
Brno, Czechoslovakia

INTRODUCTION

Contemporary commercially available ultrasound real-time sector scanners usually digitise the video signal into a matrix of about 120 columns,
each corresponding to one direction in a 90 degree sector, while each
column has about 500 samples. The resulting relatively high sampling rate
is easily available with up-to-date technology, but when considering the
possibilities of real-time post-processing, archiving in digital form or
arranging cine loops on the basis of multiple image data in RAM (and also
when the component cost and availability are of concern), keeping the sampling
rate at its natural minimum given by the sampling theorem may be worth a
certain effort. In the present study the typical point spread function,
as measured in Jan et al. (1987), is Fourier analysed and conclusions are
drawn as to the lowest acceptable sampling rate. The validity of the conclusions has been verified by analysing typical tomograms of biological
objects.

MEASUREMENT AND ANALYSIS

This chapter focuses on the case of a typical sector scanner (ADR 4000)
with a wobbling crystal 3 MHz transducer. A special interface was provided
for connecting the scanner and an 8-bit microcomputer served for controlling
the experiment and acquiring the image data in the original format in polar
co-ordinates (before the image format conversion).

As shown experimentally in Jan et al. (1987), the point spread function
(PSF) is practically constant over about 60 to 70% of the image area in this
original acquisition form so that it is this type of data that enables a
reasonable analysis to be carried out, while analysing the converted output
data in TV display format provides results which are difficult to interpret.

The results of measurements of point responses (Jan et al., 1987) from
different areas of the image field are partly summarised in Fig. 1, where
radial and lateral central profiles of responses are plotted. It can be
seen that, in accordance with the isoplanarity mentioned above, the samples
are distributed along similar curves for all the responses; at the same
time it is obvious that the original image (120 columns x 512 pixels) was
oversampled and that some reduction of the amount of data was possible.

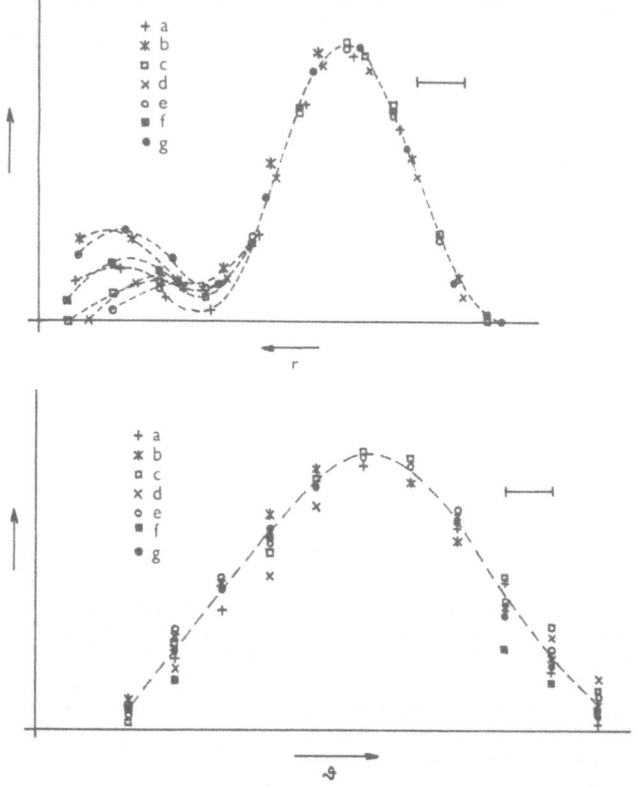

Fig. 1. Upper panel: radial (column) profiles of
"point" responses (a, b, ..., g: responses
in different locations in the image area,
marked sampling interval about 0.28 mm);
lower panel: lateral (row) profiles (marked
sampling interval 0.75 degree).

Due to the different nature of both polar co-ordinates, only 1-D
Fourier analysis on selected columns (that is, θ = const) or rows (r =
const) of images of point targets and of typical biological objects (heart,
liver, gall bladder) has been performed. The selection of columns or rows
has been made by the operator so that their content was typical for the respective type of image; the selected portions of images corresponding to the
following example results are marked in Figs. 2 - 5. Series of many rows or
columns have been analysed in order to ensure a sound basis for generalising
the conclusions and excluding artefacts.

The radial profiles are 512 samples long, but only portions of 256
samples from the most interesting part of pictures (rows 176 - 431) were analysed; the lateral profiles 120 samples long were padded by zeros to a
length of 128. The power spectra were estimated by squaring the amplitudes
of spectral components derived by DFT. In the figures from Fig. 6 onwards,
the spectra are presented in the range of spatial frequencies from 0 to the
Nyquist limit (half the sampling frequency) so that they have 128 and 64
components, respectively.

The first part of the results covers the analysis of measurements of
PSF that were carried out on a group of "point" targets arranged as a

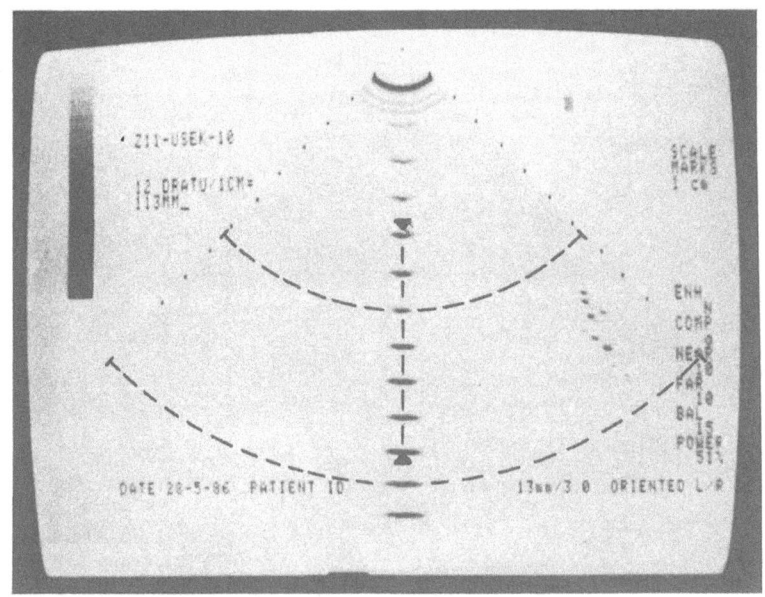

Fig. 2. Converted (output) image of a ladder-like set of
"point" targets with marked profiles that are
analysed further.

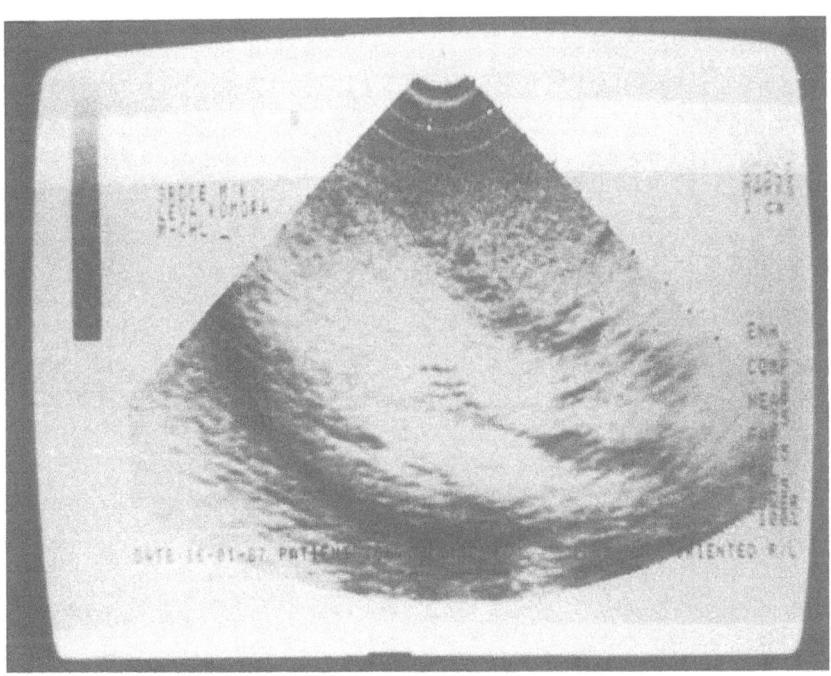

Fig. 3. Output image of the heart.

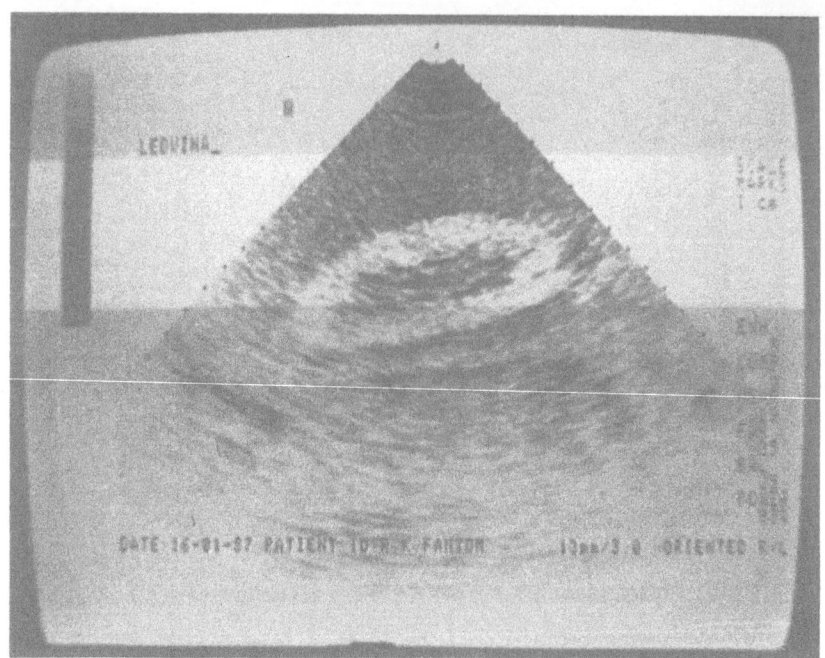

Fig. 4.   Output image of the liver.

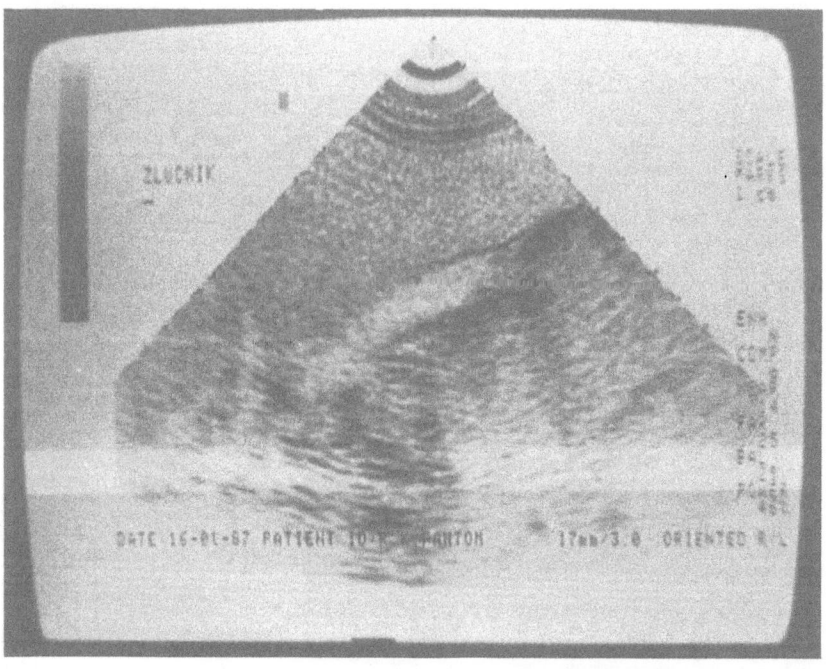

Fig. 5.   Output image of the gall bladder.

ladder-like set of thin wires perpendicular to the tomographic plane. The distance between "rungs" was 10 mm, enough to avoid any interference between neighbouring responses. Two sector depths were involved, 145 and 200 mm. Spectra of radial profiles (Figs. 6, 8, 10 and 12) are characterised by discrete lines due to periodical structure of the original functions but, from our point of view of more importance, the higher components are negligibly small. The spectra corresponding to the smaller sector depth have important components only to about $\frac{1}{3}$ of the Nyquist frequency, while the more sparsely sampled case of the greater sector depth has the spectra not exceeding about $\frac{1}{2}$ of that frequency. Higher, very weak, spectral components may be attributed to noise without information content.

The lateral (row) profiles, where the independent variable is the azimuth angle θ, describe, in fact, the directional characteristic of the ultrasonic probe distorted by non-linearities of the signal processing system. It is obvious that the lateral resolution of the system is rather low as the characteristic is quite wide (about 10 samples); this is reflected in the spectrum by a rapid decline towards the higher frequencies. The profile is similar to a cosine-like impulse which causes some weak side lobes to occur besides the main peak in the spectrum; the main lobe is substantially narrower than the Nyquist limit.

The second group of results concerns the analysis of images of real biological objects. The analysis of radial profiles (Figs. 8, 10 and 12) essentially verifies the results of PSF analysis. Typically, all the energy was concentrated in the lower third of the column spectra in images with a maximum depth of 145 mm (gall bladder and heart), while with a greater depth of 200 mm (kidney), about all of the lower half of the spectra was non-zero. Therefore, the sampling rate could have been 3 or 2 times lower, respectively, without losing any information. The row spectra (Figs. 9, 11 and 13), on the other hand, showed a less pronounced (but still obvious) concentration of the energy in the lower part of the spectra in comparison with the case of PSF measurements, clearly due to the presence of speckle textures, as the spectrum of the isolated PSF was also practically band-limited to about $\frac{1}{6}$ of the sampling frequency. This means that lowering the sampling rate along θ could cause a certain degree of distortion due to aliasing although the important information is contained in lower frequency components. Thus, simply lowering the number of radial rays (that is, the number of samples on a lateral profile) to, say, 60 cannot be directly recommended, although such a sampling would carry all the information content allowed by the given PSF of the system. Rather, filtering out the speckle components in full 120-column data can be considered before resampling, but to design such a filter would not be easy as the speckle noise has a more or less multiplicative character.

CONCLUSION

According to our experimental results, the information capacity of the image matrix used of 120 x 512 pixels is far from being fully utilised. Substantial reduction in the acquisition sampling rate (especially along r) is possible without causing any important deterioration in the resulting image quality. This may be of interest in image processing or if many frames need to be stored and reproduced, such as in archiving or making cine-loops. Typically, the amount of data reduction may be from 2:1 to 8:1. Evaluation of the quality of output images converted on the basis of the reduced data is the subject of Jan and Kubák (1987).

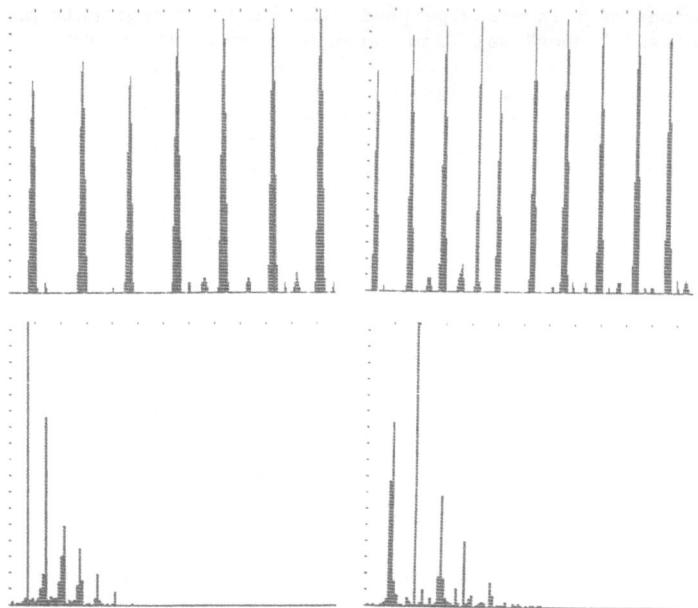

Fig. 6. Radial profiles of ladder-like set of "point" targets and their power spectra (left: max. sector depth 145 mm; right: 200 mm, both column 57).

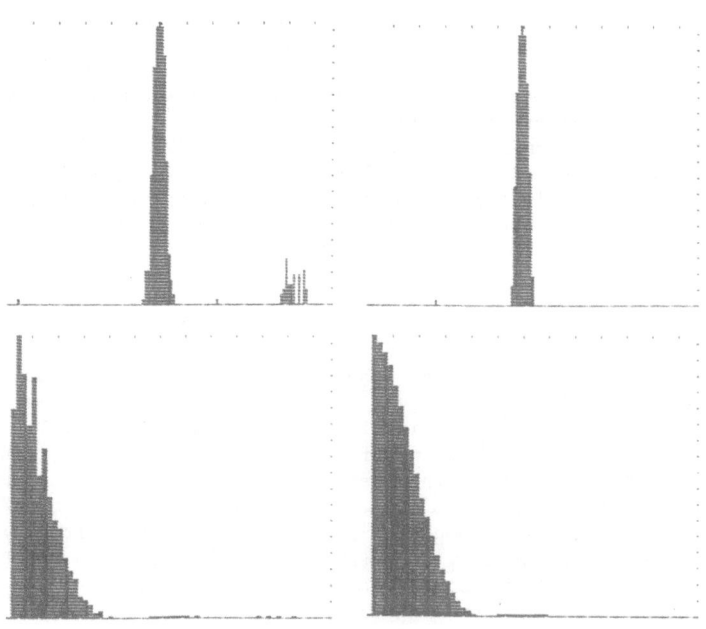

Fig. 7. Lateral profiles of "point" targets and their power spectra (left: row 307; right: row 455).

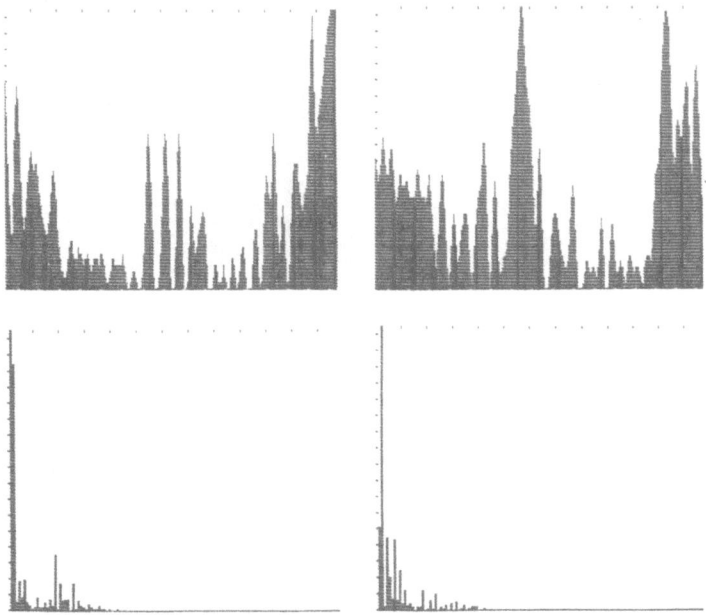

Fig. 8. Radial profiles of the heart and their power
spectra (left: column 57; right: column 96).

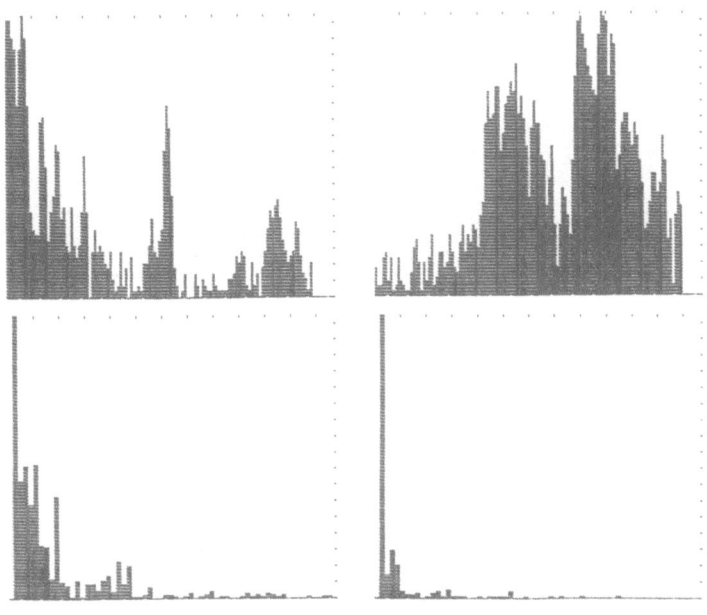

Fig. 9. Lateral profiles of the heart and their power
spectra (left: row 307; right: row 490).

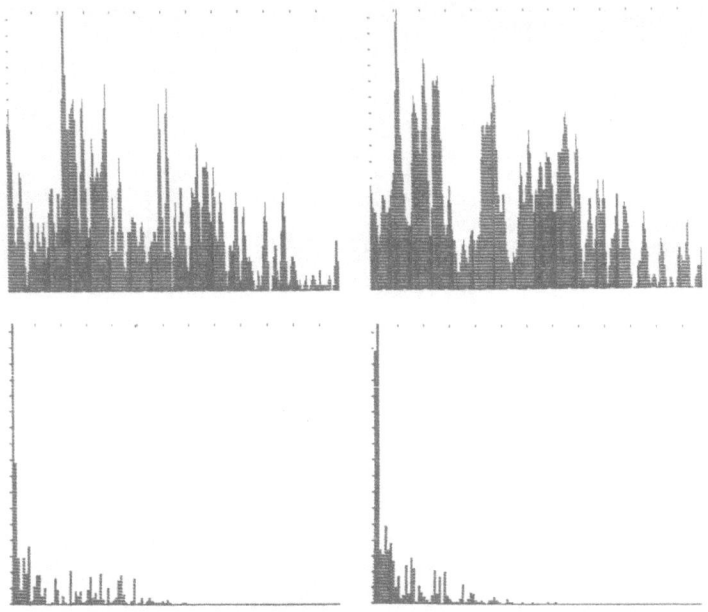

Fig. 10. Radial profiles of the liver and their power
spectra (left: column 36; right: column 93).

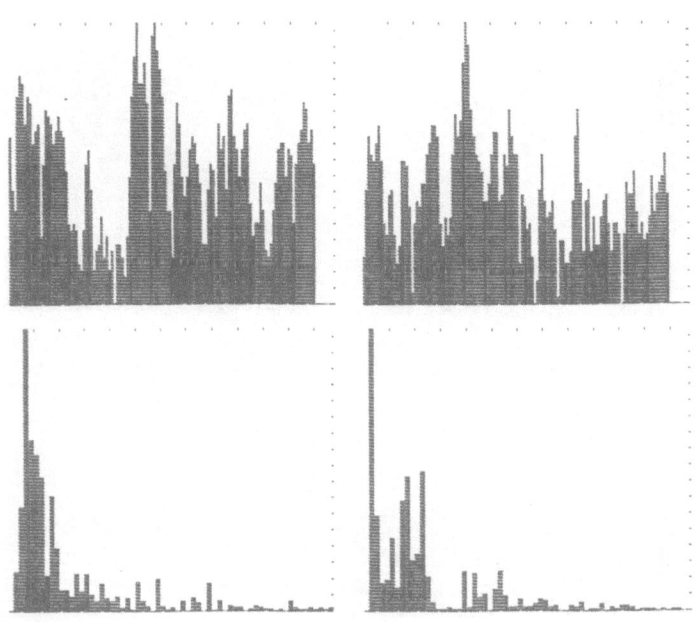

Fig. 11. Lateral profiles of the liver and their power
spectra (left: row 304; right: row 344).

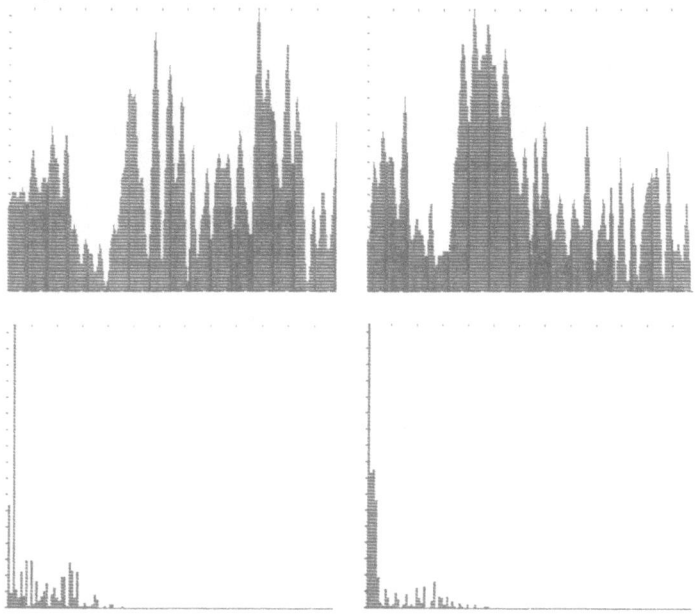

Fig. 12. Radial profiles of the gall bladder and their
spectra (left: column 50; right: column 64).

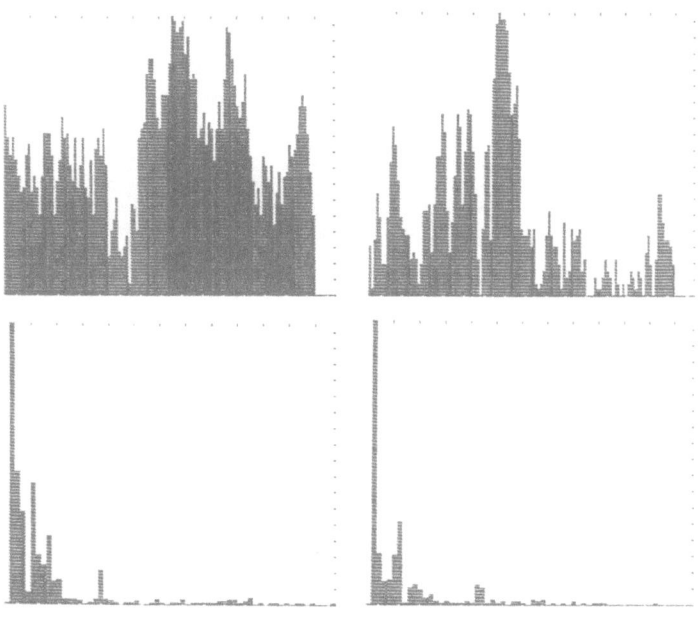

Fig. 13. Lateral profiles of the gall bladder and
their spectra (left: row 263; right: row 360).

REFERENCES

Jan, J., and Kubák, R., 1987, Minimizing data volume and conversion effort in ultrasonographic imaging, in: "Proc. 7th Congress Euroson '87", Finnish Society for Ultrasound in Medicine and Biology, Helsinki : 293.
Jan, J., Kubák, R., and Knotek, M., 1987, On isoplanarity of point spread function of ultrasound real time scanners, in: "Proc. UBIOMED VII Conference (Eisenach 1986)", M.L. Universitat, Halle-Wittemberg : 46.

DIGITAL PROCESSING OF BIOMEDICAL DYNAMICAL IMAGES

A. M. Taratorin, A. G. Kogan, E. E. Godik and Y. V. Guljaev

Institute of Radio Engineering and Electronics
USSR Academy of Sciences
Leningrad, USSR

INTRODUCTION

A great deal of attention in the field of biomedical image analysis is
now paid to the methods which are based on the study of spatio-temporal
dynamics of informative parameters. These dynamic images are very different
from static ones because they enable us to analyse transient processes and
the temporal behaviour of physiological control systems, and therefore they
characterise the functional state of the biological object. Examples of
dynamic biomedical images include: dynamic thermography sequences, spatio-
temporal mapping of heart and brain magnetic fields and dynamic tomography.

Because of the large volume of processed data the problem of dynamic
image sequence processing requires a rather powerful computing system:
high computing speed, a large amount of operating memory and some special
devices for image representation are necessary. Modern specialised image
processing systems can now process image sequences of reasonable size: up
to several hundred frames of 128 * 128 pixel resolution with 8 bits per
point depth. Our experience shows that a personal computer, compatible
with an IBM PC XT model, with an image processing board can be an appropriate solution for a large variety of dynamic image processing problems.

METHODS

For several years work has been carried out in the Institute of Radio
Engineering and Electronics of the USSR Academy of Sciences in the field
of creating programs and algorithms for the processing of biomedical image
sequences. Software developed includes a set of algorithms for dynamic
curve analysis, preliminary processing of image sequences, spatio-temporal
filtering, spatio-temporal slice analysis and the creation of functional
images.

One of the main and most common problems in the analysis of image
sequences is the comparison of temporal curves obtained in different spatial
regions of the object. These curves reflect the dynamics of the parameters
studied, parameters such as temperature, brightness and magnetic field.
The processing and analysis of these curves yields a representation of the
behaviour of the system.

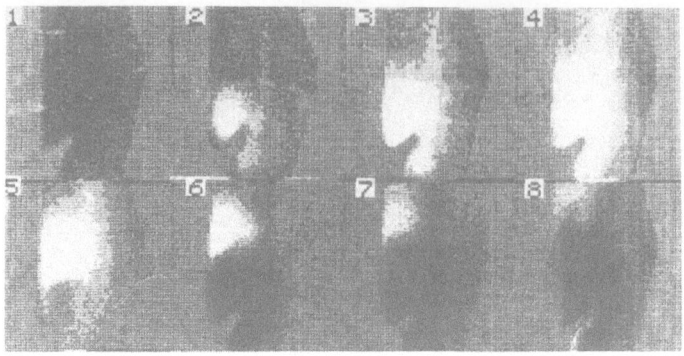

Fig. 1. Thermal autowaves of spreading depression in the
rat brain.

Preliminary processing is a transformation of the source frame sequence
after which it becomes more useful for visual analysis and investigation.
One of the main functions of preliminary processing is the suppression of
non-homogeneous spatial distribution, masking the informative changes of
brightness. In many cases the registered distribution of parameter (for
example, the temperature of brain or body) can be imagined as being two
distributions: one is stable and usually of high amplitude; the second is
weak and dynamic. To select this second distribution it is necessary to
estimate the first one and then to subtract it from the image sequence.
A method very close to this is the temporal derivation of image sequence.
After this operation, the stationary image regions are suppressed and
changing regions are enhanced. This technique is very useful for the
analysis of wave processes. For example, Fig. 1 shows the thermal autowaves
of spreading depression (SD) in the rat brain which have been selected in
this way (Guliyaev et al., 1984; Shchikalov et al., 1984).

Recorded images are always noisy to some degree. For noise suppression we used image averaging over times which were small compared with the
time constant of the process being investigated, synchronous averaging and
spatio-temporal filtration. Dynamic image filtering can also solve more
complex problems. For example, it is possible to select moving regions in
the image which have a given speed of propagation and a certain orientation
of movement.

A dynamic image is a three-dimensional signal. It has been shown that
a useful technique for the analysis of changes in the image sequence is a
method of spatio-temporal slices or projections. The slice is a two-dimensional image of the temporal dynamics of some spatial contour of the image.
Fig. 2 illustrates the spatio-temporal slices of rabbit brain temperature
dynamics. On the right the vertical line slice is shown. A vertical dark
and light structure is easily seen. This structure means that the selected
line temperature oscillates synchronously. On the horizontal slice (shown
in the upper portion of the figure) more complex structures are selected.
These distorted lines show the wave of temperature propagation in the horizontal direction, the angle of these lines corresponding to the propagation
speed.

For a quantitative description of the dynamic processes, a method of
spatio-temporal gistogram analysis was used. The image gistogram is an
estimate of the brightness density of probability in the image. Temporal
gistogram changes reflect the process dynamics. These gistogram changes
can be used to select stationary and non-stationary regions of image dynamics, laws of temporal behaviour and some other useful parameters.

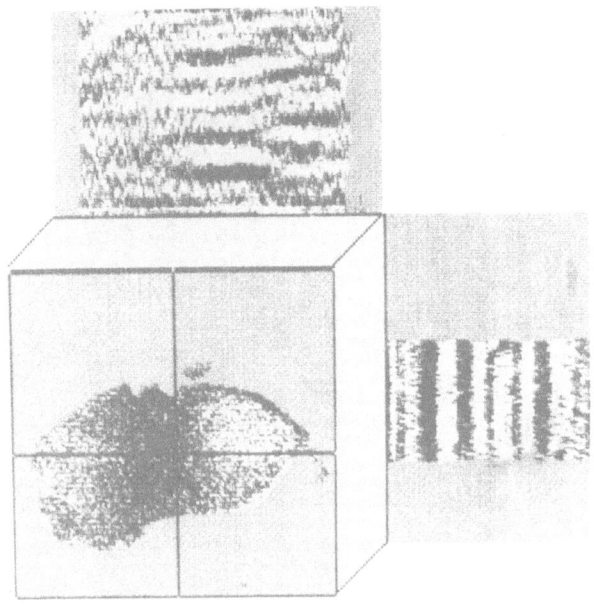

Fig. 2.   Spatio-temporal slices of the rabbit
brain temperature dynamics.

The main purpose of dynamic image processing is to obtain information
about the structure and functioning of homeostatic control systems. This
information is contained in the parameters of dynamic behaviour and its
spatial organisation. To obtain the integral representation of information
contained in image sequences, we used the method of functional imaging. A
functional image is the spatial distribution of some valuable parameters
of a dynamic image. For example, it can be the spatial distribution of
time constants, amplitudes or speed changes. We have proposed the method
of image segmentation on the regions of the close temporal behaviour. In
this method in every spatial point of the image, the temporal dynamics
curve is analysed and some set of parameters of this curve is selected.
Then the classification of the image in the spatial region is performed on
the basis of selected temporal patterns. The determination of the regions
which behave synchronously in time is very important for the determination
of the object functional structure. The synchronism of dynamics in different regions of space reflects functional homogeneity of these regions
in the structure of the object. Fig. 3 shows the functional image of the
human hand which cools after being heated by an external source. Two
regions are selected within the image: the lighter regions are cooling
monotonically, the darker regions are cooling with typical "overshoot"
behaviour. The temperature of the dark regions first falls below the
initial level and then returns to this level. These regions are, in fact,
large veins which are cooling because of the cold blood flow from the
fingertips. The changes in the spatial organisation of functional images
usually correspond to malfunctions of physiological control systems.
Therefore, functional images constitute the basis for the functional diagnosis of biological objects (Taratorin, 1986; Taratorin et al., 1986).

The processing of spatio-temporal maps of the magnetic fields of the
human heart and brain is somewhat different from the methods described.
In this case the recorded magnetic field distribution is usually generated
by several current sources. The signal from each such source approximates
to a current dipole distribution. These field sources have different

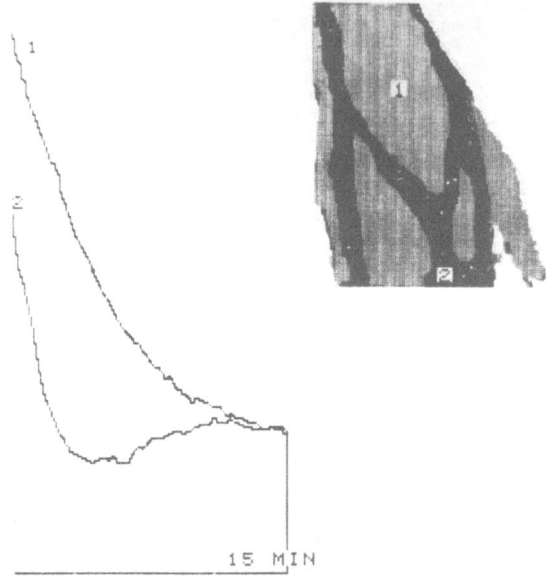

Fig. 3.   Functional image of the human hand.

localisations and different temporal dynamics. The main problem in processing a dynamic magnetic map is the separation of such sources and the selection of the dynamics of each current (the solution of the inverse problem).
It is well known that for a static field distribution, this problem cannot
be solved uniquely. Analysis of dynamic maps which are recorded during
small time intervals (milliseconds) in some cases can lead us to reliable
estimates of the amplitudes and localisation of one of the dipolar sources.
By subtracting the estimated field distribution from the source map it is
possible to estimate the remnant structure dynamics (Matloshov et al., 1986).
This algorithm is based on a multidimensional search of several dipole pairs
which best fit the experimental data. Temporal and spatial regularisation
have been used to provide stability of solution. Use of a regularisation
technique and of additional constraints on the solution behaviour provided
a significant narrowing of the region of solution indeterminacy. Parameters
of the equivalent current sources obtained in this way are used to characterise the dynamics of the heart and brain and to determine anomalies.

The software and algorithms developed have been used in studies of the
physical fields of biological objects: the surface temperatures of brain
and body, the radio-brightness temperature of internal organs, the mapping
of electrical parameters of the epidermis, and heart and brain magnetic
fields. On the basis of practical work with this software it is possible
to conclude that the methods of dynamic image processing described constitute a necessary part of any system which is designed for the research of
dynamic processes.

REFERENCES

Guliyaev, Y. V., Godik, E. E., Petrov, A. V., and Taratorin, A. M., 1984,
    Concerning the possibility of remote diagnosis of biological objects
    by their auto-infrared radiation, DANSSR (Proc. Acad. Sci. USSR),
    277(6) : 1486 (in Russian).
Matlashov, A. N., Zhuravlev, Y. E., and Valiyev, I. V., 1986, Dynamic map-
    ping of the magnetic field of the heart, DANSSR, 286(2) : 451 (in Russian).

Shchikalov, E. N., Petrov, A. V., and Taratorin, A. M., 1984, Investigation of own temperature fields connected with the excitation of the cortex of the major brain of the rat, DANSSR, 278(1) : 249 (in Russian).

Taratorin, A. M., 1986, Concerning methods of determination of the structure of dynamic biomedical images, Autometriya, no. 3 : 85 (in Russian).

Taratorin, A. M., Godik, E. E., and Gulyaev, Y. V., 1986, Functional imaging of biological objects, DANSSR, 287(5) : 1088 (in Russian).

COMPUTER SIMULATION OF THE FOURIER METHOD OF IMAGE RECONSTRUCTION FROM

PROJECTIONS IN TOMOGRAPHY

I. Bajla, S. Matej and M. Bognárová

Institute of Measurement and Measuring Technique
Electro-Physical Research Centre
Slovak Academy of Sciences
Bratislava, Czechoslovakia

INTRODUCTION

Nowadays, tomographic imaging methods based on various physical principles represent an attractive tool for medical diagnosis. In contemporary X-ray tomography systems the convolution filtered backprojection method still remains the leading mathematical approach implemented. However, continually increasing demands, namely computational speed in multiplanar or 3-D spatial methods and dynamic tomographic studies, have evoked a renewed interest in earlier proposed methods. One of the most promising of these has proved to be the Direct Fourier Inversion method (the so-called DFI method). This method was proposed by Ramachandran (Ramachandran and Lakshminarayanan, 1971), but unsatisfactory reconstruction quality was attained and, as a result, the method was not pursued further. Several papers were published from the late 1970s onwards (Mersereau, 1976; Stark et al., 1981; Niki et al., 1983), which exhibited a new interest in the DFI method. Although certain possibilities of error suppression were described in these papers, some important questions still remained to be tackled. We therefore considered it appropriate to re-evaluate discrete mathematical aspects of the DFI method and to develop effective computational algorithms. The Fourier spectral character of the measured data in Magnetic Resonance Tomography (Cho et al., 1982) made such research still more interesting.

In the next section the mathematical background underlying the problem of reconstruction from projections is introduced. A methodological basis of the DFI method is then outlined. Some problems yet to be solved if this method is to be applied in actual tomography systems are specified. Following this, the program package MODUSIRP developed at our Institute is described. This package represents a universal and flexible tool for simulation-based research aimed at investigating thoroughly the discrete reconstruction problems which have been identified. In the last section some examples of computer experiment results are presented and discussed.

IMAGE RECONSTRUCTION FROM PROJECTIONS

In general, tomographic reconstruction deals with the generation of a quantitative representation of a spatially distributed physical quantity that characterised the internal invisible structure of an examined object. The problem itself consists of determining the unknown distribution on the basis of indirect measurements. In the particular case of X-ray tomography, indirect measurements are represented by projection data.

According to the standard approach (for example, Herman (1979; 1980)), for a real valued function $f(x, y)$ which is continuous in the plane E_2, the Radon transform is given as an integral operator $\mathcal{R}$ which maps the function $f(x, y)$ into a function $p(s; \theta)$ defined through the integrals of the function f along the parallel straight lines l (perpendicular to a projection line l_p):

$$p(s; \theta) = \mathcal{R}[f(x, y)] = \int_{-\infty}^{\infty} f(s.\cos \theta - q.\sin \theta, s.\sin \theta + q.\cos \theta)dq \qquad (1)$$

Here s is the distance of a line l from the origin and θ represents the angle of the projection line l_p (see Fig. 1). The function p with the argument s, θ is called the projection. Then the object of the Radon Inversion problem is to find the unknown function f given the projection function p.

There are several equivalent methods which have been proposed for solving the Radon inversion problem in continuous formulation. In actual discrete reconstruction problems the transform methods are no longer equivalent because of discretisation errors and approximations of transform formulae. The main goal of research into image reconstruction methods is to develop a method of reconstruction from the measured discrete projection data yielding the best estimation $\hat{f}(x, y)$ of the function $f(x, y)$ with minimum computational cost. We focused our attention on developing the DFI method.

DIRECT FOURIER INVERSION METHOD

The formula for the DFI method is based on the projection theorem (Herman, 1979):

$$[\mathcal{F}_y \mathcal{R} f](S, \theta) = [\mathcal{F}_2 f](S.\cos \theta, S.\sin \theta), \qquad (2)$$

where $\mathcal{F}_y$ is the 1-D Fourier transformation (FT) operator for projections, $\mathcal{F}_2$ denotes the 2-D Fourier transformation operator (2-D FT). Then the formula for the DFI method (for CT and MR tomography) can be obtained as follows:

$$f(x, y) = [\mathcal{F}_2^{-1} \mathcal{F}_y \mathcal{R}]f(x, y) = [\mathcal{F}_2^{-1} \mathcal{F}_y]p(s; \theta) = \mathcal{F}_2^{-1}P(S; \theta). \qquad (3)$$

Let us now briefly describe the discrete version of this projective reconstruction method. For a finite number of projective angles a set of discrete projection vectors p is given. For each angle θ we have the corresponding discrete Fourier transform of p - $P(S; \theta)$. According to the Projection Theorem (2) this means that the two-dimensional Fourier spectrum of the function f is given on a polar point raster only. Further, let $F(u, v)$ denote the same Fourier spectrum obtained by means of an interpolation procedure applied to the polar grid. Then the discrete estimate $\hat{f}$ of the unknown function f is obtained as a result of 2-D inverse Fourier transformation (Fig. 2).

As we have indicated above, the discrete nature of actual tomographic reconstruction gives rise to a wide range of specific problems. We have focused our attention on some of those which have not been investigated until now. They are as follows.

First, no relevant details concerning the appropriate raster size of the 2-D inverse FT or the optimum dimension enlargement of this transformation have been mentioned in previous publications. The next unsolved

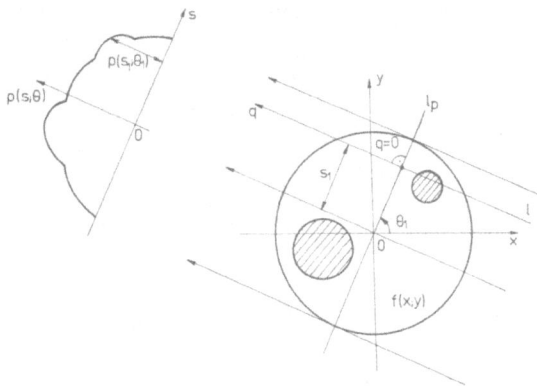

Fig. 1.   Radon transformation geometry.

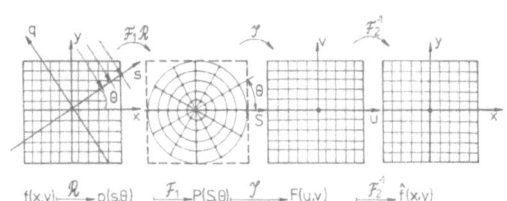

$f(x,y) \xrightarrow{\mathcal{R}} p(s,\theta) \quad \xrightarrow{\mathcal{F}_1} P(S,\theta) \quad \xrightarrow{\mathcal{I}} F(u,v) \quad \xrightarrow{\mathcal{F}_2^{-1}} \hat{f}(x,y)$

Fig. 2.   Basic steps of DFI method algorithm.

problem in the DFI method is associated with the investigation of optimum
filtering in the Fourier spectral domain.   This problem has been studied
consistently in the case of the backprojection method only.   The choice of
optimum reconstruction parameters for noisy and noiseless images, and for
low and high contrast images, represents another reconstruction issue.   The
most important such parameters to be considered are: type of filtering
functions, polar raster density in the radial direction, enlargement size
of 2-D FT, type of polar raster interpolation.   The dependence of the
acceptable level of noise present in input projective data on the prescribed
image reconstruction accuracy also had to be found out.   There was one more
interesting question to be investigated, namely, the possibility of noise
suppression by means of modification of the reconstruction algorithm.

When discussing the unsolved discrete reconstruction problems an important methodological aspect should be mentioned.   There exist theoretical and
very sophisticated approaches resulting in mathematical formulae for exact
reconstruction (Alliney and Sgallari, 1984;   Inouye, 1984;   Naterrer, 1985).
However, on the basis of these results, although they comprise some quantitative measures of the reconstruction errors, we are able to estimate only
the nature of the resulting image artifacts.   Perhaps in addition it is possible to make a prediction of how reconstruction errors will be influenced
by changing the algorithm parameters.

Unfortunately, all this does not provide quantitative information
about the appropriate values of individual reconstruction parameters, information which is necessary for correction of reconstruction errors. Obviously,
it is impossible to judge whether the final reconstruction is acceptable or
not using only the analytical estimation formulae.   Hence, computer simulation research represents the only way of searching for the optimum method
and parameters which will guarantee fast and accurate image reconstruction.

A program system which has been developed in our Institute to make simulation research possible is described below. Since special features were incorporated into the architecture of this program system, it differs from the other ones known hitherto (Herman and Rowland, 1978; Huesman et al., 1977). It therefore seems reasonable to discuss it in more detail.

MODULAR PROGRAM SYSTEM FOR COMPUTER SIMULATION OF IMAGE RECONSTRUCTION FROM PROJECTIONS (MODUSIRP)

The demands which a program system used for this kind of computer simulation has to meet can be divided into three groups:

- (i) functional demands; these correspond to the specific task to be solved and to the aims of the discrete mathematical modelling;
- (ii) user's demands comprising easy orientation in the system, well-arranged communication and simple manipulation with program and date blocks; and
- (iii) system (computer) demands, which are derived from the computer facilities (limitations of data storage, operational speed in performing arithmetic computations and data transfer).

To meet the first category of demand, modelling techniques have to make possible the following:

- generation and application of mathematical models of real objects, particularly the images or functions $f(x, y)$ defined in E_2 and E_3 respectively,
- generation of projection vectors representing in the models the source data obtained by indirect measurement procedures,
- modelling the performance of various reconstruction algorithms and studying their behaviour under different conditions (such as discretisation, filtering, dimension, noise level),
- visual and quantitative comparison of the individual results obtained with the test objects (source images or geometric phantoms).

The demands of the above category necessitate a modular structure for the reconstruction simulation software. The demands of the next two categories make this necessity even more important. Separate modules are required to perform the individual steps of source data generation, and each stage of the reconstruction algorithms. The system has also to incorporate the module which evaluates the quality of the reconstruction results. Making the individual simulation program modules interchangeable requires the input and output data structure of these modules to be strictly defined.

For discrete modelling of various reconstruction conditions, and for testing the reconstruction algorithm modification as well, it is necessary to provide the user with an interactive means for entering the important functional or auxiliary parameters in a flexible manner, for example, the number of projections, the values of discretisation steps in the spatial and Fourier spectral domain, filter type and the parameters of the filters, the noise parameters, and so on. To satisfy the requirements of the actual computer environment (storage capacity), the reconstruction parameters have to be continually checked in each module.

User-friendly programs are supposed to provide the user with clarity and simplicity of structure. Thus, the comments included in the program modules have to refer to all inevitable modifications made by the author which differ from the previously determined convention. The overall structure of each module has to be optimal from a computational point of view.

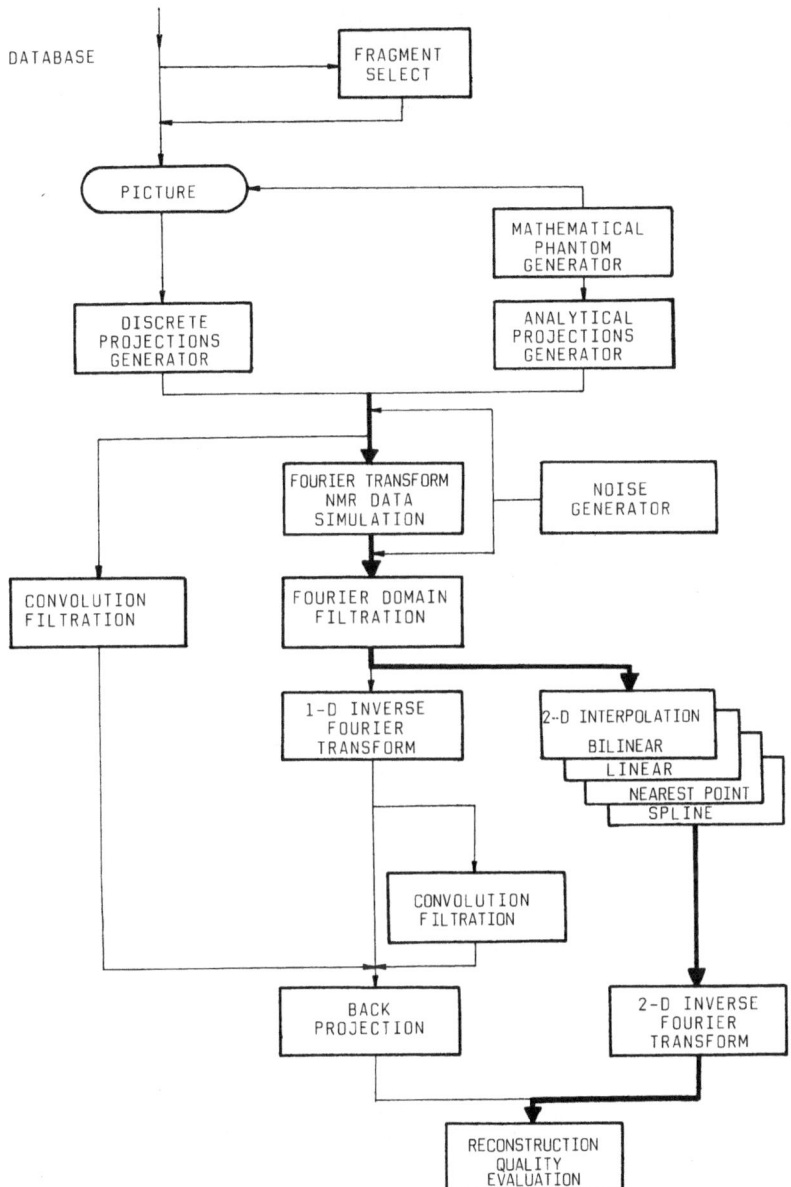

Fig. 3. Structure of the modular program system MODUSIRP.

On the basis of the aforementioned requirements, the following convention for constructing the modular program system MODUSIRP has been developed:

(i) the program module's head comprises the following information presented
in this order: a functional description of all important variables
encountered, a description of input and output data structures, the
names of all subroutines called from the main program and the accompanying data about the storage of these subroutines, a list of depar-
tures from convention, a brief description of the program module's
function, basic data concerning the compiler used, and a command file
for the task builder;

(ii)   the number of the parameters entered by the user is minimised; the
       same is true for the number of the formal parameters in the sub-
       routines;

(iii) within a period of 10 to 20s a message informing the user of how
       the processed task is proceeding is displayed (in textual form on
       the terminal or pictorially via the image display);

(iv)   the data storage is standardised in the following way: the projec-
       tion data vectors are not formatted, the arrays are segmented into
       columns or rows according to the buffer capability; there are cases
       in which the process of space and time optimisation leads to unavoid-
       able effects on the data storage convention, requiring that the
       changed data organisation be characterised explicitly within the
       program module's head; and

(v)   in order to save the limited disk data medium space (in an image re-
       construction task a huge quantity of data is being processed), the
       user is provided with an option of deleting the previous output data
       files, which is presented just at the first stage of module perfor-
       mance; the same option is included in the user's dialogue after the
       module task has been finished.

     The structure of the modular program system MODUSIRP is depicted in
Fig. 3. A detailed description may be found in Matej (1987a). The host
hardware environment for running the program system MODUSIRP that enables
the actual simulation to be carried out consists of the SM 52-11 minicomputer (equivalent to a PDP 11-60 computer) and the videographic display processor VGJ-64 (256 x 256 pixels/8 bits). The operating program system
RSX-11/M has been used.

     It should be pointed out that although the actual working version of
the system MODUSIRP reflects the limitation of hardware available, it can
be easily implemented in other image processing systems. This is due to
the fact that the computational core of the program system is written in
the high level FORTRAN-77 language and is thereby entirely computer-indepen-
dent. The only requirements are:

(i)   the minimum space reserved by the operating system for the user's
       task in the main memory should be 64 kBytes, while the total capacity
       of the main memory is expected to be not less than 248 kBytes (this
       comprises the space dedicated for the task, virtual arrays and the
       operating system); and

(ii)   the elementary drivers (subroutines) controlling the communication
       between a computer and the digital image processing system available
       have to be written additionally; then the new library of special
       drivers can be linked simply to the unchanged program modules of the
       MODUSIRP.

COMPUTER SIMULATION RESULTS

     The well-known mathematical phantoms of Shepp and Logan (1974) with
rasters of 102 x 102 and 128 x 128 pixels have been used for computer simulation. A modification of object values in the test images appeared to be
necessary in order to investigate the reconstruction behaviour in two distinct cases; namely, images including objects with 20% contrast (so-called
contrast images) and images with 2% contrast have been considered. The
result of low contrast image reconstruction (128 x 128 pixels) with the
experimentally obtained optimal parameters is depicted in the top right
part of Fig. 4. The original input phantom is displayed on the left side.
(For the sake of greater visibility all images in this chapter will be

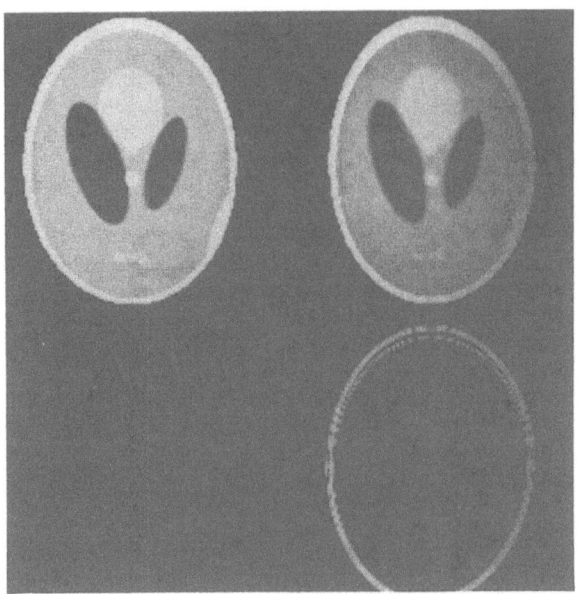

Fig. 4. A comparison of the original image with
the image reconstructed by the DFI
method using the optimal parameters
(phantom size = 128 x 128, low contrast
phantom, bilinear interpolation, $\beta = 3$,
COS filter, 150 projections, 2-D IFT
dimension = 256^2).

presented as linearly expanded in a grey scale window.) These two images
are compared and the corresponding difference image is displayed at the bottom:
absolute values on the left side, normalised values on the right side.

The experimental results have been evaluated visually and quantitatively. The following three basic mathematical measures common to the
literature, for example Herman (1980), have been used: the normalised root
mean squared distance, the normalised mean absolute distance, and the worst
distance measure defined in terms of the mean value of four adjacent raster
points. The last one has been modified to yield invariance with respect to
error shift in the image. It has been found that in the case of low contrast images the quantitative measure of reconstruction image quality indicates the actually obtained visual quality rather poorly. This is demonstrated in Fig. 5. Although the top right reconstructed image is apparently
of better visual quality, all three quantitative measures exhibit better
results in the case of the left side image. On the other hand, in the case
of contrast images, the visual reconstruction quality coincides with the
tendencies indicated by quantitative measures.

The simulation research results obtained are described in detail in
Matej (1987a). A brief summary can be found in Matej (1987b). Some of
these examples will be presented and discussed. An interpolation of point
values into a rectangular raster from the given values in a polar raster
(both situated in the Fourier spectral domain) proved to be the most intri-
cate factor which influences the DFI reconstruction accuracy. Therefore,
the presentation of some results concerning the interpolation procedures
applied in the DFI method seemed to be a very instructive example of the
worth of computer simulation research.

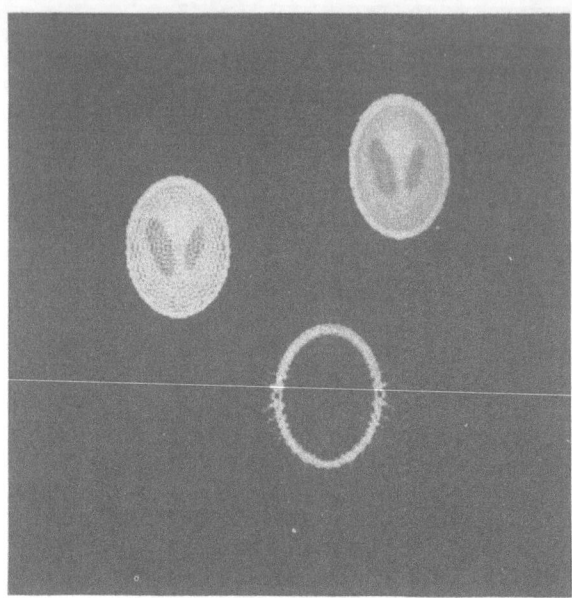

Fig. 5.   A comparison of the unfiltered recon-
           structed image (left) with the image
           filtered (right) by the Hamming filter
           α = 0.54 (phantom size = 64 x 64, low
           contrast phantom, bilinear interpolation
           β = 3, 150 projections, 2-D IFT dimension
           = 64^2 - left image/128^2 - right image.

As the nearest neighbour interpolation did not yield satisfactory reconstruction, we focused our attention on investigating how linear and bilinear interpolation would influence the optimum reconstruction parameters
in the case of high contrast and low contrast images. In particular, both
interpolation techniques have been investigated in association with the
coefficient β which represents the artificial enlargement of the projection
vector dimension. It means that the raster density in the Fourier spectral
domain has been increased 2^β times.

The picture in Fig. 6 illustrates the reconstruction error caused by
inaccurate interpolation which is in turn due to insufficient density of
the polar raster in the Fourier spectral domain. The picture on the left
side corresponds to β = 1, that on the right side to β = 3. The difference
picture is displayed below. Bilinear interpolation and a COS filter have
been used. In this case 150 projections represent the optimum number.

The next picture (Fig. 7) also demonstrates that the parameter β is
very important, especially in the case of low contrast images. Here, from
the top left to the bottom right the reconstructions with β = 0, 1, 2, 3 are
displayed. Other parameters have been preserved. On the basis of results
obtained by bilinear interpolation experiments, the following optimum parameters have been derived: COS - filter, parameter β = 3, the raster of
inverse Fourier transformation which is 125% of the reconstruction domain
size. In this case, reconstruction quality is comparable to the quality
of convolution back-projection reconstruction.

An interesting fact concerning low contrast images should be mentioned,
namely that in the case of linear interpolation, using the same parameters

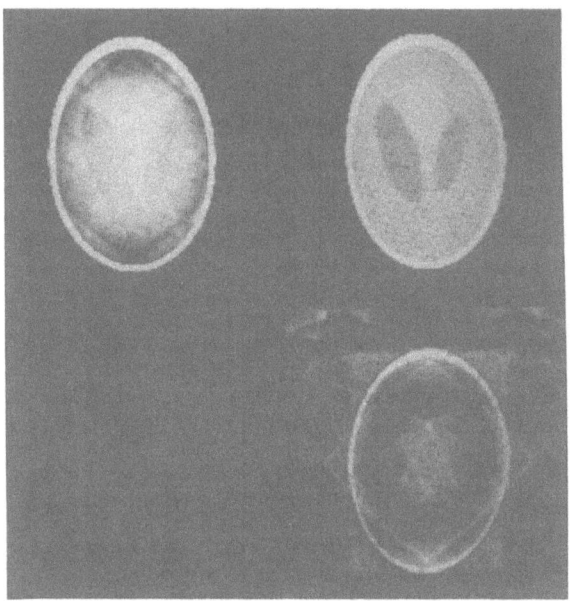

Fig. 6.   An illustration of the reconstruction
error caused by too small polar raster
density in the radial direction (phantom size = 102 x 102, low contrast
phantom bilinear interpolation, β = 1 -
left image/β = 3 - right image, COS
filter, 150 projections, 2-D IFT dimension = 128^2).

as in optimum bilinear interpolation satisfactory reconstruction quality is
not obtained.

In the case of contrast images, using distinct interpolation techniques
(Fig. 8) the reconstruction results exhibit less differences in visual
quality than in the previous case. In Fig. 8 the upper left picture
represents the result of reconstruction via the convolution backprojection
method. The other three pictures show the results of DFI reconstruction.
The upper right picture corresponds to bilinear interpolation, the lower
ones represent: left - linear interpolation, right - interpolation by the
nearest raster point. The latter one evidently demonstrates that this kind
of interpolation is unsatisfactory.

CONCLUSIONS

On the basis of our analysis of the theory of the DFI method and the
corresponding unsolved problems of discrete reconstruction, the need to
perform thorough computer simulation research has been established. The
universal program system MODUSIRP, based on methodology that was specially
elaborated for simulation research, has been developed. This paper describes the chief principles of the MODUSIRP design which allow this system
to be implemented easily in various image processing systems. Some examples
of simulation results have been presented. These results justify the use
of the program system MODUSIRP as an effective tool for computer simulation
of the projection reconstruction methods. By means of this program system,

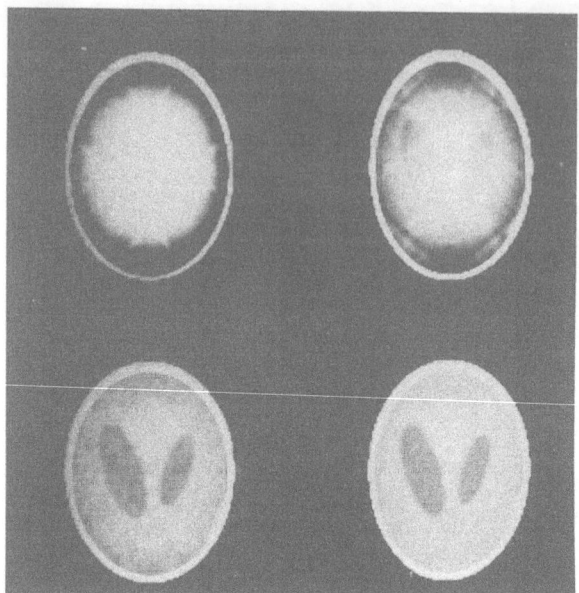

Fig. 7. A demonstration of the influence of the
parameter β on the reconstruction quality
(phantom size = 102 x 102, low contrast
phantom, bilinear interpolation, β = 0 -
top left/β = 1 - top right/β = 2 - bottom
left/β = 3 - bottom right, COS filter,
150 projections, 2-D IFT dimension = 128^2).

a reconstruction method was derived which can be characterised in the
following way: it is considerably faster than the back-projection method,
yet the reconstructed image quality obtained is comparable.

Finally, it should be pointed out that, due to the modular structure
and universal construction of the system MODUSIRP, it can be advantageously
used for more extensive simulation research in the field of reconstruction
from projections.

REFERENCES

Alliney, S., and Sgallari, F., 1984, An "ill-conditioned" Volterra integral
    equation related to the reconstruction of images from projections,
    SIAM J. Appl. Math., 44 : 627.
Cho, Z. H., Kim, H. S., Song, H. B., and Cumming, J., 1982, Fourier trans-
    form nuclear magnetic tomographic imaging, Proc. IEEE, 70 : 1152.
Herman, G. T. (ed.), 1979, "Image Reconstruction from Projections - Imple-
    mentation and Applications", Springer-Verlag, Berlin.
Herman, G. T., 1980, "Image Reconstruction from Projections - The Fundamen-
    tals of Computerized Tomography", Academic Press, New York.
Herman, G. T., and Rowland, S. W., 1978, "SNARK 77. A Programming System
    for Image Reconstruction from Projections", Technical Report No. 130,
    Department of Computer Science, State University of New York of
    Buffalo.
Huesman, R. H., Gullberg, G. T., Greenberg, W. L., and Budinger, T. F.,
    1977, "Donner Algorithms for Reconstruction Tomography RECLBL Library
    - User's Manual", Publication No. 214, Lawrence Berkeley Laboratory,
    University of California.

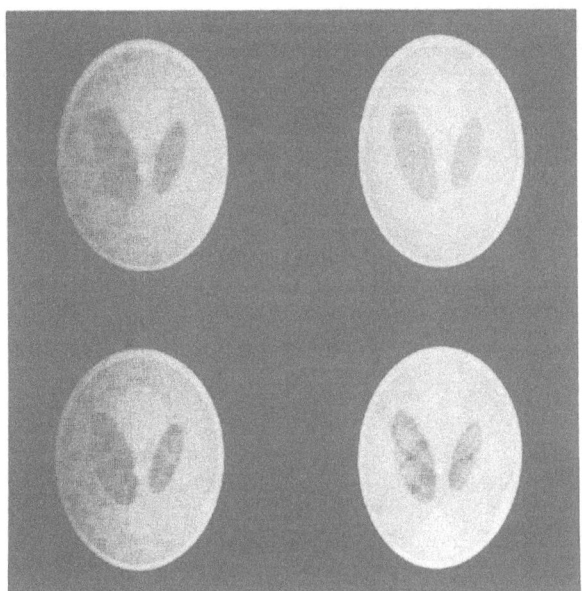

Fig. 8. Reconstruction of the contrast type of
         image using various reconstruction and
         interpolation methods (phantom size =
         102 x 102, convolution backprojection
         method - top left/DFI method: bilinear
         interpolation - top right/linear inter-
         polation - bottom left/nearest neighbour
         interpolation - bottom right, β = 2, no
         filter, 150 projections, 2-D IFT dimen-
         sion = 128²).

Inouye, T., 1984, Analysis of artifact pattern generation in two-dimensional
    Fourier transform image reconstruction algorithm for computerized
    tomography, in: "Proc. of the 1984 International Symposium on Noise
    and Clutter Rejection in Radars and Imaging Sensors" : 170.
Matej, S., 1987a, "A Contribution to Projection Reconstruction Methods in
    Tomographic Systems", Ph.D. Thesis, Slovak Academy of Sciences, Bratislava.
Matej, S., 1987b, Fast transform methods of image reconstruction from pro-
    jections and their parallel implementation on SIMD-type computer, in:
    "Computer Analysis of Images and Patterns. Proc. 11th Int. Conf. on
    Automatic Image Processing (CAIP '87)", L. P. Yaroslavskii, A. Rosen-
    feld and W. Wilhelmi, eds., Akademie-Verlag, Berlin : 40.
Mersereau, R. M., 1976, Direct Fourier transform techniques in 3-D image
    reconstruction, Comput. Biol. Med., 6 : 247.
Naterrer, F., 1985, Fourier reconstruction in tomography, Numer. Math.,
    47 : 343.
Niki, N., Mizutani, T., and Takahashi, Y., 1983, A high-speed computerized
    tomography image reconstruction using direct two-dimensional Fourier
    transform method, Systems, Computers, Controls, 14(3) : 56.
Ramachandran, G. N., and Lakshminarayanan, A. V., 1971, Three-dimensional
    reconstruction from radiographs and electron micrographs: Application
    of convolutions instead of Fourier transform, Proc. Nat. Acad. Sci.
    (USA), 68 : 2236.
Shepp, L. A., and Logan, B. F., 1974, The Fourier reconstruction of a head
    section, IEEE Trans. Nucl. Sci., NS-21 : 21.

Stark, H., Woods, J. W., Paul, I., and Hingorani, R., 1981, An investigation of computerized tomography by direct Fourier inversion and optimum interpolation, <u>IEEE Trans. Biomed. Eng.</u>, BME-28 : 496.

A MATHEMATICAL MODEL FOR CT-RECONSTRUCTION WITH UNCERTAIN DATA

S. Alliney

Department of Mathematics and Informatics
University of Udine
Udine, Italy

INTRODUCTION

It is well known that the problem of reconstructing a plane figure from its projection is ill-posed in the sense of Hadamard. This means that random disturbances on the measured data (that is, the projections) can be dramatically amplified during the solution procedure. However, the inverse problem becomes stable if we slightly modify the mathematical model which is currently used in biomedical applications, such as in computer tomography. In this chapter, after a short general discussion, some "modified" projection operators will be described, leading to a stable solution, even in the presence of disturbances on the data.

This chapter is intended to be only a preliminary account of research work in progress, and therefore most of the theoretical details will be omitted. A complete exposition, together with numerical results, will be presented in a forthcoming paper (Alliney, 1988); here, it is intended to give a general idea of the concepts underlying the method that is proposed.

SOBOLEV SPACE ANALYSIS

To begin the discussion, a short account of some recent results of Natterer (1980) are presented. First of all, some definitions are needed: for any β real, $H^\beta(R^2)$ denotes the Sobolev space of all tempered distributions such that

$$\| f \|_\beta^2 = \int \left(1 + |\underline{\xi}|^2\right)^\beta |F(\underline{\xi})|^2 \, d\underline{\xi} < +\infty \tag{1}$$

where F is the Fourier transform of f. For Ω, a sufficiently regular domain in R^2, we put

$$H_o^\beta(\Omega) = \{f \in H^\beta(R^2) : \text{supp}(f) \subseteq \bar{\Omega}\} \tag{2}$$

As shown in Natterer (1980), it is reasonable to think of the density function f of a picture in Ω as a function in $H_o^\beta(\Omega)$, with $\beta > \frac{1}{2}$. In physical terms, this means that certain smoothness constraints are imposed on the density function; on the other hand, the presence of noise during the measure process can be viewed as a violation of such constraints. More precisely, the density function f can be considered as a random process with covariance

$$E\{\left[f(\underline{x}) - E(f(\underline{x}))\right] \cdot \left[f(\underline{x}') - E(f(\underline{x}'))\right]\} = \exp(-a|\underline{x} - \underline{x}'|^\gamma) \ , \ a > 0 \qquad (3)$$

It can be shown (Alliney, 1988; Natterer, 1980) that the norm

$$E\{\ \|\ f - E(f)\ \|_\beta^2\ \} \qquad (4)$$

is finite for bounded Ω if, and only if, $\beta < \gamma/2$ and therefore the order β of the Sobolev space to be considered is conditioned by the covariance function of the image. This is a quite important point, because stability results for the inverse problem can be obtained only for $\beta > \frac{1}{2}$; on the other hand, relevant random disturbances on f may lead to a value of γ in (3) far less than one.

However, for $\beta > \frac{1}{2}$ we have the following remarkable result (Natterer, 1980): let $R_{hk}(f)$ denote a finite set of line integrals

$$R_{hk}(f) = \iint \delta(x\ \cos\theta_k + y\ \sin\theta_k - s_h) f(x, y)\ dx\ dy \qquad (5)$$

and suppose that only approximate values g_{hk} of $R_{hk}(f)$ are known, with

$$\|\ R(f)\ -\ g\ \|_N^2 = \frac{1}{N}\ \sum_{h,k} \left(R_{hk}(f)\ -\ g_{hk}\right)^2\ \leqslant\ \varepsilon^2 \qquad (6)$$

with $N = h_{max} \cdot k_{max}$ and samples equally spaced both in θ and s.

Theorem 1. If the reconstruction problem is solved by minimising the Tikhonov functional

$$j_\lambda(z) = \|\ R(z)\ -\ g\ \|_N^2 + \lambda^2\ \|\ z\ \|_\beta^2 \qquad (7)$$

then the root mean square error of the reconstruction is of the order $O\left(\varepsilon^{\beta/(\beta+\frac{1}{2})} + N^{-\beta/2}\right)$.

For further details on the regularising functional of Tikhonov, see, for example, Tikhonov and Arsenin (1977).

VARIATIONAL FORMULATIONS

The theorem of Natterer reported above cannot be used directly in numerical implementations, mainly because a noisy density function may not belong to $H_0^\beta(\Omega)$ with $\beta > \frac{1}{2}$: in that case, a possible correct definition of the transform domain is clearly $L^2(\Omega)$. From a theoretical point of view, however, we can consider an operator

$$P\ :\ L^2(\Omega)\ \longrightarrow\ H_0^\beta(\Omega) \qquad (8)$$

such that

$$\|\ P(f)\ \|_\beta^2 = c^2\ \|\ f\ \|_{L^2}^2 \qquad (9)$$

or at least

$$\|\ P(f)\ \|_\beta^2 \leqslant c^2\ \|\ f\ \|_{L^2}^2 \qquad (10)$$

As it is widely known (Alliney and Sgallari, 1984; Tikhonov and Arsenin, 1977), the regularisation method of Tikhonov can be restated in variational form as

Problem 1.   minimise $\|\ z\ \|_{L^2}^2$ $\qquad\qquad$ (11)
$\qquad\qquad$ subject to $\|\ R(z)\ -\ g\ \|_N^2 \leqslant \varepsilon^2$

By introducing a suitable Lagrange multiplier, a formulation similar to (7) can be obtained, with the difference that the smoothing part of the functional is here $\|\cdot\|_{L^2}$.

Now, in order to use Natterer's convergence result, it is necessary to project $f \in L^2(\Omega)$ into $H^\beta_o(\Omega)$, $\beta > \frac{1}{2}$. Using definition (8), Problem 1 can be restated as

Problem 2.   minimise $\| P(z) \|^2_\beta$ (12)

   subject to $\| RP(z) - g \|^2_N \leqslant \epsilon^2$

The function f which solves Problem 2 is indeed the solution of the reconstruction problem, which achieves the optimal error bound. Furthermore, using (9), it is possible to consider the equivalent:

Problem 3.   minimise $\| z \|^2_{L^2}$ (13)

   subject to $\| \tilde{R}(z) - g \|^2_N \leqslant \epsilon^2$

where $\tilde{R}$ denotes RP. Using relation (10) instead of (9), Problem 3 can be viewed as an approximation to Problem 2.

The last formulation has remarkable advantages:

   (i)   the optimal error bound of Theorem 1 is achieved;
   (ii)   only L^2 norms are present, with obvious advantages from the numerical point of view; and
   (iii) there exist iterative algorithms which determine the minimum L^2-norm solution to Problem 3.

Point (iii) is particularly relevant; for a complete discussion of the numerical details, the reader is referred to Nashed (1981), whilst a deeper theoretical discussion can be found in Nashed (1976).

MODIFIED PROJECTION OPERATORS

It remains to be shown how in practice to find the operator P defined in (8). Using formulation (13), however, an equivalent modified projection operator $\tilde{R}$ = RP can be sought. It can be noted that such an operator will contain some smoothing device, as can be seen from the definition of the space H^β_o itself. Let us begin by considering a function $k(x;\alpha)$ such that

$$\int_{-\infty}^{+\infty} k(x;\alpha)\,dx = 1$$
                                                                              (14)
$$\lim_{\alpha \to 0} \int_{-\infty}^{+\infty} k(x;\alpha)\,f(x)\,dx = f(0)$$

Then, using a function such as the kernel of the integral representation of R, a "modified" projection can be defined as

$$\hat{f}(s,\theta;\alpha) = \int\!\!\int k(x \cos\theta + y \sin\theta - s;\alpha)\,f(x,y)\,dx\,dy \qquad (15)$$

It follows from (14) that the usual definition of the projection is recovered as the limit

$$\hat{f}(s,\alpha) = \lim_{\alpha \to 0} \hat{f}(s,\theta;\alpha) \qquad (16)$$

Now the Fourier transform of $\hat{f}(s,\theta,\alpha)$ can be evaluated with respect to s:

283

$$F(\rho,\theta;\alpha) = \int_{-\infty}^{+\infty} \hat{f}(s,\theta;\alpha) \, e^{-j s \rho} \, ds$$

$$= \iint \int_{-\infty}^{+\infty} k(x \cos\theta + y \sin\theta - s;\alpha) \, e^{-j s \rho} \, f(x,y) \, ds \, dx \, dy \tag{17}$$

If it is assumed that the Fourier transform of k exists and interchanging the order of evaluation of the integrals, we obtain:

$$\int_{-\infty}^{+\infty} k(x \cos\theta + y \sin\theta - s;\alpha) e^{-j s \rho} \, ds = e^{-j \rho (x \cos\theta + y \sin\theta)} K(\rho;\alpha) \tag{18}$$

where $K(\rho;\alpha)$ denotes the Fourier transform of $k(x;\alpha)$ with respect to its first argument. Thus we have from (17):

$$F(\rho,\theta;\alpha) = K(\rho;\alpha) \iint e^{-j \rho (x \cos\theta + y \sin\theta)} f(x,y) \, dx \, dy \tag{19}$$

The integral in (19) simply defines the Fourier transform of $f(x,y)$ in polar co-ordinates and therefore we conclude that:

$$F(\rho,\theta;\alpha) = K(\rho;\alpha) \, F(\rho,\theta) \tag{20}$$

This result can be viewed as an extension of the so-called "projection slice theorem". What is relevant in the present context is that $K(\rho;\alpha)$ can be chosen in such a way that

$$\int_0^{2\pi} \int_0^{+\infty} (1 + \rho^2)^\beta \, |F(\rho,\theta;\alpha)|^2 \, \rho \, d\rho \, d\theta < +\infty \tag{21}$$

as required for the H_0^β-norm. An obvious example is

$$K(\rho;\alpha) = (1 + \alpha^2 \rho^2)^{-\beta/2} \quad ; \quad \alpha = 1 \tag{22}$$

which also satisfies the condition (9). Many other choices are possible, and some of them are reported in Table 1. A complete analysis of the constraints to be imposed on $K(\rho;\alpha)$ can be performed following Tikhonov and Arsenin (1977), and will be reported in Alliney (1988). Here we simply point out that (20) also provides an estimate of the difference between the regularised solution and the "true" density function. By computing the inverse Fourier transform of $F(\rho,\theta;\alpha)$ we obtain the convolution

$$f(\underline{x};\alpha) = w(|\underline{x} - \underline{\xi}|;\alpha) * f(\underline{\xi}) \tag{23}$$

where $w(r;\alpha)$ denotes the two-dimensional inverse Fourier transform of $K(\rho;\alpha)$. More precisely, using polar co-ordinates, we have:

$$w(r;\alpha) = \frac{1}{(2\pi)^2} \int_0^{2\pi} \int_0^{+\infty} K(\rho;\alpha) \, e^{j \rho r \cos(\phi-\theta)} \rho \, d\rho \, d\theta \tag{24}$$

$$= \frac{1}{2\pi} \int_0^{+\infty} K(\rho;\alpha) \, J_0(\rho r) \rho \, d\rho$$

where $J_0(\cdot)$ is the zero-order Bessel function.

AN EXAMPLE

In practice, the discretisation of (15) is performed exactly as in the standard case and R_{hk} is defined as R_{hk} in (5). Numerical details and experimental results will be reported in Alliney (1988); here we present a simple analytical example.

Table 1. Possible Kernels for the "Modified" Projection Operator

$k(t;\alpha)$	$K(\rho;\alpha)$	$w(r;\alpha)$	References (Gradshteyn and Ryzhik, 1980)
$\dfrac{1}{\alpha\sqrt{\pi}} e^{-t^2/\alpha}$	$e^{-\alpha^2\rho^2/4}$	$\dfrac{1}{\pi\alpha^2} e^{-r^2/\alpha^2}$	n. 6.631.4, p. 717
$\dfrac{1}{2\alpha} e^{-|t|/\alpha}$	$\dfrac{1}{1+\alpha^2\rho^2}$	$\dfrac{1}{2\pi\alpha^2} K_0(r/\alpha)$	n. 6.532.4, p. 678
$\dfrac{1}{2\alpha} P_n(|t|/\alpha) e^{-|t|/\alpha}$	$\dfrac{1}{(1+\alpha^2\rho^2)^n}$	$\dfrac{1}{2\pi\alpha^2} \dfrac{(r/\alpha)^{n-1}}{2^{n-1}(n-1)!} K_{n-1}\left(\dfrac{r}{a}\right)$	n. 3.737.1, p. 413 n. 6.565.4, p. 686
$\dfrac{1}{\alpha\pi} K_0(|t|/\alpha)$	$\dfrac{1}{\sqrt{1+\alpha^2\rho^2}}$	$\dfrac{1}{2\pi\alpha^2} \dfrac{1}{(r/\alpha)} e^{-r/\alpha}$	n. 6.671.14, p. 732 n. 6.564.1, p. 686
$\dfrac{1}{\alpha\pi} C(\mu) \dfrac{K_{\mu-\frac{1}{2}}(|t|/\alpha)}{(|t|/\alpha)^{\mu-\frac{1}{2}}}$	$\dfrac{1}{(1+\alpha^2\rho^2)^{1-\mu}}$	$\dfrac{1}{2\pi\alpha^2} \dfrac{2^\mu}{\Gamma(1-\mu)} \dfrac{K_\mu(r/\alpha)}{(r/\alpha)^\mu}$	n. 6.699.12, p. 749 n. 6.565.4, p. 686
$\dfrac{e^{-|t|/\alpha\sqrt{2}}}{2\alpha\sqrt{2}} \left|\cos\left(\dfrac{|t|}{\alpha\sqrt{2}}\right)+\sin\left(\dfrac{|t|}{\alpha\sqrt{2}}\right)\right|$	$\dfrac{1}{1+\alpha^4\rho^4}$	$\dfrac{-1}{2\pi\alpha^2} kei(r/\alpha)$	n. 3.727.1, p. 408 n. 6.537, p. 678

$K_v(\cdot)$ = modified Bessel function

$C(\mu) = 2^{\mu-\frac{1}{2}} \sqrt{\pi}/\Gamma(1-\mu)$; $0 \leq \mu < 1$

$$P_n(x) = \frac{1}{4^{n-1}(n-1)!} \sum_{k=0}^{n-1} \frac{(2n-k-2)!(2x)^k}{k!(n-k-1)!}$$

Table 2. Numerical Results for the Example Considered.

r	$\frac{\alpha^2}{4} = 0.07$	$\frac{\alpha^2}{4} = 0.08$	$\frac{\alpha^2}{4} = 0.09$	$\frac{\alpha^2}{4} = 0.10$
0.00	.9718	.9561	.9377	.9180
0.05	.9708	.9551	.9365	.9168
0.10	.9684	.9517	.9331	.9127
0.15	.9634	.9463	.9271	.9061
0.20	.9569	.9385	.9181	.8973
0.25	.9475	.9277	.9070	.8852
0.30	.9359	.9148	.8927	.8707
0.35	.9208	.8983	.8758	.8532
0.40	.9030	.8793	.8556	.8325
0.45	.8816	.8566	.8325	.8094
0.50	.8561	.8303	.8060	.7830
0.55	.8270	.8010	.7767	.7542
0.60	.7941	.7678	.7441	.7223
0.65	.7569	.7315	.7090	.6883
0.75	.6720	.6500	.6308	.6136
0.80	.6252	.6058	.5890	.5737
0.85	.5763	.5601	.5455	.5327
0.90	.5255	.5130	.5014	.4913
0.95	.4759	.4657	.4573	.4495
1.00	.4236	.4186	.4134	.4082
1.05	.3749	.3725	.3700	.3678
1.10	.3364	.3290	.3288	.3286
1.15	.3051	.2897	.2895	.2910
1.20	.2468	.2475	.2516	.2555
1.25	.2092	.2168	.2222	.2234
1.30	.1631	.1718	.1838	.1914

Let us consider a density function $f(r)$ such that $f = 1$ if $r \leqslant R$ and $f = 0$ elsewhere. The projection $f(s)$ is simply

$$f(s) = \begin{cases} 2(R^2 - s^2)^{\frac{1}{2}} & \text{if } |s| \leqslant R \\ 0 & \text{elsewhere} \end{cases} \tag{25}$$

Using the Gaussian regulariser (see Table 1) defined as

$$K(\rho;\alpha) = e^{-\alpha^2\rho^2/4} \tag{26}$$

all the computations can be performed in closed form:

$$F(\rho,\theta) = 2\pi\, R\, J_1(\rho R)/\rho \tag{27}$$

$$F(\rho,\theta;\alpha) = 2\pi\, R\, e^{-\alpha^2\rho^2/4}\, J_1(\rho R)/\rho \tag{28}$$

$$f(r;\alpha) = R\int_0^{+\infty} e^{-\alpha^2\rho^2/4}\, J_1(\rho R)\, J_0(\rho r)\, d\rho \tag{29}$$

The last integral can be evaluated in terms of hypergeometric functions (see Gradshteyn and Ryzhik, 1980), p.718, n.6.633.1); the numerical values obtained are reported in Table 2 for different choices of the parameter α.

REFERENCES

Alliney, S., 1988, Modified projection operators for CT-reconstruction from uncertain data, Riv. Mat. Pura Appl. (in press).
Alliney, S., and Sgallari, F., 1984, An ill-conditioned Volterra integral equation related to the reconstruction of images from projections, SIAM J. Appl. Math., 44 : 627.
Gradshteyn, I. S., and Ryzhik, I. M., 1980, "Tables of Integrals, Series and Products", Academic Press, New York.
Nashed, M. Z., 1976, "Generalized Inverses and Applications", Academic Press, New York.
Nashed, M. Z., 1981, Continuous and semicontinuous analogues of the iterative methods of Cimmino and Kaczmarz with application to the inverse Radon transform, in: "Mathematical Aspects of Computerized Tomography", G. T. Herman and F. Natterer, eds., Springer-Verlag, Berlin.
Natterer, F., 1980, A Sobolev space analysis of picture reconstruction, SIAM J. Appl. Math., 39 : 402.
Tikhonov, A. N., and Arsenin, V. Y., 1977, "Solutions of Ill-posed Problems", Wiley, New York.

PART 3

MODELLING AND SIMULATION

DEVELOPMENT IN MINIMAL MODELLING OF IVGTT: THE MEASUREMENT OF GLUCOSE

PRODUCTION

G. Pacini and C. Cobelli

Institute of Systems Dynamics and Bioengineering (LADSEB-CNR)
and Department of Electronics and Informatics
University of Padova, Padova, Italy

INTRODUCTION

The minimal modelling approach has been widely applied to assess insulin sensitivity in vivo from intravenous glucose tolerance test (IVGTT)
data (Bergman et al., 1985). This technique yields a measurement of the
effect that insulin exerts in the disposition of the injected glucose, by
simultaneously enhancing its tissue utilisation, Rd(t), and inhibiting its
liver production, Ra(t). Standard IVGTT data, that is, glucose and insulin
plasma concentration time course following a bolus injection of cold gluc-
cose, do not allow the separate estimation of the contribution of each of
those two factors. The use of tracer seems therefore the best candidate
for solving this problem. In a previous study (Cobelli et al., 1986) we
have developed a minimal model for the analysis of IVGTT data when tracer
was injected along with cold glucose. From the estimated insulin effect
on labelled glucose disappearance, we were able to evaluate, in quantitative terms, the true insulin effect on glucose disposal. This was possible
since tracer allowed the segregation of the confounding effect of insulin
on glucose production. Therefore, the natural pursuance of this study will
be the assessment of insulin effect on liver glucose production. To
achieve this goal, the cold glucose must also be included in the data set
to be analysed by means of the minimal model developed for labelled IVGTT.

The fundamental problem is, then, that of formulating a model which
accurately describes Ra(t) during a non-steady condition. The model of
Steele (1959) may seem the best candidate for obtaining Ra(t), since it is
general purpose and allows easy computation by means of an explicit formula.
However, a recent study (Cobelli et al., 1987) showed that, in general,
the model suffers from an error which depends both on the system being
studied and on the particular experimental conditions. In Cobelli et al.
(1987) new developments for estimating Ra(t) in the non-steady state were
proposed, but they are only applicable in a specific experimental situation,
that is, the glucose clamp. This chapter discusses the issue of the reconstruction of the glucose production time course during the IVGTT; in
particular, we point out the need for an experiment where cold and labelled
glucose are administered and measured concomitantly, and we suggest some
working hypotheses.

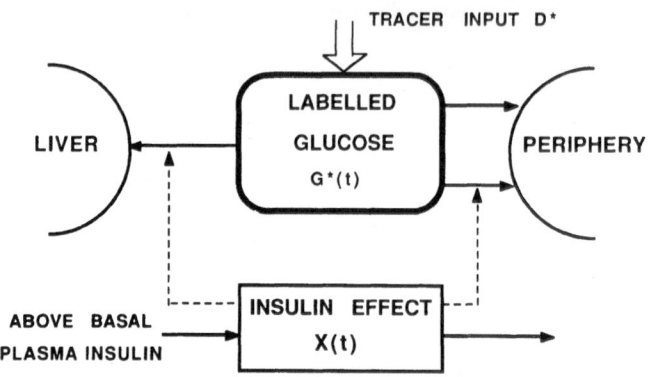

Fig. 1.   Block diagram of the minimal model of labelled
glucose disappearance.

THE MODEL

Steele's method quantifies glucose production in the non-steady state by means of the following expression (Steele, 1959):

$$Ra(t) = (Ra^*(t) - pVG(t)dSA/dt)/SA(t) \qquad (1)$$

where $Ra^*(t)$ is tracer infusion rate, pV is the distribution volume corrected by the pool-fraction factor p, $G(t)$ is glucose concentration, and $SA(t)$ the specific activity defined as $G^*(t)/G(t)$, where $G^*(t)$ is the concentration of tracer in plasma. The method requires the calculation of a derivative which becomes extremely difficult in a highly dynamic experiment. Moreover, the assumption of an a priori known value for the distribution volume is necessary. The assumption of a single compartment for the glucose distribution space is implicit in (1) and this hypothesis, too, has been the object of several criticisms (Cobelli et al., 1987).

The minimal model for the analysis of labelled IVGTT (Cobelli et al., 1986) is the following (Fig. 1):

$$dG^*(t)/dt = -(p_1 + X(t))G^*(t) + \delta(t)D^* , \quad G^*(0) = 0 \qquad (2)$$

where D^* is the injected dose of tracer, and $\delta(t)$ is the Dirac function. This expression evidences both the effect of tracer per se to enhance its own disappearance rate at basal insulin (p_1) and the effect of dynamic insulin to augment glucose utilisation ($X(t)$). The latter is assumed to be described by a first order process, that is:

$$dX(t)/dt = -p_2X(t) + p_3I(t), \quad X(0) = 0 \qquad (3)$$

where $I(t)$ is the above-basal plasma insulin concentration, and p_2 and p_3 are constant parameters. Also this model is mono-compartmental, thus it suffers from those limitations found in Steele's approach. However, it presents some advantages due to the fact that it is a physiological mode, that is, it explicitly incorporates the basic knowledge about the real system, both in terms of structure and function. In particular, behind the model of (1), there is a time-varying parameter representing the fractional glucose disappearance rate, $K(t)$. The model of (2) and (3) gives a physiological realisation to $K(t)$, that is:

$$K(t) = p_1 + X(t) \qquad (4)$$

In addition, the inverse problem, which is ill posed for (1), becomes well posed for the parametric model of (2) and (3).

Additional help for the solution of the general problem of the measurement of Ra(t) comes from the possibility of segregating the effects on plasma glucose concentration of endogenous production and exogenous administration. Once we have the cold and labelled glucose measurements, the segregation may be achieved as follows. The general model that accounts for the dynamics of labelled glucose may be written:

$$dG*(t)/dt = -Rd*(t) + \delta(t)D*, \quad G*(0) = 0 \tag{5}$$

where Rd*(t) is the rate of disappearance of tracer (that is, the utilisation by peripheral tissues and uptake by the liver). On the other hand, the dynamics of the plasma concentration of cold glucose, G(t), may be described by:

$$dG(t)/dt = -Rd(t) + Ra(t) + \delta(t)D, \quad G(0) = Gb \tag{6}$$

where D is the exogenously injected dose, and Gb is the basal (pre-injection) glucose level. From the tracer-tracee indistinguishability principle (Carson et al., 1983) we can define:

$$Rd*(t)/G*(t) = Rd(t)/G(t) = K(t) \tag{7}$$

and therefore (5) and (6) become, respectively:

$$dG*(t)/dt = -K(t)G*(t) + \delta(t)D*, \quad G*(0) = 0 \tag{8}$$

$$dG(t)/dt = -K(t)G(t) + Ra(t) + \delta(t)D, \quad G(0) = Gb \tag{9}$$

Let us now make the assumption that G(t) can be split into two components: one due to the exogenous glucose administration, $G_E(t)$, and the other to endogenous glucose production, $G_p(t)$, such that $G(t) = G_E(t) + G_p(t)$. This yields the possibility of dividing (9) into two expressions (Fig. 2):

$$dG_E(t)/dt = -K(t)G_E(t) + \delta(t)D, \quad G_E(0) = 0 \tag{10}$$
and
$$dG_p(t)/dt = -K(t)G_p(t) + Ra(t), \quad G_p(0) = Gb \tag{11}$$

Comparing (10) with (8) and considering the specific activity of the infusate, defined as $SA_{INF} = D*/D$, the time course of the exogenous component of plasma glucose can be computed as:

$$G_E(t) = G*(t)/SA_{INF} \tag{12}$$

where G*(t) is measured and SA_{INF} is known. Consequently, the time course of the endogenous component $G_p(t)$ can also be easily obtained as:

$$G_p(t) = G(t) - G_E(t) \tag{13}$$

Up to this point, we have shown how it is possible to isolate the endogenous component of plasma glucose. Equation (11) will allow the estimation of the time course of Ra(t), once we are able to describe mathematically both K(t) and Ra(t). K(t) represents the fractional glucose disappearance which is a function of insulin. We have already seen that, from the minimal modelling technique, a specific description can be given to K(t), that $K(t) = p_1 + X(t)$ (see (4)), and, in the previous study on labelled IVGTT (Cobelli et al., 1986), we have shown that the analysis of tracer data by means of the system of (3), (4) and (8) yields the true

293

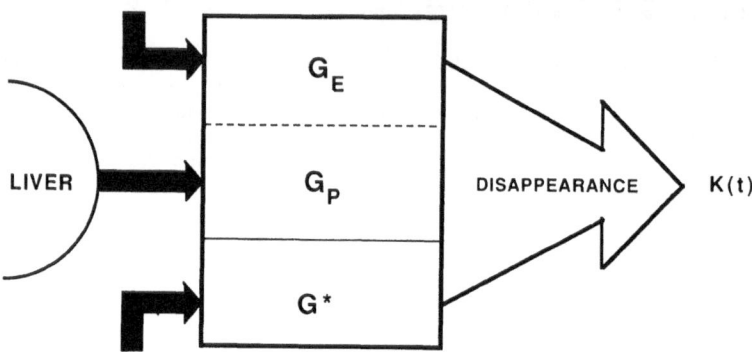

COLD GLUCOSE D

LIVER

G_E

G_P

G^*

DISAPPEARANCE K(t)

LABELLED GLUCOSE D*

Fig. 2.   Diagram depicting the rationale for the segregation
of endogenous and exogenous glucose components.

value of the parameters p_1, p_2 and p_3. If we apply this system to our
$G^*(t)$ and $I(t)$ data, we can estimate parameter p_1 and the time course of
$X(t)$. Therefore, in (11) the only unknown is Ra(t) which must be described
by mathematical models to be proposed and validated.

MODELS OF GLUCOSE PRODUCTION

Some mathematical models which explicitly include Ra(t) already exist
(see the discussion on them in Cobelli et al., 1987).   However, they were
conceived to be used with data obtained with glucose clamp type experiments,
and their application to non-steady state situations was limited to the
transition period between two steady states. Furthermore, their aim was
that of describing a physiological process which is known to be only inhibited under the circumstances of a glucose clamp type experiment (eu- or
hyperglycaemia and hyperinsulinaemia). As a matter of fact, one of the
major points upon which the validity of the models was assessed concerned
the answer to the question as to whether the estimated Ra(t) fell to zero
and maintained this level, or became negative, or just decreased remaining
low, but greater than zero. The conditions during the IVGTT in a normal
subject are, on the contrary, quite different (Fig. 3). When the glucose
level rises to its peak, inducing a high supra-basal insulin release, Ra(t)
is expected to be inhibited, but when the insulin basal level is re-estab-
lished and glucose slowly returns to its fasting pre-injection value, Ra(t)
is also raised to its pre-injection level. Moreover, since the glycaemia
often exhibits a moderate, but well detectable, undershoot, it is likely
that in such a phase Ra(t) is the major component responsible for making
plasma glucose concentration re-attain the fasting value.

Modelling a pattern of behaviour such as the one described in the
previous paragraph is a difficult task. Physiological knowledge says that
Ra(t) depends mainly upon glucose and insulin plasma concentrations, thus a
structured model which accounts for these controls seems the most appropriate candidate. We have described Ra(t) with a series of models where
glucose and insulin were explicitly included according to several different
mathematical relationships:

$$q_1 Gb/(1 + q_2(G - Gb)X) \qquad q_1 Gb/(1 + q_2(G - Gb)X + q_3 X)$$
$$q_1 Gb/(1 + q_2(G - Gb) + q_3 GX) \qquad q_1 Gb(1 - q_2(G - Gb))/(1 + q_3 GX)$$
$$q_1 Gb(1 - q_2 X)/(1 + q_3(G - Gb)X) \qquad q_1 Gb(1 + q_2 Gb\ Ib)/(1 + q_2 G\ I)$$
$$q_1 Gb/(1 + q_2(G - Gb)I) \qquad q_1/(1 + q_2 G\ I)$$

(14)

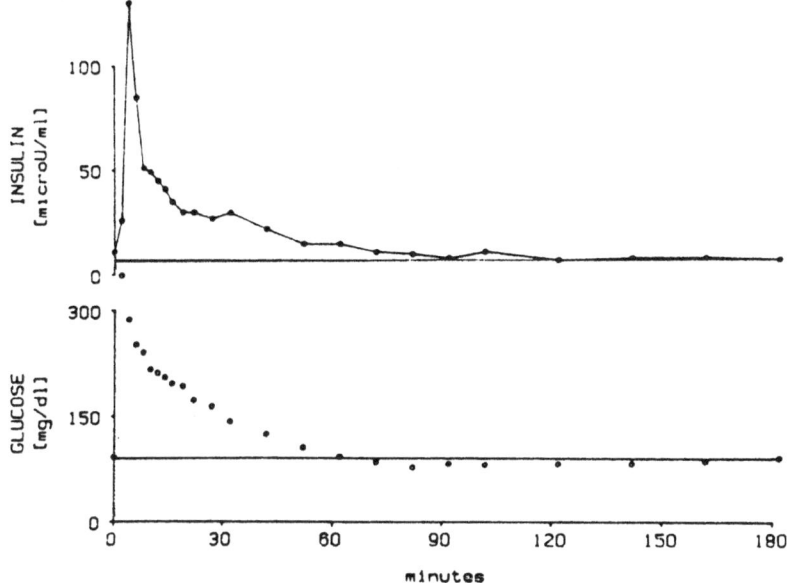

Fig. 3. Example of IVGTT (glucose 0.33 g kg^{-1} is injected as a
bolus at t = 0).

We took into account insulin both as "action" (X = X(t)), and as measured
(smoothed) concentration (I = I(t), Ib = basal). In the above expressions
q_1, q_2, q_3 are constant parameters. The application of these models to
our IVGTT data did not yield satisfactory solutions both in terms of parameter estimates (high coefficient of variation) and of data fit (high resi-
duals, elevated sum of squares, systematic errors).

The other possibility is that of describing glucose production by
means of models with time-varying parameters or by means of models with no
direct dependence upon any physiological process, that is <u>parametric</u> models.
The latter need a mathematical representation which describes the expected
pattern of Ra(t) without giving any specific meaning to the parameters. A
combination of exponentials seems to be the most suitable description, but
a polynomial function can also accomplish this purpose. As a preliminary
result, we describe a function of Ra(t) that we have applied to describe
glucose production in four normal dogs which underwent labelled IVGTTs.
The expression is the following:

$$Ra(t) = Rab - A t^2 e^{-Bt} \tag{15}$$

where Rab is the basal (pre-injection) production and A and B are constant
parameters. From (11) it is easy to show that Rab = p_1 Gb, thus A and B
are the only unknown parameters to be estimated. In Fig. 4 the average
time course of the endogenous component of plasma glucose, $G_p(t)$, as calculated using (13), is shown in the upper panel. Given this pattern and the
values of p_1, p_2 and p_3, estimated from the system of (3), (4) and (8),
parameters A and B have been estimated in every single dog, yielding an
average time course of Ra(t) such as that shown in the bottom panel of
Fig. 4. As expected, glucose production, after a short period (6 - 10 min)
during which it remains constant, falls to a nadir of approximately 50% of
its fasting level, reached at t = 30 min. Then it remains inhibited for
another 20 - 30 min, after which it begins rising to reach again the basal
value after 2 h. It should be noticed that the Ra(t) does not reach zero

295

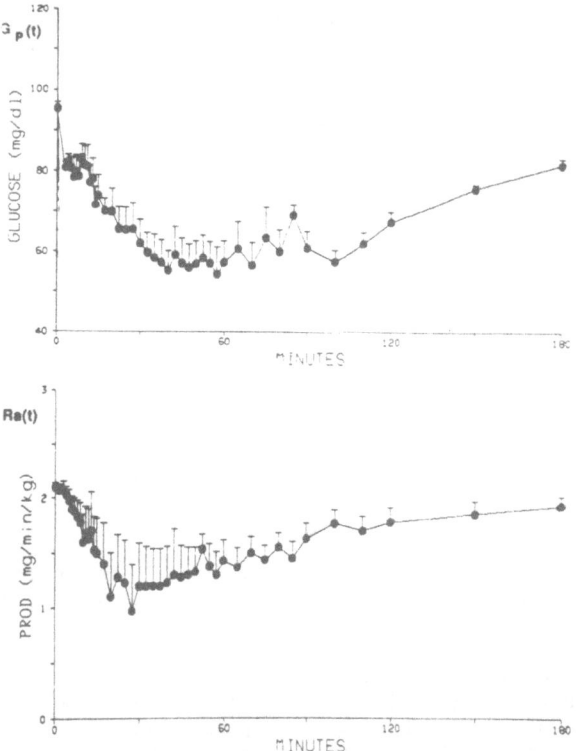

PARAMETRIC MODEL : Ra(t) = Rab - A t² exp(-Bt)

Fig. 4.   Endogenous glucose component (upper
panel) and reconstructed glucose production (bottom panel) in four dogs
which underwent the labelled IVGTT.

or unphysiological negative values as in studies employing Steele's method,
but remains always greater than zero as recent evidence has demonstrated
that it should be (Cobelli et al., 1987).   Of course, the validity of such
a model must be thoroughly assessed, both by comparing model predictions
with known measurements (when and if available), and by evaluating the
overall physiological plausibility of the Ra(t) pattern in normal and
pathological states.

REFERENCES

Bergman, R. N., Finegood, D. T., and Ader, M., 1985, Assessment of insulin
       sensitivity in vivo, Endocr. Rev., 6 : 45.
Carson, E. R., Cobelli, C., and Finkelstein, L., 1983, "The Mathematical
       Modeling of Endocrine and Metabolic Systems: Model Formulation,
       Identification and Validation", Wiley, New York.
Cobelli, C., Pacini, G., Toffolo, G., and Sacca, L., 1986, Estimation of
       insulin sensitivity and glucose clearance from minimal model: new
       insights from labeled IVGTT, Am. J. Physiol., 250 : E591.
Cobelli, C., Mari, A., and Ferrannini, E., 1987a, Non-steady state: error
       analysis of Steele's model and developments for glucose kinetics,
       Am. J. Physiol., 252 : E679.
Steele, R., 1959, Influence of glucose loading and of injected insulin on
       hepatic glucose output, Ann. NY Acad. Sci., 82 : 420.

MODELLING THE CONTROL PROPERTIES OF THE BASAL GANGLIA

S. S. Hacisalihzade, M. Mansour and C. Albani

Institute of Automatic Control, Swiss Federal Institute of
Technology (ETH); and
Neurology Department, University Hospital
Zurich, Switzerland

INTRODUCTION

The precise role of the basal ganglia in human motor activity is not
yet fully resolved. A popular school of thought advocates the notion of
learned motor "programs" which are hypothesised to be recorded in the basal
ganglia (Tatton and Bruce, 1981; Marsden, 1982). The complexity of these
"programs" is, however, a subject of controversy. This chapter is an
attempt to analyse simple motor activity and to infer the possible controller role of the human basal ganglia.

The tool used for the quantitative assessment of motor ability, a
computer-based visual tracking device, will be described. A mathematical
model of the relevant control loops involved in the control of simple motor
activity will then be introduced. Finally, the results of measurements
with a control group and with patients suffering from Parkinson's disease
will be presented and discussed.

QUANTITATIVE ASSESSMENT OF MOTOR ABILITY

The standard neurological examination is the classical method of
assessing the sensory and motor system functions (De Jong, 1967). This
examination consists of subjective evaluations of strength, sensory and
motor co-ordination, equilibrium, gait, sensory perception, mental state,
language, reflexes and the cranial nerves. The examiner studies the patient
for signs and symptoms of disease and judges the extent of impairment.
The overall impression the examiner acquires from each neurological category is the so-called neurological function of the patient.

It has to be stressed that the judgments of factors like speech, facial
expression, touch and pressure sensation are necessarily subjective. A
given function is then assigned a vaguely defined category, such as "mildly
abnormal" or "moderately abnormal", to assess sensory and motor functions
better and each category is enumerated. This kind of assessment is called
an ordinal scale rating. It is simple and very quick to perform. Although
such a scale can be used for initial evaluation and differential diagnosis
of a neurological disorder, the rating categories are often too broad and
vaguely defined to detect effects of medication or small changes in the

patient over a longer period of time. It also has to be mentioned that, in addition to differences in the rating techniques of different neurologists, the rating style of the same examiner may change over time and with experience. Therefore, direct comparisons - one of the primary motivations behind rating scales - may become worthless or even impossible to carry out.

Many rating scales have been developed and employed to assess parkinsonian deficits over the years. The most commonly used rating scales include the evaluation methods introduced by Hoehn and Yahr (1967), Webster (1968), Duvoisin (1971), and Marsden et al. (1973).

One of the methods of measuring the neurological and motor functions is to evaluate the tracking performance of patients (Velasco and Velasco, 1973; Flowers, 1976; Baroni et al., 1983; Wing and Miller, 1984). Tracking can be defined generally as the problem of making the output of a controlled process correspond as closely as possible to the reference input (Poulton, 1974; Sheridan and Ferrel, 1974). Two principal types of tracking task can be defined:

(i) closed loop systems, where the behaviour of the process which is to be controlled is fed back to the human operator visually and/or through somatosensory feedback; and

(ii) open loop systems, where the process output is not fed back to the human operator. In many cases the absence of visual information about the behaviour of the process is enough to interrupt this feedback.

Differences between closed loop and open loop tracking performance can represent valuable diagnostic data in many cases. Therefore, these two principal types of tracking tasks reflect the neurological tests used in clinical examination to assess sensory and motor function as well as co-ordination.

A VISUAL TRACKING DEVICE

The configuration depicted in Fig. 1 is a simple device with which open and closed loop pursuit tracking tests can be performed (Hacisalihzade, 1986).

Since motor tasks can be performed using different motor strategies, that is the co-ordinated activation of different muscles, measurements using movements with several degrees of freedom are usually difficult to analyse. The measurements discussed in this chapter use the displacement of the end phalanx of the thumb to perform the tracking tasks. This movement has a single degree of freedom and the peripheral control mechanisms have been analysed in detail (Marsden et al., 1976).

A single digital process computer (PDP 11/23) is used in the measurements
- to drive the stimulus of the tracking task (real time/on line);
- to record the tracking performance of the subject (real time/on line); and
- to analyse the recorded tracking data (off line).

The end phalanx of the thumb is tightly strapped to the shaft of a direct current motor. The angular position of the shaft is measured by a potentiometer. An individual plaster of Paris cast is used to ensure that the hand rests in its most relaxed form. Potentials of relevant muscles can be checked by means of EMG to ensure that the tracking is performed by the flexion and extension movements of the thumb. Through adjustment

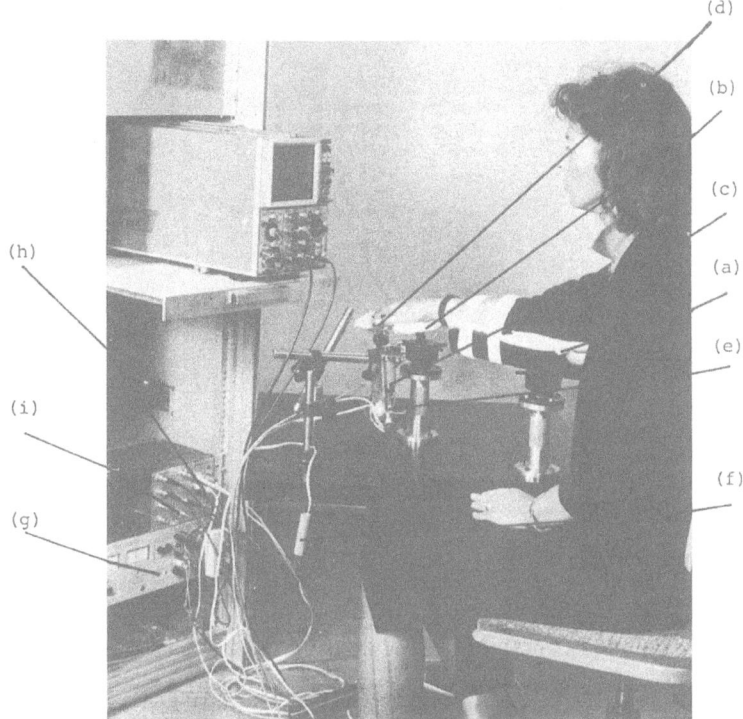

Fig. 1.   Picture of the device with which tracking tasks can
be performed. (a) armrests, (b) hand platform, (c)
motor, (d) thumbscrew, (e) potentiometer measuring
the position of the thumb, (f) button to start the
experiment, (g) power amplifier, (h) overheating
surveillance, (i) D/A and A/D converters.

of the shaft's position with respect to the fixed position of the hand in
the cast, it is made sure that this movement is executed through the activation of the extensor and flexor pollicis longus muscles alone and not by
thenar and forearm muscles.

The reference input of the tracking task is a small target which moves
on the vertical axis of an oscilloscope screen. The task is to move a
point of light up and down on the screen so that it stays in the middle of
the target. The position of the point of light corresponds to the angular
position of the shaft which can be rotated by the movements of the end
phalanx of the thumb.

The motor can produce disturbances of the thumb position by applying
a variable torque. It is possible to perform measurements with this configuration with or without disturbing torque as well as measurements in
which the actual position of the thumb is visible only for a fraction of
the time (visual or somatosensory open loop conditions). Fig. 2 shows the
functional diagram of the measurement configuration.

A MATHEMATICAL MODEL OF HUMAN MOTOR CONTROL MECHANISMS

The model of the major afferent and efferent connections shown in
Fig. 3 assumes a control loop which has the block-diagram depicted in

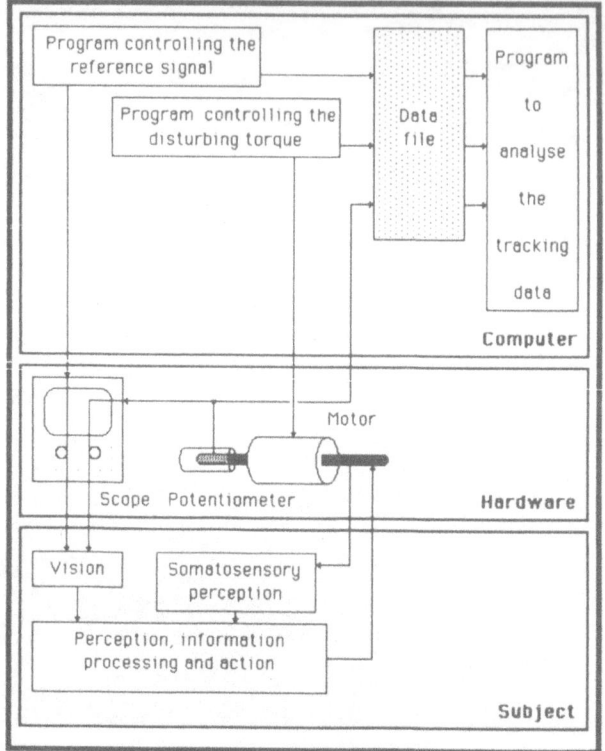

Fig. 2. Schematic diagram depicting the experiment configuration.

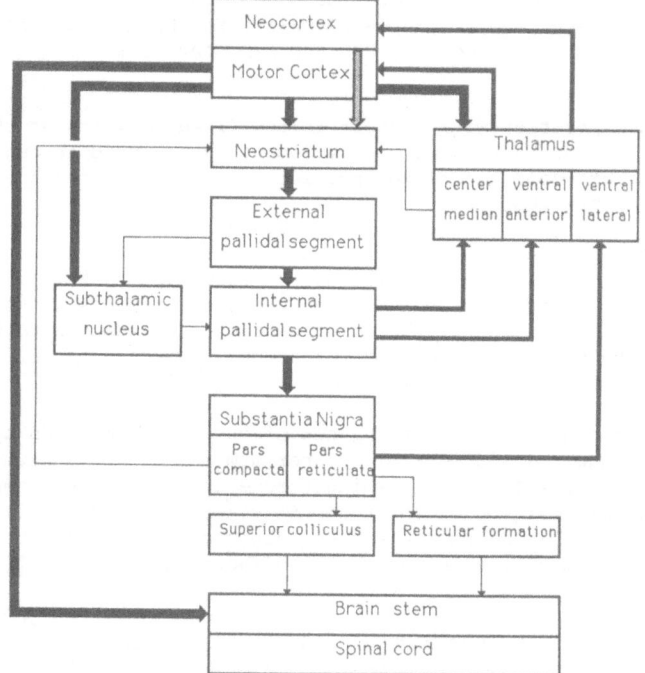

Fig. 3. Major afferent and efferent connections in the brain.

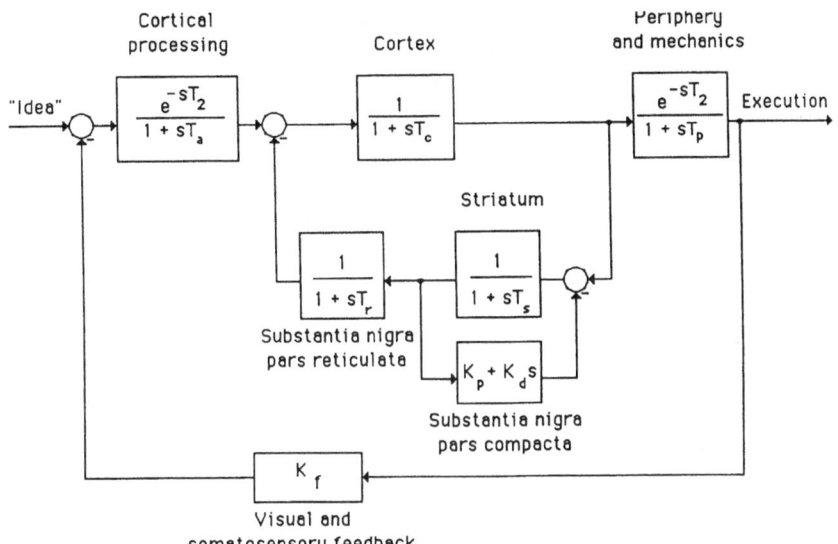

Fig. 4. Block diagram of the control loop which is affected in Parkinson's disease.

Fig. 4. In this model the "idea" for a motor action is processed in the cortex with a time delay and first order dynamics. It is then further processed in the basal ganglia and conveyed to the periphery (which is also modelled as a delay with a first order low-pass) for the execution of the desired action. The input of the cortical processing is modified continuously during the execution of the action by comparing the "idea" with the visual and somatosensory feedback of the present state of the execution. The basal ganglia itself is also modelled as a control loop, where both the pars reticulata of the substantia nigra and the striatum are modelled as first order systems and where the pars compacta of the substantia nigra has a PD-controller role over the striatum.

Obviously this highly complex and high order model cannot be verified. Therefore, it makes sense to simplify it. The result of this simplification is shown in Fig. 5. In this simplified model, the "idea" is processed with a delay in the cortex and then further processed in the basal ganglia before being sent to the periphery, which is also modelled as a simple delay. The input of the cortical processing is modified continuously also in this model during the execution of the action by comparing the "idea" with the visual and somatosensory feedback of the present state of the execution. The basal ganglia itself is again modelled as a control loop, where the substantia nigra acts as a PD-controller on the striatum.

RESULTS

It is now possible to verify the model in Fig. 5 and to identify the parameters of this system. A series of measurements was conducted with nine controls and eleven parkinsonians. In each measurement session, tracking performances of the subjects were recorded as step responses to pseudo-random square waves. The resulting step responses (42 for each subject) were then averaged for every individual.

Fig. 6 shows the comparison of the measured and modelled step responses of (a) an average control and a typical parkinsonian (b) with and (c) without

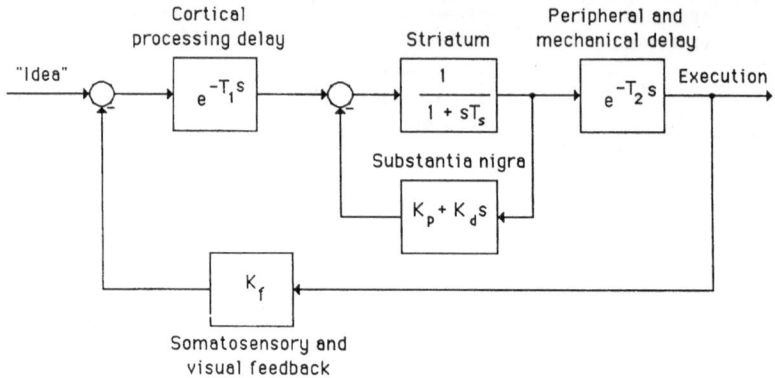

Fig. 5. Simplified model showing the controller characteristic
of the substantia nigra.

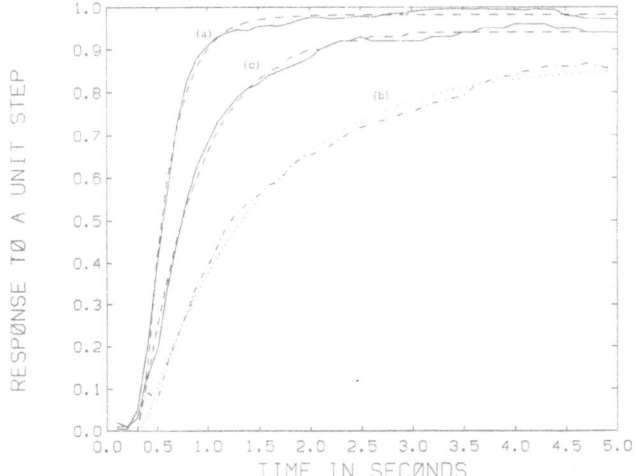

Fig. 6. Modelled and measured step responses of
(a) a control, (b) a parkinsonian in "off"
state and (c) a parkinsonian in "on" state.

medication ("on" and "off" states). The dashed curves in (a) and (c), as
well as the dotted curve in (b), are the results of simulation with identified values of parameters substituted in the model shown in Fig. 5. The
continuous curves in (a) and (c), as well as the dash-dotted curve in (b),
are the results of averaged measurements.

Table 1 shows the results of identification of parameters of the model
shown in Fig. 5. The standard deviations are shown in brackets below the
mean values of the identified parameters.

It was observed that the time constant of the striatum (T_S), which
was modelled as a first order low-pass, was not significantly different in
the controls and the parkinsonians with or without medication. This is in
agreement with the fact that the striatum is not influenced in Parkinson's
disease. It was also observed that the total delay in cortical processing
and in the periphery ($T_1 + T_2$) (which cannot be identified separately) is

302

Table 1. Identified Parameters of the Model Depicted in Fig. 5 for Nine
Controls and Eleven Parkinsonians

Parameter		Controls	Parkinsonians	
			"off"	"on"
T_s	[s]	0.069 (0.019)	0.071 (0.018)	0.067 (0.022)
$T_1 + T_2$	[s]	0.251 (0.087)	0.257 (0.094)	0.243 (0.117)
K_p	[-]	0.018 (0.007)	0.157 (0.054)	0.049 (0.027)
K_d	[s]	0.195 (0.071)	1.243 (0.347)	0.277 (0.121)
K_f	[-]	0.001 (0.001)	-	-

about the same in all groups, which is also in agreement with previous
findings about the reaction time in parkinsonians not being affected by the
disease.

However, the significant differences in the proportional and the derivative feedback gains (K_p and K_d) between the control group and the parkinsonian population supports the conclusion that the basal ganglia actually
have a PD-controller characteristic and that the neuronal depletion which
causes Parkinson's disease actually increases the values of these gains
significantly (5 to 10 times). This makes fast tracking impossible, thus
rendering the parkinsonians bradykinetic. Furthermore, the effect of the
medication is such that the controller gains get closer to their normal
values with adequate medication, which is also reflected in Fig. 6.

Two important questions which have to be answered by further research
concern whether a correlation exists between the increase in the values of
controller gains and the stage of the disease, and whether a decoupled
parallel controller model - in which a separate simple model, as described
above, can be employed for every single degree of freedom - can be used
for movements combining several degrees of freedom.

CONCLUSIONS

Starting with a physiological model, a mathematical model of the
afferent and efferent connections relevant to the control of motor activity
has .been derived. This model has then been simplified. Measurements based
on simple visual tracking tasks with a single degree of freedom have been
performed on controls and parkinsonian patients with defective basal ganglia.
These measurements have been used to verify the derived mathematical model
which explains the control properties of the basal ganglia. Parameter
identification of this model has shown no significant difference in the
parameters of the striatum, cortical processing and periphery between both
groups, whereas significant differences were observed in the substantia
nigra parameters between the controls and the parkinsonians. These results
confirm a simple PD-controller role of the substantia nigra over the striatum during motor action with a single degree of freedom.

ACKNOWLEDGEMENT

This work was partially supported by the Swiss National Science Foundation grant 3.889-0.85.

REFERENCES

Baroni, A., Benvenuti, F., Fantini, L., Pantaleo, T., and Urbani, F., 1983, Human ballistic arm abduction movements, Neurology, 34 : 868.
De Jong, R. N., 1967, "The Neurologic Examination", Harper & Row, New York.
Duvoisin, R. C., 1971, The evaluation of extrapyramidal disease, in; "Monoamines Noyaux Gris Centraux et Syndrome de Parkinson", J. Ajuriaguerra and G. Gauthier, eds., Georg & Cie., Geneva.
Flowers, K., 1976, Visual "closed-loop" and "open-loop" characteristics of voluntary movement in patients with parkinsonism and intention tremor, Brain, 99 : 269.
Hacisalihzade, S. S., 1986, "Optimization of Levodopa Therapy in Parkinson's Disease", Doctoral Dissertation No. 8201, Swiss Federal Institute of Technology (ETH), Zurich.
Hoehn, M. M., and Yahr, M. D., 1967, Parkinsonism: onset, progression and mortality, Neurology, 17 : 427.
Marsden, C. D., 1982, The mysterious motor function of the basal ganglia, Neurology, 32 : 514.
Marsden, C. D., Barry, P. E., Parkes, J. D., and Zilkha, K. J., 1973, Treatment of Parkinson's disease with levodopa combined with L-alphamethylhydrazine, J. Neurol. Neurosurg. Psych., 36 : 10.
Marsden, C. D., Merton, P. A., and Morton, H. B., 1976, Servo action in the human thumb, J. Physiol., 257 : 1.
Poulton, E. C., 1974, "Tracking Skill and Manual Control", Academic Press, London.
Sheridan, T. B., and Ferrel, W. R., 1974, "Man-Machine Systems", MIT Press, Cambridge, MA.
Tatton, W. G., and Bruce, I. C., 1981, A schema for the interactions between motor programs and sensory input, Can. J. Physiol. Pharmacol., 59 : 691.
Velasco, F., and Velasco, M., 1973, A quantitative evaluation of the effects of levodopa on Parkinson's disease, Neuropharmacology, 12 : 88.
Webster, D. D., 1968, Critical analysis of disability in Parkinson's disease, Mod. Treatment, 5 : 257.
Wing, A. M., and Miller, E., 1984, Basal ganglia lesions and psychological analyses of the control of voluntary movement, in: "Functions of the Basal Ganglia, Ciba Foundation Symposium 107", Pitman, New York.

IDENTIFICATION PROBLEMS OF A NON-LINEAR RLC LUNG MODEL DURING

ARTIFICIAL VENTILATION

G. Bogányi

Kandó Kálmán College MITI
Budapest
Hungary

INTRODUCTION

Mathematical modelling of physiological systems has shown a substantial advance in recent years. As a general trend, models are able to describe more and more detailed properties of the physiological system. Consequently, the question of parameter estimation arises during modelling to a greater extent since this is a fundamental factor in verifying and applying the model.

The aim of this chapter is to present a model-based identification method which is suitable for the identification of mechanical properties of the lungs in a special field of application - during positive pressure-controlled ventilation. It is known that the mechanical properties of the lungs fundamentally influence the strategy to be adopted in adjusting the operational parameters of a respirator.

Many research groups have already examined the changes in the mechanical properties of the lungs which occur during low frequency ventilation ($f \leqslant 120$ min^{-1}), assuming linear models (Avanzolini and Barbini, 1984; Bates et al., 1985; Bogányi, 1985; Rossi et al., 1985). Non-linear models are, however, more exact in describing the lung mechanics, especially in pathological conditions (Bogányi, 1986; Kovács and Bogányi, 1983; Nunn, 1977). Some simulation results have shown that it is expedient to use a non-linear model even in the case of respiratory volume changes which are small compared to the value of functional residual capacity (FRC) (Lutchen et al., 1982). As a result, parameter estimation is assuming ever greater importance.

MODEL DERIVATION

The airways and the lungs can be considered as a non-linear distributed parameter system. The model described here is non-linear and is a one-compartment lumped parameter system. It consists not only of the non-linear properties of the airways, but of the lungs, too. The operating point of the lungs is the actual FRC value, which can be measured, and as such is assumed to be known (Mitchell et al., 1982). The respiratory system has two sub-systems, namely the conducting airways and the lungs.

The first sub-system is essentially of rigid structure and therefore its pressure (p) - flow ($\dot{v}$) properties can be described by the following equation:

$$p_m(t) - p_a(t) = R_1 \dot{v}(t) + R_2 \dot{v}^2(t) \tag{1}$$

where R_1 represents the resistance of the airways in the case of a laminar flow pattern and R_2 the resistance for the case of turbulent flow. Building non-linear flow properties into the model is necessary because the flow pattern is in fact mixed, even in the case of spontaneous quiet breathing (Kovács and Bogányi, 1983; Nunn, 1977). The non-linear properties of the airways become more dominant in the case of chronic obstructive pulmonary diseases as a result of the constriction of the airways.

The following equation describes the complex mechanical properties of the lungs. The inertial effect is modelled by inductance (L), which plays a more and more important role if the ventilatory volume or frequency is increasing:

$$p_a(t) - p_p(t) = L\ddot{v}(t) + \frac{\alpha}{v}\dot{v}(t) + f(v) \tag{2}$$

Another possibility in order to obtain a more detailed model is to consider the effect of the resistance of the lung tissue. In this case α/v corresponds to the value of the specific resistance characterising the given operating point.

The last term in the equation describes the static non-linear pressure-volume relations of the lungs as a sub-system, where β is the pressure coefficient and ε the volume coefficient:

$$f(v) = \beta \exp\left(\frac{v}{\varepsilon}\right). \tag{3}$$

Their size is influenced by the patient's age and physiological condition. This non-linear relation concerning the lung sub-system can be approached in terms of a Taylor series about the operating point FRC = V_o.

$$f(v) \cong f(v_o) + \frac{\beta}{\varepsilon}\exp\left(\frac{v_o}{\varepsilon}\right) + \frac{\beta}{2\varepsilon^2}\exp\left(\frac{v_o}{\varepsilon}\right) + \dots \tag{4}$$

The first term in the Taylor series is the value characterising the pressure at the operating point. Tidal volume is the volume change around the FRC. The dynamic elasticity of the non-linear system is given by the following equations:

$$E_1 = \frac{f(v) - f(v_o)}{v - v_o} = \frac{\beta}{\varepsilon}\exp\left(\frac{v_o}{\varepsilon}\right)$$

$$\tag{5}$$

$$E_2 = \frac{f(v) - f(v_o)}{(v - v_o)^2} = \frac{\beta}{2\varepsilon^2}\exp\left(\frac{v_o}{\varepsilon}\right)$$

According to these relations, the volume change must be an indirect function of time. It can be seen that both values, those of E_1 and E_2, are functions of the operating point and of the parameters characterising the patient's patho-physiological condition. Differentiation between these effects is very important from the point of view of exact diagnosis.

The sum of (1) and (2), making use of (5), leads to the differential equation of the non-linear model characterising the airways and the lungs:

$$p(t) = p_m(t) - p_p(t) = R_1\dot{v}(t) + R_2\dot{v}^2(t) + L\ddot{v}(t) + \frac{\alpha}{v_o}\dot{v}(t) + E_1v(t) + E_2v^2(t). \tag{6}$$

According to the structure of the model, the linear resistive members characterising the conductive airway and the lung tissue can be lumped together as the total resistance R_1^* of the system.

$$p(t) = R_1^* \dot{v}(t) + R_2 \dot{v}^2(t) + L\ddot{v}(t) + E_1 v(t) + E_2 v^2(t) \qquad (7)$$

where $R_1^* = R_1 + \dfrac{\alpha}{v_0}$.

IDENTIFICATION

The aim of identification is to determine the parameters of the differential equation. Identification will be carried out first assuming an ideal measuring and data processing system. The sets of independent ideal input signs are the directly measurable ones, namely:

(i)   driving pressure at the mouth; and
(ii) time function of flow.

The following derived quantities must be determined in order to achieve model identification:

(i)   time function of the first derivative of flow;
(ii)   time function of quadratic flow;
(iii) time function of volume; and
(iv)   time function of quadratic volume.

The boundary conditions of identification are as follows:

(i)   the input signs of the system have to be periodic;
(ii)   the mean values of $p_m(t)$, $\dot{v}(t)$ and $v(t)$ have to be zero; and
(iii) the characteristic parameters of the lungs as a mechanical system should not change during the period of measurement.

This last condition assumes, of course, that FRC remains constant.

The differential equation characterising the physiological system is multiplied by the primary and derived time functions and their mean value is taken. According to the second boundary condition, using the orthogonality of some product functions, the following simpler linear equation system can be obtained.

$$
\begin{bmatrix}
M[\dot{v}^2] & 0 & 0 & 0 & 0 \\
0 & M[\dot{v}^4] & 0 & 0 & 0 \\
0 & 0 & M[v^2] & 0 & M[\ddot{v}v] \\
0 & 0 & 0 & M[v^4] & M[\ddot{v}v^2] \\
0 & 0 & M[v\ddot{v}] & M[v^2\ddot{v}] & M[\ddot{v}^2]
\end{bmatrix}
\cdot
\begin{bmatrix}
R_1^* \\
R_2 \\
E_1 \\
E_2 \\
L
\end{bmatrix}
=
\begin{bmatrix}
M[p\dot{v}] \\
M[p\dot{v}^2] \\
M[pv] \\
M[pv^2] \\
M[p\ddot{v}]
\end{bmatrix}
\qquad (8)
$$

The matrix containing the coefficients shows that the solution of R_1^* and R_2 is given directly, but the determination of the parameters E_1, E_2 and L requires many calculations.

If all the boundary conditions are fulfilled, it can be proved, both for the periodic and complex periodic cases, that parameter identification is based on moment estimation. It can further be proved that the identification obtained is, in fact, unambiguous. If E_1 and E_2 are determined, the parameters ε and β corresponding to the operating point which is assumed to be known can be derived primarily from the solution of (5):

$$\varepsilon = \frac{E_1}{2E_2} \qquad\qquad \beta = \frac{E_1^2}{2E_2} \exp\left(-\frac{2v_o E_2}{E_1}\right) \qquad\qquad (9)$$

However, it must be clearly stated that this method cannot be used directly to separate the components of R_1^*, namely the resistance of the airways and of the tissue. This arises, of course, from the fact that the model contains only a single compartment. This problem can be solved by examining the mechanical properties of the lungs at two different FRC values. Therefore, the parameters α and R_1 can be determined as well, in such an indirect manner:

$$R_{11}^* = R_1 + \frac{\alpha}{v_{o1}} \qquad\qquad R_{12}^* = R_1 + \frac{\alpha}{v_{o2}} . \qquad\qquad (10)$$

The solution of (10) is:

$$\alpha = \left(R_{11}^* - R_{12}^*\right) \frac{v_{o1} \, v_{o2}}{v_{o2} - v_{o1}} \qquad\qquad R_1 = \frac{R_{11}^* \, v_{o2} + R_{12}^* \, v_{o1}}{v_{o2} - v_{o1}} \qquad\qquad (11)$$

REALISATION

The block-scheme of the respirator-patient model and of the measuring system is shown in Fig. 1. This summarises the experimental set-up. The time function of flow is transformed into a pressure difference by the Fleisch-tube and this, in turn, into an electric sign. The mouth pressure is measured according to the ambient pressure. The set of independent input parameters is fed to the input of the data processing system.

Expedient placing of the measuring instruments shows that they do not cause any systematic error from the viewpoint of parameter estimation, because the mouth pressure transducer is not measuring the presssure drop at the Fleisch-tube. Another important conclusion is that the inner impedance of the applied generator does not influence parameter identification.

The digital data processing system has to fulfil the following functions:

(i) making derived input signs out of the directly measurable ones;
(ii) producing the zero mean value of the time functions of pressure, flow and volume; and
(iii) calculation of the moments necessary for parameter estimation.

It is evident that parameter estimation can be carried out over a finite averaging time only. It is a well-known fact that moment estimation is undistorted, independent of the rates of the moments. Therefore R_1^*, R_2, E_1, E_2 and L must be undistorted, too. If the measuring time is the reciprocal of the ventilatory frequency, the parameters of the respiratory mechanical model will be determined on a breath-by-breath basis. In this case the variance of parameter estimation will be influenced in a negative manner by the measuring time, which is short compared to the spectrum of the input signs.

It must be realised that both the choice of model and the flow differentiation can cause measurement errors. Partitioning of R_1^* occurring in the model can, in principle, be realised, but it has no useful meaning in practice, since its variance can be as great as 100%. This awkward effect has to be taken into account in every case of application. It is impossible to measure the inductive term without deriving the flow. The

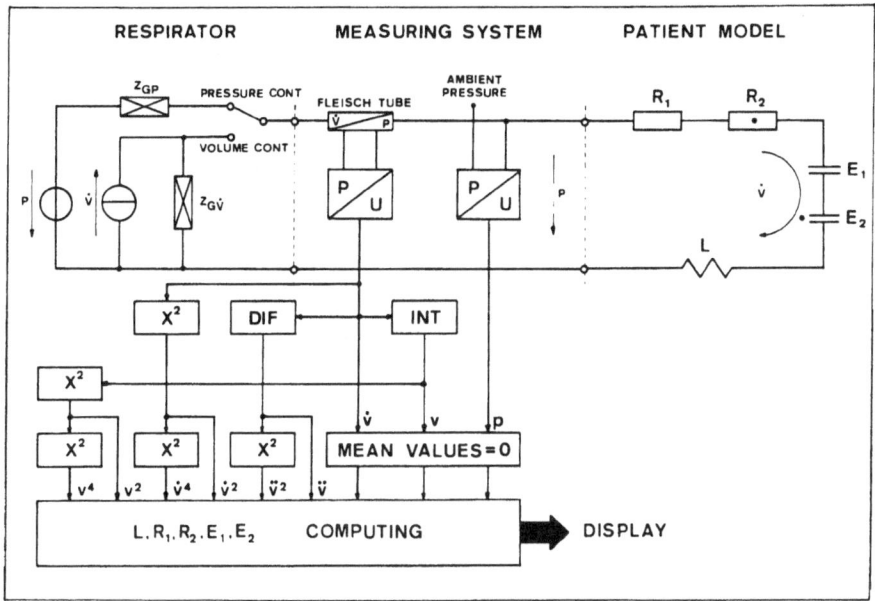

Fig. 1. The experimental set-up.

errors arising from digital differentiating algorithms are well-known. The solution is on the one hand to decrease in a rational manner the band-width of the input signs, and on the other hand to increase the sampling frequency. The latter is a very effective method, but it might influence the price of the data processing system enormously.

SUMMARY

The sophisticated estimation of the mechanical features of the respiratory system is based on the measurement of the input impedance using a forced oscillation method. Fitting the input impedance function to the parameters of the model is another source of difficulty. Therefore, this complicated method will not be used in the near future in measurements obtained during artificial ventilation. Equally, the method described by Bates et al. (1985) and Rossi et al. (1985) is sensitive to noise, and therefore the variance of parameter measurement is high. It can be used both in pressure and volume control. It must be mentioned, however, that this method is not suitable for taking into account non-linearities.

In contrast, some of the advantages of the method described in this chapter are as follows:

(i) It is suitable for characterising lung mechanics in the case of both pressure and volume controlled artificial ventilation.
(ii) The application of positive end-expiratory pressure does not influence parameter estimation.
(iii) In applying this method the problem of parameter fitting does not arise. Its consequence is that the computing time is less than in other cases.
(iv) Its sensitivity to errors, which is an advantage of the measuring system, is based on the averaging effect.
(v) The non-linear static pressure-volume relation, which characterises

the lungs can be determined exactly, provided that E_1 and E_2 are known.

REFERENCES

Avanzolini, G., and Barbini, P., 1984, A versatile identification method applied to analysis of respiratory mechanics, IEEE Trans. Biomed. Eng., BME-31 : 520.

Bates, J. H. T., Rossi, A., and Milic-Emili, J., 1985, Analysis of the behaviour of the respiratory system with constant inspiratory flow, J. Appl. Physiol., 58 : 1840.

Berend, N., and Thurlbeck, W. M., 1982, Exponential analysis of pressure volume relationship in excised human lungs, J. Appl. Physiol., 52 : 838.

Bogányi, G., 1985, Identification of the mechanical properties of the lungs during artificial ventilation, in: "Proc. 14th International Conference on Medical and Biomedical Engineering and 7th International Conference on Medical Physics", Helsinki, A.-L. Kairento, T. Katila, N. Saranummi and H. Seitsonen, eds., IFMBE : 334.

Bogányi, G., 1986, Exponential model-based estimation of the mechanical properties of the lungs during artificial ventilation, in: "Proc. 8th Annual IEEE Conference of the Engineering in Medicine and Biology Society", G. V. Kondraske and C. J. Robinson, eds., IEEE, New York : 885.

Kovács, B., and Bogányi, G., 1983, Changes of the mechanical properties of the lungs following abdominal surgery, Acta Chir. Hung., 24 : 251.

Lutchen, K. R., Primiano, F. P., and Saidel, G. M., 1982, A non-linear model combining pulmonary mechanics and gas concentration dynamics, IEEE Trans. Biomed. Eng., BME-29 : 629.

Mitchell, R. R., Wilson, R. M., Holzapfel, L., Benis, A. M., Sierra, D., and Osborn, J. J., 1982, Oxygen wash-in method for monitoring functional residual capacity, Crit. Care Med., 10 : 529.

Nunn, F. J., 1977, "Applied Respiratory Physiology with Special Reference to Anaesthesia", Butterworths, London.

Rossi, A., Gottfried, S. B., Higgs, B. D., Zocchi, L., Grassino, A., and Milic-Emili, J., 1985, Respiratory mechanics in mechanically ventilated patients with respiratory failure, J. Appl. Physiol., 58 : 1849.

EVALUATION OF CARDIORESPIRATORY FUNCTIONS DURING HEART CATHETERISATION

THROUGH SIMULATION MODEL IDENTIFICATION

J. Kofránek, M. Munclinger, B. Šerf, M. Fusek, J. Kautzner,
V. Duchác, Z. Pokorný, Z. Brelidze and J. Gondzhilasvili

Department of Pathological Physiology and 2nd Department
of Medicine, Charles University,
Prague, Czechoslovakia; and
Central Science Laboratory, Ministry of Health,
Tbilisi, USSR

BASIC RELATIONSHIP

The development of personal computers has made it possible for complex computations to be performed during clinical examination in catheterisation laboratories. For this purpose we have developed a mathematical model of blood gas transport, which takes as its input the data gained in the heart catheterisation presented in Table 1. The mathematical realisation of the physiological laws in the simulation model is presented in Appendix I, with

Table 1. Input Data

BASIC

- body weight,
- body height,
- rectal temperature.

VENTILATION

- barometric pressure,
- room temperature,
- expired volume (ATPS),
- time for measurement of expired volume,
- fractional concentration of oxygen in
  inspired and expired gas,
- fractional concentration of carbon dioxide
  in inspired and expired gas (optional).

BLOOD

- haemoglobin concentration,
- pH, PCO_2, PO_2 in arterial and mixed venous
  blood (at 37°C),
- oxygen saturation in arterial and mixed
  venous blood (optional).

Table 2. Output Data

ARTERIAL AND MIXED VENOUS BLOOD

- pH, PO_2, PCO_2 corrected for rectal temperature,
- base excess,
- actual and standard bicarbonate concentrations,
- oxygen saturation,
- blood O_2 and CO_2 contents.

END PULMONARY CAPILLARY BLOOD

- O_2 and CO_2 contents.

ALVEOLAR GAS

- PO_2 and PCO_2.

RESPIRATION

- minute ventilation,
- O_2 consumption (BTPS),
- CO_2 production (BTPS),
- alveolar ventilation (BTPS),
- respiratory quotient.

CIRCULATION

- systemic blood flow (cardiac output and index)
- pulmonary blood flow,
- right-to-left shunt blood flow.

VENTILATION-TO-PERFUSION RATIO

(alveolar ventilation/cardiac output)

---

the symbols used listed in Appendix II. Incorrect data are rejected and, if the measured data are consistent with physiological relationships, the clinically relevant data presented in Table 2 are calculated.

In the first step of model identification, the total blood oxygen and carbon dioxide contents in arterial and mixed venous blood are calculated from pH, PCO_2 and PO_2 (measured at a temperature of 37°C) using (4), (5) and (6) (see Appendix I). Then the values of pH, PCO_2 and PO_2 are corrected for the patient's temperature using (8) - (10), and arterial and mixed venous blood acid-base parameters (base excess, base excess in virtually fully oxygenated blood - BEox, actual and standard bicarbonate concentrations) are calculated using (2), (3), (7), (11) and (12).

Furthermore, the respiratory quotient (RQ) is determined (by (14)) from the O_2 and CO_2 contents in arterial and mixed venous blood. If the RQ value obtained is beyond the physiologically feasible range, the inconsistent nature of the measured input data is indicated.

In the second step, the minute ventilation is calculated using (17) and (18). For the correct evaluation of the oxygen consumption we have to know the volume fractions of oxygen and carbon dioxide in dry expired gas ($FeCO_2$) - see (19). If the value of $FeCO_2$ is not measured, the oxygen consumption can be evaluated indirectly from alveolar ventilation, using (22) - (24).

If FeCO$_2$ is measured, the comparison between the direct and indirect
calculations of oxygen consumption can be used for testing the consistency
of the measured data. Moreover, it is possible to calculate the carbon
dioxide consumption directly from (20), and the respiratory quotient, which
can be compared with the indirectly obtained value, from (14), for checking
the consistency of the measured data.

For indirect estimation of alveolar ventilation it is necessary to have
knowledge of the value of alveolar oxygen tension. This can be obtained
from the Bohr equation (24). However, for this calculation we need the
value of alveolar CO$_2$ tension, which can be obtained using an iterative
procedure. At the start of this iteration, the equivalence of the arterial
and alveolar carbon dioxide blood tensions is assumed. Then, oxygen alveolar tension is determined from carbon dioxide tension using the Bohr equation (24). If the value of alveolar PO$_2$ obtained is lower than arterial
PO$_2$, the inconsistency of the measured data is indicated. Now, supposing
that the blood gaseous tensions in alveoli and in the pulmonary end-capillary blood are in equilibrium, the oxygen and carbon dioxide contents in
the pulmonary end-capillary blood are calculated by a computer procedure
(using (1), (3) - (6) and (8) - (11)) from the values of PO$_2$, PCO$_2$, BEox (this
value of base excess in virtually fully oxygenated blood is the same in
arterial and pulmonary end-capillary blood), haemoglobin concentration and
blood temperature.

Subsequently, using (16) the new value of carbon dioxide content in the
end-pulmonary capillary blood is determined. Afterwards, the new value of
PCO$_2$ in pulmonary-end capillary blood is calculated by another computer
procedure from the values of BEox, blood gas contents, haemoglobin concentration and blood temperature, using (1), (3) - (6) and (8) - (11). Assuming equivalence of carbon dioxide tensions in alveoli and pulmonary end-
capillary blood, we obtain the new value of alveolar CO$_2$ tension for the
next iteration.

The values of alveolar oxygen and carbon dioxide tensions (which are
essential for the calculation of alveolar ventilation and, subsequently,
exact determination of oxygen consumption) are obtained as the result of
this iterative procedure. A further result of this iterative sub-routine
is the value of oxygen content in pulmonary end-capillary blood, which is
necessary for determination of right-to-left shunt perfusion (by (15)).

In contrast to the classical procedure for calculating the oxygen
content in the pulmonary end-capillary blood from arterial and mixed venous
blood gases (Kelman, 1966b; Ruiz et al., 1975; Newell et al., 1980), no
assumptions are needed regarding the identity of PCO$_2$ and pH in arterial
and pulmonary end-capillary blood. Finally, cardiac output, pulmonary perfusion and the ventilation to perfusion quotient can be determined.

Through model identification, not only were relevant clinical data
obtained, but also the consistency of the input measured data was checked.
The program informs us whether the instrumentally measured data are consistent with physiological relationships. Incorrect data are rejected in the
following program steps:

(i)   if the respiratory quotient value calculated by (14) from O$_2$ and
      CO$_2$ contents in arterial and mixed venous blood deviated from the
      predicted physiological limit (0.15 - 1.7: incorrect data);

(ii) if the alveolar O$_2$ pressure computed by Bohr's equation did not exceed
      the measured arterial pressure (discrepancy of blood gases).

In the case of measured FeCO$_2$, the oxygen consumption and carbon dioxide production can be calculated directly using (19) and (20). This

operation leads to an additional control on the validity of the measured data:

(iii) if the oxygen consumption calculated directly by (19) deviated more than 10% from the value calculated indirectly by (22);

(iv) if the carbon dioxide production estimated by (20) diverged more than 10% from the value obtained by (21) from directly calculated oxygen consumption and indirectly calculated respiratory quotient;

(v) if the respiratory quotient obtained by directly calculated O_2 consumption and CO_2 production diverged more than 10% from the value of respiratory quotient calculated indirectly from O_2 and CO_2 arterial and mixed venous blood contents.

RESULTS

The simulation model implemented on a PMD-85 microcomputer was evaluated by identification using catheterisation data from 131 patients. Cardiac output was computed by the model using the Fick principle from paired samples of arterial and mixed venous blood obtained in the 5th minute of breathing into a Douglas bag. The fractional concentration of oxygen in inspired and expired gas was measured with an OA 101 Mk II oxygen analyser (Servomex, Great Britain), pH and blood gases pressure values were determined by means of a 1302 I.T.S. analyser (USA). The fractional concentration of carbon dioxide in expired gas was not measured. Cardiac output was subsequently assessed in 37 patients by the dye-dilution method (Cariogeen Hynson, Westcott and Dunning, USA; cardiodensitometer Beckman, USA).

The input data were rejected as incorrect (respiratory quotient exceeded the physiologically feasible limit) in 9 patients (7%) and further data in another 30 patients (23%) were taken out due to the discrepancy of the blood gases with Bohr's equation of alveolar gases. Blood gas values were accepted as correct data in accord with physiological principles in 92 patients (70%).

The reliability of the program was verified by a good correlation of cardiac index values calculated by the model (using the Fick principle) with those determined by dye dilution ($y = 0.612 * x + 1.144$, $r = 0.734$, $p < 0.001$). The difference between cardiac index values assessed by both methods did not exceed $1.0 \, l \, min^{-1} \, m^{-2}$ in 78% of the patients. The reproducibility was confirmed by close correlation between the maximal and the minimal cardiac index values computed by the simulation model from different combinations of blood samples after these values had been randomly divided into two groups ($y = 0.884 * x + 0.447$, $r = 0.891$, $p < 0.001$). The difference between maximal and minimal cardiac index values did not exceed $0.5 \, l \, min^{-1} \, m^{-2}$.

DISCUSSION

The application of the simulation model identification procedure in the evaluation of the cardiorespiratory functions provides a test of the validity of laboratory data obtained by instrumental measurements against basic physiological principles. The rapid delivery of results by the program enables invalid data to be detected in the course of cardiac catheterisation, to be eliminated immediately, and to be substituted with correct data gained from the re-examination of blood samples. These arrangements undoubtedly add significantly to the quality of parameters derived from the haemodynamic investigation.

The limits for the respiratory quotient values in the program were set rather wide so as to reject only the very extreme values arising either from incorrect sampling of blood specimens, from their inappropriate handling during transportation, or caused by faulty instrument processing. Smaller deviations in the respiratory quotient, caused for instance by irregular breathing of the patient,were accepted by the program.

The control functions of our model, as well as the accuracy of its output parameters, could be further enhanced through the direct measurement of the fractional concentration of CO_2 in inspired and expired gas and by direct estimation of arterial and mixed venous oxygen saturation. Clinical exploitation of some output data computed by simulation model identification (Table 2) have not so far been used in routine practice in haemodynamic laboratories. This is something which still remains to be explored.

Copies of the computer program used (Basic and Turbo-Pascal versions) are available on request from the author by sending a blank IBM-compatible diskette and self-addressed envelope to Dr. Jiří Kofránek (see Address List of Contributors).

APPENDIX I.   LIST OF EQUATIONS USED IN THE SIMULATION MODEL

$$pH37 = A1 + A2*Y + (A3 + A4*Y) * log\ PCO_237)/(A5 + A6*Y) \tag{1}$$

where:

$$
\begin{aligned}
Y &= (A7 + SQRT\ (A8 + A9*(BEox + 0.3*Hb*(1 - SO_237))))/A10 \\
A1 &= 9.963500*10^2 - 1.03500*10^1 * Hb \\
A2 &= 3.516875*10^1 + 2.58750*10^{-1} * Hb \\
A3 &= -8.241000*10^1 + 2.01000 * Hb \\
A4 &= -5.276250 - 5.02500*10^{-2} * Hb \\
A5 &= 1.210000*10^2 - Hb \\
A6 &= 2.625000 + 2.50000*10^{-2} * Hb \\
A7 &= -2.556000 - 9.44000*10^{-2} * Hb \\
A8 &= 1.387634*10^1 + 1.86653*10^{-1} * Hb + 5.34936*10^{-3} * Hb^2 \\
A9 &= 5.480000*10^{-1} + 1.37000*10^{-2} * Hb \\
A10 &= 2.740000*10^{-1} + 1.37000*10^{-2} * Hb
\end{aligned}
$$

[reference: Kofranek, 1980]

$$BE = (A10*X - A7)^2 - A8)/A9 \tag{2}$$

where:

$X = (A1 + A3 * log\ PCO_237 - A5 * pH37)/(A6 * pH37 - A2 - A4 * log\ PCO_237)$
Coefficients A1 - A10 (see (1))
[reference: Kofranek, 1980]

$$BEox = BE - 0.3 * Hb * (1 - SO_237) \tag{3}$$

[reference: Siggaard-Andersen, 1974]

$$
\begin{aligned}
SO_237 = 0.9995 &- 1.0000/(1 + ((P + 7)/33.7)^{3.3}) \\
&- 0.0050/(1 + ((P - 130)/35)^2) \\
&+ 0.0045/(1 + ((P - 68)/12)^6) \\
&- 0.0050/(1 + ((P - 35)/3)^4) \\
&- 0.0050/(1 + ((P - 15)/4)^4) \\
&+ 0.0035/(1 + ((P - 26)/3)^6) \\
&+ 0.0020/(1 + ((P - 53)/8)^4) \\
&- 0.0040/(1 + ((P - 40)/0.9)^4) \\
&- 0.0020/(1 + ((P - 200)/65)^8) \\
&+ 0.0040/(1 + ((P - 9)/3)^2)
\end{aligned}
\tag{4}
$$

where:

$$P = PO_237 * 10^{0.40} * (pH37 - 7.4) + 0.06 \log (40/PCO_237)$$

[references: Kelman, 1966b; Ruiz et al., 1975]

$$O_2tot = 3.0473*10^{-5} * PO_237 + 1.39 * 10^{-2} * Hb * SO_237 \qquad (5)$$

[reference: Siggaard-Andersen, 1974]

$$CO_2tot = 0.02226 * PCO_237 * (C1 + (C2*aH + C3)/aH^2 + C4*D/aH \qquad (6)$$
$$+ C5 * (1 - SO_237)/((aH/D)^2 + C6*aH/D + C7*PCO_237)$$
$$+ C8 * SO_237/((aH/D)^2 + C9*aH/D + C10*PCO_237))$$

where:

$$aH = 10^{(9-pH37)}$$
$$D = XO * SO_237^{X1} * pH37^{X2} * e^{X3*SO_237 + X4*pH37}$$
$$XO = 7.0388002*10^{-3}$$
$$X1 = 3.6447450*10^{-4}$$
$$X2 = 7.9099077$$
$$X3 = -2.0113444*10^{-1}$$
$$X4 = -1.4790526$$
$$C1 = 3.0700000*10^{-2} - 2.2580645*10^{-4} * Hb$$
$$C2 = 2.3038631*10^{1} - 6.7561967*10^{-1} * Hb$$
$$C3 = 4.7648549*10^{1} - 1.3973181 * Hb$$
$$C4 = 5.5042129*10^{-1} * Hb$$
$$C5 = 1.3735969*10^{1} * Hb$$
$$C6 = 3.9800000*10^{1}$$
$$C7 = 2.2152680*10^{1}$$
$$C8 = 3.2420343 * Hb$$
$$C9 = 1.7900000*10^{1} * Hb$$
$$C10 = 5.2285900*10^{1}$$

[reference: Kofranek, 1980]

$$HCO_3st = 1.2 * 10^{pHst} - 6.1008 \qquad (7)$$

where:

pH37 = (A1 + A2 * Y + 1.60206 * (A3 + A4*Y))/(A5 + A6*Y)
Y = (A7 + SQRT (A8 + A9*BE)/A10
Coefficients A1 - A10 (listed in (1))
[reference: Kofranek, 1980]

$$pHt = pH37 - (0.0146 - 0.0065 * (7.4 - pH37) \qquad (8)$$
$$- 0.00003 * BEox * (t - 37)$$

[reference: Severinghaus, 1966]

$$PCO_2t = PCO_237 * 10^{0.0185*(t-37)} \qquad (9)$$

[reference: Kelman and Nunn, 1966]

$$PO_2t = PO_237 * 10^{(t-37)*(0.0049+0.0261*(1-e^X))} \qquad (10)$$

where:

$$X = 52 * (SO_237 - 1)$$

[reference: Severinghaus, 1979]

$$SO_2t = (O_2tot - aO_2t * PO_2t)/(1.39*10^{-2} * Hb)$$ (11)

where:

$$aO_2t = 5.9519*10^{-5} - 1.266*10^{-6} * t + 1.3*10^{-8} * t^2$$
[reference: Kelman, 1966a]

$$HCO_3t = 10^{pHt - pKt} + log (aCO_2t * PCO_2t)$$ (12)

where:

$$pKt = a1 + a2 * t + a3 * t^2 + a4 * t^3$$
$$aCO_2t = b1 + b2 * t + b3 * t^2 + b4 * t^3$$
$$a1 = 6.3852 \qquad b1 = 0.0907$$
$$a2 = -1.3288*10^{-2} \quad b2 = -3.3730*10^{-3}$$
$$a3 = 1.7364*10^{-4} \quad b3 = 6.7490*10^{-5}$$
$$a4 = -6.0084*10^{-7} \quad b4 = -5.4076*10^{-7}$$
[reference: Reeves, 1976]

$$Qt = VO_2/(O_2tot_a - O_2tot_v) = VCO_2/(CO_2tot_v - CO_2tot_a)$$ (13)

$$RQ = VCO_2/VO_2 = (CO_2tot_v - CO_2tot_a)/(O_2tot_a - O_2tot_v)$$ (14)

$$Qsh/Qt = (O_2tot_c - O_2tot_a)/(O_2tot_c - O_2tot_v)$$ (15)

$$CO_2tot_c = CO_2tot_a - RQ * (O_2tot_c - O_2tot_a)$$ (16)

$$VE = Vexp/time * (Patm - PH_2Otlab) * (273.15 + t)$$
$$/((Patm - PH_2Ot) * (273.15 + tlab))$$ (17)

where:

PH_2Ot and PH_2Otlab were calculated by (18).

$$PH_2Ot = 2.4225 + 0.67734*t - 0.0082*t^2 + 0.00061*t^3$$ (18)

[reference: Kofranek, 1980]

$$VO_2 = (VI*FiO_2 - VE*FeO_2) * kBTPS\_STPD$$ (19)

where:

$$VI = VE * FeN_2/FiN_2$$
$$FeN_2 = 1 - FeO_2 - FeCO_2$$
$$FiN_2 = 1 - FiO_2 - FiCO_2 \quad (FiCO_2 = 0)$$
$$kBTPS\_STPD = (Patm - PH_2Ot) * 273.15/(760 * (273.15 + t))$$

$$VCO_2 = (VE*FeCO_2 - VI*FiCO_2) * kBTPS\_STPD$$ (20)

$$VCO_2 = VO_2 * RQ$$ (21)

$$VO_2 = VA * ((FAN_2/FiN_2) * FiO_2 - FAO_2) * kBTPS\_STPD$$ (22)

where:

$$FAN_2/FiN_2 = (PACO_2 + RQ * PAO_2)/(FiCO_2 * (Patm - PH_2Ot)$$
$$+ RQ * FiO_2 * (Patm - PH_2Ot))$$
$$kBTPS\_STPD \text{ (see (19))}$$

$$VA = VE - VD = VE * (1 - (FAO_2 - FEO_2)/(FAO_2 - FiO_2)) \qquad (23)$$

where:

$$FAO_2 = PAO_2/(Patm - PH_2Ot)$$

$$PAO_2 = (FiO_2 * (Patm - PH_2Ot) + PACO_2 * (FiO_2 * (1 - RQ) - 1)/RQ \qquad (24)$$

APPENDIX II. LIST OF SYMBOLS

aCO_2t	[mmol/torr]	Carbon dioxide solubility coefficient in plasma
aO_2t	[1-STPD/torr]	Oxygen solubility coefficient in plasma
BE	[mmol/l]	Blood base excess concentration
BEox	[mmol/l]	Base excess concentration in virtually oxygenated blood
CO_2tot	[1-STPD/1 blood]	Total blood carbon dioxide content
HCO_3st	[mmol/l]	Standard bicarbonate concentration
HCO_3t	[mmol/l]	Actual bicarbonate concentration at given temperature
$FACO_2$		Volume fraction of carbon dioxide in dry alveolar gas
FAN_2		Volume fraction of nitrogen in dry alveolar gas
FAO_2		Volume fraction of oxygen in dry alveolar gas
$FeCO_2$		Volume fraction of carbon dioxide in dry mixed expired gas
FeN_2		Volume fraction of nitrogen in dry mixed expired gas
FeO_2		Volume fraction of oxygen in dry mixed expired gas
$FiCO_2$		Volume fraction of carbon dioxide in dry inspired gas
FiN_2		Volume fraction of nitrogen in dry inspired gas
FiO_2		Volume fraction of oxygen in dry inspired gas
O_2tot	[1-STPD/1 blood]	Total blood oxygen content
$PACO_2$	[torr]	Carbon dioxide tension in alveoli
PAO_2	[torr]	Oxygen tension in alveoli
Patm	[torr]	Atmospheric pressure
PCO_2t	[torr]	Blood carbon dioxide tension at given temperature
PCO_237	[torr]	Blood carbon dioxide tension at 37°C
PH_2Ot	[torr]	Vapour pressure at given temperature
PH_2Otlab	[torr]	Vapour pressure at room temperature
PH_2O37	[torr]	Vapour pressure at 37°C
pHt		Plasma pH at given temperature
pH37		Plasma pH at 37°C
PO_2t	[torr]	Blood oxygen tension at given temperature
PO_237	[torr]	Blood oxygen tension at 37°C
Qsh	[1/min]	Right-to-left shunt perfusion
Qp	[1/min]	Lung perfusion
Qt	[1/min]	Cardiac output
RQ		Respiratory quotient

SO₂t		Oxygen haemoglobin saturation at given temperature
SO₂37		Oxygen haemoglobin saturation at 37°C
t	[°C]	Given temperature
tlab	[°C]	Room temperature
time	[min]	Time for measurement of expired volume
VA	[1-BTPS/min]	Alveolar ventilation
VD	[1-BTPS/min]	Death volume ventilation
VE	[1-BTPS/min]	Minute ventilation
VI	[1-BTPS/min]	Inspired minute ventilation
Vexp	[1-ATPS]	Expired volume
VCO₂	[1-STPD/min]	Carbon dioxide production
VO₂	[1-STPD/min]	Oxygen consumption

Indices:

a	Arterial blood
c	Pulmonary end-capillary blood
v	Mixed venous blood

REFERENCES

Kelman, R. G., 1966a, Digital computer subroutine for the conversion of oxygen tension into saturation, J. Appl. Physiol., 21 : 1375.

Kelman, R. G., 1966b, Calculation of certain indices of cardio-pulmonary function using a digital computer, Resp. Physiol., 1 : 335.

Kelman, R. G., and Nunn, F. J., 1966, Nomograms for correction of blood pO₂, pCO₂, pH and base excess for time and temperature, J. Appl. Physiol., 21 : 1484.

Kofránek, J., 1980, "Simulation of Acid-base Control", Ph.D. Thesis, Charles University, Prague (in Czech.).

Newell, J. C., Stratton, H. H., Deno, D. C., Gisser, D., and Ostrander, L. E., 1980, On-line blood and respiratory gas analysis, IEEE Trans. Biomed.Eng., BME-27 : 523.

Reeves, R. B., 1976, Temperature-induced changes in blood acid-base status: pH and pCO₂ in a binary buffer, J. Appl. Physiol., 40 : 752.

Ruiz, B. C., Tucker, W. K., and Kirby, R. R., 1975, A program for calculation of intrapulmonary shunts, blood-gas and acid-base values with a programable calculator, Anesthesiology, 42 : 88.

Severinghaus, J. W., 1966, Blood gas calculator, J. Appl. Physiol., 21 : 1108.

Severinghaus, J. W., 1979, Simple, accurate equation for human blood O₂ dissociation computation, J. Appl. Physiol., 46 : 599.

Siggaard-Andersen, O., 1974, "The Acid-Base Status of Blood", 4th edn., Williams and Wilkins, Baltimore.

COMPUTER MODELLING OF ATRIOVENTRICULAR NODAL PROPERTIES

M. Malik and A. J. Camm

Department of Computer Science, Charles University
Prague, Czechoslovakia; and
Department of Cardiological Sciences
St. George's Hospital Medical School
London, UK

INTRODUCTION

From the clinical point of view, the AV node is one of the most complex cardiac structures. Its external behaviour, that is, the function of the whole structure, has previously been investigated in clinical and laboratory studies. Based on such laboratory data, different approaches have been used to describe the nodal properties in an exact way and different mathematical formalisms of nodal functions have been suggested (Heethaar et al., 1973; Dorveaux et al., 1985a).

On the other hand, the internal behaviour of the node, that is, the properties and functions of its individual elements and fibres as well as their mutual influence, is known only approximately. Although laboratory studies enable separate nodal fibres to be examined, the interactions between intra-nodal cells and parts are still not accessible by current laboratory techniques. Hence, our belief concerning the internal nodal function is mostly based on unproved hypotheses and logical speculations.

Since direct laboratory testing of hypotheses which have been suggested is beyond the frame of our possibilities, computer modelling offers a useful tool for improvement of our knowledge. The principles and features expected by different hypotheses can be incorporated into computer simulation experiments. Certainly, the comparison of laboratory and clinical data with the results of computational models does not offer exact verifications of our ideas, but it can, to a great extent, highlight the problems.

COMPUTER MODEL OF CARDIAC CONDUCTION

Simulation studies of AV nodal properties require a "complete" heart model incorporating both atrial and ventricular myocardia, sinus and AV nodes and possible bypass tracts. Despite this, the model structures representing these parts of the heart can be very concise. We have therefore constructed a simulation model introducing either one or several elements for each relevant part of the heart (Malik et al., 1987a). The actions of these elements are also simplified. The model supposes that some elements are able to generate the excitation depolarisation signal and to transmit it to their neighbouring elements, while other elements are only able to convey the depolarisation wavefront. Further, the model distinguishes depolarisation and resting states for all elements, assuming the changes between them to be instantaneous. The properties of elements creating the excitation include a periodicity of signal production which dictates the cycle length; for other elements, a conduction delay is introduced.

The elements of the model are linked together using special connectors. Each transmits the excitation wavefront from one element to another. Conduction delay is also introduced for these connectors, although depolarisation and resting states are not considered. The connection between two elements can be either uni- or bidirectional. For bidirectional connections, the "anterograde" and "retrograde" delays may differ. The computer implementation of the current version of the model makes it possible to create different structures incorporating different numbers of elements and/or connectors. The properties of the elements and connectors (for example, conduction delay, refractory period duration) are fully programmable and are introduced into each model computation in the form of input data.

The procedure of depolarisation wavefront transmission between the model elements mirrors the natural process: the depolarisation signal created in one of the "generation" elements is transmitted along all connectors leading from this element. A target element accepts the signal transmitted along a connector only if it is in its resting state. On accepting the excitation, the element changes its state to become depolarised.

A computer model acceptable for studies of supraventricular tachycardia must also incorporate the cycle length dependences of the duration of repolarisation of different cardiac structures and cycle length dependence of intra-AV nodal conduction delay. In the model reported here, the length of the repolarisation phase of each element is not set to a certain value, but it is automatically obtained from a numeric table, representing a multilinear curve, which describes how the refractoriness of the given element depends on the cycle length. These curves may be different for different elements. Further, the model supposes, in accordance with reality, that the cycle length's influence on refractoriness is not immediate. Therefore, an average cycle length of several recent cycles is computed for each element and used to update the duration of its repolarisation phase. Intra-AV nodal conduction is treated in the same manner, but its dependence is only upon the last cycle.

Each computational experiment with the model gives results in several formats. A complete tracing of polarisation changes of all or of selected heart model elements is produced in the form of a timetable. Using this table, the way in which the depolarisation wave radiated through the artificial heart may be followed. For a more realistic display of results, a simulated one-lead ECG record is also produced. Since the simplified nature of the heart model image makes it impossible to mirror the ECG generation process and to compute the exact form of ECG waves and complexes, the model uses several predefined prototypes of P waves and QRS complexes distinguishing different types and directions of atrial and ventricular depolarisation. Similar prototypes are introduced for T and Ta waves.

The experiments presented here were performed using two versions of the model. The original version was written in FORTRAN 77 and implemented on a NORD-100 minicomputer, while a new version has been implemented on an IBM PC AT microcomputer.

MATHEMATICAL DESCRIPTION OF AV NODAL CONDUCTION

Special attention should be given to the prolongation of AV nodal
conduction in response to premature atrial depolarisation which may be
essential for junctional tachycardia initiation. We have modelled the
initiation of AV re-entry tachycardia (AVRT) by an atrial premature beat
(APB) during pre-excited sinus rhythm in which the AV conduction is predominantly via an accessory bypass tract (Malik et al., 1987b). An APB
occurring shortly after a regular atrial event may not be conducted via
the accessory path because of its refractoriness. Then, the AV node is the
only available AV conduction and its prolonged conduction enables both the
atrial myocardium and the accessory pathway to recover and to complete the
tachycardia circuit.

Initial oscillations of the tachycardia rhythm may sometimes be recorded. Some hypotheses explain these oscillations by "recursive" influence
of the AVN tissue (Schamroth and Sareli, 1986). However, the AV node is
not a simple structure. It is composed of a large number of separate elements which mutually influence each other. Nevertheless, we can mathematically elaborate the hypothesis that a short tachycardia cycle causes prolongation of AV nodal conduction which results in a longer cycle, making
subsequent AV nodal conduction faster so that a shorter cycle follows.

To enhance the problem of mutual influence between nodal elements, we
consider a linear structure consisting of conduction elements. Each element transmits the excitation impulse and the transmission velocity depends
on the current cycle length. Hence, we can represent the linear image of
the AV node by a real interval $\langle A, H \rangle$. Each conduction through the node may
then be described by a function $\underline{d}$ which is the reciprocal of the conduction
speed. It describes how the signal is proportionally delayed during nodal
transmission. The total nodal delay corresponds to

$$\int_A^H \underline{d}(e)\,de \tag{1}$$

where the variable e is used to denote the elements of the $\langle A, H \rangle$ interval.
The conduction through a fully recovered node (during physiological sinus
rhythm) is probably not homogeneous. We will describe it by a function
$\underline{d}_0$. If an APB with the prematurity interval $\underline{P} = A_1 A_2$ is transmitted
through the node, the cycle length is not the same for all nodal elements.
This is because the transmission of the premature event can be delayed
within the node in a manner other than the regular atrial depolarisation.
The cycle length of an element $\underline{u} \in \langle A, H \rangle$ can be expressed by the formula

$$\underline{P} + \int_A^u \underline{d}(e)\,de - \int_u^H \underline{d}_0(e)\,de \tag{2}$$

where $\underline{d}$ is the delay function of the nodal transmission of the given premature atrial event.

The way in which an element's delay is affected by its shorter coupling interval is probably different for different elements. To describe
the dependence, we introduce a function $\underline{D}$ assigning to each nodal element
and to its coupling interval the resulting delay of that element. Then,
we obtain the following equation for the delay function $\underline{d}$ describing the
conduction of an APB with prematurity $\underline{P}$:

$$\underline{d}(u) = \underline{D}\left(u, \left(\underline{P} + \int_A^u \underline{d}(e)\,de - \int_u^H \underline{d}_0(e)\,de\right)\right) \tag{3}$$

A similar approach can be applied to AVRT modelling. We can suppose

323

that the re-entry interval (the time necessary to conduct the depolarisation wavefront through the ventricular conduction system, ventricular myocardium, bypass tract and atrial myocardium) is constant and does not depend
on tachycardia cycles. Therefore, we introduce the constant r describing
this interval. Then it is possible to model all cycles of AVRT following
an APB. The tachycardia can be represented by a succession of functions
d_1, d_2, ..., where d_i denotes the delay function of the ith path through
the node. The following equation system models the tachycardia following
an APB with prematurity interval P:

$$d_1(u) = D\left(u, \ (P + \int_A^u d_1(e)\,de - \int_u^H d_0(e)\,de)\right) \tag{4}$$

$$d_{i+1}(u) = D\left(u, \ (r + \int_A^u d_{i+1}(e)\,de - \int_u^H d_i(e)\,de)\right) \tag{5}$$

for i = 1, 2, ...

Here, (4) describes the AVRT initiation cycle, whilst (5) corresponds
to proper tachycardia cycles.

COMPUTATIONAL EXPERIMENTS

Both the computer model of the cardiac conduction process and the
differential-integral equation mathematical model described above have
been employed to study some selected aspects of the AV nodal properties.
The computational experiments addressed four aspects: (i) the rhythm
alternations during AVRT; (ii) the dependence of nodal conduction delay
on cycle length and the effects of this dependence on AVRT rhythm; (iii)
the conduction mechanism producing the Wenckebach periodicity during the
second degree AV block; and finally (iv) the possible effects of collisions
of excitation wavefronts within the node.

Rhythm alternations during AVRT are occasionally observed in clinical
records. Hypotheses reported by Vohra et al. (1974) expect that the cycle
length oscillations are in the main caused by treatment which affects the
nodal conduction. Other authors suppose that two different conduction
pathways within the AV node can be responsible for the intra-tachycardia
oscillations in cycle length without changing dynamically the conduction
properties of the nodal tissue (Schamroth et al., 1981).

To simulate the later hypothesis, we used the cardiac conduction model
and examined the behaviour of the conduction structure shown in Fig. 1.
Different properties of the elements of the two AV nodal branches were
introduced. One branch was given faster conduction and longer recovery
time than the other. For elements of both branches, the model expected
prolongation of refractory period and slowing of excitation transmission
with decreasing cycle length.

When modelling an APB initiating an AVRT episode, nodal conduction is
performed via the faster branch which also activates, retrogressively, the
slower branch. This mechanism is repeated until the short cycle length
prolongs the refractory period of the faster branch. Then the nodal conduction is performed only along the slower branch. Depending on the next
impulse conducted along the faster branch, this may result in several
different patterns (Fig. 2).

The next conduction via the faster branch may not activate the terminal part of the slower branch because of its refractoriness. Then the
nodal transmission occurs simultaneously via both branches and may

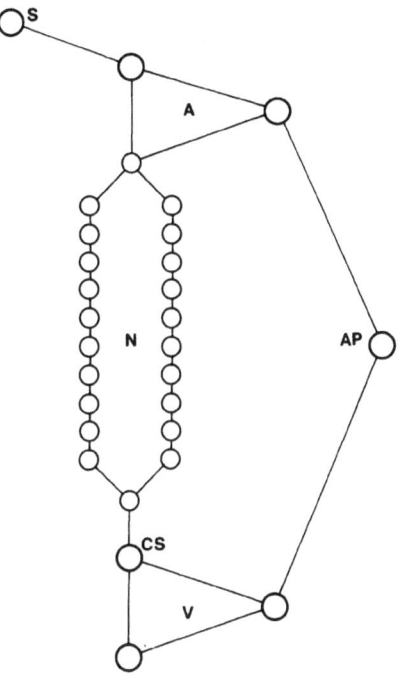

Fig. 1. The artificial heart image
employed in the first ex-
perimental series with the
cardiac conduction model.
The circles denote model
elements, the lines mark
connectors between them.
S - sinus node, A - ele-
ments of the atrial myo-
cardium, N - elements of
the AV node, CS - ventri-
cular conduction system,
V - elements of the ven-
tricular myocardium, AP -
accessory bypass tract.

(depending on the refractory period of the ventricular myocardium) result
in two closely coupled ventricular events which terminate the tachycardia
(Fig. 2a).

The slow conduction via the slower branch may also allow the terminal
part of the faster branch to recover. Then, the faster branch is activated
retrogressively and the AVRT changes its rhythm (Fig. 2b).

When supposing both the nodal branches to be unidirectional, regular
alternation of tachycardia rhythm can be modelled (Fig. 2c). A cycle conducted via the faster branch is followed by a cycle conducted via the
slower one because of the longer refractoriness of the faster branch.

However, the prolonged conduction via the slow branch can affect the
cycle length of the elements of the faster branch so that the following
conduction is performed via the faster branch too quickly to capture the
ventricular myocardium. The elements of the slower branch may also be

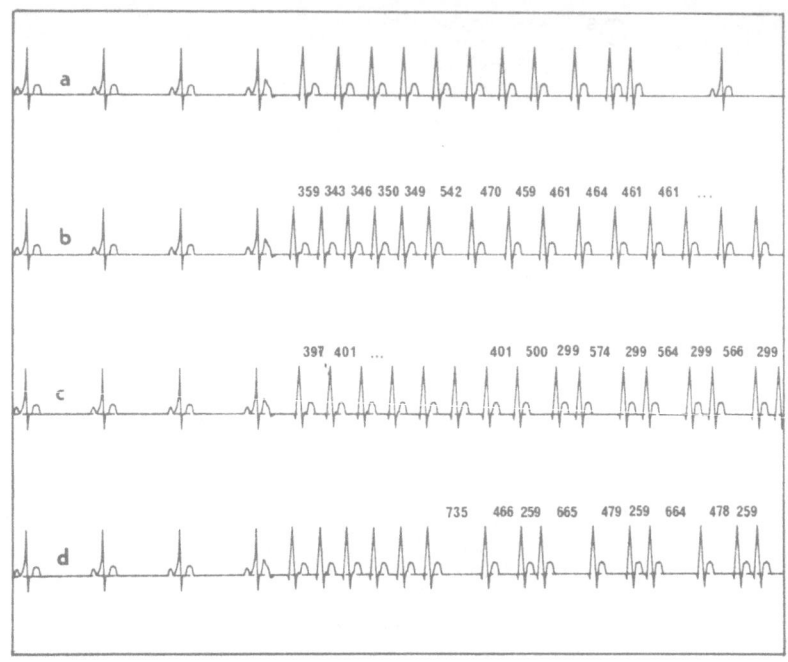

Fig. 2.   Simulated rhythm oscillations in AVRT using two different branches of the AV node (see the text for
details).   In cases b – d the numbers show the VV intervals in milliseconds (measured in excitation tables
produced by the model).

affected and conduct slightly faster than in the previous loop. This again
affects the faster branch which conducts more slowly and captures the ventricles in the next cycle. This very short cycle may not capture the
faster branch when re-entering from the atria. The slow branch prolongs
its conduction and the mechanism with the period of three different cycles
is repeated (Fig. 2d).

Using the time tracing of computer experiments we were able to construct the ladder diagrams explaining the AV conduction in modelled rhythms
(Fig. 3).

The series of experiments presented in Figs. 2 and 3 indicates that
the histological topology of nodal structure can cause very complicated
phenomena of nodal behaviour. Hence, for judging the mutual influence of
nodal elements and the dependence of nodal conduction delay on cycle length,
we should simplify the nodal image employed in computational experiments.
The linear representation of the node introduced in the differential-integral equation model can be used for such a purpose.

In experimenting with the equation model, we simplified the model
concept and introduced a constant delay function d_0 of the conduction
through the fully recovered node. In addition to this, we supposed that
the response function D does not depend on a particular element within the
linear node, that is, that all nodal parts respond to a changing cycle rate
in the same way.

Then the first experimental task was to establish an appropriate function D. However, the only laboratory and clinical data at our disposal

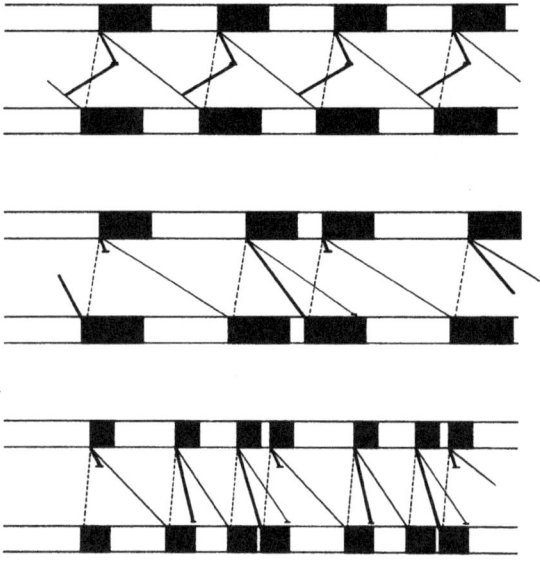

Fig. 3. Schematic ladder diagrams (derived from the model's timetables) explaining the last three cases of rhythm disorders presented in Fig. 2. The bold lines mark the conduction via the faster nodal branch, the fine lines the conduction via the slower branch, and the dashed lines show the retrograde conduction via the accessory path. Top strip - rhythm episode recorded in Fig. 2b; middle strip - Fig. 2c; bottom strip - Fig. 2d.

relate to the behaviour of the complete nodal structure derived from laboratory experiments measuring the nodal conduction of premature atrial depolarisations with different prematurity intervals (Dorveaux et al., 1985a,b). Reflecting such measurements, we require a function $\underline{D}^*$ such that, if $\underline{d}^*$ is a solution of the equation

$$\underline{d}^*(\underline{u}, \underline{P}) = \underline{D}^*\left(\underline{P} + \int_{\underline{A}}^{\underline{u}} \underline{d}^*(e, \underline{P})\,de - \underline{d_0}(\underline{u} - \underline{A})\right), \tag{6}$$

then the function

$$\underline{Q}(\underline{P}) = \int_{\underline{A}}^{\overline{H}} \underline{d}^*(e, \underline{P})\,de \tag{7}$$

has the exponential form corresponding to clinical data. Here, the function $\underline{D}^*$ is the simplified form of $\underline{D}$, and $\underline{d_0}$ is the value of the constant function $\underline{d}_0$.

To solve this problem and for computational experiments with the model, we approximated the interval $\langle\underline{A},\underline{H}\rangle$ by a finite number of individual elements (1000 elements were employed in the experiments presented here). Obviously, many different functions $\underline{D}^*$ may approximate the behaviour of nodal elements and lead to a satisfactory form of the resulting function $\underline{Q}$.

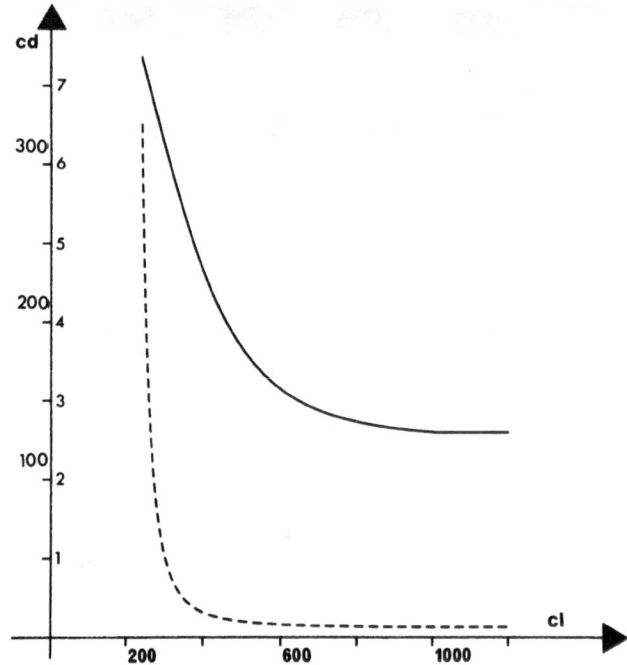

Fig. 4.   Graphs of the function $\underline{D}^*_{1000}$ (dashed line)
          and of the resulting function $\underline{Q}$ (see the
          text for details).   cl - cycle length,
          cd - conduction delay. Scaling is in milli-
          seconds, on the vertical scale. The capital
          numbers correspond to the function $\underline{Q}$, while
          the small numbers correspond to the function
          $\underline{D}^*_{1000}$.

However, the exact form of the function $\underline{D}^*$ is not too important and the use
of slightly different approximation functions probably does not affect the
results in a significant manner. The form of the function $\underline{D}^*_{1000}$ (reflecting
the number of discrete nodal elements used) which we used in the experiments,
and the form of the corresponding function $\underline{Q}$ are shown in Fig. 4.

We experimented with the model and simulated the beginnings of different cases of AVRT with different re-entry intervals and different prematurity levels of initiating APBs. The basic form of the results is given
in Fig. 5. This shows how the mutual influences of nodal elements compensate the initial differences and how the tachycardia is stabilised.

Other series of computational experiments had the aim of presenting
more evidence that complicated AV nodal function can be explained by assuming that its individual elements behave in a simple way in the context of
a complex structure. Therefore, we experimented with complicated nodal
images without introducing cycle rate dependences of parameters of individual nodal cells, that is, the refractory period and excitation transmission
delay of each AV nodal element were supposed to be constant during the experiment.

The first of these computational series used an AV nodal image suggested by the Rosenblueth concept. Rosenblueth (1958a,b) suggested that the
prolongation of the AV nodal conduction delay with shorter length was not

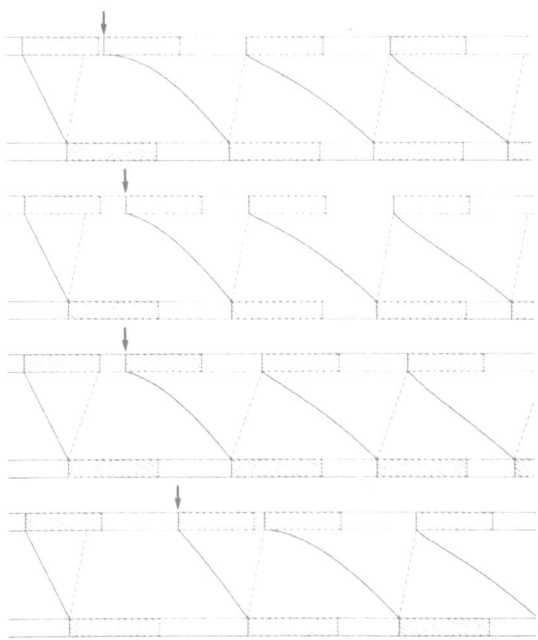

Fig. 5. Modelled activation diagrams of AVRT
by a concealed accessory pathway and
initiated APB (indicated by arrows).
The full lines correspond to the AV
nodal conduction simulated by the
differential-integral equation model;
the dashed lines mark the retrograde
conduction. The four cases presented
differ in the prematurity level of the
initiating APB and in the re-entry
interval.

due to the incremental delay exhibited by more or less all the nodal fibres,
but that the excitation wavefront was delayed by a single special layer.
This layer is expected to exhibit not only an effective refractory period,
but also a longer relative one. It is supposed to be able to accept the
excitation wavefront during its relative refractory period and delay it
until the resting state is restored. The differences between the effective
and relative refractory period of the layer can explain such phenomena as
Wenckebach periodicity observed in the 2nd degree AV nodal block. Some
experimental evidence supporting Rosenblueth's idea has been reported (Levy
et al., 1974; Young et al., 1986b). However, no detailed explanation of
the electrophysiological mechanism of Rosenblueth's layer has been suggested.

     According to theoretical considerations (Malik et al., 1987c), the
layer required by Rosenblueth can be composed of elements having much
longer repolarisation than the rest of the node. If the anisotropy of
intra-nodal conduction causes the excitation impulse to reach the "long
repolarisation layer" at many different moments, and if the excitation of
any part of the layer is transmitted within the layer with a very high
speed causing the whole layer to depolarise nearly instantaneously, the
structure will satisfy the Rosenblueth concept. The difference between
effective and relative refractory periods needs not to be introduced for
separate elements; the differences between these periods of the whole

Fig. 6. The anisotropic nodal structure
employed to model the Rosenblueth
hypothesis. The open circles denote the elements with short repolarisation period (250ms), the
solid circles denote the "long
repolarisation layer" (repolarisation period of 450ms). The conduction within the elements is nearly
instantaneous (1ms delay per element).
The conduction between neighbouring
elements parallel with the long axis
is set at 10ms per connector, and
30ms per connector in other directions. The conduction within the
Rosenblueth layer is nearly instantaneous (1ms per connector). s -
sinus node; a - atrial myocardium;
v - ventricular myocardium.

nodal structure are achieved by the nodal anisotropy which causes the excitation to be transmitted to the "long repolarisation layer" via pathways
with different total conduction times.

This principle has been incorporated into the AV nodal image shown in Fig. 6. The image introduces transmission delay differences between pathways along and cross the node. The ventricular border of the node represents Rosenblueth's layer: its repolarisation period is prolonged and the excitation conduction within the layer is nearly instantaneous.

The nodal image has been tested with the computer model of cardiac conduction. A second degree AV block with basic patterns of Wenckebach periods can be modelled (Fig. 7) when introducing a sufficient difference between repolarisation intervals of the internal nodal elements and of the Rosenblueth layer. When introducing a heterogeneous conduction anisotropy within the "long repolarisation layer", even more realistic patterns of Wenckebach periods (including such phenomena as successive shortening of RR intervals in each period (Sandoe and Sigurd, 1984) can be simulated (Malik and Camm, 1987).

The previous experimental series shows how the conduction anisotropy and repolarisation interval heterogeneity of nodal layers can contribute to the complexity of the nodal function. We have therefore examined another artificial nodal image introducing conduction and repolarisation differences between several nodal layers (Fig. 8).

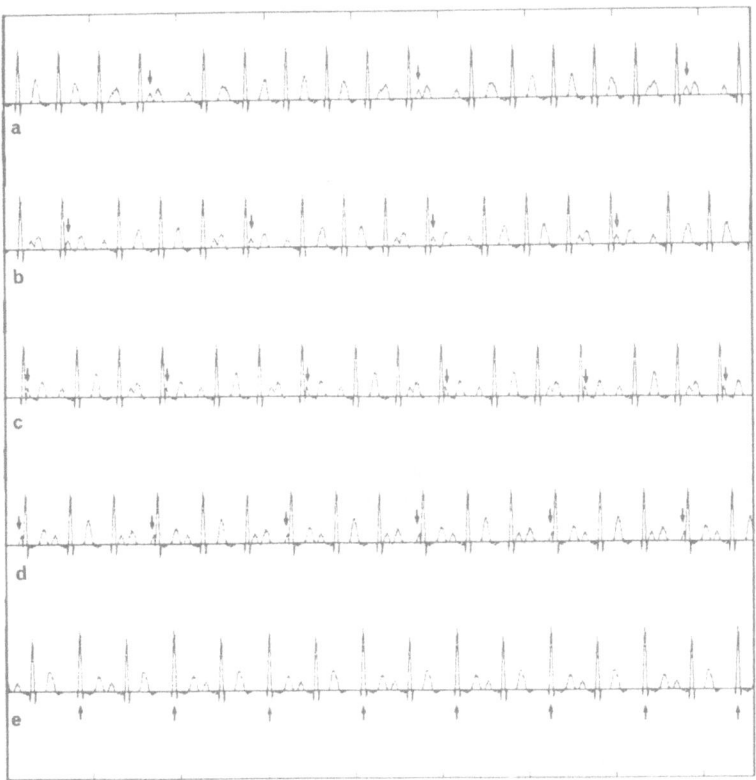

Fig. 7. Experimental results obtained with the structure shown in Fig. 6. A basic pattern of the 2nd degree (Wenckebach) AV block is produced at each of five atrial rates: a - sinus coupling intervals of 440ms, b - 420ms, c - 400ms, d - 380ms, e - 360ms. The arrows indicate the P waves which have not been transmitted to the ventricles.

331

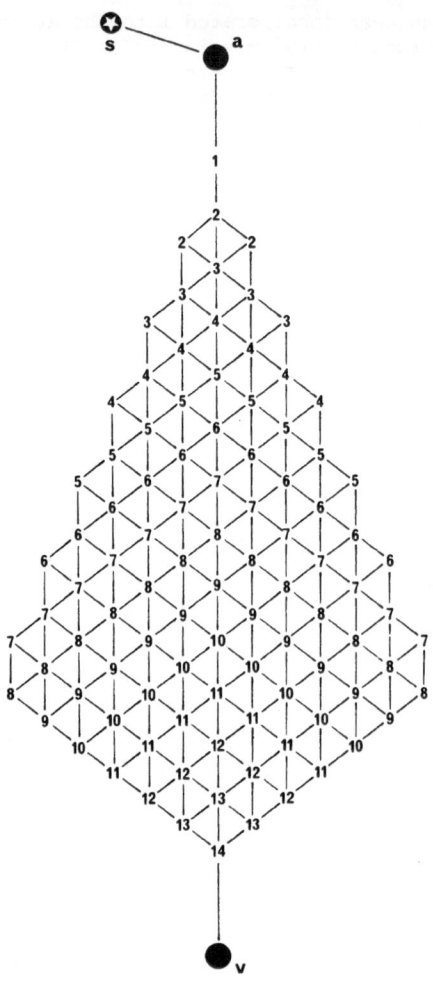

Fig. 8. The structure of the AV nodal image
used in the experiments modelling the
collisions of excitation wavefronts
within the node. The node is composed
of 14 layers of elements with different
properties. The element '1' transmits
the signal with intra-element delay of
16ms and its repolarisation period has
been set to 200ms; the elements '2'
transmit the signal with the delay of
15ms and have the repolarisation period
216ms; and so on: '3' - 14+231ms; '4' -
13+245ms; '5' - 12+258ms; '6' - 11+270ms;
'7' - 10+281ms; '8' - 9+291ms; '9' -
8+300ms; '10' - 7+308ms; '11' - 6+315ms;
'12' - 5+321ms; '13' - 4+326ms; '14' -
3+330ms.   s - sinus node, a - atrial
myocardium, v - ventricular myocardium.

This nodal image has also been examined with different frequencies of
the base sinus rhythm. Fig. 9 shows that for some sinus node frequencies

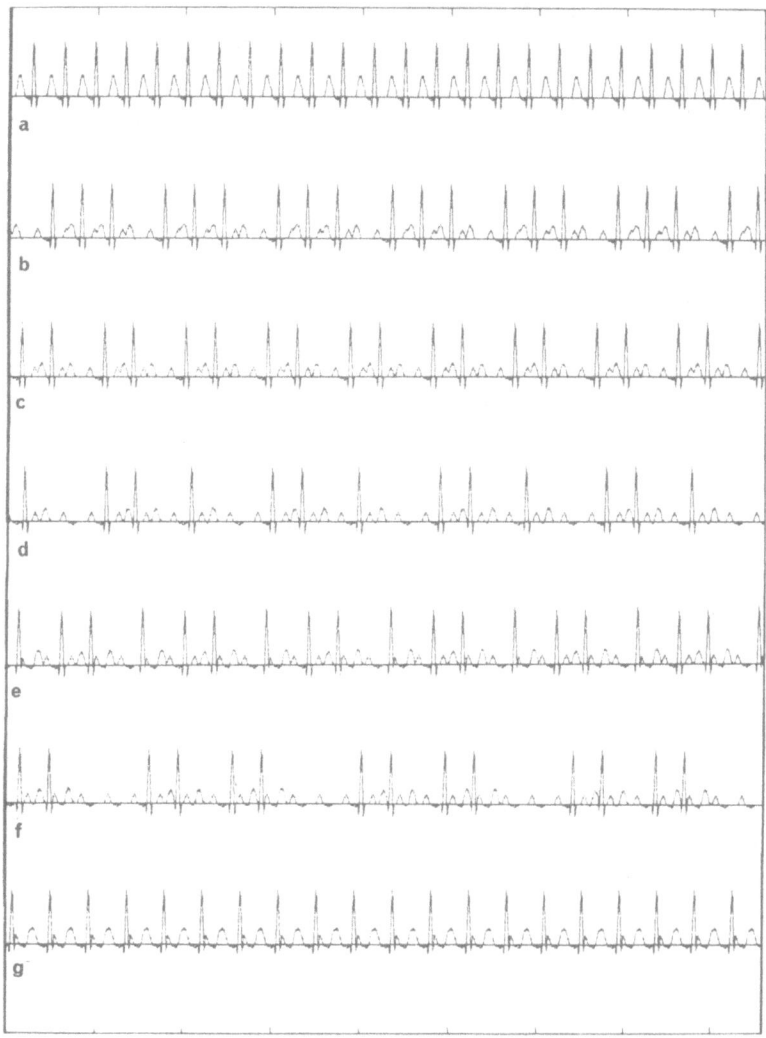

Fig. 9.   The results of experiments employing the AV nodal
          structure shown in Fig. 8. The sinus nodal coupling
          intervals differ in seven cases: a - 350ms; b -
          320ms; c - 310ms; d - 315ms; e - 280ms; f - 300ms;
          g - 215ms. The traces demonstrate phenomena which
          might be interpreted as being due to "concealed
          conduction".

the AV nodal behaviour produces satisfactory ECG traces, while for other
frequencies the results do not resemble clinical records. These complicated
patterns are caused by delays and blocks of the excitation wavefronts at
different nodal levels. Although some results model distinctly unrealistic
patterns of the AV block, a mechanism of similar principle may cause those
phenomena which we observe in clinical situations.

DISCUSSION

     Both the basic computer model and the mathematical description of intra-nodal behaviour which we used for the series of computational experiments have several important limitations. Similarly, the nodal images introduced in particular experimental studies have serious drawbacks and restrictions.

In the computer model of cardiac conduction processes, the restriction of each element to only two polarisation states and the assumption that the changes between these states are instantaneous are the most important simplifications which may affect the validity of the experiments presented. It is very likely that the different performance of nearly, but not fully, recovered elements is very important in determining the AV nodal response to coupled atrial excitations. The current version of the model can also only cope with statically anisotropic and heterogeneous structures. Dynamic anisotropy and dynamic heterogeneity should be examined since they may certainly contribute to the complexity of AV nodal properties. These aspects, as well as others, such as random conduction disturbances and disorders, have to be taken into account when developing the next version of the model.

The most important objection to our mathematical differential-integral equation model lies in the fact that it introduces a linear, one-dimensional representation of the node. This makes it impossible to consider and elaborate the effects caused by radiation of excitation impulses across nodal tissue. Nodal images similar to those presented in Figs. 6 and 8 should be described analytically to make the mathematical model more realistic. Then some medically relevant theoretical conclusions may be obtained from the continuous equation model.

The images of artificial AV nodes which we examined by the cardiac conduction model are very simplistic. The histology of the node (James, 1984) should be mirrored more closely, and more detailed structures introducing not several tens but several thousands of elements should be employed. It is likely that the reality can be more closely approximated by combining many different mechanisms (Watanabe and Dreifus, 1965; 1967; Wit et al., 1970; Schaffer and Depasquale, 1973; Friedeman et al., 1975; Young et al., 1986a) and by considering anatomical facts, for example, possible differential atrial input points (Mazgalev et al., 1984).

Despite all these important restrictions and simplifications, the experimental results reported here provide evidence that the mutual influences of individual nodal elements are very important and that the complexity of the AV nodal structure may itself be responsible for its complex behaviour. It is not necessary to assume a complicated function for the individual AV nodal cell when seeking to explain many aspects of AV nodal function.

Hence, we might propose that the complexity of AV nodal external behaviour is unlikely to be caused by an unknown single simple mechanism but that combinations of several mechanisms, together with the histological topology of the nodal tissue, can play a very important role. All possible combinations of different reported and expected intra-nodal mechanisms can hardly be examined by simple logical speculations. However, mathematical modelling and computer simulation of complex nodal images offer a powerful tool for future investigations.

REFERENCES

Dorveaux, L., Heddle, W., Jones, M., and Tonkin, A., 1985a, Examination of an exponential model of conduction through the human atrioventricular node, PACE, 8 : 646.
Dorveaux, L., Heddle, W., Jones, M., and Tonkin, A., 1985b, Comparison of exponential and hyperbolic models of conduction through the atrioventricular node, Int. J. Bio-Med. Comput., 17 : 227.

Friedeman, H. S., Gomes, J. A. C., and Haft, J. I., 1975, An analysis of
Wenckebach periodicity, J. Electrocardiol., 8 : 303.

Heethaar, R. M., van der Gon, J. J. D., and Meijler, F. L., 1973, Mathematical model of A-V conduction in the rat heart, Cardiovasc. Res., 7 : 105.

James, T. N., 1984, Sir Thomas Lewis redivivus: from pebbles in a quiet
pond to autonomic storms, Br. Heart J., 52 : 1.

Levy, M. N., Martin, P. J., Zieske, H., and Adler, D., 1970, Role of positive
feedback in the atrioventricular nodal Wenckebach phenomenon, Circ. Res.,
36 : 697.

Malik, M., and Camm, A. J., 1987, Complexity of AV nodal function: complex
nodal structure or complex behaviour of nodal elements?, PACE (in press).

Malik, M., Cochrane, T., Davies, D. W., and Camm, A. J., 1987a, A clinically
relevant computer model of cardiac rhythm and pacemaker - heart interaction, Med. & Biol. Eng. & Comput., 25 : 504.

Malik, M., Davies, D. W., Cochrane, T., and Camm, A. J., 1987b, A one dimensional model of atrioventricular nodal conduction, Int. J. Biomed.
Comput., 21 : 13.

Malik, M., Ward, D., and Camm, A. J., 1987c, Theoretical evaluation of the
Rosenblueth hypothesis, PACE (submitted for publication).

Mazgalev, T., Dreifus, L. S., Iinuma, H., and Michelson, E. L., 1984, Effects
of the site and timing of atrioventricular nodal input on atrioventricular conduction in the isolated perfused rabbit heart, Circulation,
70 : 748.

Rosenblueth, A., 1958a, Mechanism of the Wenckebach-Luciani cycles, Am. J.
Physiol., 194 : 491.

Rosenblueth, A., 1958b, Two processes for auriculo-ventricular and ventri-
culo-auricular propagation of impulses in the heart, Am. J. Physiol.,
194 : 495.

Sandoe, E., and Sigurd, B., 1984, "Arrhythmia - Diagnosis and Management",
Fachmed AG, St. Gallen : 300.

Schaffer, A. I., and Depasquale, N. P., 1973, Mechanism of Wenckebach A-V
block and the obligatory ventricular echo, Cardiovasc. Res., 7 : 696.

Schamroth, L., and Sareli, P., 1986, Compensating conduction times as a
mechanism for alternation during reciprocating tachycardia, J. Electrocardiol., 19 : 291.

Schamroth, J. M., Myburgh, D. P., and Schamroth, L., 1981, Reciprocating
tachycardia with only odd-numbered beats in the Wolff-Parkinson-White
syndrome, Chest, 80 : 643.

Vohra, J., Hunt, D., Stuckey, J., and Sloman, S., 1974, Cycle length alternation in supraventricular tachycardia after administration of vera-
pamil, Br. Heart J., 36 : 570.

Watanabe, Y., and Dreifus, L. S., 1965, Inhomogeneous conduction in the AV
node: a model for reentry, Am. Heart J., 70 : 505.

Watanabe, Y., and Dreifus, L. S., 1967, Second degree atrioventricular
block, Cardiovasc. Res., 1 : 150.

Wit, A. L., Weiss, M. B., Berkowitz, W. D., Rosen, K. M., Steiner, C., and
Damato, A. N., 1970, Patterns of atrioventricular conduction in the
human heart, Circ. Res., 27 : 345.

Young, M.-L., Gelband, H., and Wolff, G. S., 1986a, Atrial pacing induced
alternating Wenckebach periodicity and multilevel conduction block in
children, Am. J. Cardiol., 57 : 135.

Young, M.-L., Wolff, G. S., Castellanos, A., and Gelgand, H., 1986, Application of the Rosenblueth hypothesis to assess atrioventricular nodal
behaviour, Am. J. Cardiol., 57 : 131.

SIMULATION OF BIOLOGICAL PROCESSES AND OF HEALTH CARE SYSTEMS:

METHODOLOGICAL PROBLEMS

M. Kotva

Institute for Social Medicine and Organisation of Health
Services,
Prague, Czechoslovakia

INTRODUCTION

The simulation of systems undoubtedly had its beginnings in technical
fields. It soon spread, however, to the sphere of the natural sciences
and finally it also began to penetrate into the social sciences. In these
beginnings, therefore, its application in the field of designing objects
predominated, together with the development of methods only for the synthesis of simulation models and their optimisation. Here also lay the source
of understanding of the concept of systems simulation in the narrower sense,
that is as a method of experiment with the simulation model (possibly including the very creation of simulation models as an application of the method
of modelling, but more often spoken of in this context as "modelling and
simulation"). This narrowed conception of systems simulation often led,
however, during the application of experiments with the simulation model to
investigate objects, to certain misunderstandings, mostly of a methodological character. These difficulties also appear in the application of systems
simulation to the investigation of biological processes and of systems of
health care.

It became evident that it is necessary to view the simulation of systems more broadly - as a specific form of the process of cognition* which
in its basic form passes in an iterative manner through the following
phases (Kotva, 1986):

  (i)   definition of the object of cognition;
  (ii)   definition of the simulated system on the object of cognition;
  (iii) formulation of current ideas about the simulated system and its
         motion;
  (iv)   creation of the simulation model;
  (v)   testing the correctness of the simulation model;
  (vi)   testing the validity (verification of truthfulness) of the simula-
         tion model; and
  (vii) further use of the validated simulation model in the process of
         cognition instead of the simulated system.

---

* Here we have in mind the process of cognition in the broader sense, that
  is as a dialectical unity between the cognition of objective reality and
  social and productive practice.

The experiments with the simulation model themselves constitute a method which is used in virtually all of the above-mentioned phases and which distinguishes the simulation of systems from other forms of the cognition process. Therefore, we shall consider this method as the simulation of systems in the narrower sense.

The concept of systems simulation as a specific form of the cognition process not only allows but directly forces us into the creation of a general methodology of systems simulation as a part of the general methodology of science. This methodology provides, in addition to those things described above, all the basic orientation relating to the correct use of relevant methods, procedures and techniques during the solution of cognitive tasks. It is true that it will not teach anyone to simulate (just as logic will not teach anyone to think), but it should contribute to a correct use of experiments with the simulation model in conjunction with other methods used in the process of cognition without exaggerating or elevating to dogma the role of any of these methods.

TESTING THE VALIDITY (VERIFICATION OF TRUTHFULNESS) OF THE SIMULATION MODEL

Although the phases of the process of systems simulation described above are formally identical both during the designing of objects and during the investigation of objects, their material contents are - with the exception of the phases "creation of the simulation model" and "testing the correctness of the simulation model" - rather different. From this standpoint particularly delicate is the question of the "verification of truthfulness" of the simulation model. During the designing of objects we can definitively declare that the simulation model is "true" (or also "untrue") only after the realisation of the project and the verification of the conformity or non-conformity of the model to reality. During the investigation of objects this phase of the process of systems simulation in fact represents the phase of the "verification of the hypothesis about the simulated system and its motion", because the simulation model constitutes in this case a representation of this hypothesis. During the designing of objects the verification of truthfulness therefore lies, as a matter of fact, outside the process of the simulation of systems proper (usually because we compare the realised project directly with the "initial requirements" and not with the simulation model to which we return only if we want to find out, for instance, "where the mistake occurred"). During the investigation of objects it constitutes, on the contrary, its summit (which, certainly, we do not always reach). If we transfer our experience insensitively without deeper methodological reflection and also without appropriate methods, procedures and techniques, this difference can constitute a source of serious errors during the investigation of objects by means of system simulation.

We often encounter, for example, such cases in which the phase of the verification of truthfulness of the simulation model is simply omitted from the process of system simulation. This is sometimes caused by an excessively exacting character or impossibility of its realisation, and if we are aware of the fact that the simulation model represents only a mere hypothesis, there need not occur any incorrectness in its further utilisation. Far more frequent and far more dangerous, however, are those cases in which we believe that we have carried out the verification of truthfulness of the simulation model although in reality this is not, in fact, so. The cause is here mostly a very common misunderstanding (which must be seen more as a euphemism) whose sources can be found even in the literature - for example, Mervart (1977): "The deductive method ..... consists in the inference of new knowledge from the original premises with the justification that if the original premises are true, the knowledge which is being inferred will also be true, and vice versa."

The underlined part of this otherwise correct assertion is, at least, misleading. In deductive reasoning which proceeds in accordance with the principles of formal logic there obtain, it is true, the following schemes:

(1) From A follows B*
     A is true
     B is true

(2) From A follows B
     B is not true
     A is not true

but, on the other hand,

(3) From A follows B
     B is true
     ?

(4) From A follows B
     A is not true
     ?

which means that in neither case is it possible to infer any conclusion from the given premises. We can unfortunately very often find this in simulation practice, for example such cases when it seems as if the scheme (3) was complemented by the conclusion "A is true". Even though the experiment with the simulation model itself has in its own way the character of deductive reasoning, and its results can be formulated by statements of the type "from A (that is from the hypothesis represented by the simulation model) follows B (that is the phenomenon observed during the experiment with the simulation model or the judgement inferred from the experiment)", we must rather apply during the verification of truthfulness of the simulation model probabilistic reasoning which proceeds inductively (see, for example, Polya, 1954):

(3)' From A follows B
      B is true
      It is more probable that A is true

In this connection it is worth mentioning the general philosophical principle of relativity of our cognition of objective reality. However, as we very often cannot even verify the truthfulness of B with complete certainty, we must consider also the "weakened" scheme

((3))' From A follows B
        It is probable that B is true
        It is somewhat more probable that A is true

Another important scheme of probabilistic inductive reasoning is the scheme

(4)' From A follows B
      A is not true
      It is less probable that B is true

The inductive scheme (3)', or ((3))', in effect says that the confirmation of truthfulness of the consequence increases the probability that the premise is true (in the case of the weakened scheme the probability is increased less than in the non-weakened form). To put it less exactly and more subjectively - the ascertainment of truthfulness of the consequence increases our conviction that the simulation model, or the hypothesis

---

* By using the expression "from A follows B" instead of the usual "if A, then B" which is used in formal logic for the coupling of the statements A and B with the character of so-called material implication, we want to indicate that we have in mind here not only a formal connection but also a substantial connection between the contents of the statements A and B, that is that "from the premise A follows the consequence B".

represented by it, is true. The direction of the change of probability is here objective; the answer to the question "how much?" is, to be sure, necessarily subjective. Nevertheless, there are some clues at least of relative character for answering this question.

Let us suppose, for example, that we have inferred from the experiment with the simulation model several judgements which we shall designate B_1, B_2 ... B_n and that we have already verified the truthfulness of these judgements. This has increased our conviction about the probability of the hypothesis represented by the simulation model regardless of whether we could apply the basic inductive scheme (3)' or only its "weakened form", that is, the scheme ((3))'. Let us suppose further that by another experiment we have inferred the judgement B_{n+1} and succeeded again in verifying its truthfulness. The weight of this ascertainment will depend upon the degree to which the judgement B_{n+1} is similar to the preceding judgements, that is to B_1 ... B_n. In accordance with this we can then subordinate to the following scheme:

(5)'   From A follows B_1 ... B_n
B_1 ... B_n are true
_____
It is probable to the degree m that
A is true
From A follows also B_{n+1}
B_{n+1} is true
_____
It is probable to the degree m+p that
A is true

which can also be interpreted in such a way that the verification of truthfulness of another judgement inferred on the basis of experiments with the simulation model has a much greater weight for the increase of our conviction about the probability of the present hypothesis in those cases when this judgement significantly differs from judgements which have already been used earlier within the framework of the verification of the truthfulness of the hypothesis. The scheme (5)' could, of course, be presented also in a "weakened form" with consequences which are perhaps already quite evident.

Anyway, there is a direct link to the scheme (3)' here - we could speak about "chaining" of this basic inductive scheme. We would therefore obtain the "weakened form" of the scheme (5)' by chaining the "weakened" basic inductive scheme ((3))'. In a similar way, however, we could obtain by chaining the "weakened" scheme of deductive reasoning ((2)) the scheme suitable for the disproval of a hypothesis with a similar characteristic as the scheme (5)'. That is to say that we come nearer to the conclusion that a hypothesis is untrue in the case that the judgements inferred from it and evaluated as "less probable" significantly differ from each other as far as their character is concerned than in the case that they differ only a little. (It, of course, makes no sense to chain the scheme of deductive reasoning (2) because according to it a single judgement inferred from the hypothesis and evaluated as untrue suffices for the disproving of the hypothesis - this at the same time shows one of the great dangers which await us if we choose as our approach "disproving" rather than "verification"). In the situation when the object of cognition is readily accessible to our observation and measurement and it is possible to experiment with it, it is feasible to plan and to carry out experiments with the simulation model precisely with the aim of arriving at conclusions whose character would be as different as possible. In this way - in the case of verification of truthfulness or "greater probability" - our conviction about the truthfulness of the hypothesis will be increased significantly more than in the case when we are dependent upon the results of observation and measurement of the object in native conditions not differing too much. This is similarly true for the alternative of disproval of a hypothesis if we cannot evaluate the judgements which are being verified and which have been inferred from it through experiments with the simulation model as "certainly untrue".

Neither will there be practically any changes in the preceding reflections in the case when, with the help of experiments with the simulation model, we do not infer literally some judgements but verify the truthfulness of the hypothesis represented by the simulation model practically only by the mere comparison of the course of its motion (for example, behaviour) with the motion observed in the object investigated under comparable conditions. In such a case we can also apply a scheme which we could call the broadened scheme of the inference from analogy:

(6)'   From A follows B
        B is analogous to B'
        B' is true
        It is more probable that B is true
        It is somewhat more probable that A is true.

In relation to this scheme of reasoning it is necessary above all to call attention to the fact that we should not regard the truthfulness of B' as automatically proven only because it is a motion observed on the objective reality. In connection with the problem of the black box, we have to pay attention to the imperfection of our senses and measuring devices, and to the possibility of influencing the manifestations of the features of the objects of investigation by the measurement process itself. If we cannot guarantee that we have succeeded in eliminating completely these negative phenomena in the observation and measurement of the motion of the investigated object, we should use a weakened scheme of the reasoning from analogy:

((6))'   From A follows B
         B is analogous to B'
         It is more probable that B' is true
         It is somewhat more probable that B is true
         It is slightly more probable that A is true.

It is precisely the comparison of the motion of the simulation model with the motion observed on the investigated object which is the approach very often used to test the validity of the simulation model, especially as far as models of behaviour of a functional type are concerned. Therefore, let us devote some more remarks to the schemes mentioned above. Let us above all take notice of the conclusion of the weakened scheme ((6))', where by the word "slightly" we mean that in this case our conviction about the truthfulness of the hypothesis which is being verified has been strengthened still less than is indicated by the expression "somewhat" in the non-weakened form (6)'. We can certainly chain both schemes in a similar way to the basic scheme of inductive reasoning (3)', or ((3))', and come to schemes similar to (5)' or to ((5))' (which we have not even mentioned here). What we have said would again be true in this connection – in the case of a significant difference of the character of the individual pairs $B_1 - B_1'$, $B_2 - B_2'$ our conviction about the truthfulness of the hypothesis which is being verified will increase more rapidly than in the case in which the character of motion during the individual experiments will not differ too much. It is true that we must take into consideration also the fact that in the schemes (6)' and ((6))' we have reasoned on the basis of the conviction that B is "assuredly" analogous to B'. Experience hitherto, however, shows that we cannot always use such an expression with complete justification and that rather it is advisable to use the statement

"B is probably analogous to B'", and so on. In this way we could pass to other forms of weakened or even twice weakened schemes of probabilistic reasoning from analogy.

Let us add, for the sake of completeness, that we could in a similar way infer analogous (broadened) schemes of probabilistic reasoning from analogy for the disproving of a hypothesis.

All the schemes of probabilistic reasoning described above can be used for quantification of the starting hypothesis during testing the correctness of the simulation model, that is during verification of the correspondence of the model to the current hypothesis about the simulated system and its motion. This hypothesis was inferred from hitherto obtained results of observations, measurements and experiments with the object of cognition. However, a generally usable starting point of the verification of scientific hypotheses is the question "what should <u>still</u> be observable if the hypothesis is true?" on which we provide an answer in the process of systems simulation precisely on the basis of experiments with the simulation model. The schemes (3)', (5)' and (6)' (or their weakened forms) indicate that after the inference of the judgement B on the basis of an experiment with the simulation model, there must follow the verification of its truthfulness. We can do this only by returning to the performance of observations and measurements, or to experiments with the object of cognition (or also to the knowledge base if we are concerned with such a judgement whose truthfulness can be verified by some already verified scientific theory or explanation, or in a negative sense - a hypothesis which has already been disproved earlier).

The results obtained hitherto from observation and measurement of the object of cognition are not, in fact, relevant here (in the same way as scientific theories and explanations or disproved hypotheses which we have already used) because we have inferred the present hypothesis from them and have ascertained the conformity of the simulation model with them within the framework of testing its correctness. Nevertheless, they are precisely the schemes (3)', (5)' and (6)' which show that, already, during testing of the correctness of the simulation model we can regard the successful comparison of the results of experiments with the simulation model with the results of observation and measurements of the investigated object as the first step towards testing the validity of the simulation model. However, we cannot regard this success (as is sometimes done under the influence of the erroneous opinions quoted at the beginning of this paragraph) as the final verification of the truthfulness of the starting hypothesis. Rather it provides confirmation of the fact that the hypothesis represented by the simulation model has a non-zero probability (and we can also affirm that we have not - for the time being - succeeded in disproving it).

In conclusion, let us recall that all schemes of probabilistic reasoning make possible only a rational estimate of the level of probability that the hypothesis (that is the simulation model) which is being verified is (or is not) true (it can also happen that this rational estimate will in some concrete case prove ultimately to be incorrect - also an estimate made to the best of one's knowledge and belief will always remain only an estimate and we can never confuse it with the "ascertainment" of truthfulness of the starting hypothesis). Anyway, they represent a significant heuristic tool which makes it possible to strengthen or, on the contrary, to weaken our conviction about the truthfulness of the starting hypothesis during the application of systems simulation to the investigation of objects. Certainly in real cases this character of the probabilistic reasoning can lead us into very complex situations. We cannot generally expect - especially during the first passages through the process of the simulation

of systems - that we shall be able to evaluate universally our judgements
about the simulated system inferred from the experiments with the simulation
model as "more probable" or on the contrary as "less probable"; and therefore
our conviction about the truthfulness of the present hypothesis will always
increase with the advance of verification of the individual judgements or,
on the contrary, will always become weaker. We must rather expect greater
or smaller oscillations in the level of this conviction of ours, that is
that some judgements will prove "less" and some "more probable". Because,
as already indicated, we can only estimate the increase or lowering of this
level in the individual steps, the final estimate of the level of our conviction will be very difficult. In any case, such a course of testing the
validity of the simulation model, and consequently also the truthfulness
of the present hypothesis represented by it, will necessarily create in us
a conviction which, approximately, is that the hypothesis is only "partially
true (untrue)" or "incomplete".

This evaluation, even if it is somewhat vague, suggests how we should
further proceed. Let us divide the judgements which are being verified
into two groups, depending on whether they lead to a strengthening or, on
the contrary, to a weakening of our conviction about the truthfulness of
a hypothesis. Then, on this basis, we can try to infer conclusions which
show what in the present hypothesis is "true" (more exactly - what is
"more probable") and what is "untrue" ("less probable"), or in which sphere
the hypothesis appears as "more" and in which sphere as "less probable".
(We could also possibly use the expression "determination of the sphere of
adequacy" or "validity" of a hypothesis or of a simulation model). This
can serve at the same time as a basis for the planning of other experiments using the simulation model for the definition (with greater precision) of the "frontiers" of these areas, and in any case it is a basis
which should help us during the necessary return to the formulation of the
hypothesis in its modified form.

OTHER METHODOLOGICAL PROBLEMS

The verification of the truthfulness of the simulation model or of
the hypothesis represented by this model, if carried out correctly, should
lead to the detection of all mistakes which we have made in the process
of simulation of biological processes or of systems of health care. It
turns out that the sources of these mistakes have again, first and foremost, a methodological character. They are caused by the insensitive
transfer of experience and procedures from the applications of simulation
of systems which concern, above all, the designing of technical objects
to the sphere of the investigation of objects.

So, for example, the definition of the object of cognition during the
designing of objects is a question of the determination of demands made on
the designed object and of the determination of the partial objects which
are capable of use. For objects of a technical character it mostly means,
in practice, that the object which is being designed is identical with the
object of cognition upon which we focus our attention during systems simulation. During the application of system simulation to the investigation
of objects, however, it is often not so. To put it in a different way,
we must often define the object of cognition for systems simulation more
broadly than the object being investigated. It should not be forgotten
that we cannot, for example, observe or measure the properties of the investigated object directly; we can do it only by means of their manifestations in the interaction with other objects. To these objects can belong our
measuring devices or other objects from the environment of the investigated
object, or possibly the investigator himself. This can be precisely the
reason why we should broaden the definition of the object of cognition so

that it will include these objects and then simulate the system defined by the viewpoint of the investigation on this "broadened" object, and not only on the object of investigation proper. Another reason can be the necessity to exclude feedback between the "outputs" and "inputs" of the system via its environment. The object of cognition must be defined in such a way that any feedback will lie within the simulated system defined on this object.

The definition of the simulated system on the object of cognition represents by itself a frequent source of failures during system simulation. The definition of a system on the object of cognition is under the influence of experience gained in technical fields, very often (though for the most part unconsciously) understood purely subjectively. That is, it is as if the determination of elements and properties of the system, of properties of its elements, of the relationships among the elements and of the relationships between the system and its environment, and so on, depended only on our decision. In reality by this procedure we create only our idea of those aspects of the system which we have defined on the object of cognition already by the choice of the viewpoint from which we investigate it. The existence of the simulated system is objective in the sense that it belongs to the object of cognition and its content is independent of man and mankind. It is probably the subjective conception of the system which gives rise to the omission, mentioned above, of the phase of the "verification of truthfulness of the simulation model" or of its identification with the phase of the "testing of correctness of the simulation model". The correctness of the simulation model is verified by us as its adequacy in relation to our idea about the simulated system. If we do not admit the objective existence of this system, we no longer, of course, need to occupy ourselves with the adequacy of the simulation model to the objective reality, that is, with the verification of its truthfulness, either. We shall then certainly feel the consequences of this approach during the further use of the simulation model in the process of cognition.

CONCLUSION

In this chapter we have simply outlined the most serious errors which can arise from a methodological standpoint during the investigation of objects in general. These methodological problems also manifest themselves to a large extent during the simulation of biological processes and of systems of health care. If we neglect them it could justify the points of view of those people who today still believe that the simulation of systems only helps "to confirm what its user desires".

REFERENCES

Kotva, M., 1986, Dohoda o chápání pojmu "simulace systémů" (The agreement on understanding the notion "The simulation of systems"), Automatizace (Prague), 29 (12): 299.

Mervart, J., 1977, "Základy Metodologie Vědy: Aplikace na Ekonomické Vědy" ("The Bases of the Methodology of Science: Applications in the Field of Economic Sciences"), Svoboda, Prague.

Polya, G., 1954, "Mathematics and Plausible Reasoning", Princeton University Press, Princeton, NJ.

MODEL OF ELECTROMAGNETIC FIELD DISTRIBUTION IN THE HUMAN BRAIN DURING

ELECTRIC STIMULATION

M. Mikuláš and J. Miertušová

Department of Radioelectronics
EF SVŠT
Bratislava, Czechoslovakia

INTRODUCTION

The objective control of the regular operation of the neurostimulator is one of the most important problems in the direct stimulation of the brain by the implanted stimulator. This objective control requires the measurement of several parameters. They include stimulation artefacts which are detected on the surface of the head during stimulation. Experimental measurements have shown that it is possible to measure these signals by means of a sensor system suitable for detecting EEG signals. The final recorded signal is the superposition of the EEG signal and stimulation artefacts. It has been shown that the value of such signals depends on the distance between the stimulating and detecting electrodes. However, it is not possible to observe such dependences by direct experiment. For this reason we have solved this problem theoretically using appropriate models.

THE PHYSICAL MODEL

In our laboratory the physical model for the observation of the electromagnetic field distribution was elaborated. Such a model enables measurements to be made of the dependences of stimulation artefacts on the distance between the stimulating and detecting electrodes in the space of the human head during electric stimulation. This model was made in the Department of Automation and Regulation SjF SVST, directed by Ing. L. Dedík.

The model of the human head is formed of a plastic sphere, 23 cm in diameter, in which there are a number of holes. Through these holes it is possible to fill the sphere with liquid (physiological solutions, blood, blood plasma, and so on). In the bottom part of the sphere, the holes for the stimulating electrodes are arranged symmetrically, 36 holes in total. There is the possibility to change the depth of the electrodes introduced and to change the angle between the axis of the hole and the electrode.

MODEL ANALYSIS AND RESULTS

It can be shown theoretically that, with some simplifying assumptions, with the help of this model it is possible to observe the electromagnetic field distribution in the space of the head as a function of the electrode position and the type of liquid.

As a first step we have supposed that the space in the sphere is 50% liquid and 50% air. For this condition the potential on the surface of the sphere is given by

$$\psi = k \sum_{i=1}^{4} \left(\frac{1}{a_{i1}} - \frac{1}{a_{i2}} \right) \tag{1}$$

where a_{i1}, a_{i2} are the distances of electrodes from the measured point. Numerical solution and graphic representation were performed by computer in the Research Institute of Medical Bionics by Ing. M. Hrubý. Some results are shown in Fig. 1.

The first measurements were carried out for the case when half the space was full of liquid and they confirmed the theoretical predictions. All the measurements were done by d.c. and impulse excitation with a real neurostimulator. The scheme of interconnection for the two cases is shown in Fig. 2.

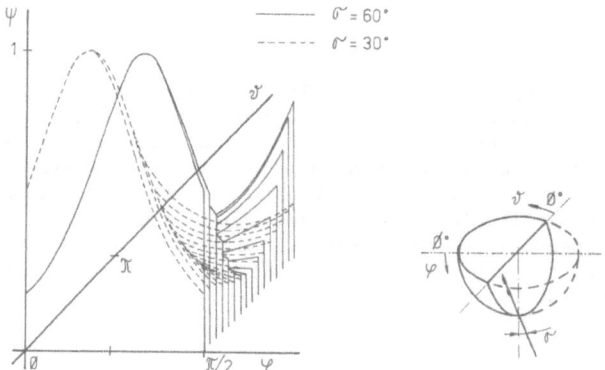

Fig. 1. Typical computer simulation results

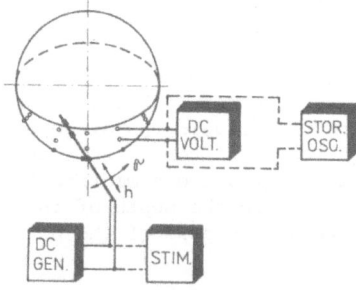

Fig. 2. The measurement system

Measurements on the physical model verified the need to elaborate the model for the measurement of the electromagnetic field distribution in the space of the human head during stimulation. The theoretical and experimental results are not exactly equal, the difference being about 30% at the margin of the space. However, in the real condition, this difference is only a few per cent.

There needs to be taken into account the fact that some simplifications were made in the theoretical solution. Experimental measurements on the physical model enable the mathematical model to be simulated. Further, it will be possible to calculate the map of the electromagnetic field distribution by computer and to define the optimal position of the detecting electrode system when the localisation of the stimulating electrodes is known. In this way the sensor system determined by computer will be used in detecting stimulation artefacts. The objective evaluation of stimulation efficiency will be performed by the method of selecting and measuring the value of the stimulation artefact.

This work is a partial problem solved in a complex programme of neurostimulation control systems. Recent results show that it is possible to make the computer model usable in clinical practice.

PART 4

DECISION SUPPORT

INTELLIGENT INSTRUMENTATION IN CRITICAL CARE MEDICINE

E. R. Carson, A. Shamsolmaali, R. Summers, M. S. Leaning
and D. G. Cramp

Centre for Measurement and Information in Medicine and Depart-
ment of Systems Science, City University; and
Departments of Chemical Pathology and Medical Informatics
Royal Free Hospital School of Medicine
London, U.K.

INTRODUCTION

Advances in information technology have resulted in the embedding of
computer-based information handling within the measurement and instrumentation processes, giving rise to what is termed intelligent measurement.
A major area for the application of intelligent measurement and instrumentation is clinical medicine, particularly in relation to the management of
the critically ill patient. This paper examines the nature of intelligent
measurement and instrumentation in the context of clinical medicine, out-
lines the management problem in critical care medicine and then illustrates
these concepts through two case studies. The first relates to the development of a predictive knowledge-based system for the interpretation of
patient data from a critical care unit, whilst the second focuses on an
artificial intelligent respirator system for ventilator management.

INTELLIGENT MEASUREMENT AND INSTRUMENTATION

The term intelligent measurement tends to be applied to a range of
activities involving machine information processing. At the simplest level
are signal enhancement, including the extraction of signal from noise and
linearisation, and basic signal processing. Such functions can be performed by smart instruments which are sometimes erroneously called intelligent. The measurement processes truly involving significant machine
intelligence are those of inferential measurement, pattern cognition and
measurement as a part of an integrated information system (Finkelstein
and Carson, 1986).

In critical care medicine, many crucial variables and parameters are
not directly measurable and hence have to be identified, using appropriate
mathematical models, from other directly observed variables. This use of
mathematical models together with appropriate parameter estimation schemes,
termed inferential measurement, generally involves substantial computing,
on- or off-line, and embodies a degree of machine intelligence. Inferential measurement can also be used to plan treatment. The parameters estimated from observable input/output measurements or from other physiological

knowledge provide a representation of a particular patient; a model which can be used to assess alternative therapies by simulation thus providing an additional aid to clinical decision making.

Where the physiological process models are not well understood, they may be induced through pattern cognition. In this way combinations of laboratory test values and clinical observations which are most discriminative of patient state can be identified. For instance, data obtained from on-line monitoring of cardiac output or blood pressure clearly contain information which is relevant to circulatory stability. Since the process models are not well understood, they are induced by pattern cognition in order to determine those signal features which are most discriminative of circulatory function.

The process of managing the patient relies on measurement as part of an integrated information system. Having been acquired, enhanced and processed into an appropriate form, measurement information forms input to decision and action or control processes. Much of the information handling is carried out by machine and hence can be viewed as the highest level of intelligent measurement.

THE MANAGEMENT PROBLEM IN CRITICAL CARE MEDICINE

In critical care management, the patient enters the critical care unit in an abnormal state and the task of the unit is to return the patient to a normal state by as smooth a path as is possible. The more irregular or "stepped" this path, the longer the return will take, and the higher the possibility of permanent or fatal injury to the patient.

Treatment within the critical care unit is based on information extracted from data on the patient state gathered from automatic on-line monitors, from laboratory results and from clinical observations. Machine intelligence is generally limited to the extraction and processing of monitored physiological signals. The major problem is that there is a larger volume of data available which is presented in varied formats. The relationships between the monitored variables are very complex and it is only with highly skilled interpretation that the underlying process of disease or trauma can be understood and hence the most appropriate management effected.

In managing the critically ill patient, the maintenance of body fluid volume, circulatory stability and metabolic homeostasis, together with optimisation of oxygen delivery, is of great importance. An intelligent measurement system which operates upon the variables to these physiological processes and their derangements would be of great value in order to provide an intelligent decision support system for the clinician. The development of such an intelligent measurement system involves fusion of data from sensors of various types attached to each patient with nursing observation and laboratory data in order to provide a report, including a flexible display of integrated patient data, which can be accessed at any time by medical and nursing personnel (Carson et al., 1986). Progress on two aspects of this work is described in the case studies presented below.

A PREDICTIVE KNOWLEDGE-BASED SYSTEM FOR THE INTERPRETATION OF PATIENT DATA FROM A CRITICAL CARE UNIT

In this first example, a prototype knowledge-based system for the enhanced interpretation of fluid-electrolyte and acid-base balance data is being implemented and evaluated. These laboratory data when integrated

with clinical (bedside) data contain the information which is required by
the clinician for the decision-making process (Shamsolmaali et al., 1987).
In terms of the laboratory data, providing an analytical result is not an
end in itself; analytical results need transformation into useful information within the clinical context in which the tests are performed.

Rule-based systems encapsulating elements of knowledge and reasoning
have considerable potential as aids to interpreting the causal relationships between disease, treatment and the evolving patient state as measured
using laboratory-generated information. Such systems are not, however, the
only way in which expert knowledge can be expressed. Dynamic mathematical
models, which are formulated as differential or difference equations and
are soundly based on the underlying physiological processes, have been
developed and tested in a number of areas of medicine which are heavily
reliant upon laboratory data including fluid-electrolyte, acid-base balance,
to the extent that they constitute useful representations of the time-
course of physiological events. As such, they can be used to track and
predict the evolving patient state in response to pathological change, and
the therapeutic input, by which the clinician seeks to return the patient
to a more normal state.

Intelligent knowledge-based systems and dynamic mathematical models,
in the context of intelligent instrumentation, together offer the prospect
of computer-based aids to the interpretation of laboratory data and hence
to patient management, yielding benefits in excess of those which either
type of representation could provide on its own. The basic structure of
the prototype system comprising patient specific database, diagnostic and
treatment modules and dynamic mathematical model, together with the interfaces to the clinical user and to the patient record file, is shown in
Fig. 1.

Patient Specific Database

The patient specific database comprises both clinical (bedside) and
laboratory data; both current data input to the system by the clinical

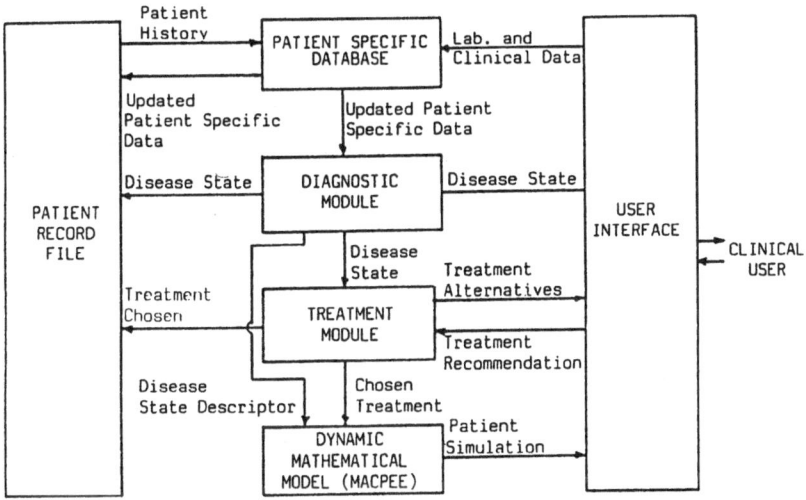

Fig. 1. Predictive knowledge-based system for the interprtation of
patient data from a critical care unit.

Table 1.   Variables in the Patient Specific Database

Laboratory data:	plasma concentrations of -	sodium
		potassium
		albumin
		creatinine
		urea
	urine concentrations of -	sodium
		albumin
	urine osmolality	
	haemoglobin	
Clinical data:	blood pressure	
	central venous pressure	
	pulmonary capillary wedge pressure	
	temperature difference (core - periphery)	

user and those contained in the record file for the specific patient.   The variables included are listed in Table 1.

The input of the patient identifier by the clinical user automatically triggers the computer system to look for the corresponding patient record and retrieve any existing data (for example, patient history).   Where no file already exists, a new one is opened.   The patient database is updated with the newly input data and returned to the patient record file for later encounters.

Diagnostic Module

The patient specific data are then input to this module which yields an output in the form of the most probable diagnosis expressed as a disease state.   This module comprises a rule-based system which defines the conditions required for a disease-state to be present.   All the states are "candidate states" unless there are some data which contra-indicate a particular state.   Currently there are 15 disease states defined by the system, all of which are water and salt related disorders including: water overload;   syndrome of inappropriate antidiuretic hormone release (SIADH);   congestive heart failure;   and dehydration.

Treatment Module

Based on the responses from the clinical user to questions asked regarding the cardiac, renal and dietary status of the patient and details of currently prescribed drugs, together with the disease state as diagnosed, one or more treatment strategies will be suggested to the user.   Facilities will be provided for the user to choose one such suggested treatment, or indeed any other treatment, and implement this on the dynamic mathematical model (see below) in order to observe the changes in the patient's condition over time which are predicted by the model simulation.

The Dynamic Mathematical Model

Using the data output from the diagnostic module, the parameters of the dynamic mathematical model are adjusted so as to tune it to the diagnosed disease state.   It is then used to track the dynamic changes in the patient state over time, in response to the chosen therapeutic regimen.

The model used is MacPee which simulates the systemic circulation, kidneys and body fluids (Dickinson et al., 1985). MacPee is designed around MacMan, an earlier model which simulates circulation. While MacMan is designed to examine and simulate rapid fluctuations in haemodynamic function and has a two-second computing interval, MacPee is not concerned with such rapid fluctuations, and has a normal computing interval of 60 minutes (changeable between 5 and 120 minutes). It is designed to illustrate the much slower changes in haemodynamic function related to alterations in the size and distribution of the contents of body fluid spaces.

The output of the model is in the form of a 6-hourly graphical representation of blood pressure and heart rate, together with a selection of 8 variables. At the end of each simulation period values are printed for plasma sodium, potassium, albumin, urea and creatinine concentrations, haemoglobin, right atrial pressure, body weight and so forth. Adjustable factors include dietary sodium, protein, potassium, cardiac contractility, renal artery pressure, glomerular function, tubular sodium loss, aldosterone function, and a range of fluids either administered or withdrawn.

User Interface

The clinical user communicates with this intelligent system by means of a menu-driven interface. Facilities are provided to: input or update data; display data; make a diagnosis; generate alternative therapies; and output predictions of therapeutic outcome (via model-based intelligent measurement).

Implementation

This predictive knowledge-based system for the interpretation of patient data is being developed on an IBM PC with a minimum of 512k RAM, running PC DOS. MacPee is written in FORTRAN 77, whilst the other modules, including the user interface, are being developed in LPA-PROLOG version 1.4.

AN ARTIFICIAL INTELLIGENT RESPIRATOR (AIR) - A SYSTEM FOR VENTILATOR MANAGEMENT

Ventilator management is another area of critical care medicine where the application of an intelligent knowledge-based system can assist the clinician. In situations where the physiological system which controls respiration is not intact, an alternative external controller is required. This is the role of an artificial ventilator - it takes over from the respiratory physiological control system so as to sustain life. The purpose of an artificial ventilatory system can therefore be defined as to keep designated respiratory variables within pre-set desirable limits. This enables the production of a patient-specific management plan, the desired end-point of which is to return the patient to a state where spontaneous ventilation occurs. This is the point when the internal physiological control system can retake command from the external "artificial" system.

The Need for Intelligent Instrumentation

The management strategies for mechanical ventilation adopted in clinical practice are rarely optimal. The difficulty faced by the clinician is that of integrating and interpreting a range of variables spanning the respiratory, circulatory and metabolic systems of the patient. The problem is one of seeking to minimise the mismatch between the ventilator setting chosen and the needs of the patient. This requires development of a knowledge-based system which provides advice for ventilator management based on respiratory data interpretation in the context of the patient's clinical state. Ventilator management involves manipulation of minute

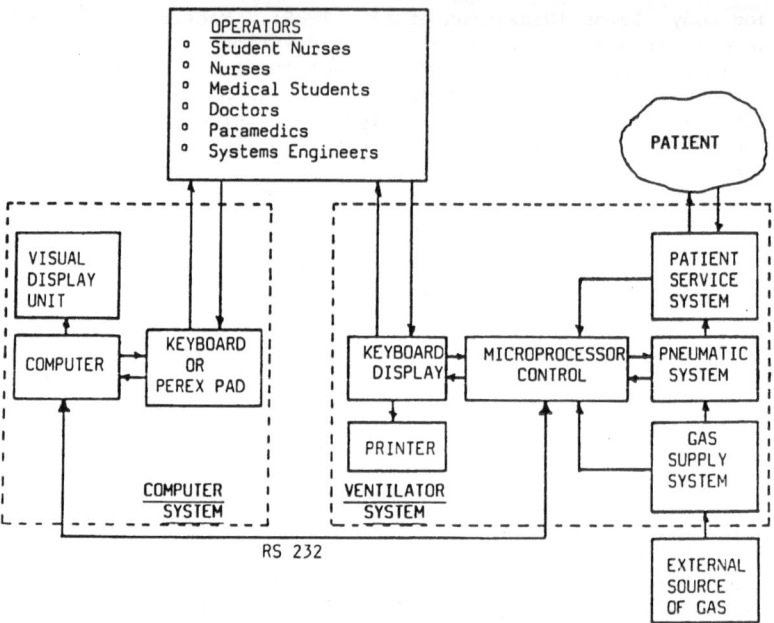

Fig. 2. Artificial intelligent respirator system (AIRS)

volume, inspired oxygen concentration and inspired CO_2 concentration. The patient's state is typically defined by end-tidal CO_2 concentration (as an indirect measure of arterial CO_2 concentration), cardiac output and pH. These variables together provide an index of the respiratory, circulatory and metabolic states of the patient.

The System

The Artificial Intelligent Respirator System (AIRS) described below consists of four functional units, each of which can be further sub-divided into their constituent elements (see Fig. 2). These functional units are the patient, the operator, the ventilator system and the computer system.

The patient can be described as a complex physiological super-system, whose constituent systems are interlinked in such a way as to optimise the process for sustaining life. Of interest is the respiratory system where reasons for introducing mechanical ventilatory support are numerous. These range from depression of the respiratory centre in the brain (for example, due to drug action or brain tumour), to muscle weakness (for example, caused by muscle relaxants administered to patients prior to surgery or pathological syndromes such as myasthenia gravis).

The operators catered for in an AIR system interaction include doctors and nurses (both fully qualified and students), paramedical staff and the system engineer. Differing levels of system interaction are envisaged depending on the status of the operator.

The ventilator system consists of six functional sub-units which can be divided into two sub-systems: the pneumatic sub-system and the electrical sub-system. Within the pneumatic sub-system the functional sub-units are the gas supply system, the pneumatic system controller and the patient service system. The purpose of the pneumatic sub-system at its highest level is to deliver gas (air, oxygen) to the patient at a pressure and flow

Table 2. Patient Data Set Obtained from Ventilator

Respiratory Rate
Minute Volume
Mean Airway Pressure
Inspiration : Expiration
Tidal Volume
Spontaneous Minute Volume
Peak Airway Pressure
Plateau Pressure

---

rate which are in the physiological range. Pressure and gas flow are monitored continuously so that adjustments can be made which enable the delivery
of the desired pressure and flow rate characteristic. The electrical sub-
system consists of the microprocessor control unit, the keyboard display
panel, and a printer. The microprocessor control unit is the central pivot
on which the ventilator system operates. It receives inputs from the keyboard display unit and from pressure, flow and temperature sensors found
in the pneumatic system. A controlled output port leads to the pneumatic
system controller which generates and controls gas flow. Interaction with
the keyboard display panel indicates patient and ventilator performance;
these indicants are transferred to the computer system via an RS-232 cable.
A printer is included in this system so that a hard-copy of a patient report
can be produced.

The computer system consists of the system unit, a keyboard or PEREX
(pressure-sensitive graphics pad) pad for data entry, and a visual display
unit so that the operator can observe the output of the system. An RS-232
connection exists between the system unit and the microprocessor control
unit housed in the ventilator. This connection allows patient data collected
by the ventilator to be captured by the computer system, thereby initiating
the process of intelligent data processing.

Typical data produced by the ventilator system are listed in Table 2.
If these are complemented by data derived from haemodynamic and metabolic
monitoring, and a knowledge-base of relevant associations between data
findings and management options available is embodied within the computer,
then a powerful data-driven computer-aided decision support system emerges
(Summers et al., 1987).

Although a closed-loop system is not envisaged for AIR, Katona (1983)
has described two reasons for developing systems which automatically control
the designated respiratory variables. First, ventilator settings may have
to be continually adjusted so as to allow optimal gas exchange to take
place for the metabolic requirements of the patient. (The available ventilator settings are shown in Table 3). Secondly, it is sometimes desirable
to keep some of the monitored respiratory variables constant so that interpretation of respiratory manoeuvres is made easier. This enables patient
state to be inferred, and hence the state trajectory of the patient can be
monitored.

CONCLUSIONS

The nature of intelligent measurement and instrumentation has been
examined and its applicability in critical care medicine outlined. It has
been shown that intelligent measurement within an overall clinical measurement information system constitutes a knowledge-based decision support

Table 3. Set of Ventilator Settings

Respiratory Rate
Tidal Volume
Peak Inspiratory Flow
FIO_2
Sensitivity
PEEP

Mode (CMV, SIMV or CPAP)
Flow Waveform (Square, Ramp or Sine)

100% O_2 Suction (Off or On)
Nebuliser   (Off or On)
Automatic Sigh (Off or On)

system which can assist in the management of the critically ill patient. Two particular examples of intelligent instrumentation have been examined: a predictive knowledge-based system for the interpretation of patient data, and an artificial intelligent respirator system for ventilator management. In both cases the system in question provides clinical assistance and a consultative facility, thereby offering the prospect of enhanced patient care.

ACKNOWLEDGEMENTS

This work was supported by the U.K. Science and Engineering Research Council; by a Department of Health and Social Security Grant; and by NATO Grant No. RG85/0207.

REFERENCES

Carson, E. R., Cramp, D. G., and Finkelstein, L., 1986, Towards intelligent measurement in critical care medicine, in: "Proc. 8th Annual Conference of the IEEE Engineering in Medicine and Biology Society", G. V. Kondraske and C. J. Robinson, eds., IEEE, New York : 799.
Dickinson, C. J., Ingram, D., and Ahmed, K., 1985, The Mac family of physiological models, ATLA, 13 : 107.
Finkelstein, L., and Carson, E. R., 1986, Intelligent measurement in clinical medicine, in: "Proc. 5th IMEKO Symposium on Measurement Theory", Jena, DDR, 1986.
Katona, P. G., 1983, Automated control of physiological variables and clinical therapy, Crit. Rev. Biomed. Eng., 8 : 281.
Shamsolmaali, A., Carson, E. R., and Cramp, D. G. , 1987, A knowledge-based clinical laboratory data interpretation system incorporating a mathematical model, in: "Proc. 9th Annual Conference of the IEEE Engineering in Medicine and Biology Society", IEEE, New York : 377.
Summers, R., Leaning, M. S., Cramp, D. G., and Carson, E. R., 1987, A knowledge-based approach to ventilator management, in: "9th Annual Conference of the IEEE Engineering in Medicine and Biology Society", IEEE, New York : 379.

ICAR - INTENSIVE CARE EXPERT SYSTEM

M. Matvejev, N. Kasabov and O. Hinkov

Institute of Biomedical Engineering
Medical Academy
Sofia, Bulgaria

INTRODUCTION

There are now many reasons for considering that of all the artificial
intelligence products of medical application, expert systems for intensive
care and emergency care offer the greatest potential (Blum, 1985). Bearing
this in mind, we decided to try to build a modular expert system of an open
type, with the possibility of adding knowledge bases oriented towards the
specific tasks of the departments of resuscitation and intensive care.

THE SYSTEM

The general structure of the system is given in Fig. 1. The connecting
link in this structure is the programming system, GESMI - Generator of
Expert Systems for 16-bit MIcrocomputers, compatible with IBM PC/XT/AT personal computers. The generator includes:

(i) An editor program (RULEMAKER) used to transform expert knowledge into
a knowledge base describing a tree-like logic structure by a set of
production rules. The editor has extended possibilities for text
editing, dialogue control to conform with the semantics and syntactic
rules for writing the text, switching between modes and commands
through menus, display or printout of the structure of the knowledge
base in a form of enumerated texts of the conditions and conclusions
included in the rules, tables of the logical connections between
rules, succession of rules entered or a list of the possible final
conclusions (Figs. 2 and 3).

(ii) Inference engine - program for the logical deduction of the rules
(EXPERT), designed for building specific knowledge bases and obtaining
final conclusions, corresponding to the current values of the parameters included in the base. Proving is done by the principle of
backward-chaining.

The functional modules of the expert system generator are written in
MICROSOFT PASCAL (version 3.30), under the control of the MS/PC DOS operating system. The conditions and the conclusions representing the rules of
the knowledge bases, implemented by GESMY, can be represented in the form
of texts, input and output manageable text and graphic files and any type
of arithmetic operations allowable with PASCAL syntax. The names of other

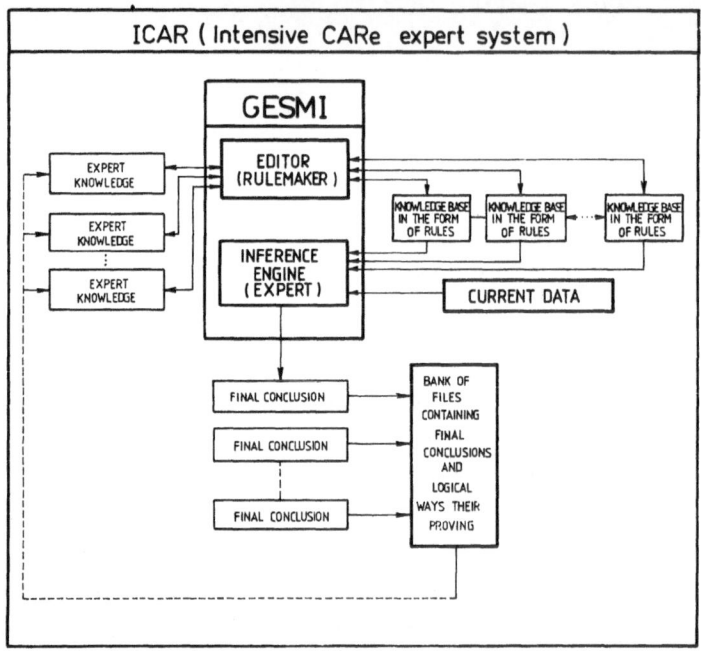

Fig. 1.   General structure of ICAR.

knowledge bases, executable programs and operational system commands can
also be final conclusions, thus ensuring modularity (and/or hierarchy) of
the expert system structure, with the possibilities of transition from one
knowledge base to another and back, and of ending with output to other
program products, and so on.

Values can be assigned to the variables in the knowledge base by one
of the following means:

(i)   entering via the keyboard as answers to dialogue questions during
      the process of reaching a conclusion;

(ii) setting of a cursor on a curve on the screen (the curve is obtained
      from points with co-ordinates stored in the graphic file); in par-
      ticular, the data file could be taken after analog-to-digital con-
      version;

(iii) using a function of correspondence between variables and the order
      number of data in a set of real numbers of the data file.

The input, output and processing of graphic information in the knowledge bases is provided by direct use of graphic primitives available in
the software for the high-resolution graphic display PC board of Hercules
Computer Technology, Inc.

STRUCTURE OF THE KNOWLEDGE BASES

The specific structure of knowledge bases obtainable using GESMI will
be demonstrated by means of two sub-systems included so far in the expert
system ICAR. The first sub-system, comprising three knowledge bases
(Fig. 4), is designed to provide electrocardiological interpretation based
on the Frank orthogonal lead system. The data (Fig. 5) are stored in a
file after primary signal processing and measurement of their morphological

```
 1: \pm<<'Pacemaker? Type YES (Y) or NO (N) ==>
 2: \(pm=y)
 3: >?pminf
 4: \prdur:=getf(51)
 5: \(prdur>=210)
 6: >?av-i
 7: \(pr<120)
 8: \qrsdur:=getf(53)
 9: \(qrsdur>=120)
10: >?shortprd
11: \hrythm<<'Normal rythm? Type YES (Y) or NO (N) ==>
12: \(hrythm=y)
13: >?abnrythm
14: \PAMPLXPS:=getf(7)
15: \pamplzps:=getf(39)
16: \pamplps:=pamplxps+pamplzps
17: \(pamplps>1.5)
18: >?latrover
19: \pamplyps:=getf(23)
20: \(pamplyps>2.0)
21: \pamplzng:=getf(40)
22: \pamplz:=-pamplzng
Arrow keys controls the scroll direction ( up , down ). ESC exits
```

```
Rule    1:      1,      2,      0,      0,      0,      0,   =>      3
Rule    2:     -2,      4,      5,      0,      0,      0,   =>      6
Rule    3:     -6,      7,      8,     -9,      0,      0,   =>     10
Rule    4:     -2,     11,     12,      0,      0,      0,   =>     13
Rule    5:    -12,     14,     15,     16,     17,      0,   =>     18
Rule    6:    -12,     19,     20,     21,     22,     23,   =>     24
Rule    7:      9,      7,      0,      0,      0,      0,   =>     25
Rule    8:    -25,      9,     26,     27,      0,      0,   =>     28
Rule    9:     29,     30,     31,      0,      0,      0,   =>     26
Rule   10:     32,     33,     34,      0,      0,      0,   =>     27
Rule   11:    -25,      9,     35,      0,      0,      0,   =>     36
Rule   12:     37,     38,     39,      0,      0,      0,   =>     35
Rule   13:     40,      0,      0,      0,      0,      0,   =>     35
Rule   14:    -25,      9,    -28,    -36,      0,      0,   =>     41
Rule   15:     28,      0,      0,      0,      0,      0,   =>     42
Rule   16:     36,      0,      0,      0,      0,      0,   =>     42
Rule   17:     41,      0,      0,      0,      0,      0,   =>     42
Rule   18:    -42,     43,      0,      0,      0,      0,   =>     44
Rule   19:     45,     46,     47,      0,      0,      0,   =>     43
Rule   20:     48,     49,     50,     51,      0,      0,   =>     43
Rule   21:     52,     53,      0,      0,      0,      0,   =>     43
Rule   22:     54,     55,      0,      0,      0,      0,   =>     43
Arrow keys controls scroll direction ( up , down ). ESC exits
```

Fig. 2.   Rules within the knowledge base.

parameters using a microcomputer electrocardiograph developed in the Institute of Biomedical Engineering of the Medical Academy (Daskalov, 1988). The logic of the first knowledge base is built according to the interpretation system developed by Pipberger et al. (1982). The transition to the specialised knowledge base for diagnosis and localisation of myocardial infarction (MI) is accomplished when the Pipberger system issues a statement involving myocardial infarction signs. The need for detailed interpretation of the MI data is due to the fact that the detection of MI from the Q/R ratios in three leads only can induce a significant number of false statements, for example in the presence of ventricular hypertrophy. This knowledge base logic was built using the discrimination rules of Abreu-Lima et al. (1983), these having a much greater information content than those of the Pipberger criteria. Further, a transition is possible from this knowledge base to a system for the determination of the size of the MI lesion, after the method of Cowan et al. (1982).

The second sub-system is intended to give assistance to the diagnosis and treatment in severe disbalance of the cardiovascular, cardiopulmonary

```
No    No    IF                        Rule No    6
Cond  Text  DDDDDDDDDDDDDDDDDDDDDDDDDDDDDDDDDDDDDDDDDDDDDDDDDDDDDDDDDDDDDDDDDDDDDDDDDDDDDL
  1:    12.  \(hrythm=y) - NOT

  2:    19.  \pamplyps:=getf(23)

  3:    20.  \(pamplyps>2.0)

  4:    21.  \pamplzng:=getf(40)

  5:    22.  \pamplz:=-pamplzng

  6:    23.  \(pamplz<-1.0)


DDDDDDDDDDDDDDDDDDDDDDDDDDDDDDDDDDDDDDDDDDDDDDDDDDDDDDDDDDDDDDDDDDDDDDDDDDDDDDDDDD
            THEN:
       24.  >?ratrover


Arrow keys controls scroll direction ( up , down ). ESC exits



   3:  >?pminf
 137:  >?normecg
Arrow keys controls scroll direction ( up , down ). ESC exits
```

Fig. 3. List of possible final conclusions.

and oxygen transport systems. The structure of this knowledge base is
built in four levels. The first level consists of rules defining the clinical state (SYMPTOMS) from the set of quantitative parameters included in
the base and derived data (ratios, indexes, and so on). The second level
is built of rules, whose conclusions are defined by specific combinations
of clinical symptoms and quantitative data. At this level, expert conclusions of haemodynamic changes, of pulmonary ventilation and perfusion, of
tissue gas exchange, and so on, are deduced, related to the basic clinical
SYNDROMES in resuscitation: general body reaction to acute stress, hypo-
volaemia, cardiac insufficiency, pulmonary insufficiency, pulmonary inflam-
mation, cardiac tamponade and sepsis. A typical example of rules of the
first and second levels is given in Fig. 6. The third level consists of
rules whose conclusions represent consulting texts of the necessary therapeutic measures for control of the left ventricular preload, myocardial

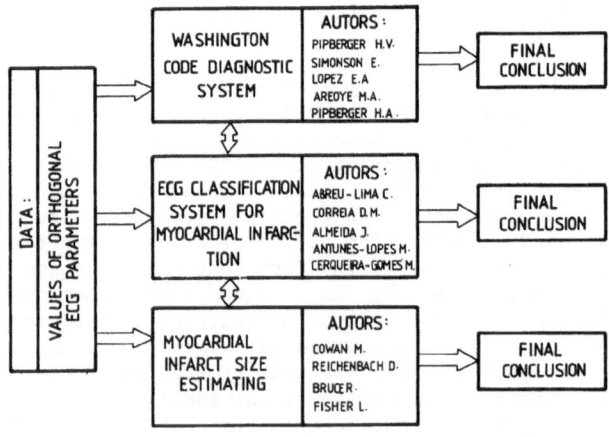

SUBSYSTEM TO ECG DIAGNOSTIC

Fig. 4. The first sub-system comprising three
         knowledge bases.

COMPUTED FRANK

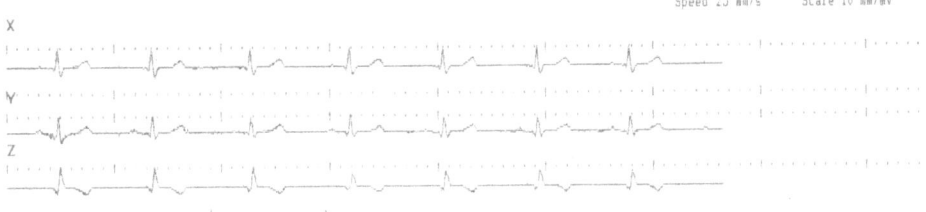

Speed 25 mm/s   Scale 10 mm/mV

X

Y

Z

DATA REPORT

Heart Rate	71	bPm
R-R interval	845	ms
P-Q interval	140	ms
P duration	60	ms
QRS duration	117	ms
Q-T interval	365	ms
QTr	.97	
QRS axis	50	deg
T axis	55	deg

	Q aV	R aV	S aV	ST aV	T+ aV	T- aV	P+ aV	P- aV	J aV	Rdur ms	Rp-k ms	Qdur ms	P/R -	Q/R -	T/R -	R/S -
X	.053	.444	.180	.039	.190		.053		.019	35	45	5	.11	.11	.42	2.46
Y	.107	.366	.117	.048	.180		.058		.019	35	47	12	.15	.29	.49	3.12
Z	.146	.395		-.019		.170	.024	-.014		65	67	40	.06	.36	.43	>10.00

Fig. 5. Data which are stored following primary signal processing.

contractility, pulmonary ventilation and perfusion. All the three levels
are based on quantitative parameters. A fourth level, having a higher
reliability, is built using the personal clinical impressions of the physician and their combination with the results of additional clinical investigations, recommended by the expert system.

CONCLUSIONS

The expert system ICAR, after loading it with functional knowledge
bases, should offer the user the possibility of detecting changes in normal
body functions, of displaying information on functional groups and of interpreting data of the patient state (by issuing conclusions on the semantics
of quantitative parameters). It should also be able: to offer exertise
based on syndromes of the patients in the resuscitation and intensive care
units, in order to give consultation on the basic therapeutic actions; to
provide for retrospective assessment of diagnosis and therapy by storing
information on the patient state and the sequence of logical reasoning,
brought to the expert conclusion; and to promote study of the accumulated
expert knowledge and clinical experience, by storing characteristic cases
and modelling clinical situations by entering corresponding values for the
parameters.

REFERENCES

Abreu-Lima, C., Correia, D. M., Almeida, J., Antunes-Lopes, M., and Cer-
    queira-Gomez, M., 1983, A new EGC classification system for myocardial

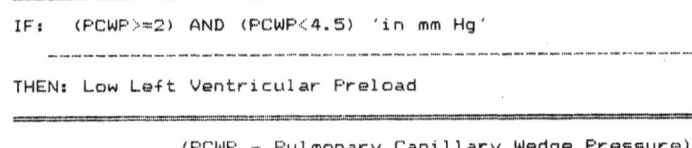

A first level rule

```
IF:  (PCWP>=2) AND (PCWP<4.5) 'in mm Hg'
-----------------------------------------------------------------

THEN: Low Left Ventricular Preload
```

(PCWP - Pulmonary Capillary Wedge Pressure)

A second level rule

```
IF: 1. (Moderately Tachycardia) OR (Heavy Tachycardia) OR
       (Critical Tachycardia) OR (Bradycardia) OR (Criti-
       cal Bradycardia)
    2. (Normal Respiratory Rate) OR (Moderately Tachypnea)
       OR (Heavy Tachypnea)
    3. .......................................................
-----------------------------------------------------------------
THEN: Low Heart Preload (O.XX Probability)
```

Fig. 6.   A typical example of rules of the first and
second levels.

infarction based on receiver operating characteristic curve analysis
and information theory, Circulation, 67 : 1252.

Blum, B. I., 1985, Artificial intelligence and medical informatics, J. Clin.
Eng., 10 : 109.

Cowan, M. J., Reichenbach, D. D., Bruce, R. A., and Ficher, L., 1982,
Estimation of myocardial infarct size by digital computer analysis of
the VCG, J. Electrocardiol., 15 : 307.

Daskalov, I. K., 1988, Family of microcomputer electrocardiographs, in:
"Advances in Biomedical Measurement", E. R. Carson, P. Kneppo and
I. Krekule, eds., Plenum, London : 3.

Pipberger, H. A., Simonson, E., Lopez, E. A., Araoye, M. A., and Pipberger,
B. A., 1982, The ECG in epidemiologic investigations. A new classification system, Circulation, 65 : 1456.

AN EXPERT SYSTEM FOR THE DIAGNOSIS AND TREATMENT OF ISCHAEMIC HEART

DISEASE : CARDEXP

E. Kékes, M. Kakas, J. Aszalos, I. Préda, J. Barcsák,
J. Kovács and Z. Antalóczy

Postgraduate Medical School, 2nd Medical Department
(Cardiology), and
Computing Applications and Service Company
Expert System Department
Budapest, Hungary

INTRODUCTION

Ischaemic heart disease (IHD) is one of the most common causes of death in many countries of the world. This is why the development of a widely accepted expert system for the evaluation of the diagnostic and therapeutic decision model of ischaemic heart disease constitutes a general need.

The diagnostic algorithms were based on the case history and on routine medical methods, such as physical and non-invasive and invasive cardiological examinations (Diamond et al., 1980; 1983).

THE DIAGNOSIS AND TREATMENT OF IHD

In evaluating a patient presenting with chest pain, the case history, physical examination and the resting ECG are the key-elements in determining the extent and direction of diagnostic procedures and the possibility of medical or surgical intervention.

The medical aims of our expert system are:

1. differential diagnosis of chest pain;
2. diagnosis of different forms of IHD;
3. analysis of heart function;
4. determination of the sequence of diagnostic procedures: and
5. determination of the optimal therapeutic decisions.

Physicians need a systemic approach to actual clinical problems. Such a strategy ensures the maximum diagnostic accuracy at minimum risk and expense to the patient. Most cardiologists pursue a line of questioning to establish whether the actual chest pain is typically cardiac, atypical or non-cardiac in origin (Greenberg et al., 1984). It is therefore important to determine the sequence of the steps in the diagnosis of IHD. The first diagnostic step is to eliminate the chest pain syndrome of non-cardiac origin and to verify the different forms of cardiac pain (Table 1).

Table 1. Differential Diagnosis of Cardiac Pain.

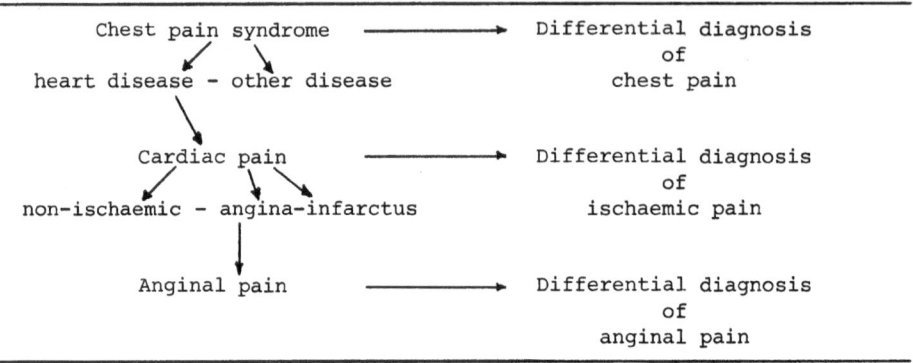

The other aspect was to estimate the likelihood of IHD using so-called
risk factors including genetic factors, smoking, alcohol-intake, hypertension, and so on (Table 2). Each of the factors was weighted and evaluated
as per cent. The total risk was calculated by the following formula:

$$V = 1 - \prod_{i=1}^{n} x(1 - p_i \times v_i) \tag{1}$$

where V is the total risk value, v_i are the individual risk factors and
p_i is the factor weight.

The next step was the selection of patients for a more detailed analysis of their disease. The schedule of the non-invasive and invasive investigations being necessary for the exact diagnosis of IHD, the assessment
of the functional state of the heart as well as the evaluation of the parameters influencing the final therapeutic decision should be determined.

The patients' data are acquired by questioning or from previously
stored data. The line of questioning follows a predetermined network
structure with only relevant questions being asked (Table 3).

The probability of the establishment of the exact diagnosis in IHD is
related to the actual level of health service. Using this hypothetical
assumption, three diagnostic and therapeutic models were elaborated:

1. The first steps of the diagnostic procedures should be evaluated at
    the level of Primary Medical Care in the general practitioner's office

Table 2. Risk Factors

Genetic factors
Smoking
High blood pressure
Hyperlipidaemia
Diabetes
Alcohol
Stress
Personality
Contraceptives
Overweight
Age
Sex

Table 3. Questioning Following a Predetermined
          Network Structure

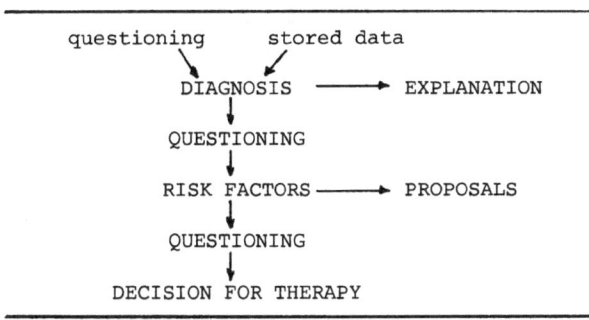

questioning   stored data

DIAGNOSIS ⟶ EXPLANATION

↓

QUESTIONING

↓

RISK FACTORS ⟶ PROPOSALS

↓

QUESTIONING

↓

DECISION FOR THERAPY

or in an ambulance during the first interview with the patient.

2. The second place is usually the cardiologist's office or cardiac
    ambulance, where the non-invasive examinations make it possible to
    select those patients requiring detailed invasive study or where the
    therapeutic decision can reasonably be made if the anatomic or haemo-
    dynamic diagnosis is not obligatory.

3. The last diagnostic steps should be organised at the Cardiac Centre
    where the definitive diagnosis can be made by means of special methods
    and the final therapy can be initiated or performed including invasive
    interventions or cardiac therapy.

All the diagnostic schemes proposed have an increasing degree of diagnostic power and they should be adjusted properly to the actual level of
Health Care. Decision algorithms were elaborated according to the possibility and requirements of the actual diagnostic or therapeutic problems.
For example, the first interview and examination of a patient with unstable
angina pectoris can involve decisions either in the direction of drug
therapy, further special investigations or even an urgent admission into
an intensive care unit (CCU), see Table 4.

THE EXPERT SYSTEM - CARDEXP

The CARDEXP is an Expert System using the tools and methods of Artificial Intelligence (Hunter, 1986). Its main characteristics are the following:

1. Tree-like problem spaces;
2. Backtracking;
3. Backward-chaining;
4. Rule and network-based knowledge;
5. Application of necessary, possible, accessory, excluding and positive
    preconditions;
6. Probability factors; and
7. Menu-based interaction.

The architecture of CARDEXP reflects the paradigm commonly accepted
by the most expert system, but there is some difference in the problem area
and in the approach of understanding the expert's task (Fig. 1).

The CARDEXP is a rule-based system in which the knowledge for making
diagnostic and therapeutic decisions is represented in rules (Knoebel, 1986;
Pauker and Kassirer, 1981):

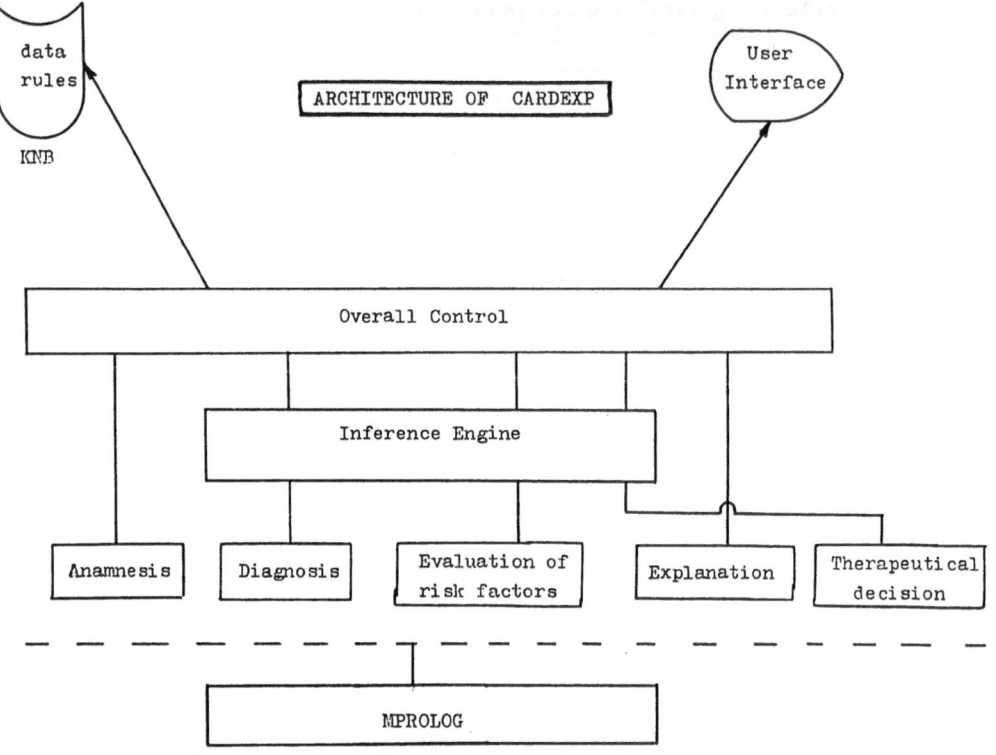

Fig. 1. The architecture of the expert system CARDEXP.

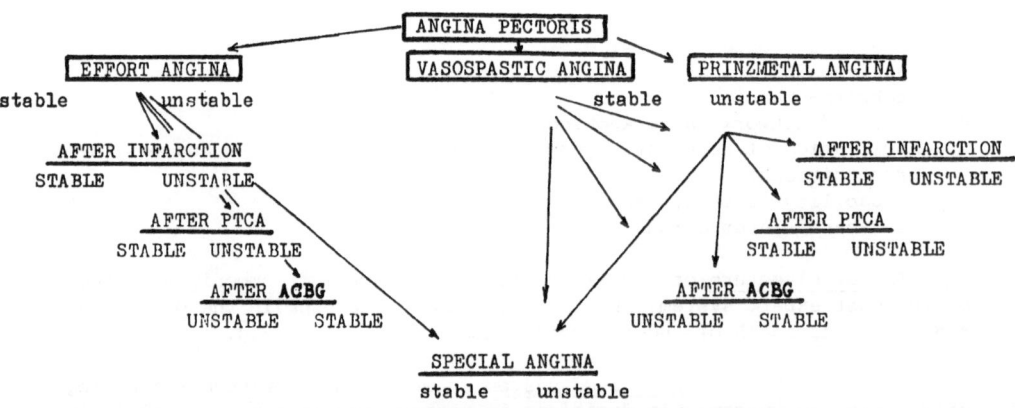

Fig. 2. Representation of IHD as a network.

Table 4. Decision-making at First Interview with a Patient Exhibiting
          Unstable Angina Pectoris

C A R D E X P

DECISIONS IN EFFORT UNSTABLE ANGINA	PRIMARY CARE
SITUATION	DECISION
IF	

RESTING ECG	ST depression type 2 or 3	
	ST depression rate 1 or 2	
	and NTG effective	direct to the specialist
	NTG not effective	drug + CORONARY CARE

RESTING ECG	ST depression type 2 or 3	
	ST depression rate 3 or 4	
	and NTG effective	take to the HOSPITAL
	NTG not effective	drug + CORONARY CARE

RESTING ECG normal ....

RESTING ECG was not performed ....

....

....

....

....

If a precondition excluding the syndrome can be realised, the conditional probability becomes zero per cent just as if a positive precondition
does not occur. If all the necessary preconditions can be realised, the
probability of the syndrome will be 100%. Otherwise the CARDEXP calculates
the probability percentage according to the necessary, possible and accessory preconditions (Table 5).

For diagnostic purposes, during the evaluation of a syndrome, CARDEXP
applies the inference strategy in a goal-driven fashion. The different
forms of IHD represent a network: Fig. 2 shows this network structure in
angina pectoris. The network is traversed by the program at the different
levels utilising at each cross-point all the information obtained until
now. A given constellation can be manifested in some diagnosis on the display.

In the algorithms for therapeutic decisions, the CARDEXP uses the same
backward chaining method using the data obtained by questioning and also
the internal results of the diagnostic algorithms.

The CARDEXP works in <u>three</u> different <u>modes</u>:

Table 5.   Calculation of Probability Percentage
           According to Necessary, Possible and
           Accessory Preconditions

CARDEXP

PRECONDITIONS

FOR VENTRICULAR ARRHYTHMIA
    IN ACUTE MYOCARDIAL INFARCTION

Q - ST
    RHYTHM - a

        NECESSARY

                ECG   ES TYPE "2"   100%

                ES CODE 2   100%

            or ES CODE 3   100%

            or ES CODE 4   100%

            or ES CODE 5   100%

            or ES CODE 6   100%

            or ES CODE 7   100%

            or ECG rit 4   100%

        POSSIBLE
                RHYTHM   2   60%

        ACCESSORY

        EXCLUDING
                ECG mut 1

            or ES type 1

        POSITIVE
                ECG mut 3

---

1. as a system helping in operative medical work or screening the population for IHD;
2. as a teaching model in the university and postgraduate education schools;
   and
3. as an examination testing model for university students and physicians.

IMPLEMENTATION

     CARDEXP is implemented in MPROLOG and runs on an IBM/PC XT or AT or
other compatible configurations. The minimum storage requirement is 512-
bytes; one floppy disk is required and a Winchester disk is recommended
for storing patient data.

370

REFERENCES

Diamond, G. A., Forrester, J. S., Hirsch, M., Staniloff, H. M., Vas, R., Berman, D. S., and Swan, H. J. C., 1980, Application of conditional probability analysis to the clinical diagnosis of coronary artery disease, J. Clin. Invest., 65 : 121.

Diamond, G. A., Staniloff, H. M., Forrester, J. S., Pollock, B. H., and Swan, H. J. C., 1983, Computer assisted diagnosis in the non-invasive evaluation of patients with suspected coronary artery disease, J. Am. Coll. Cardiol., 1 : 444.

Greenberg, P. S., Ellestad, M. H., and Clover, C. C., 1984, Comparison of the multivariate analysis and CADENZA systems for determination of the probability of coronary artery disease, Am. J. Cardiol., 53 : 493.

Hunter, J. R. W., 1986, Artificial intelligence in medicine: a tutorial survey, Parts 1 and 2, Biomed. Meas. Infor. Contr., 1 : 64 and 147.

Knoebel, S. B., 1986, "Perspectives in Clinical Decision Making", Futura, New York.

Pauker, S. G., and Kassirer, J. P., 1981, Clinical decision analysis by personal computer, Arch. Intern. Med., 141 : 1831.

SEARCH FOR THE MOST DISCRIMINATIVE FEATURES OF CORONARY HEART DISEASE USING

A BRANCH AND BOUND METHOD

A. Bartkowiak, S. Łukasik, K. Chwistecki and M. Mrukowicz

Institute of Computer Science, University of Wrocław, and
Cardiological Institute, Medical Academy.
Wrocław, Poland

THE MEDICAL PROBLEM

Data have been collected in the Wrocław Coronary Heart Disease (CHD)
Prevention Study, being part of the ERICA programme co-ordinated by the WHO
Collaborating Centre in Heidelberg. After 4 years of follow-up a sample
from the monitored cohort of men working in industrial plants of Wrocław
was taken. It comprised 2433 men, of whom 39 showed CHD symptoms. The
number of variables considered is p = 12: age, body weight, systolic and
diastolic blood pressure, measured casually and in standard conditions,
number of cigarettes smoked daily, cholesterol, high density lipoprotein
(HDL), triglyceride, uric acid and glucose. The goal is to find variables
which differentiate between men with and without CHD symptoms. The task is
performed using methods of discriminant analysis. Linear (in p variables)
and quadratic (in p + p variables) discriminant functions have been used.
A subset of variables is sought which best discriminates between the two
groups of men (with and without CHD symptoms). As the criterion of discriminative power the Mahalanobis distance is adopted.

THE BRANCH AND BOUND METHOD IN THE SEARCH FOR THE OPTIMAL SUBSET

Discrimination between two groups can be performed using regression
methods (see, for example, Lachenbruch, 1975). Also the Mahalanobis distance can be evaluated as the residual variance in a special case of
normal equations (see, for example, Bartkowiak, 1984). Therefore the
methods of search for an optimal subset in regression analysis can be transferred directly to the search for an optimal set in discriminant analysis.

In this case, the algorithm used is one elaborated recently by Bartkowiak (1987). It permits the finding of the subset yielding the minimal
residual sum of squares in a linear regression problem. The task is to
find the subset yielding the maximal Mahalanobis distance. It can be shown
(see, for example, Bartkowiak, 1984, p. 59) that the Mahalanobis distance
can be obtained as a residual sum of squares after reversing the sign of
this residual sum of squares. Therefore the problem of finding the subset
with the largest Mahalanobis distance is equivalent to the problem of
finding the subset with the smallest residual sum of squares in a properly
set linear regression problem. The method of building the regression equation is described, for instance, by Lachenbruch (1975), or Bartkowiak (1984).

Let us define an artificial predictor variable y

$$
y = \begin{cases} \dfrac{n_2}{n_1 + n_2} & \text{for individuals belonging to the first group of data comprising } n_1 \text{ individuals} \\[2em] -\dfrac{n_1}{n_1 + n_2} & \text{for individuals belonging to the second group of data comprising } n_2 \text{ individuals} \end{cases}
\tag{1}
$$

and consider the regression

$$
y = b_0 + b_1 x_1 + \ldots + b_p x_p + e
\tag{2}
$$

To find the optimal subset we proceed as follows:

First, a new order of the variables 1, 2, ..., p is introduced according to the magnitude $Q(-1)$, ..., $Q(-p)$ of their residual sums of squares $Q(-i)$, $i = 1, \ldots, p$, defined as the residual sum of squares obtained when introducing all, but the ith, variables into the regression set. Ordering the set of $Q(-i)$, the following sequence is obtained:

$$
Q(-1) \geq Q(-2) \geq \ldots \geq Q(-p)
\tag{3}
$$

The variables are then relabelled so that their $Q(-i)$s satisfy the inequalities (1).

Next, the search for the best subset is performed considering $k + 1$ "branches" of generated subsets of size k:

The $(k + 1)$th branch comprises only one set: 1, 2, ..., k
The kth branch comprises the integers 1, 2, ..., k - 1 and one of the integers $p - k + 1, \ldots, p$
The $(k - 1)$th branch comprises the integers 1, 2, ..., k - 2 and two of the integers $p - k + 2, \ldots, p$
... and so on ..., up to
The 1st branch which comprises all k-tuples that can be chosen from the integers 2, 3, ..., p.

It follows from the definition of the subsets comprising a branch that the number of subsets contained in the jth branch is equal to $\binom{p - j}{k - j + 1}$:

$$
N_j = (\text{number of subsets in the } j\text{th branch}) = \binom{p - j}{k - j + 1}
\tag{4}
$$

Seeking the optimal subset we proceed as follows:

(i) We start from the $(k + 1)$th branch and evaluate the residual sum of squares RSS for this subset. If RSS < $Q(-k)$, we have found the optimum subset and the search is terminated. Otherwise we retain RSS as RSS_0 and we proceed for $j = k, k - 1, \ldots, 1$ considering the subsequent "branches".

(ii) For the jth branch we ask whether the current RSS_0 is smaller than $Q(-j)$. If yes, then the subset yielding RSS_0 is the optimal subset and we stop the search. Otherwise, we generate sequentially the subsets belonging to this branch. Finding an RSS smaller than RSS_0 the current RSS is relabelled as RSS_0.

(iii) Having considered all subsets belonging to the jth branch, we diminish j by one. If $j > 0$, we start (ii), otherwise RSS_0 is the minimal value, and the subset which yielded RSS_0 is the optimal subset.

THE CONCEPT OF AN ε-OPTIMAL SUBSET

The residual sum of squares $Q(-i)$ defined by (1) is a lower bound for the RSSs evaluated for all subsets belonging to the ith branch and simultaneously for branches with a smaller index than i. If for a particular subset found in the $(i+1)$th branch the RSS is smaller than $Q(-i)$, then no subset belonging to the branches 1, 2, ..., i can give a smaller RSS.

Instead of considering the inequality

$$RSS < Q(-i) \tag{5}$$

we could verify another inequality

$$RSS < Q(-i) + \varepsilon \tag{6}$$

In practice we consider instead of (6) a modified inequality

$$RSS < Q(-i) + \varepsilon SS(y), \tag{6a}$$

where $SS(y)$ is the total adjusted sum of squares of the variable y introduced in (1):

$$SS(y) = \sum_{i=1}^{n} (y_i - \bar{y})^2, \quad n = n_1 + n_2 \tag{7}$$

A subset with RSS satisfying (6) or (6a) is said to be ε-optimal (its RSS differs from the RSS of the optimal subset not more than by ε).

RESULTS FOR THE CHD DATA

We consider first $p = 12$ and next $p = 24$ variables with the artificial regression functions (see (2))

$$y = b_0 + b_1 x_1 + \ldots + b_{12} x_{12} + e, \tag{8a}$$

$$y = b_0 + b_1 x_1 + \ldots + b_{12} x_{12} + b_{13} x_1^2 + \ldots + b_{24} x_{12}^2 + e \tag{8b}$$

The two regressions lead appropriately to linear and quadratic discriminant functions, respectively. We have been looking for optimal subsets of size $k = 3, 4, 5$. The total number of subsets to be evaluated when using the traditional all-subset search is given in Table 1. The branches and the N_js, the number of subsets in these branches, are given in Table 2 (the N_js were evaluated using (4)).

Table 1. The Number of Subsets to be Evaluated in the All-subset Search. (p - number of variables, k - size of the subset)

k	p = 12	p = 24
3	$\binom{12}{3} = 220$	$\binom{24}{3} = 2024$
4	$\binom{12}{4} = 495$	$\binom{24}{4} = 10626$
5	$\binom{12}{5} = 792$	$\binom{24}{5} = 42504$

It can be seen that the totals of the numbers of subsets in the branches numbers k + 1, k, k - 1, ..., 1 are equal to the numbers of subsets given in Table 2.

Table 2. Branches and the Numbers of Subsets in These Branches When Seeking Subsets of Size k = 3, 4, 5 Out of p = 12 and p = 24 Variables

Number of the branch j	Considering p = 12 Variables N_j - number of subsets		Considering p = 24 Variables N_j - number of subsets	

size of the subset: k = 3

Number of the branch j	Considering p = 12 Variables		Considering p = 24 Variables	
j = 4	$\binom{8}{0}$ =	1	$\binom{20}{0}$ =	1
j = 3	$\binom{9}{1}$ =	9	$\binom{21}{1}$ =	21
j = 2	$\binom{10}{2}$ =	45	$\binom{22}{2}$ =	231
j = 1	$\binom{11}{3}$ =	165	$\binom{23}{3}$ =	1771
Total		220		2024

size of the subset: k = 4

j = 5	$\binom{7}{0}$ =	1	$\binom{19}{0}$ =	1
j = 4	$\binom{8}{1}$ =	8	$\binom{20}{1}$ =	20
j = 3	$\binom{9}{2}$ =	36	$\binom{21}{2}$ =	210
j = 2	$\binom{10}{3}$ =	120	$\binom{22}{3}$ =	1540
j = 1	$\binom{11}{4}$ =	330	$\binom{23}{4}$ =	8855
Total		495		10626

size of the subset: k = 5

j = 6	$\binom{6}{0}$ =	1	$\binom{18}{0}$ =	1
j = 5	$\binom{7}{1}$ =	7	$\binom{19}{1}$ =	19
j = 4	$\binom{8}{2}$ =	28	$\binom{20}{2}$ =	190
j = 3	$\binom{9}{3}$ =	84	$\binom{21}{3}$ =	1330
j = 2	$\binom{10}{4}$ =	210	$\binom{22}{4}$ =	7315
j = 1	$\binom{11}{5}$ =	462	$\binom{23}{5}$ =	33649
Total		792		42504

Table 3. Results of the Search for an ε-Optimal and Optimal Subset Considering p = 12 Variables

k, Size of the Subset	Branches Considered	Subset Chosen (notation as above)	Fraction of Subsets considered
ε-optimal subset with ε = 0.02 according to (6a)			
3	j = 3	1, 11, 12	10:220 = 0.045454
4	j = 4	1, 8, 11, 12	9:495 = 0.018182
5	j = 5	1, 4, 6, 11, 12	8:792 = 0.010010
optimal subset			
3	j = 3	1, 11, 12	10:220 = 0.045454
4	j = 4 v j = 3	1, 8, 11, 12	45:495 = 0.090909
5	j = 5 v j = 4	1, 4, 6, 11, 12	36:792 = 0.045454

From Table 3 it can be seen that the ε-optimal sets and optimal sets comprise the same variables. Generally, the optimal algorithm worked for p = 12 variables very rapidly and the difference in time between the two algorithms is not very large.

When considering p = 24 variables the ε-optimal subset was found very speedily. When seeking the optimal subset we were not lucky and had to evaluate all subsets. The subset of size k = 3 is the same for both methods. The subsets of size k = 4 and k = 5 differ by one variable. Performing the calculations for p = 24 variables and seeking a good subset, the great advantage of searching for a ε-optimal subset can be seen: the gain in time here was tremendous! The details are given in Table 4.

Table 4. Results of the Search for an ε-Optimal and Optimal Subset Considering p = 24 Variables

k, Size of the Subset	Branches Considered	Subset Chosen (notation as above and (8b))	Fraction of Subsets considered
ε-optimal subset with ε = 0.02 according to (6a)			
3	j = 3	1, 12, 23	22: 2024 = 0.010870
4	j = 4	1, 4, 12, 23	21:10626 = 0.001976
5	j = 5	1, 4, 12, 23, 24	20:42504 = 0.000471
optimal subset			
3	j = 3 v j = 2 v j = 1	1, 12, 23	all = 1.0
4	j = 4 v j = 3 v j = 2 v j = 1	1, 12, 23, 24	all = 1.0
5	j = 5 v j = 4 v j = 3 v j = 2 v j = 1	1, 11, 12, 23, 24	all = 1.0

DISCUSSION OF THE RESULTS

The method described shows a great advantage over the traditional one. The application of the branch and bound method for discriminant analysis was possible only because the Mahalanobis distance, taken as the criterion of the discriminative power of variables under investigation, has the property of being a monotonic function of the number of variables in the discriminative set being considered: adding a new variable to this set, the Mahalanobis distance between the two groups of data can never be decreased, but it may possibly increase.

The results obtained for the detailed medical problem are, at first sight, a little surprising. The most discriminative features here are: age, uric acid and glucose. These (except perhaps for age) are not judged as the most important risk factors for Coronary Heart Disease (Multiple Risk Factor Intervention Trial Research Group, 1982). In particular, the variable "uric acid" is somehow questionable, although there are some reports on the importance of this variable when considering CHD (Persky et al., 1979). It is surprising that the feature "smoking" was not revealed by the search procedure. One possible explanation could be that the question, "How many cigarettes do you smoke per day?", was not precise enough and did not give much information on the smoking history of the interviewed individual. Another possibility is that the groups of data considered are truly not differentiated at all and that the results obtained are spurious.

It should be remembered that the people considered were still in the working age group and were employed in industrial plants. It can be concluded that the CHD symptoms stated in these people (men) were not very advanced because they were still able to carry out their professional activities. Therefore, the final (medical) conclusion that these people with stated CHD symptoms do not differ statistically from those with no CHD symptoms - at least with regard to the 12 parameters considered - is not surprising.

REFERENCES

Bartkowiak, A., 1984, "SABA - An Algol Package for Statistical Data Analysis on the ODRA 1305 Computer", Universitas Wratislaviensis, Wrocław.
Bartkowiak, A., 1987, Experience in computing optimal regression by branch and bound, Zastosowania Matematyki/Applicationes Mathematicae, 20(2): (in press).
Lachenbruch, P., 1975, "Discriminant Analysis", Hafner Press, Macmillan, London.
Multiple Risk Factor Intervention Trial Research Group, 1982, Multiple risk factor intervention trial. Risk factor changes and mortality results, J.A.M.A., 248:1485.
Persky, V. W., Dyer, A. R., Idris-Soren, E., Stamler, J., Shekelle, R. B., Schoenberger, J. A., Berksom, D. M., and Lindberg, H. A., 1979, Uric acid: a risk factor for coronary heart disease?, Circulation, 59:969.

APPLICATION OF EXPERT SYSTEMS AND SIMULATION PROCESSING IN THE PHYSIOLOGY

OF EXERCISE

R. Mader, J. Potucek and E. J. H. Kerckhoffs

Institute for Sport Medicine
Prague, Czechoslovakia; and
Technical University.
Delft, Holland

INTRODUCTION

Knowledge-based expert systems can be used in modelling and simulation both as advisory systems and linked to conventional simulation models for decision making and for intelligent control (Holmes, 1985; Kerckhoffs and Vansteenkiste, 1986; Kerckhoffs et al., 1986; Luker and Adelsberger, 1986; Luker and Birtwistle, 1987).

The task of knowledge-based advisory systems in, for instance, continuous systems simulation could be one of the following:

(i)   acting as a support for the user in the synthesis of differential models (general knowledge regarding the modelling process; choice and elaboration of the candidate model; selecting and lumping domain primitives; defining differential models from schematic representations such as flow diagrams used in compartmental analysis, symbolic notation of chemical reactions, block diagrams used in engineering, Bond graphs, and so on);

(ii)   providing knowledge of the known mathematical properties of a model;

(iii) formal manipulations, such as symbolic differentiation or symbolic solution of equations (to calculate stationary solutions, sensitivity functions, and so on);

(iv)   providing support for the user in the choice of algorithms, such as numerical integration algorithms, parameter estimation algorithms, validation procedures;

(v)   assisting the user in the interpretation of mathematical expressions (to overcome the drawback that equations are often semantically disconnected from the application domain and they are much less informative than, for instance, schematic representations, mathematical equations could be translated into such schematic representations); and

(vi)   providing an intelligent man/machine interface, also in experimenting on the models and analysing the simulation results.

Interfacing knowledge-based systems or other intelligent systems with simulation models will frequently occur. Qualitative models of dynamic systems may be coupled to numerical simulations to reach a behavioural representation without requiring complete quantitative specification of the model. Furthermore, linking knowledge-based systems to simulation models is needed in simulation with multiple decision making agents, each performing tasks for which AI is well-suited (such as planning, scheduling, hypothesis formulation, performed in an environment full of uncertainty and incomplete and distorted information). For such systems those sub-processes which are clearly heuristic and symbolic (for example, decision making) could be modelled by expert systems, while the sub-processes which are, for example, physical processes are conventionally modelled as dynamic systems. The requirement for knowledge-based expert systems in such simulations is the ability to operate in a rapidly varying environment and to do so speedily. These requirements could be met by the use of parallel techniques (Uhr, 1987; Wise, 1987).

PHYSIOLOGY OF EXERCISE AS THE APPLICATION AREA

In the following we shall deal with the application of expert systems linked with simulation models for intelligent control and diagnostics (evaluation of results) of spiro-ergometric stress tests.

Spiro-ergometry now represents a method, which is recognised worldwide, suitable for non-invasive investigation and examination and for objective assessment of the performance of the cardio-respiratory system. It makes possible an assessment of the adaptability of the organism to physical load and an objective evaluation of sports training and/or physical rehabilitation. It is an experimental method. Its accuracy depends on the quality of the instruments, on standardised examination and on the methods used to evaluate the results (Horak, 1984). One of the main tasks, nowadays, is to establish the quality control of ergometric measurements (Mellerowicz, 1984).

A standardised stress test is a commonly used ergometric measurement of performance capability (fitness) and respiratory insufficiency. During the test, which is usually run on a treadmill or a bicycle ergometer, the response of the physiological parameters is observed during exercise when the load is being changed in an appropriate manner. Some parameters can be acquired on-line during the test (parameters of the ECG signal, concentration and volume of expired oxygen and carbon dioxide, total volume of expired gases, temperature and blood pressure). Data from the biochemical analysis of samples taken during the test are usually acquired off-line. From the analysis of the physiological response during exercise the performance capabilities can be ascertained, for example, a physiological diagnosis. Possible deficiencies detected can lead, after special diagnostics, to a pathophysiological diagnosis. Therefore, stress tests are used both in clinical practice, including screening, and in sports medicine for testing of performance capabilities of athletes (Lollgen, 1983; Lollgen and Mellerowicz, 1984).

INTRODUCTORY ANALYSIS

Let us consider the diagnostic system as a whole, including the technical equipment as well as the medical personnel and the subject (athlete) under test. When introducing a computer-based system here, the aim is to improve the functioning of the whole system. The criteria used include:

- safety of the test;
- objectivity of measurement;
- accuracy; and
- the possibility of obtaining most of the results on-line, including the suggested diagnosis.

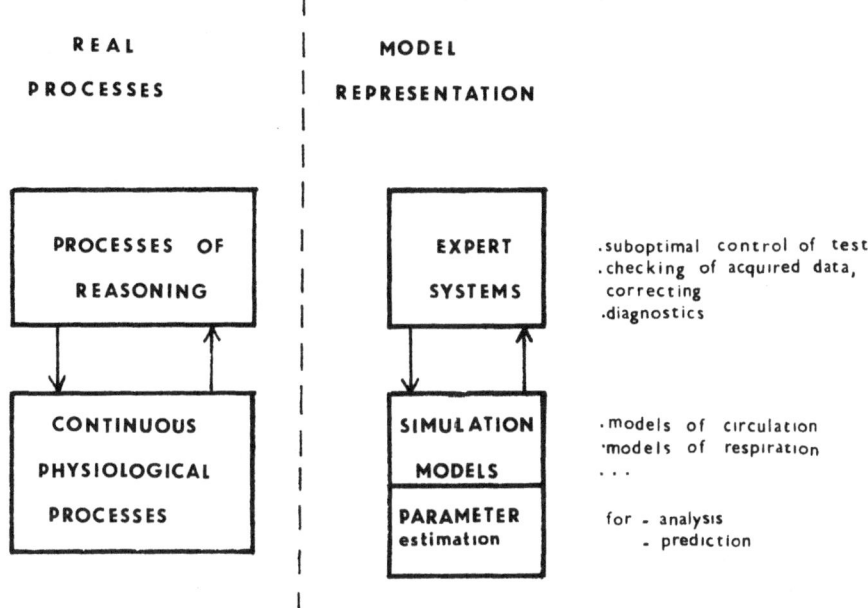

REAL   MODEL

PROCESSES   REPRESENTATION

PROCESSES OF   EXPERT   .suboptimal control of test
REASONING   SYSTEMS   .checking of acquired data,
                                         correcting
                                        .diagnostics

CONTINUOUS   SIMULATION   .models of circulation
PHYSIOLOGICAL   MODELS   .models of respiration
PROCESSES   PARAMETER   . . .
                       estimation   for - analysis
                                            - prediction

Fig. 1. Hybrid structure of the diagnostic system showing the conti-
         nuous physiological processes working in parallel with heuris-
         tic processes of reasoning.

We take into account the role and the relations among the main components
of the diagnostic system: that is the patient, physician and technical
equipment.

     From our experience, the limiting factor for the functioning of the
system appears to lie in the relations between the physician and the technical equipment. The more complicated the technical procedures that we use,
the greater is the risk that the physician will not use them properly and/or
will not believe them and, finally, that he would not use them at all if
they were not necessary. Thus the physician should participate in the
design of the system. The system should be interactive, with possibilities
of explanation of the suggested decisions and providing a suitable form of
communicating the information, for example, in graphical form as far as
possible, adopting common standards.

     The diagnostic system is characterised by its hybrid structure: con-
tinuous physiological processes are working in parallel with heuristic processes of reasoning (see Fig. 1). There exist strong interactions among
the processes. An example can be the problem of control of the test. The
respiratory and circulatory systems are working continuously, both influenced
by changes of load during exercise as a result of decision making by the
physician, these being based on the test schedule and the feedback of his
experience as to how to maintain test safety and achieve the required goals
(reference points). Let us simplify the situation and represent this system
using an expert system (ES) to model the physician's reasoning and a set of
simple models of the chosen physiological processes such as models of heart
rate regulation (Hajek et al., 1980; Potucek et al., 1980) and a model of
respiration. Each of the models can be called by the expert system to run
at any sampling frequency when needed for decision-making about the next
changes of load, circulatory warnings, and so on.

We prefer the use of a set of simple models, each of them being valid over a limited range of respiratory state variable behaviour, instead of a complex model in this application. There should exist an optimal relation between the robustness of the simulation models and the robustness of the managing expert system. This problem seems to be more sensitive in the case of time-critical tasks solved by parallel running of the control system on a multiprocessor system of MIMD structure. We have the possibility of using the Delft Parallel Processor (developed at the Technical University of Delft, Holland) for this purpose.

The use of an expert system linked to simulation models in the frame of the stress test control system can offer new possibilities for the conducting of the tests; for example, the task of following a required heart rate slope by intelligent control of the load. This can be useful in the case when a limitation in the function of the circulatory system must be respected, or where the observed physiological response of subjects should be "normed" by their heart rate response.

The following task for the diagnostic system is the representation of the final reasoning process (after the end of the run of the exercise) leading to the formulation of the final diagnosis. This process can be modelled by a rule-based expert system. We suggest running this system using a reduced knowledge base (excluding rules working with information acquired off-line) to have the possibility of obtaining the first suggestion of the diagnosis on-line. It must be stressed that the physician is responsible for the final diagnosis which is not valid without his agreement.

A necessary part of the system is the database, containing source data for the expert systems (values of physiological parameters and data on anamnesis) and the output information of the system (diagnoses). The systematic storage of tests results seems to be necessary for diagnosis, because we are interested in the longitudinal dynamics of the adaptability to stress and intra-individual comparison with earlier test results can provide useful information. The database can also be used for research purposes and can be useful in the context of implementation of the knowledge base of an expert system.

STRUCTURE OF THE DESIGNED DIAGNOSTIC SYSTEM

The structure of the system is shown in Fig. 2 and it can be regarded as a multi-paradigm computer-based system. It consists of five main blocks which are described below:

1. Data Acquisition and Monitoring System

   Spiro-ergometric data and information about the ECG are acquired on-line. They are tested logically to see whether they lie within the appropriate range. Some artefacts are automatically corrected. Warnings concerning technical failure or danger are included. The graphical output provides for the representation of the slope of eight chosen physiological parameters in parallel.

2. Intelligent Control System

   This includes an expert system ESO linked to a set of simulation models (Fig. 3) for the control of the load during the test. Adaptive simulation models of heart rate regulation and respiration are incorporated. The models work with data acquired on-line. Their parameters are currently estimated during the test. The models are used for prediction of the behaviour of the subject under test. Some of the estimated parameters are finally used for classification of the observed response.

The operation of the simple models is performed by a simple rule-based expert system.

3. Computation of Derived Parameters of Physiological Response

This block provides a source information for the expert system.

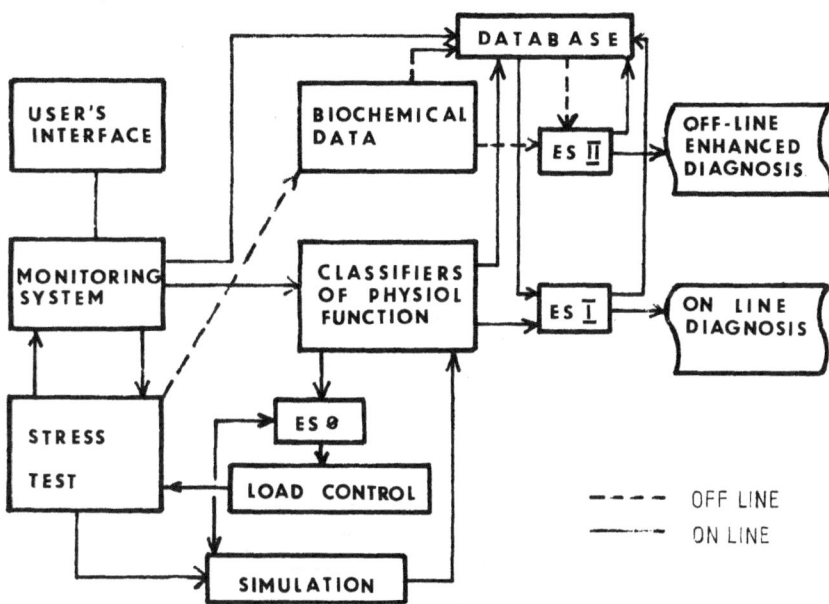

Fig. 2. Structure of the designed diagnostic system.

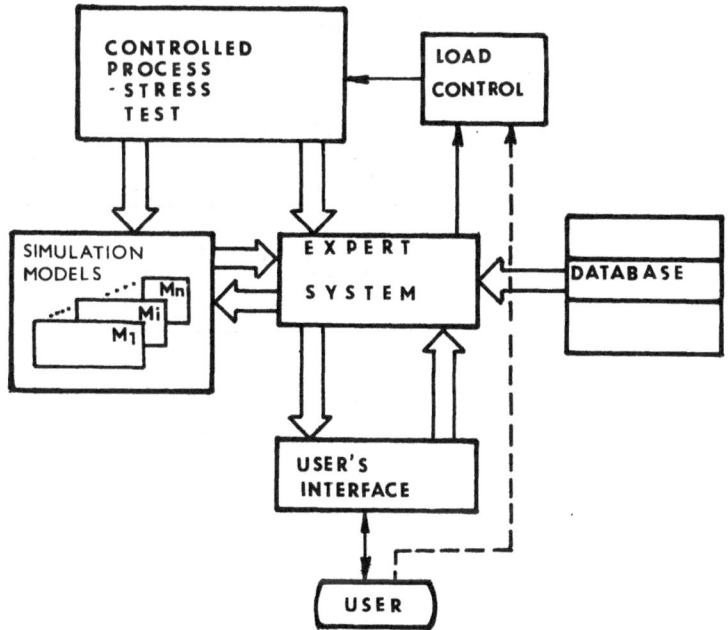

Fig. 3. Intelligent control system, including an expert system ESO, linked to a set of simulation models.

4. Expert Systems ES I and ES II

These are rule-based expert systems used for on-line and off-line derivation of the diagnosis. Their input information consists of:
- parameters of the physiological response;
- information about anamnesis, training and feeding; and
- information about the effect of environmental factors.
The ES II also works with the biochemical data. All the data for the
expert systems are taken from database files.

5. The Database

This is linked with all the other blocks and contains data for the
expert systems and the output information of the systems (diagnoses).
Use of the data base for research purposes can also lead to the actuali-
sation of knowledge bases for the expert systems (Fig. 4).

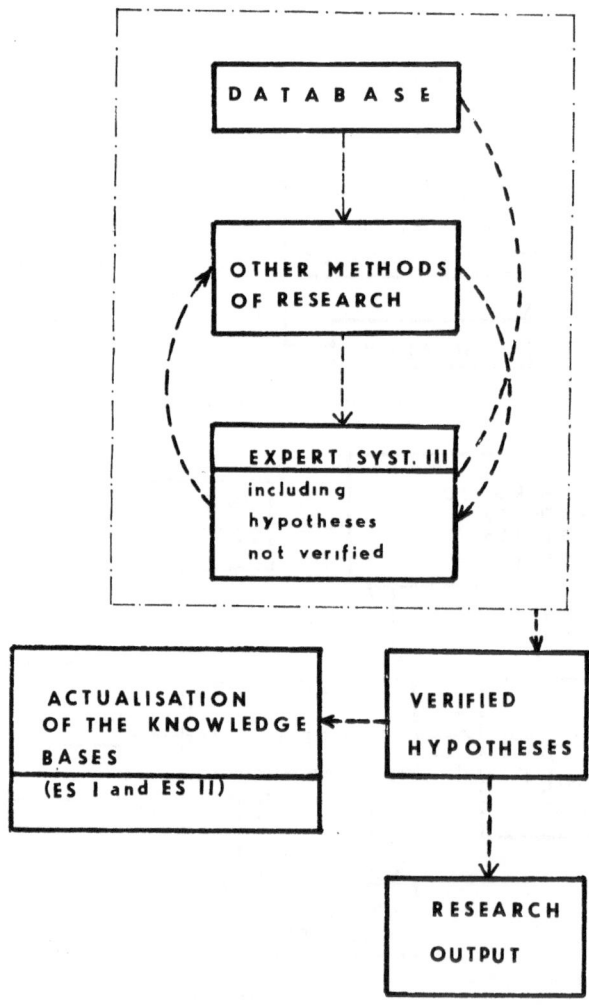

Fig. 4.   Linkage of the database to all other
blocks containing data for the expert
system and output information of the
system.

TECHNICAL EQUIPMENT USED

The system is being implemented in a computer network consisting of an IBM PC in the physiological laboratory and a DEC 11/23 as the host computer of the Delft parallel processor (DPP 84, MIMD/Multiple Instruction Multiple Datastream/structure). The tasks are distributed as follows:

time-critical tasks (simulation)	DPP	(macroassembler, FORTRAN
operational database	IBM PC AT	(TURBOPASCAL)
monitoring and control	IBM PC AT	(TURBOPASCAL)
central database	DEC 11/23	
diagnostic ES	IBM PC AT	(TURBOPASCAL)

The equipment used for measurement is:

OXYCON4 : spiro-ergometry (30 s sampling interval)
Hellige : ECG monitoring
Electronically controlled ergometers.

DISCUSSION

Standardisation of the stress tests is necessary to enable application of the system which has been described. Nowadays, stress tests are generally standardised with a given time-course of load. The heterogeneity in the capabilities of the subjects tested causes dispersion of their physiological response. This fact can cause serious loss of objectivity in the subsequent analysis of the test results. Application of intelligent control of the test facilitates standardisation of the behaviour of, for example, the circulatory and/or respiratory system of the subject tested. For instance, a test can be standardised in terms of the required heart rate response. A 30 s sampling interval, as implemented using the OXYCON 4, will not be sufficient in the case of short, highly dynamic tests. Breath-by-breath measurement is applied in such cases. Then the intelligent control system will have to be speeded up according to the variable sampling interval, which can be as short as 300 ms. We believe that this task can be solved only by the implementation of decision-making procedures, together with simple simulation models implemented on the parallel processor.

The system is being developed in stages and its validity is currently being examined in daily practice.

CONCLUSION

The approach suggested in this chapter can lead to the following benefits in relation to the analysis of the physiology of exercise:

- higher validity of acquired data (parameters) reached by the quality control of measurement;
- higher objectivity in the conclusions reached;
- higher efficiency of the physician's work in the laboratory (printout of the suggested diagnosis); and
- systematic storage and easy accessibility of constrained information about tests for longitudinal analyses and for research purposes.

REFERENCES

Hajek, M., Potucek, J., and Brodan, V., 1980, The mathematical model of heart rate regulation during physical exercise, Automatica, 16 : 191.

Holmes, W. M., ed., 1985, "Artificial Intelligence and Simulation", Simulation Councils, Inc. (Society for Computer Simulation), San Diego.

Horak, J., 1984, Introductory remarks, in: "Progress in Ergometry: Quality Control and Test Criteria", H. Lollgen and H. Mellerowicz, eds., Springer Verlag, Berlin : 6.

Kerckhoffs, E. J. H., and Vansteenkiste, G. C., 1986, The impact of advanced information processing on simulation. An illustrative review, Simulation, 46(1) : 17.

Kerckhoffs, E. J. H., Vansteenkiste, G. C., and Zeigler, B. P., eds., 1986, "AI Applied to Simulation. Simulation Series, Vol. 18, No. 1", Simulation Councils, Inc. (Society for Computer Simulation), San Diego.

Lollgen, H., 1983, "Kardiopulmonale Funktiondiagnostik", CIBA-GEIGY GmbH, Wehr/Baden.

Lollgen, H., and Mellerowicz, H., eds., 1984, "Progress in Ergometry: Quality Control and Test Criteria", Springer-Verlag, Berlin.

Luker, P. A., and Adelsberger, H. H., eds., 1986, "Intelligent Simulation Environments. Simulation Series, Vol. 17, No. 1", Simulation Councils, Inc. (Society for Computer Simulation), San Diego.

Luker, P. A., and Birtwistle, G., eds., 1987, "Simulation and AI. Simulation Series, Vol. 18, No. 3", Simulation Councils, Inc. (Society for Computer Simulation), San Diego.

Mellerowicz, H., 1984, Preliminary remarks on the present state and future tasks of ergometry, in: "Progress in Ergometry: Quality Control and Test Criteria", H. Lollgen and H. Mellerowicz, eds., Springer Verlag, Berlin : 8.

Potucek, J., Brodan, V., and Hajek, M., 1980, Hybrid computer analysis of the heart rate regulation during physical load, in: "Simulation of Systems '79", L. Dekker, ed., North-Holland, Amsterdam : 685.

Uhr, L., 1987, "Multi-computer Architectures for Artificial Intelligence - Towards Fast, Robust, Parallel Systems", Wiley, New York.

Wise, M. J., 1987, "Prolog Multiprocessors", Prentice-Hall, Englewood Cliffs, NJ.

A PROGRAM SYSTEM FOR COMPUTER-AIDED DRUG DOSAGE

E. Světlovská, A. Krechňáková and M. Oravcová

Medical Bionics Research Institute
Bratislava
Czechoslovakia

INTRODUCTION

In most diseases there is the requirement that the effective drug concentration in its place of efficacy should be maintained over a period of
time. This is achieved by repeated drug administration to offset its elimination from the organism. The doses and the time intervals between their
administration are determined by the concentration required, as well as by
the pharmacokinetic parameters of resorption, distribution, metabolism and
elimination.

The so-called standardised drug dosage assigned to common clinical
practice by clinical pharmacological research assumes that the pharmacokinetic processes follow their physiological course. This dosage, however,
cannot be used in the sick patient when pathological changes influence some
of the pharmacokinetic parameters of the administered drug (Dzúrik and
Dzúriková, 1979). The changes which are of greatest significance in disease
are those of drug elimination. In fact, the drug elimination mechanism and
its dosage have the greatest influence on the drug concentration in the
body. From the viewpoint of elimination, the liver and kidneys play the
decisive role. The liver eliminates the drug into the gall-bladder by means
of metabolism and secretion. The kidneys effect elimination more significantly, producing both the drug and its metabolites. During renal impairment, the production of renally-eliminated drugs is reduced, their biological half-life is prolonged and their elimination rate constant is lowered
(Dettli, 1977a).

If in such states a standard drug dosage is adopted, then the drug is
accumulated in the body and its concentration in plasma rises, thus risking
damage to the patient. For the majority of drugs, the kidneys are the only
major organs of elimination. Therefore, in cases of patients suffering from
renal impairment, adjustment of the common dosage regimen is usually required. Our program system called "Dosages", which is used in a physician-
computer interactive mode, is capable of determining the reduced drug
dosage schemes for patients with renal impairment.

METHODS

The underlying assumption of our computational scheme is that the
total amount of drug eliminated by the organism in a period of time is the

387

sum of the elimination by extra-renal processes and by the kidneys. Further, it is also supposed that lowering the total elimination rate in patients with renal impairment is caused by lowering the rate of renal elimination, while the rate of extra-renal elimination remains unchanged. The starting point for the solution was the basic postulate that the dosing scheme must be modified so as the resulting steady-state drug concentration in patients with renal impairment was the same and was achieved in a similar period of time as in patients without renal impairment (Dettli, 1977a,b).

The method for solving the task given is based on the following:

(i)   the pharmacokinetic parameters of drug are determined by compartmental analysis using one compartment only,

(ii) the rate of renal elimination of a drug is related to the creatinine clearance which correlates well with changes of biological half-life of many drugs, and

(iii) the patient is in a state of dynamic balance.

There are no theoretical and computational objections to using a multicompartment model in our system. There are, however, practical and ethical limitations whereby the variables of the patient which can be measured are severely restricted resulting in it not being possible to make unique estimates of all the model parameters. We are aware of the restrictions imposed by the one-compartment model; however, the primary aim of our system development was its use for practical needs in real clinical settings. The system is being gradually extended to cover all relevant drugs registered in Czechoslovakia and to be used in a wide range of our health-care facilities.

The model parameters have been determined using data in the literature based on population studies set up with a one-compartment model (pharmacokinetic parameters of drugs). Having carried out a review of the accessible literature we have not, as yet, encountered any other approaches to drug dosage reduction with regard to renal impairment based on multi-compartmental analysis without taking multiple blood samples.

The computation of a drug dosage reduction scheme requires the establishment of a database covering drugs allowing flexible selection, maintenance, and modification of required drug pharmacokinetic parameters and related data with regard to treatment modes for each drug.

RESULTS

The system "Dosages" has been developed for the SMEP computer family in versions for OS FOBOS, DOS RV, and DIAMS II. The programming languages used are FORTRAN IV (OS FOBOS and OS DOS RV) and MUMPS Standard (OS DIAMS), respectively.

The system can be logically divided into two main components: one serving the needs of the database administrator and the other for the end-user. The database administrator has at his disposal programs for setting up the database, for its maintenance and for its implementation.

The database is arranged in the form of independent data files. These are organised as non-formatted direct access files (OS FOBOS and DOS RV) or as B-trees (OS DIAMS). The basic file is a catalogue of drug groups. Each group is composed from drugs each having an identical sequence of the first three characters of its generic name. A pointer figured out from the first three characters enables the subsequent directory of drugs belonging to a group to be addressed. At the same time, the pointer serves as an access key to the file of drug parameters.

Programs are available which facilitate:

  (i)   incorporation of a new drug into the database;
 (ii)   removal of an outdated drug from the database;
(iii)   modification of parameters of already registered drugs;   and
 (iv)   production of well-arranged extracts of drugs and their parameters
        from the database.

These programs are dedicated to the database administrator (clinical pharmacist). Programs making extracts from the actual list of drugs are available also for the end-user, the consulting physician.

The end-user has at his disposal programs for dosing schemes. These programs are designed to enable the consulting physician to obtain the calculation of the reduced dosage scheme immediately. Such a scheme for the given drug and patient is inferred from data embodied in the database, so that there is no need for the physician to have knowledge of the drug pharmacokinetic parameters. The system performance is based on the formulae for calculation of the initial dose, the reduced maintenance dose, the prolonged interval between consecutive drug administrations and the drug biological half-life (Dettli, 1977a).

The role of the initial dose is to make sure that the drug therapeutic concentration in plasma is reached straightaway in those cases where a gradual concentration increase up to the steady-state is unsuitable (for example, in the case of antibiotics). The initial dose is obligatory also for patients without renal impairment, just as it is a prerequisite for the rational treatment by antibiotics. The importance of the initial dose is increased by severity of renal impairment as the decrease in drug elimination prolongs the time needed for reaching the steady-state.

The programs adjust the dosing scheme as follows:

  (i)   by reduction of the standardised maintenance dose, the drug administration interval is preserved;
 (ii)   by prolonging the administration interval, the standard maintenance dose is preserved;   and
(iii)   when the drug administration interval exceeds 24 hours, which is not suitable, for example in the case of antibiotics, then both of these modes are combined into one scheme;   this modification is particularly dependent on the parameters assigned to each drug.

The creatinine clearance is the measure of renal function and its value is used in computations. The constant for normal renal function is 1.67 ml s^{-1}. In the so-called dynamic balance states, with balanced and little altered renal function, the system allows the dependence of creatinine concentration in the patient's serum with his own creatinine clearance to be used in order to judge the function of the kidneys directly (Cockcroft's relation) (Cockcroft and Gault, 1976). It is not necessary in this case to determine the creatinine clearance in the laboratory, so speedy introduction of treatment is thus possible.

Basic information about a drug and supplementary text giving treatment information are provided. This is illustrated in a scheme for Gentamycin, as shown in Fig. 1.

In its operation the system makes use of the structure of drug parameters created for each medication. Relations between the given parameters and their values determine, for a particular drug, the corresponding functional scheme of individual programs in the system. A flexible computational structure is being set up in this manner, whilst the actual scheme

Multiple-drug dosage in a patient with renal impairment

Last name: Smith
First name: John
Date of birth: 12th May, 1934

Gentamycin - Gentamycin Inj I.M., I.V.
Ampoules: 40 mg/1 ml,
80 mg/2 ml

Creatinine clearance: 1.10 ml/s
Body weight: 100 kg

Standardised dosage:

Dose:	170 mg	Interval:	8 hours

Reduced dosage scheme:

Initial dose:	245 mg		
dose:	116 mg	Interval:	8 hours
dose:	170 mg	Interval:	11.5 hours

The dosage is to be correlated with the clinical state of the patient.

Date: 20th July, 1987

Fig. 1. Illustration of a print-out of results for a particular patient. The upper part contains the patient identification data, followed by the generic name of the selected drug to which its commercial name, form of application and administration mode, as well as the amount of effective substance, are attached. Then the required patient parameters are listed, followed by the standardised and recommended reduced dosage scheme. In the reduced scheme there are listed the initial dose and two types of alternative maintenance administration: a reduced dose with the standard administration interval and the standard dose with prolonged administration interval. The selection is made by the physician. Finally, advice is included drawing the physician's attention to various circumstances.

which results (recommended reduced drug dosage scheme) depends only on determining the parameters of a particular drug.

REFERENCES

Cockcroft, D. W., and Gault, M. H., 1976, Prediction of creatinine clearance from serum creatinine, Nephron, 16 : 31.

Dettli, L., 1977a, Elimination kinetics and dosage adjustment of drugs in patients with kidney disease, in: "Progress in Pharmacology", Stuttgart, vol. 1, no. 4 : 1.

Dettli, L., 1977b, Individualization of drug dosage in patients with renal disease, Med. Clin. N. America, 58 : 977.

Dzúrik, R., and Dzúriková, V., 1979, Drug dosage in patients with renal impairment, in: "Materia Pharmaceutica 3", 1st edn., M. Salava and O. Horáková, eds., Martin, Osveta : 70.

GRAPHSEARCH A* ALGORITHM IN INSULIN THERAPY

R. Hovorka, Š. Svačina and M. Malík

Third Medical Department
Charles University
Prague, Czechoslovakia

INTRODUCTION

Dietary therapy or the use of oral hypoglycaemic agents provides sufficient treatment for the majority of diabetic patients. Those patients
with a deeper disorder of glycoregulation, however, require insulin therapy.
Usually, insulin is injected subcutaneously. Nowadays, various types of
insulin preparation are available, differing in their duration of action.
Originally only insulin in solution was used. Its clinical effect starts
very shortly after the injection, but it lasts only for about 6 hours.
This original short-acting preparation is now termed regular insulin.
Later, the so-called depot insulin preparations were developed. The onset
of their effects is delayed after the injection, but the effect is longer
lasting. Amorphous insulin acts for about twice as long, and crystallised
insulin for about 3 to 4 times as long as the regular preparation. The
quantity of insulin needed varies considerably from patient to patient,
and the individual insulin need also varies during the day. For example,
it rises at mealtimes and decreases during physical activity. At the same
time, defining the exact dosage of insulin and its rate of delivery during
the day are the most important aspects of insulin therapy. To achieve such
an optimum, preparations with different durations of action are combined
within a single insulin injection. This makes it possible to reduce the
number of injections. Originally, regular insulin had to be injected about
four times a day. Currently, two injections, or sometimes one, are sufficient in most cases.

The establishment of proper insulin therapy, that is the quantitative
dosage and its daily profile, is obviously rather complicated. Moreover,
development of the disease requires that the therapy should be updated
according to the blood glucose level of the patient. Some algorithms for
establishing insulin therapy have been reported (Skyler et al., 1979; 1982)
and different mathematical models of the insulin – blood glucose relationship have been developed (see, for instance, Bergman et al., 1979, and
Carson et al., 1983).

Some years ago, a new method of insulin therapy was introduced. A
small special technical device, called the insulin pump, is fastened to
the patient's body. It outputs insulin continuously into the subcutaneous
tissue. During the whole day the pump delivers insulin at a constant basal
level and it also injects additional doses (termed insulin boluses) at

either preprogrammed or manually selected times. In our patients, for
instance, we use 6 boluses during the whole day, one before each meal.
The insulin pump enables quick and safe changes of insulin therapy to be
made. For many patients the permanent pump treatment is not acceptable
for social, psychological or other reasons. Therefore, the optimum insulin
dosage and its proportional delivery by the pump have to be converted into
the optimum combination of injections. To help physicians to perform such
a regimen conversion, and in order to standardise it, a computer system
assisting the regimen transformation has been developed. The aim of this
chapter is to describe the theoretical and computational background of the
system.

SYSTEM DESCRIPTION

The system developed consists of a mathematical model of insulin pharmacodynamics, and of a control algorithm directing the performance of simulation experiments using the mathematical model. The model is simple but
it is able to simulate the pharmacodynamics of any combination of insulin
preparations in a sufficiently realistic way. Therefore, the simulation
experiments enable various injections to be compared, and the pharmacodynamic effect of any injection combination to be matched with the desirable
effect which has been established by the pump treatment (we shall return
to this point when we consider the model description). To direct and
evaluate the simulation experiments, we postulated that the patient has to
be treated by that injection combination, the effect of which differs mini-
mally from the optimum pump effect. However, the search for such a combination is a complex combinatorial problem. Therefore, a heuristic algorithm
controls and schedules simulation experiments using the model.

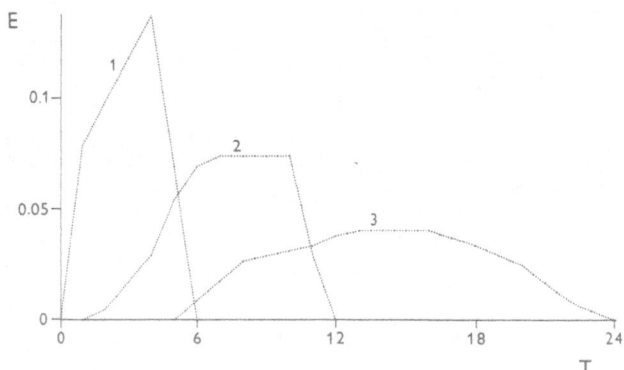

Fig. 1.   Pharmacodynamic effects of the three
          insulin preparations considered.
          Each curve corresponds to one preparation
          and describes the time evolution of its
          effect. All curves consider an injection
          containing 1 IU of the given preparation,
          being delivered at the moment O.
          1 - regular preparation PN (Pur-Insulin-
          Neutral SPOFA), 2 - amorphous preparation
          PD (Pur-Insulin-Dep SPOFA), 3 - crystallic
          preparation PS (Pur-Insulin-Superdep
          SPOFA), E - pharmacodynamic effect in
          symbolic units, T - elapsed time from the
          insulin application in hours.

Mathematical Model of Insulin Pharmacodynamics

A simple model of the effect of insulin on blood glucose level was proved to be sufficient for our purposes. The model is based on the pharmacodynamic characteristics of three Czech insulin preparations: Pur-Insulin-Neutral SPOFA (PN - regular insulin), Pur-Insulin-Dep SPOFA (PD - preparation with a prolonged action) and Pur-Insulin-Superdep SPOFA (PS - preparation with a very prolonged effect). Each preparation considered is described in the model by its pharmacodynamic curve (Fig. 1). Each of these curves expresses the pharmacodynamic effect of 1 international insulin unit (IU) as a function of the time elapsed since the insulin application. The effect is expressed in symbolic units and the model supposes the total effect of 1 IU of each insulin preparation is the same. This means that the areas under all three curves are the same. The forms of the curves which we use are estimated from published data (Smolek, 1982). Further,

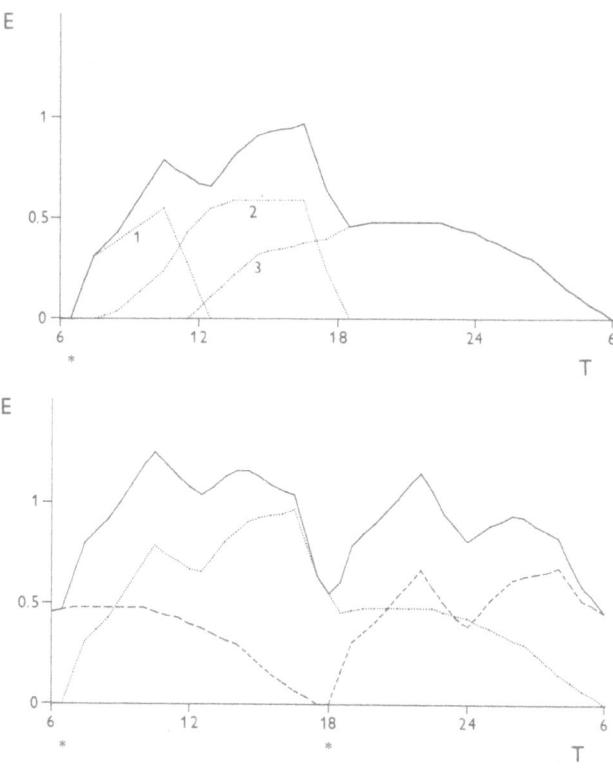

Fig. 2.   Sample of the pharmacodynamic effect of an injection combination.

The dotted line describes the morning injection, the dashed line the evening injection, and the full line corresponds to the total effect of both injections.

E - pharmacodynamic effect in symbolic units, T - time of day, the asterisks indicate the moments of injection delivery.

In this particular case, the morning injection contains 4 IU of PN, 8 IU of PD and 12 IU of PS, while the evening injection consists of 4 IU of PN, 4 IU of PD and 12 IU of PS.

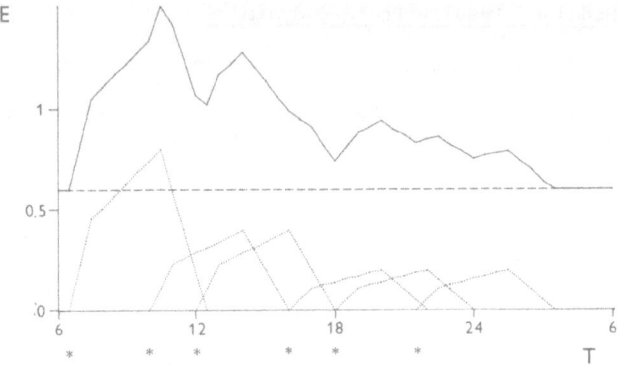

Fig. 3.   Sample of the pharmacodynamic effect of a
          pump regimen.
          The dashed line corresponds to the constant
          basal level of the pump. The six dotted
          lines show the effects of pump boluses (trig-
          gered at 6.30 a.m., 10.00 a.m., 12 noon,
          4.00 p.m., 6.00 p.m. and 9.30 p.m.). The
          full line, which is the sum of the basal
          level and the effects of the boluses, des-
          cribes the total effect of the pump therapy.
          E - pharmacodynamic effect in symbolic units,
          t - time of day, the asterisks indicate the
          times of the application of the boluses.
          In this particular case, the basal level cor-
          responds to the constant delivery of 1.19 IU
          of PN per hour, while the boluses contain
          5.81, 2.91, 2.91, 1.45, 1.45 and 1.45 IU of
          PN preparation, respectively.

the pharmacodynamic effect is assumed to depend linearly on the quantity
of insulin. The model also assumes different insulin preparations to be
mutually independent. Therefore, the total pharmacodynamic effect at a
given time is considered to be a simple sum of the pharmacodynamic effects
of all injected preparations at that time. Thus, to construct the total
pharmacodynamic curve, a partial pharmacodynamic curve is created for each
preparation applied during the day, and its beginning is placed on the time
axis according to the delivery time. Then the total pharmacodynamic curve
is the sum of these partial curves (Fig. 2). Since some preparations may
also act the next day, those parts of the partial curves, which lie behind
the current day interval, are inserted at the beginning of the total day
curve. This fully corresponds to the continuous process of insulin treatment.

     The optimum pump therapy established during the first stage of the
treatment enables the optimum pharmacodynamic curve to be computed. Since
the effect of the pump basal level does not vary during the day, its partial
curve is a straight line parallel with the time axis. Insulin boluses are
evaluated as micro-injections of the same insulin quantity. The total
pharmacodynamic curve is the sum of the basal and bolus partial curves
(Fig. 3).

     For practical computational purposes, the time axis is divided into
separate intervals of 30 minutes, and the pharmacodynamic effects considered
are assumed to be constant within each of these intervals.

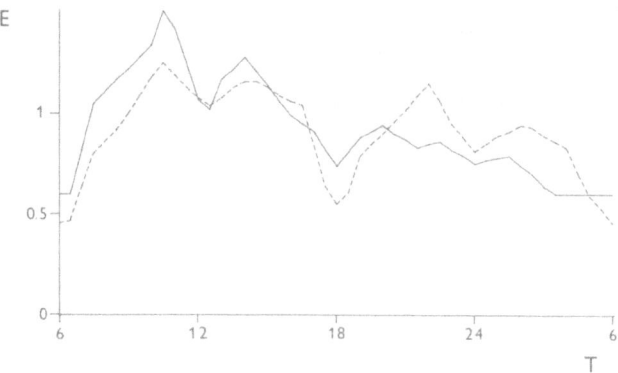

Fig. 4. Approximation of the established pump
optimum effect (full line) by the effect
of a combined injection (dashed line).
Both curves correspond to the cases which
are shown in detail in Figs. 2 and 3.
E - pharmacodynamic effect in symbolic
units, T - time of day.

The difference in any injection combination from the established pump
regimen may be expressed as the deviation of its curve from the pump curve
(Fig. 4). The best injection therapy exhibits the smallest curve deviation.
The difference criterion uses the minimal sum of squares of differences
over all 30-minute intervals.

The simple mathematical model described can be used to compute and
compare the pharmacodynamic curves. Therefore, the main problem is to
find the optimum injection combination in an effective manner.

Heuristic Searching

Generally, morning and evening injections consist of PS, PD and PN preparations. For practical purposes, the quantity of one preparation in any
injection is limited to 40 IU. Obviously, the minimum dose of each preparation is 0 IU. The visual check on insulin injections usually allows 0.1 ml
quantities to be distinguished, corresponding to 4 IU in standard concentration. Thus, 11 possible quantities corresponding to 0, 4, 8, ..., 36,
40 IU have to be considered for each of the 3 available preparations in
both morning and evening injections.

The 2 injections represent a combination of 6 dose preparations.
Since there are 11 possibilities for each of these doses, the total number
of conceivable combinations is equal to 11^6 (= 1,771,561). Such a large
number of possibilities makes it impracticable to test each of them. Considering, for instance, 0.1 second for evaluation of one proposal (this
corresponds roughly to the CPU time of our computer MYCRON 2200, Techonor,
Norway), the total time is more than 2 days. For this reason a heuristic
searching algorithm is used to reduce the number of proposals evaluated;
the resulting demand on CPU time has been decreased to several minutes.

We have already mentioned that the selection criterion "crit" of an
injection combination has the form of the sum of squares of differences
between its pharmacodynamic curve and the optimum pump curve. Both curves
are approximated discretely into 30-minute intervals (48 values on the one-
day scale). Therefore, the criterion can be expressed by the formula:

$$crit = \sum_{i=1}^{48} \left(T_c(i) - T_p(i)\right)^2 \tag{1}$$

where $T_c(i)$ is the value of total pharmacodynamic effect of the injection therapy considered, and $T_p(i)$ is the value of the pharmacodynamic effect of the pump optimum at time i.

All of the total of 11^6 possibilities of injection combinations can be ordered in a hypothetical tree with 7 levels of nodes. Each node on the internal levels of the tree has eleven sons; the leaves of the tree on its last (6th) level represent the conceivable injection combinations.

Our heuristic searching algorithm is based on cutting out some branches of the tree without evaluating their internal nodes. Before giving details on this algorithm, we define the tree in an exact way.

Each node of the tree is associated with 6 integer numbers $<q_1, \ldots, q_6>$ representing one injection combination of the insulin preparation quantities:

q_1 - morning PS dose q_2 - evening PS dose
q_3 - morning PD dose q_4 - evening PD dose
q_5 - morning PN dose q_6 - evening PN dose

For a node $<q_1, \ldots, q_6>$ at the ith tree level ($0 \leqslant i \leqslant 6$), its preparation quantities are zero for the preparations having an index higher than the level of the node, that is $q_j = 0$ for $i < j \leqslant 6$. This means that the root of the tree corresponds to a "dummy" injection: $<0,0,0,0,0,0>$.

Each node at an internal tree level $i = 0, \ldots, 5$ has 11 sons, which have the same preparation quantities except for the q_{i+1} dose. This quantity ranges from 0 to 40 IU. This means that the sons of the node $<q_1, \ldots, q_i, 0, \ldots, 0>$ at the ith level ($0 \leqslant i \leqslant 5$) are the nodes:

$$<q_1, \ldots, q_i, 0, 0, \ldots, 0>$$
$$<q_1, \ldots, q_i, 4, 0, \ldots, 0>$$
$$\cdot \quad \cdot \quad \cdot \quad \cdot$$
$$<q_1, \ldots, q_i, 40, 0, \ldots, 0>$$

For the purposes of the heuristical algorithm, we consider the selection criterion "crit" for any node, that is, the selection criterion of its injection combination, and we also assign to each node an estimated criterion "est", which is calculated by a similar formula to the selection criterion, but only sums squares of positive differences:

$$est = \sum \left(T_c(i) - T_p(i)\right)^2 \tag{2}$$
$$(1 \leqslant i \leqslant 48) \text{ and}$$
$$\left(T_c(i) - T_p(i) > 0\right)$$

where $T_c(i)$ is the value of the total pharmacodynamic effect of the injection combination corresponding to the node considered, and $T_p(i)$ is the value of the total pharmacodynamic effect of the pump optimum at time i.

The estimated criterion of any node is less than or equal to the criterion of that node and to the criteria of all nodes within its sub-tree. This can be proved simply since the quantity of each insulin preparation represented by a node is less than or equal to the corresponding quantities of nodes in its sub-tree.

The estimated criteria of the nodes are the control values in the following algorithm, which searches for the optimum injection combination:

1. Create a list A and put the root <0, ..., 0> into this list. Set an integer variable SolutionNo to 1.

2. Take the first node $<q_1, ..., q_I, 0, ..., 0>$ from the list A, remove it from the list and assign its level in the tree to variable I.

3. If I = 6 then:
    Output SolutionNo[th] best injection therapy $q_1, ..., q_6$.
    Set SolutionNo to SolutionNo+1.
    If SolutionNo > 10, then exit else go to step 2.

4. Expand the note $<q_1, ..., q_I, 0, 0, ..., 0>$ by creating its sons:
    $$<q_1, ..., q_I, 0, 0, ..., 0>$$
    $$<q_1, ..., q_I, 4, 0, ..., 0>$$
    . . .
    $$<q_1, ..., q_I, 40, 0, ..., 0>$$
    at the (I+1)[th] level.

5. If the created nodes are at the 6th level, assign to each of them the value of its criterion crit, otherwise assign to each of them the value of its estimated criterion est.

6. Add each created node to the list A and sort the whole list in ascending order.

7. Go to step 2.

   This algorithm is a modified special case of the algorithm A* described by Hart et al. (1968) and Nilsson (1982), which has been proved to output the optimum combination.

   However, this general purpose algorithm is not able to reduce the number of the nodes evaluated to a reasonable number and to satisfy our computation speed demands. Therefore, the following three restrictions of the expansion step 4 are employed:

   (i)   Do not expand any node at level 2 (morning and evening PSs are evaluated) when the sum of the morning and evening PSs is less than 20% of the insulin quantity delivered by the pump during the whole day.

   (ii) Do not expand any node at level 4 (morning and evening PSs and PDs are evaluated) when the sum of the morning and evening PSs and PDs is less than 50% of the insulin quantity delivered by the pump during the whole day.

   (iii) Do not expand any node at any level, when the sum of all insulin preparation quantities in the node exceeds 120% of the insulin quantity delivered by the pump during the whole day.

   These additional restrictions may result in the algorithm not finding the mathematically optimum node in the sense of the smallest criterion value. However, they are based on the suggestions of experts in insulin therapy. Their clinical experience postulates that suitable injection combinations have to agree with these rules.

DISCUSSION

   Although there exists a direct analytical solution to the problem, we have employed this relatively time-consuming algorithm. The function $T_c(i)$, which is used in "crit", is a linear function of insulin quantities. Setting the partial differentiations of the function "crit" with respect to each insulin quantity equal to zero leads to six linear equations and should directly offer the optimum solution. However, for practical reasons, the solution has to be expressed in integer values and, even more, values are to be multiples of four. There is no evidence that such a solution lies near (in the analytical sense of the word) the global or local minimum obtained from the linear equations. Furthermore, the A* algorithm offers several solutions together with semi-graphically drawn pharmacodynamic curves and, as the model neglects individual factors, this output arrangement helps the clinician to select the solution which accords with the characteristics of the patient, for instance, higher insulin consumption during lunch and during the night.

The optimum solutions depend on the criterion type. In this case, square differences were used. Certainly, other possibilities such as absolute differences or a non-symmetric criterion should be considered, because short but marked overdosage of insulin may have a more damaging result than a short underdosage. During development of the project the absolute and square criteria were tested. There was no significant difference between solutions obtained using these criteria and, as we wanted to avoid high variations, the square criterion was adopted. The non-symmetric criterion was not introduced since there was no obvious model as to the form of the asymmetry.

During clinical testing of the system the same square and absolute criteria applied to integrals of pharmacodynamic curves instead of the pharmacodynamic curves were also proposed. While the criteria employed minimise actual variances in insulin pharmacodynamics, the integral criteria prefer solutions where the actual variances are adequately balanced by variations in the opposite direction over a short time. This will be considered in future versions of the system.

The system employs a very simple model of insulin pharmacodynamics. However, the insulin pharmacodynamics system is extremely complex. More sophisticated mathematical models might consider such aspects as the absorption process of the injected insulin and functions of insulin receptors (Kolendorf et al., 1978; Kobayashi et al., 1983; Binder et al., 1984; and Lauritzen, 1985). Some of these features are still not fully understood and the existing mathematical models (Bergman et al., 1979; Carson et al., 1983) of the relevant sub-systems have limited validity in their predictive abilities, let alone the problems of finding the individual parameters of these models. Our simple model neglects some aspects of contemporary knowledge on insulin pharmacokinetics. When improving the whole system, the non-linear dependence of insulin absorption on the injected quantity and concentration, the dependence of pharmacodynamic effects on the disease compensation, and so on, have to be taken into account.

Despite all these limitations, the system developed has proved its clinical relevance. It normalises and standardises the treatment transformation process and helps the clinician to select the correct doses. By offering several nearly optimum possibilities, the system also enables the personal clinical characteristics of each patient to be taken into account.

REFERENCES

Bergman, R. N., Ides, Y. Z., Bowden, C. R., and Cobelli, C., 1979, Quantitative estimation of insulin sensitivity, Am. J. Physiol., 236 : E667.
Binder, C., Lauritzen, T., Faber, O., and Pramming, S., 1984, Insulin pharmacokinetics, Diabetes Care, 7 : 188.
Carson, E. R., Cobelli, C., and Finkelstein, L., 1983, "The Mathematical Modelling of Metabolic and Endocrine Systems", Wiley, New York.

Hart, P. E., Nilsson, N. J., and Raphael, B., 1968, A formal basis for the heuristic determination of minimum cost paths, <u>IEEE Trans. Syst.Sci. Cybern.</u>, SSC-4 : 100.

Kobayashi, T., Sawano, S., Itoh, T., Kosaka, K., Hirayama, H., and Kasuya, Y., 1983, The pharmacokinetics of insulin after continuous subcutaneous infusion or bolus subcutaneous injection, <u>Diabetes</u>, 32 : 331.

Kolendorf, K., Aaby, P., Westergaard, S., and Deckert, T., 1978, Absorption, effectiveness and side effects of highly purified porcine NPH-insulin preparation, <u>Eur. J. Clin. Pharmacol.</u>, 14 : 117.

Lauritzen, T., 1985, Pharmacokinetics and clinical aspects of intensified subcutaneous insulin therapy, <u>Danish Med. Bull.</u>, 32 : 104.

Nilsson, N. J., 1982, "Principles of Artificial Intelligence", Springer-Verlag, Heidelberg : 53.

Skyler, J. S., Ellis, G. J., Skyler, D. L., Lasky, I. A., and Lebovitz, F. L., 1979, Instructing patient in making alterations in insulin dosage, <u>Diabetes Care</u>, 2 : 39.

Skyler, J. S., Miller, N. E., O'Sullivan, M. J., Reeves, M. L., Ryan, E. A., Seigler, D. E., Skyler, D. L., and Zigo, M. A., 1982, Use of insulin in insulin-dependent diabetes mellitus, <u>in</u>: "Proc. Insulin Update", Key Biscayne : 125.

Smolek, J., 1982, Insuliny Spofa (Spofa insulins), <u>in</u>: "Diabetes Mellitus a Jeho Lecba Insulinem" ("Diabetes Mellitus and its Insulin Treatment"), O. Dub, ed., Aviceum, Prague : 258 (in Czech).

AN EXPERT SYSTEM FOR THERAPY MANAGEMENT OF TYPE I DIABETICS - DIABETEX

G. Zahlmann, S. Oranien, G. Henning and W. Müller

Department of Biomedical Engineering, Ilmenau Institute of
Technology, Ilmenau; and
Clinic for Eye Diseases, Medical Academy, Erfurt,
GDR

INTRODUCTION

Over the last few years we have seen the development of a new methodological approach to problem-solving and decision-making in complex and complicated regulatory processes, namely the development of so-called "expert
systems". The knowledge of experts for decision making is necessary because
these processes can hardly be described completely and exhaustively by
mathematical methods. In our opinion, expert systems make possible decision
proposals for a given task based on a vast amount of information about the
controlled process, the logical connections thereof, and the organisation
of the man-machine dialogue. The real decision maker is always man, in
our case the physician (Wernstedt, 1985).

The working base of such systems is shown in Fig. 1. This new method
for problem-solving involves working with knowledge consisting of various
kinds of information, methods and data.

Fig. 1. Methodology and knowledge
base.

The designer of a medical expert system has to take into consideration some peculiarities of items of information and their processing:

(1) The systems are very complex and change with time.
(2) They are disturbed.
(3) There are often only a few items of information available about the system. Most of them describe a point in the past.
(4) The information is not exact and is often subjective.
(5) There are large inter- and intra-individual differences.
(6) In most cases there exist no algorithms for decision-making, but the experience and intuition of the physician are very important.
(7) The assessment of the quality of the decision is often possible only after weeks, months or years. It will always be a subjective evaluation by an individual expert.

Today we can develop an expert system for a special problem in two different ways: either using a so-called "general-purpose knowledge-based development tool-frame system"; or designing a special system for only that one problem. Examples of these two approaches are shown in Fig. 2.

The following steps for developing an expert system are in both cases the same. The notes appended to these six points give recommendations for the work to be performed.

(1) Making oneself familiar with the specific set of problems and formulation of the goal:
    - selection of experts
    - questionnaire
    - interviewing and/or observation of experts.

(2) System conception:
    - hardware conception
    - structure (parts) of the expert system.

(3) Selection of the decision route and the methods of decision-making:
    - decision tree, decision table
    - rules
    - pattern
    - classification
    - model.

(4) Software conception and realisation.

(5) Testing and correcting.

(6) Application in medical practice.

```
           ┌─────────────────────────┐
           │    EXPERT  SYSTEM        │
           ├─────────────────────────┤
           │ KNOWLEDGE        BASE    │
           └─────────────────────────┘

           FRAME-SYSTEM    SPECIAL SYSTEM

             · MYCIN        · GLUCON
             · ONCOCIN      · DIABETEX
             · EXPERT
             · GENIE
               ⋮              ⋮
```

Fig. 2.   Examples for different developments of expert systems.

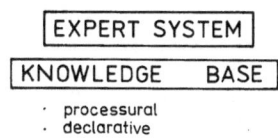

Fig. 3. Main structure of an expert system.

The main structure of all these expert systems, shown in Fig. 3, is the same.

Expert systems are used in medicine for diagnosis, therapy selection, therapy management or prognosis. The peculiarity of expert systems for therapy management is that with only a few items of information about the system in the past and with consideration of the probable effect of a physician's planned action, the prediction of the future state of the system is calculated or else a proposed course of action is calculated, so as to realise a favourable system state in the future. In practice we find such expert systems for short-term prediction (for example, one hour (Kiesewetter et al., 1985)), but not for therapy management of out-patients. However, this is an important field for further activities in expert system design.

THE EXPERT SYSTEM DIABETEX

The Medical Problems

The improvement of insulin therapy plays an important role in diabetes research, because a physiological application of insulin is the pre-requisite for a normal blood glucose level and therefore for the delay or prevention of diabetes-related diseases such as retinopathy and nephropathy.

In Fig. 4 the two complexes of research activities in this field are shown. The subject of our work is a special system for the therapy management of out-patient type I diabetics.

In practice, we find three methods of therapy management: conventional therapy with syringes, intensified conventional therapy with additional

Part of Diabetes mellitus Research

medical	technical
· transplantation	· artificial organ
– pancreas	· expert systems
– segment of pancreas	in several stages
– isolated islet	

aim: adequate insulin delivery
correlated to glucose intake
under all conditions of life

Fig. 4. Research activities for the
improvement of insulin therapy.

blood glucose self-monitoring, and the continuous infusion of insulin with portable insulin pumps. The main goal of all therapy variants is the maintenance of a relatively stable blood glucose level without hypo- or hyperglycaemic excursions. This is not easy to put into practice because there are many regulatory processes which affect the blood glucose level. The best results have been obtained with pump therapy. However, the majority of type I diabetics are treated today with syringes. The improvement of conventional therapy involves new dosage regimes with more injections and makes use of various kinds of insulin.

So we have a wide range of possibilities of treatment with a large variety of conditions for a decision. The selection of the dosage profile for a day for one patient is a very serious task which needs a large amount of knowledge - the knowledge and the experience of experts (Kerner, 1987; Menzel, 1984; Menzel and Bruns, 1982).

Development and Structure of DIABETEX

First we analyse the problem. From a cybernetic point of view we have an open control loop (Fig. 5). There is no adequate information for every time-point. We therefore obtain the majority of our information in a discontinuous manner.

For building up the knowledge base and the decision module, we needed more information about the treatment of diabetes. From the cybernetic methods available for the acquisition of information we used interviews with, and the observation of, experts. Here the experts are leading diabetologists, nationally and internationally, and also some intelligent and cooperative patients, because they have the most experience of managing their diseases by regulation day in, day out. So we also received several action patterns and evaluated reactions of the physician and/or the patient. The patient's activities also revealed a new problem: the intelligent patient as an opponent of the expert system. In a future version of the system game theory will be considered.

The structure of the expert system DIABETEX is shown in Fig. 6. With the help of the input unit all data and information have to be fed into the knowledge base. The information input is separated into three parts:

(1) Information about the carbohydrate metabolism (blood glucose values, insulin units, carbohydrate units);
(2) Information about patient's state (anamnesis, laboratory findings);
(3) Information about life situations (special situations) of the patient (exercise, illness, stress, and so on).

The knowledge base consists of the compressed and condensed expert knowledge, all data and all cybernetic methods necessary for decision making

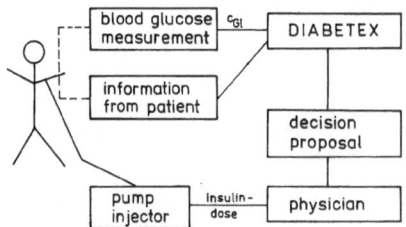

Fig. 5. Open-loop system with DIABETEX.

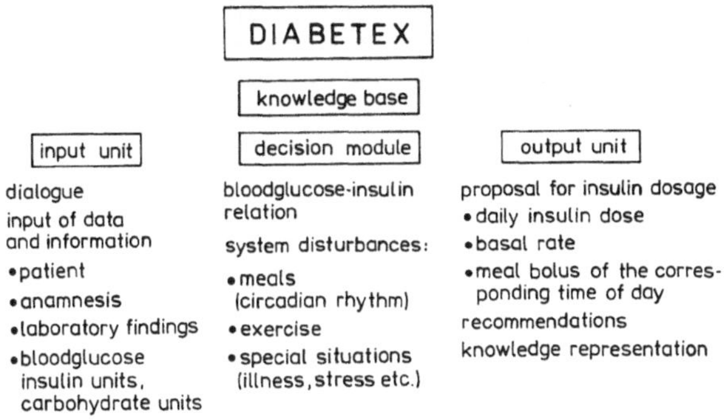

```
                    ┌─────────────┐
                    │  DIABETEX   │
                    └─────────────┘

                  ┌──────────────────┐
                  │  knowledge base  │
                  └──────────────────┘

┌────────────┐    ┌──────────────────┐    ┌─────────────┐
│ input unit │    │ decision module  │    │ output unit │
└────────────┘    └──────────────────┘    └─────────────┘
```

input unit	decision module	output unit
dialogue	bloodglucose-insulin	proposal for insulin dosage
input of data	relation	• daily insulin dose
and information		• basal rate
• patient	system disturbances:	• meal bolus of the corres-
• anamnesis	• meals	ponding time of day
• laboratory findings	(circadian rhythm)	recommendations
	• exercise	knowledge representation
• bloodglucose	• special situations	
insulin units,	(illness, stress etc.)	
carbohydrate units		

Fig. 6. Structure of the expert system DIABETEX.

and also for knowledge representation. In the decision module we find a
decision tree which is data-driven. The criterion of quality is the haemoglobin finding HbAl which characterises the stability of the patient's
blood glucose level over the last 3 - 5 weeks. This represents the first
decision step. The branches of the tree are distinguished by ranges of
laboratory parameters and other findings.

The insulin-blood glucose and the carbohydrate-blood glucose relations
are described by individual models. These models need only a few values
obtained from the patient's records. The diabetologist can affect the calculations of the model by varying the diet, the range of blood glucose,
time parameters and other information. The information of normal and
special life situations allows a situation classification and also the
learning of individual reaction patterns by the system. Under consideration
of uncertainties and possible reactions (consent or rejection) of the
patient, these steps lead to the decision proposal.

The main output is a dosage or dosage-profile proposal under a given
therapy variant for all situations, which the system "knows". This proposal
is evaluated by the diabetologist, the real decision maker, by his consent
or rejection. The selected information is the new base for the system's
learning. The physician can obtain the basic data of the decision process
and all other information as a record in an adequately condensed form.

CONCLUSION

Our goal was the development of an expert system in its first version
for managing the treatment of type I diabetics, a system to be used in a
clinical environment to support the decision making of the physician in
different therapy regimes. In the course of this year the system will be
tested in practice.

REFERENCES

Kerner, W., 1987, Neue Wege der Insulintherapie, Der Internist, 19(28) : 236.
Kiesewetter, M., Wernstedt, J., Lemke, K., and Möricke, R., 1985, Entwurf
    und Erprobung einer rechnergestützten Steuerung des Glucoseniveaus des
    Menschen mit dem Beratungssystem "GLUCON", 30.IWK,TH Ilmenau, vol. B : 149.
Menzel, R., 1984, Aktuelle Therapie des insulinabhängigen Diabetes mellitus,
    Medicamentum, 25(7) : 155.
Menzel, R., and Bruns, W., 1982, Die moderne Diabetestherapie des Diabetes
    mellitus - ein Beitrag zur Prävention diabetesspezifischer Spätkompli-
    kationen, Dt. Gesundheitswesen, 37(34) : 1473.

Wernstedt,J., 1985, Prozesssteuerung und Entscheidungsfindung durch den
    Menschen auf der Grundlage von Beratungs-/Expertensystemen - Methoden
    und Erfahrungen, 30.IWK, TH Ilmenau, vol. A : 367.

PROJECTION OF THE PREVALENCE OF TYPE 1 AND TYPE 2 DIABETES

F. Hauser and M. Anděl

Institute of Social Medicine and Health Services Organisa-
tion, and
Institute of Clinical and Experimental Medicine
Prague, Czechoslovakia

INTRODUCTION

The epidemiology of chronic diseases has attracted more and more
interest in recent years. There are two main reasons for this interest.
First, advances in medical science have, with just a few important excep-
tions, conquered communicable infectious diseases which were the main source
of morbidity and mortality in the past. Secondly, the process of demographic transition culminates in the industrially developed countries where
the earlier high birth and death rates have been replaced by low ones, and
the proportion of inhabitants in the older age groups, which are the most
afflicted by chronic and degenerative diseases, continues to rise to an
all time high.

Looking at the social importance of chronic disease, we find that diabetes mellitus occupies one of the foremost places. Although it is at
present only very rarely the immediate cause of death, diabetes is the
source of health problems leading to the main causes of death. In addition,
diabetes is an important cause of invalidity and one of the prime risk
factors of many serious diseases such as atherosclerosis and renal failure.

The prevalence of diabetes has risen in most countries, but the causes
of this rise have not yet been completely explained. The prevalence of diabetes differs considerably for males and females and for various age groups.
In the German Democratic Republic (GDR), where the age structure of diabetics is routinely monitored, the prevalence in females in the age group
of 75 to 79 years was 17.8% in 1983, while the mean prevalence in the whole
population was 3.6% that year (Institute of Social Medicine and Health Care
Organisation, 1984). The importance of age distribution is further strengthened by the close connection of clinical complications of diabetes with
the duration of the disease.

In order to infer judgements about the needs for specialised health
services, drugs, instruments and facilities for diabetics, it is therefore
necessary to have estimates of future prevalence of diabetes in the individual age groups. Hence, a projection of the diabetic population is the
first task of a wider problem of epidemiological modelling in relation to
diabetes.

MODEL DESCRIPTION

The multi-state population model MULTISPOM (Mader, 1986) was used to describe the evolution of age specific prevalence and incidence. The model represents the evolution of the population age structure in several phases of the health state. These health state phases must be defined in such a way that they are linked to each other and that every individual from the population considered can be uniquely classed as being in just one phase at every time instant. Transitions take place between the phases. The transition rates depend on age, duration of the disease and the effects of various external and internal factors, generally variable in time.

The modelled process is described approximately by the set of equations:

$$Q_{k(i,j)s} = \sum_{l=1}^{K} Q_{l(i-1,\ j-1)s} \cdot P_{l,k(i-1,\ j-1)s},$$

$$i = 2, 3, \ldots, I; \quad j = 2, 3, \ldots, J; \quad k = 1, 2, \ldots, K$$

where $Q_{k(i,j)s}$ is the number of persons in health state phase k and age group i at the time step j for sex s (s = 1, 2 for males and females, respectively), $P_{l,k(i,j)s}$ is probability of transition from phase l to phase k in age group i and time step j for sex s, K is the number of health state phases, I is the number of age groups, and J is the number of time steps.

To preserve the balance the following equality must hold for each l = 1, 2, ..., K:

$$\sum_{k=1}^{K} P_{l,k(i,j)s} = 1 \tag{2}$$

The number of persons in the youngest age group is given by

$$Q_{k(1,j)s} = P_k \cdot P_s \cdot \sum_{l=1}^{K} \sum_{i=2}^{I+1} Q_{l(i-1,\ j-1)} 2 \cdot f_{k(i-1,\ j-1)} \tag{3}$$

where $f_{k(i,j)}$ is the mean number of live births for each woman in health state phase k and age group i at the time step j, p_k is probability that the new-born is in health state phase k, and p_s is probability that the new-born has the sex s.

From the above description it follows that every simulation run requires the following data: initial population age structure divided into health state phases, probabilities of transition between health state phases in given age groups and a hypothesis about their evolution, fertility coefficients for given age groups, probabilities of birth in a certain health state phase, ratio of new-born boys and girls, and hypotheses about their evolution.

Routine statistics and epidemiological studies often only provide data for five or ten year age groups. In model calculations, however, one year age groups and one year time steps are used. For this reason it was necessary to solve the problem of reconstruction of the initial yearly data from the measured five or ten year age structures and age-dependent transition probabilities. For this purpose the interpolation program KVANTAPROX (Mader et al., 1982) has been developed. The given step function is interpolated by a continuous curve so that for each group the equality is preserved between the integral under the interpolating curve and the measured number of persons in the age group. The one year structure is then derived from the calculated interpolation curve. This reconstruction of the initial

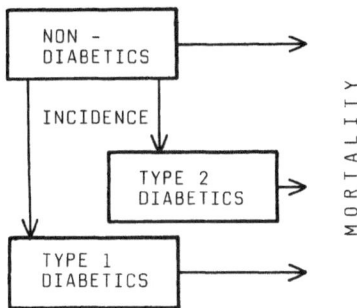

Fig. 1.   Structure of the model
for the projection of
the prevalence of dia-
betes.

yearly data is performed with a given accuracy so as to prevent significant distortion of the measured information.

Diabetes is defined as a state of chronic hyperglycaemia which may result from many environmental and genetic factors (WHO, 1985). It is in reality only a symptom of several rather distinct diseases corresponding to several ethiopathogenetic mechanisms. Although there is no completely satisfactory classification, the World Health Organisation (WHO, 1985) has suggested an interim one and recommended it for use. The WHO classification distinguishes basically insulin-dependent (type 1) diabetes, non-insulin-dependent (type 2) diabetes, impaired glucose tolerance, gestational diabetes and other types of diabetes.

Although a more detailed classification would be desirable, the available statistical data for the Czech Socialist Republic (CSR) make it barely possible to determine the initial model variables and coefficients just for a very simple case in which the population is divided only into type 1 diabetics, type 2 diabetics and non-diabetics, that is when the model represents only three health state phases (Fig. 1).

QUANTIFICATION OF THE MODEL

The model shown in Fig. 1 operates with the following data:

- age-sex structure of the whole population,
- age-sex structure of type 1 and type 2 diabetics,
- age and sex specific incidence of type 1 and type 2 diabetes,
- age and sex specific overall mortality,
- age and sex specific mortality of type 1 and type 2 diabetics,
- age specific fertility rates of non-diabetic mothers,
- age specific fertility rates of diabetic mothers,
- relative number of new-born boys, and
- peri-natal mortality of boys and girls.

The model was quantified for the CSR. The population age structure and overall mortality are contained in the demographic yearbook issued by the Czech Statistical Authority (1980). The age-specific fertility rates of non-diabetic mothers were for the time being approximated by overall fertility rates provided by the Czechoslovak Health Care Yearbook (1980) which also presents the relative number of new-born boys and the peri-natal mortalities. The fertility of diabetic mothers was roughly estimated as one-half of the overall fertility.

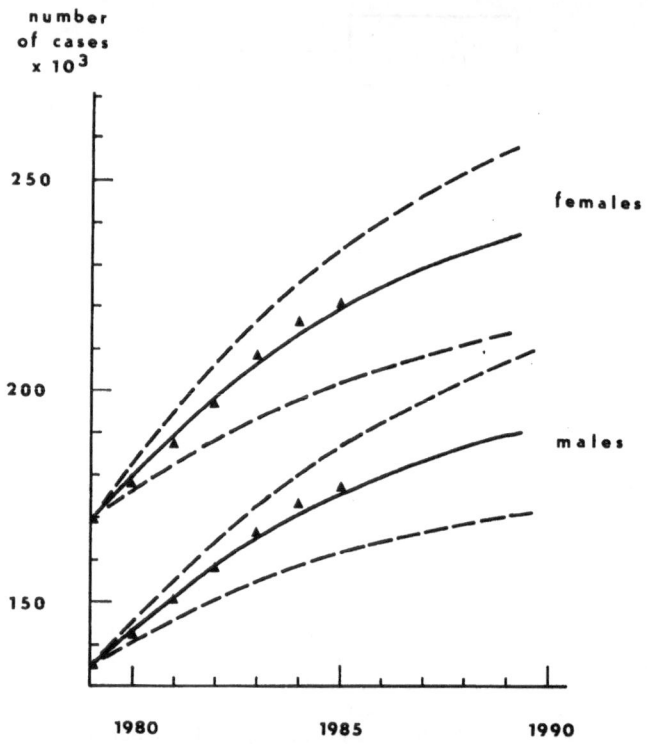

Fig. 2. Basic variant of the evolution of the prevalence of diabetes. Real numbers of diabetics ▲, basic variant of the projection
————, projections with incidence increased
and decreased by 15% ----.

The age-sex structure of diabetics was obtained from an unpublished
study conducted in 1979. The division between type 1 and type 2 patients was
performed by indirect standardisation to data from the GDR Health Care Yearbook (1984). Although the methodology of data collection does not correspond to the WHO recommendations but rather follows the type of therapy, no
better data are available at present, and genetic, environmental and social
conditions in the GDR and CSR are fairly similar. For the same reasons the
age and sex specific incidences were also taken from the GDR Health Care
Yearbook (1984) for age groups below 20, and were estimated for older age
groups to correct for the difference in classification.

There exist only very few studies on the mortality of diabetics,
though it is generally accepted that the disease shortens life expectancy.
Age and sex specific mortality data for type 2 diabetics were taken from
the study of North-American caucasians published by Kessler (1971). The
mortality of type 1 diabetics was taken from Dorman et al. (1984) for age
groups below 35 years, which was the limit of that study, and from Deckert
et al. (1978) for age groups over 45. The work presents the mortality of
patients diagnosed before 1933 and, as care for young diabetics is incom-
parably better now, the data presented there for the younger age groups
cannot be used.

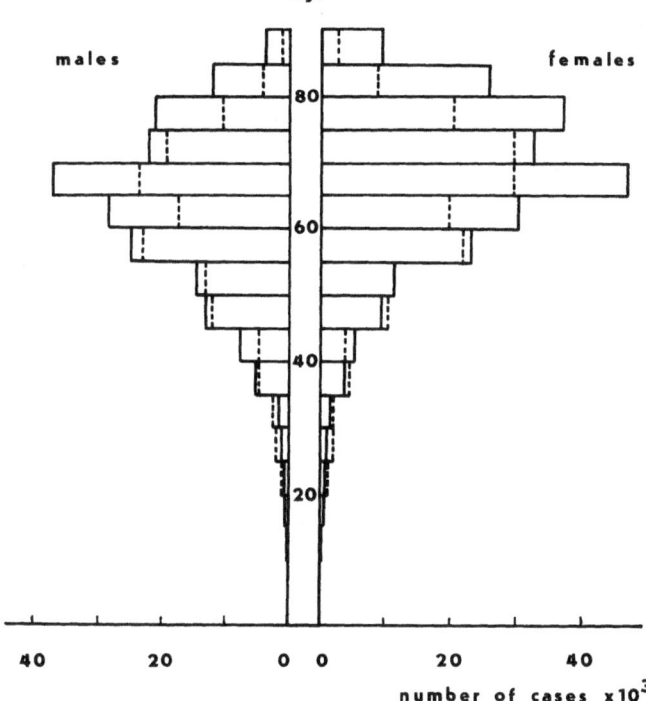

Fig. 3. Age structure of the diabetic population.
The situation in 1979 ----; in 1990 ——.

RESULTS

     With regard to the low level of knowledge about the epidemiology of
diabetes in Czechoslovakia, the mere establishment of all the necessary
input data was a difficult task. These data can thus be safely considered
as the first result of modelling.

     The basic variant of the evolution of the prevalence of diabetes is
shown in Fig. 2. It assumes that the incidence which caused the rapid increase of diabetics in the late 1970s and early 1980s will prevail and that
the demographic characteristics (birth rates, death rates, and so on) will
remain at the level of the year 1980. Fig. 2 presents the total numbers of
male and female diabetics of both types projected under these assumptions
(full lines). The triangles indicate the actual numbers of registered diabetics. Also presented (dashed lines) are the projected numbers for incidence increased and decreased uniformly in all age groups by 15%. It is
apparent that as a first approximation the basic variant can explain the
observed changes. Even when the curves represent the sum of type 1 and
type 2 patients, they describe the increase attributed to type 2 patients
who constitute more than 96% of cases.

     The age structure of the diabetic population is shown in Fig. 3. The
dashed lines represent the situation in 1979, the initial year of the projection; the full lines are for the year 1990. The most prominent feature
of the projection is the large increase in age groups over 60 years of age.
This rising prevalence is superimposed here over the increase in the older
age groups. As population aging itself cannot account for such an increase
of type 2 diabetics and mortality in the older age groups does not decrease,
the main cause has to be high incidence.

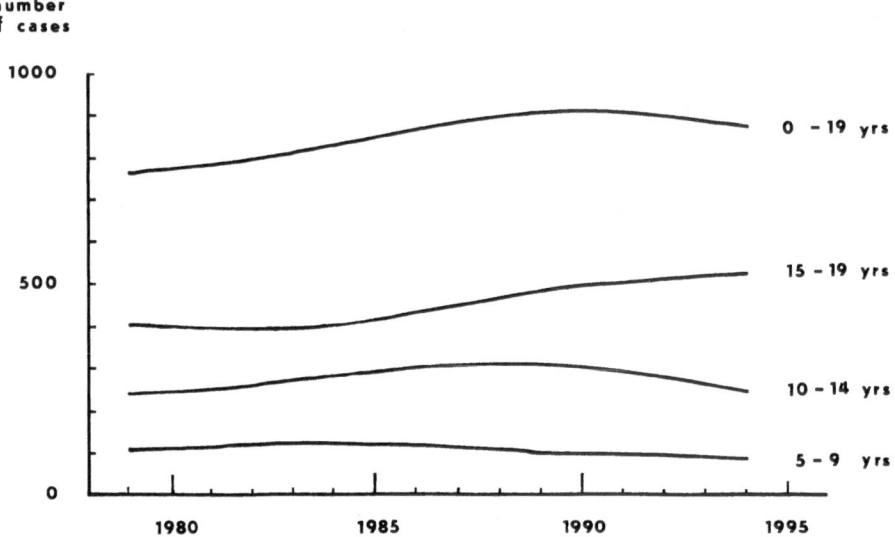

Fig. 4. Projection of the number of boys with type 1 diabetes - basic variant.

Let us turn now to type 1 diabetics. Fig. 4 indicates the basic variant of the projection for boys up to 19 years of age. Unlike type 2, the late 1970s' incidence of type 1 does not cause any change of prevalence. This is, of course, only a first approximation because epidemiological studies from several countries have shown that the incidence of type 1 diabetes is also rising slowly. The changes in numbers of patients in Fig. 4 thus reflect changes in the numbers of new-borns which, in turn, reflect changes in the numbers of women of reproductive age (the demographic characteristics remain constant).

An age structure projection for type 1 diabetics is presented in Fig. 5. The dashed lines represent an estimate of the situation in the year 1979, with the full lines corresponding to the year 1989. Here again, changes in patient numbers correspond to changes in the demographic composition.

CONCLUSION

Medico-demographic models are a valuable tool which can support the investigation of epidemiological processes. However, in the absence of an advanced health information system, the usefulness of such models is limited by poor quality, incoherence or even total lack of data. On the other hand, such models can, and in fact should, identify what data are missing or are to be improved in order to make reasonable predictions of future development.

This model is just the first step in an effort to study the epidemiology of diabetes by computer simulation. It has provided a new view of the current epidemiological situation and enabled the value of the imperfect data which are at our disposal to be increased in advance of the emergence of results from rapidly developing medical information systems.

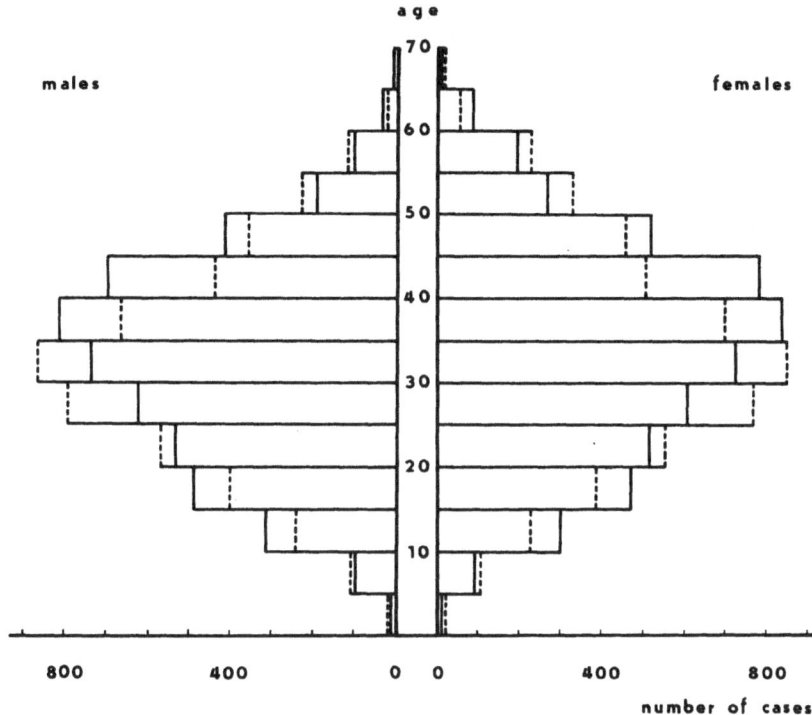

Fig. 5. Age structure of type 1 diabetics. The situation in 1979
----; in 1989 ———.

REFERENCES

Czech Statistical Authority, 1980, "Pohyb Obyvatelstva v ČSR", Czech Statistical Authority, Prague.
Deckert, T., Poulsen, J. E., and Larsen, M., 1978, Prognosis of diabetics
with diabetes onset before the age of thirty one, Diabetologia, 14 :
363.
Dorman, J. S., Laporte, R. E., Kuller, L. H., Cruickshanks, K. J., Orchard,
T. J., Wagener, D. K., Becker, D. J., Cavender, D. E., and Drash, A. L.,
1984, The Pittsburgh insulin-dependent diabetes mellitus (IDDM) morbi-
dity and mortality study, Diabetes, 33 : 271.
Institute of Health Care Information and Statistics, 1980, "CSSR zdravot-
nictví", Institute of Health Care Information and Statistics, Prague.
Institute of Social Medicine and Health Care Organisation, 1984, "Das
Gesundheitswesen der DDR", Institute of Social Medicine and Health
Care Organisation, Berlin (in German).
Kessler, I. I., 1971, Mortality experience of diabetic patients, Am. J.
Med., 51 : 715.
Mader, R., 1986, Model for the prognosis of the population age structure
evolution in several health state phases, Cs. zdravotnictví, 34 : 218
(in Czech).
Mader, R., Mandys, F., and Mikšovic, P., 1982, Regression of a variable
expressed in steps by values of the moving integral, in: "Systems
Simulation and Scientific and Technical Computation", DT ČSVTS Ostrava,
Ostrava : 58 (in Czech).
WHO, 1985, "Diabetes Mellitus", WHO Technical Report Series 727, WHO,
Geneva.

COMPUTER-AIDED CONSULTATION IN GYNAECOLOGICAL PRIMARY HEALTH CARE BASED

UPON MEDICAL EXPERT KNOWLEDGE

J. Ruszkowski

Medical Computer Laboratory
Department of Biophysics and Biomathematics
Medical Centre of Postgraduate Education
Warsaw, Poland

INTRODUCTORY REMARKS.   THE AIM OF THE STUDY

As is apparent from the title, this chapter deals with topics falling
into the area of the support of medical reasoning by means of methods of
artificial intelligence. Studies on this and related problems have been a
subject of interest to many scientists for more than twenty years (Wagner
et al., 1978). The theoretical foundations of these works are currently
based mainly upon mathematical logic, modern algebra, set theory, mathematical linguistics and the theory of programming. In recent years strong
efforts have been made to provide medical staff with facilities derived
from studies on computer-assisted modelling of medical decision-making,
medical knowledge acquisition and management (Miller, 1984; Wagner et al.,
1978). An important stimulus for these attempts seems to be the clearly
visible, increasing gap between the increased availability of the patient's
clinical and laboratory data and their use in clinical practice. The
common efforts of researchers to overcome this tendency are resulting in
the development of new, unconventional methods of knowledge acquisition
(Doroszewski et al., 1981; Ruszkowski, 1986) and effective ways of processing it aimed at aiding clinicians in their every-day practice. It should
be noted, however, that to date they have not progressed beyond the experimental prorotype (Ruszkowski and Skałzynski, 1985). This is caused by a
number of reasons, foremost amongst which are the great complexity of medical reasoning and the indeterministic flavour of medical knowledge, together
with extraordinary difficulties in acquiring it in a sufficiently reliable
form.

This chapter presents part of a research project concerned with computer-
assisted consultation in gynaecological primary health service to aid the
not very highly experienced, junior gynaecologist or the general practitioner
working in small centres, rural areas and the like. The basic needs for
such facilities coming from that group of medical staff members (physicians,
nurses and so on) may vary considerably, dependent on their experience, independence of thinking, clinical skills and local diagnostic equipment, so
that it seems proper to present the system's assumptions by answering the
question: what does the consultation consist of and what kind of assistance
may the user be expected to have during the consultation by means of the
particular system proposed in this chapter. Accordingly, let us specify
the principles which are expected to be supported by the premises of the
system:

(i)   the doctor in the primary health care unit often puts forward diagnostic hypotheses which should be proven by senior physicians in the
specialised health care unit; the gynaecologist stays here in the
exceptional position of working in the primary care system and being
a specialist as well;

(ii)   the diagnostic consultation should aid the doctor to explain the
diagnostic problem, that is the reason for the patient's visit to the
gynaecologist; it should facilitate the process of formulating
working hypotheses as well as planning the diagnostic steps to make
this process efficient and as reliable as possible;

(iii)   the physician should also be aided in taking more general decisions
in the course of managing the patient, that is which case is he able
to look after himself, and which one should he refer to the specialised unit for consultation or treatment; and

(iv)   the physician takes any decision on his own responsibility; he
decides in which situations and to what extent he should take into
account the suggestions of the computer-aided consultation.

Finally, we cannot pass over the fundamental question on premises regarding the applicability of consultation systems in the clinical environment
at all. One of them consists of embracing the majority of clinical problems
usually met in the given branch of medicine or in the specific type of health
care unit, within the range of the consultation offered. Ambulatory gynaecology gives us the chance to build a system which can satisfy such demands;
this is one of the reasons why a consultation system has been proposed for
this particular area of medicine.

THE CLINICAL FOUNDATIONS OF THE SYSTEM: COMMON DIAGNOSTIC PROBLEMS IN
THE GYNAECOLOGICAL OUT-PATIENT CLINIC

In accordance with the system's purpose, the problems and the assumed
extent of the medical consultation should now be explained further. From
an analysis of cases met in gynaecological primary health care units it is
apparent that several symptoms are dominant as the reasons why the patient
asks the gynaecologist for help. There are complaints, which patients have
assessed themselves as coming from their genital organs, even if it is not
clear that this is, in fact, the case. Thus, the first step the doctor
should take is to check the correctness of the patient's self-assessment
and so start properly with further management which is now at its very
beginning. This involves efficient recognition by the gynaecologist of
non-gynaecological reasons which could have caused the patient's complaints;
something which often causes serious difficulties even to experienced doctors, so that the chance of mis-diagnosis appears relatively high.

Apart from the sufferings mentioned by women during the course of pregnancy, the following symptoms appear to be the most frequent reasons for
consultation at the gynaecological out-patient clinic:

(i)   irregular vaginal bleeding,
(ii)   lower hypogastric pain,
(iii)   sacrodynia,
(iv)   certain menstrual irregularities, and
(v)   vaginal discharge.

Although such manifestations are known as obvious reasons for women's
visits to the out-patient clinic, it is clear that some of these symptoms
could also be caused, as already indicated, by other non-gynaecological
diseases. As a result, a number of such conditions have to be considered.
So whereas "irregular vaginal bleeding", "menstrual disorders" or "vaginal

discharge" are undoubtedly meant to be symptoms of gynaecological diseases, the other two - "lower hypogastric pain" and "sacrodynia" - may, however, also have their roots in pathologies other than those of the female genital organs, particularly in organs or systems situated within the pelvis. In the case of "sacrodynia" the aetiological factors arise mostly from: the osseo-muscular system; the nervous system: the digestive system; and the urinary system. In the case of "lower hypogastric pain" the corresponding factors arise from: the urinary system; the digestive system; the nervous system; and the circulatory system.

The aetiological factors originating out of the genital organs were involved only as aggregated groups of diseases, without differentiating them from other diseases of the particular system they are emanating from, although the simultaneous occurrence of the gynaecological and non-gynaecological conditions could never be excluded. The consultation system presents its suggestions in such a way to account for those situations as well.

THE FOUNDATIONS OF THE METHOD

Before approaching the project, two basic problems (among others) have had to be considered:

(i) the range and type of consultation provided, and
(ii) the type of medical expert knowledge and how to acquire it in a comprehensive and reliable manner.

Both these tasks have been solved paying close attention to the real clinical environment and to the physician's demands as well as to his way of thinking. Furthermore, we assumed that diagnostic problem-solving could be modelled as a step-by-step running explanatory process. Analysing the possible set of factors (diagnostic hypotheses) suspected as those having caused the symptom considered as the diagnostic problem, the doctor proceeds to confirm some of them and considers the others to be of minor importance. In his natural reasoning the physician keeps making comparisons between diagnostic hypotheses in the light of the patient's data, using his clinical skills and experience. Thus, he arranges the set of hypotheses by putting them in order with the intention of finally distinguishing one of them as the diagnosis. Furthermore, the doctor usually seeks to reach this aim in the minimum number of diagnostic steps. Although the ways of a doctor's reasoning are of a very complex character, in the light of the assumptions described above two inter-related phases may be distinguished in the diagnostic process:

(i) the evaluation and choice of the suspected reasons for the patient's complaints, that is the formation of working hypotheses, and
(ii) the choice of diagnostic actions (most suitable in relation to the next steps of the process).

Even if the above scheme is to be seen as a simplification, it is possible to apply it in the majority of clinical situations.

Given these general remarks on the basic features of the consultation system described in this chapter, let us now consider the form of medical expert knowledge required as well as the appropriate course of the consultation procedure.

MEDICAL EXPERT KNOWLEDGE

Following the assumptions described above, an ELSA-method (Expert Lattice Structured Acquirements) has been applied for expert knowledge

acquisition and management (Ruszkowski, 1986). It generally consists of
the total decomposition of the diagnostic problem, to ask expert questions
in as simple a manner as possible, so as to elicit answers which are as
helpful as possible in complex clinical situations. The philosophy of the
ELSA-method can also be expressed by the following sentence: "If medical
knowledge acquired from experts who have been asked about complex situations does not appear to be sufficiently reliable, let us simplify the
enquiries as far as possible and then compose the answers in order to reach
reliable evaluations (for example, the ordering of possible diagnoses) related to any syndrome which has been defined". Diagnostic hypotheses,
clinical manifestations (symptoms, signs, laboratory findings), medical
tests and medical knowledge structures are the objects in the application
of the ELSA-method.

Diagnostic problem - clinical fact (for example, a sign or symptom being
a subject of explanation in the diagnostic process; the explanation of
the diagnostic problem consists of finding its cause (or causes) in the
light of patient data which has been acquired,

Diagnostic hypotheses - a set of factors assumed to be taken into account
as possible causes of the occurrence of the diagnostic problem,

Clinical manifestations - a set of clinical events (symptoms, signs, laboratory findings), which should be helpful in the process of selecting a
diagnosis from the set of diagnostic hypotheses,

Diagnostic tests - a set of any diagnostic actions (subjective and objective examinations, laboratory tests, auxiliary examinations, and so on),
which aim at the acquisition of clinical manifestations,

Medical knowledge structures - a set of ordered relations within the set
of diagnostic hypotheses placed in order according to the individual
clinical manifestations considered in succession. The order relation
applied in the method is of the type: "... preferred rather than ..."
in the light of a certain assumed manifestation. For each clinical manifestation, an order relation within the set of diagnostic hypotheses has
to be acquired.

According to the principles of the ELSA-method, the following sequence
of system objects must be set up:

- the diagnostic problem (DP),
- the diagnostic hypotheses (DH),
- the symptoms, signs and laboratory findings (S),
- the diagnostic steps (tests) (DT),

what further allows:

- expert knowledge acquisition and its logical vertification.

In the procedure of knowledge acquisition two basic steps could be distinguished:

(i) in the defined diagnostic problem (DP), for each pair of diagnostic
     hypotheses (DH) and for each manifestation (S) considered, the experts
     are asked to answer the question:
     "Which of the two hypotheses d_1 and d_2 should be preferred, if mani-
     festation S is being introduced?"

(ii) the experts can select from the following:
     (a) "I do prefer d_1 rather than d_2 (or vice versa)",
     (b) "I strongly prefer d_1 to d_2 (or vice versa")",
     (c) "S is of no importance for differentiation between d_1 and d_2,
         that is no preference could be judged",
     (d) "the importance of S in preferring d_1 or d_2 is unknown".

418

Each of the expert's answers is then confronted with the previous ones to check the logical correctness of these opinions, which must be consistent as a whole.

Let us now consider an example of the acquisition of medical expert knowledge in the case of given objects defined according to the ELSA-method rules:

<u>diagnostic problem</u>:   "irregular vaginal bleeding"

<u>diagnostic hypotheses</u>: d_1 - uterine cervix carcinoma,
                           d_2 - carcinoma of the endometrium,
                           d_3 - uterine sarcoma,
                           d_4 - hyperplasia,
                           d_5 - intra-uterine polyp,
                           d_6 - uterine myoma,
                           d_7 - metritis,
                           d_8 - imminent abortion,
                           d_9 - extra-uterine pregnancy,
                           d_{10} - hydatid mole,
                           d_{11} - Stein-Loewenthal syndrome,
                           d_{12} - endometriosis,
                           d_{13} - ovarian tumour,
                           d_{14} - hyperoestrogenism,
                           d_{15} - uterine cervic erosion,
                           d_{16} - adnexitis,
                           d_{17} - psychological factors.

<u>manifestations</u>:   "nervousness observed",
                       "leukocytosis occurrence"

<u>diagnostic tests</u>:   "nervousness?",
                       "leukocytosis?"

Some of the answers of experts whose opinions have been sought in the case of "leukocytosis occurrence" are listed in Table 1.

The same enquiries have been put to experts in the case of "nervousness observed". The order relations within the set of the selected six diagnostic hypotheses obtained in this way are shown in Figs. 1 and 2. The opinions of the experts as to the full set of 17 hypotheses - possible aetiological factors of irregular vaginal bleeding - are presented in

Table 1. Expert Opinions for a Case of "Leukocytosis Occurrence"

imminent abortion	"preferred rather than"	ectopic pregnancy
imminent abortion	"no preference"	endometriosis
imminent abortion	"preferred rather than"	hyperplasia
metritis	"strongly preferred to"	ectopic pregnancy
metritis	"strongly preferred to"	endometriosis
metritis	"strongly preferred to"	hyperplasia
metritis	"no preference"	imminent abortion
adnexitis	"no preference"	imminent abortion
adnexitis	"strongly preferred to"	ectopic pregnancy
adnexitis	"no preference"	metritis
adnexitis	"strongly preferred to"	endometriosis
adnexitis	"strongly preferred to"	hyperplasia
ectopic pregnancy	"no preference"	endometriosis
ectopic pregnancy	"no preference"	hyperplasia
endometriosis	"no preference"	hyperplasia

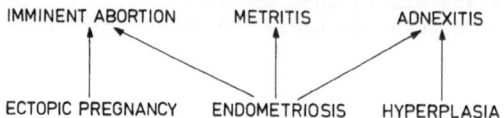

Fig. 1. Diagnostic hypotheses ordered
according to the experts' knowledge in case of "leukocytosis
occurrence" when reasons for
"irregular vaginal bleeding" are
considered.

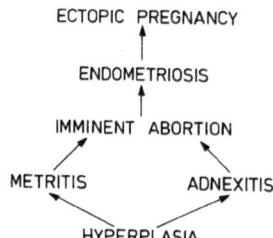

Fig. 2. Diagnostic hypo-
theses ordered
according to the
experts' know-
ledge in the case
of "nervousness
observed" when
reasons for
"irregular vagi-
nal bleeding"
are considered.

Figs. 3 and 4. As has already been mentioned, the ELSA-method also provides
a way of composing medical knowledge structures to have expert suggestions
in the light of any syndrome of manifestations which has been assumed. A
special operator for this aim has been defined (Ruszkowski, 1986). It
allows the composition of relations concerned with individual manifestations
in order to reach a new relation, reflecting the arrangement of hypotheses
performed with respect to the conjunction of all the assumed manifestations.
A detailed explanation of the procedure of knowledge composition has been
given in Ruszkowsky (1986). The numerals assigned to the lines and columns
of the matrices in Figs. 3 and 4 are the code numbers of the diagnostic
hypotheses - reasons for irregular vaginal bleeding, already specified in
this chapter. The arrows in the matrices indicate the directions of pre-
ferences for each pair. The "blank", an empty box, is equivalent to a lack
of preference.

In Fig. 5 the result of the composition of these two pieces of knowledge (represented by the relations shown in Figs. 3 and 4) is visible as
a new order relation within the set of diagnostic hypotheses reflecting the
new clinical situation: the simultaneous occurrence of "leukocytosis" and
"nervousness". When in the case of "nervousness observed" disorders of
early pregnancy have been supported in the light of the opinions of the
experts (Figs. 2 and 4), the "leukocytosis occurrence" has given preference
to inflammatory conditions of the genital organs (Figs. 1 and 3) over other
ones (Ruszkowski, 1987a). The relation obtained for these two findings
considered jointly (Fig. 5) brings with it a small preference for inflam-

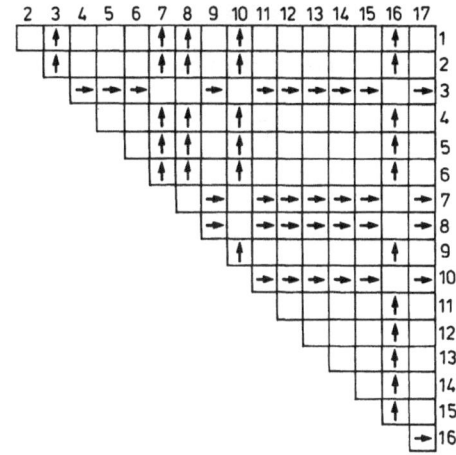

Fig. 3. Matrix of binary relations
defined within the set of
diagnostic hypotheses in the
case of "leukocytosis occurrence" when reasons for
"irregular vaginal bleeding"
are considered.

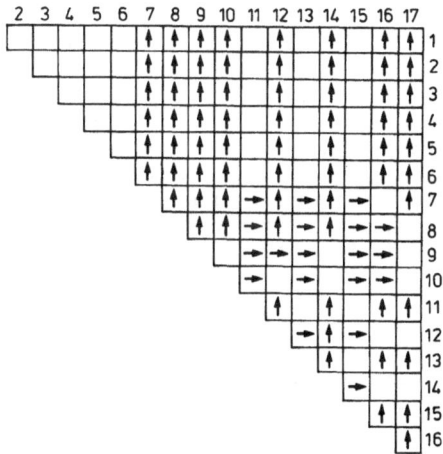

Fig. 4. Matrix of binary relations
defined within the set of
diagnostic hypotheses in the
case of "nervousness observed"
when reasons for "irregular
vaginal bleeding" are considered.

matory diseases, but in this case let us start with differentiation between
these groups of hypotheses. This is to emphasise, however, that the diagnostic hypotheses situated at the top of the graph in Fig. 5 ("adnexitis")
cannot by any means be thought of as a diagnosis, but rather it should be
considered carefully together with the other hypotheses placed on the top
levels of the structure in the next steps of the diagnostic process. This
interpretation expresses well the specific attitude to the approach to

421

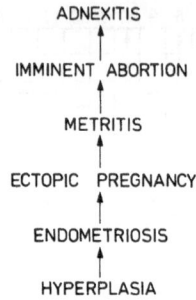

Fig. 5. Diagnostic hypo-
        theses ordered in
        the case of conjunc-
        tion of "leukocyto-
        sis occurrence" and
        "nervousness observed"
        when reasons for "ir-
        regular vaginal blee-
        ding" are considered.

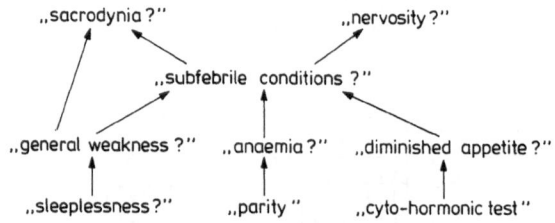

Fig. 6. Diagnostic tests ordered according
        to their diagnostic significance for
        differentiation within the group of
        diagnostic hypotheses mentioned in
        Figs. 1, 2 and 5.

consultation which is assumed in the foundations of the method. As a consequence, on the top levels of the graph in which the hypotheses are
arranged, there remain those for which there is less pressure for downward
movement in the light of the knowledge acquired from the experts.

        As has already been mentioned, the other advantage the user could
take in the course of the consultation is the assistance rendered to him
for planning the next diagnostic steps in order to differentiate most effi-
ciently between the working hypotheses which have previously been assumed
by him. Fig. 6 ' shows the system's suggestions concerning the selection of
new tests in the case of the hypotheses specified in Figs. 1 and 2. The
graph presénts a group of diagnostic tests ordered according to their diagnostic value for differentiation within the group of the selected hypotheses.
The items: "sacrodynia?", "nervousness?", "sub-febrile conditions?", and
so forth, denote questions to be answered simply as "yes" or "no", while
the others may result in:

"parity":   1. "one childbirth or more",
                         2. "no childbirth";

"cyto-hormonic test":   1. "normal oestrogenic activity",
                         2. "diminished oestrogenic activity",
                         3. "increased ectrogenic activity".

422

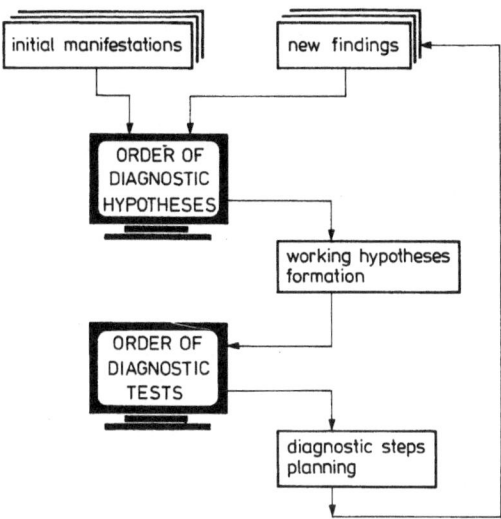

Fig. 7. The scheme of the course of con-
         sultation.

The principle of ordering the diagnostic tests in the light of the knowledge
of the experts consists of a procedure of binary comparisons of the structures of the diagnostic hypotheses (for example, as in Figs. 1, 2 and 5)
concerned with the respective symptoms, signs or laboratory findings. The
better ordered are the hypotheses in the light of certain manifestations,
the more valuable it is for differentiation within that group of hypotheses.
A good ordering in this case means a close similarity (or equivalence) to a
linear order relation. Based upon the results of such comparisons between
manifestations, the diagnostic tests could now be compared with regard to
their expected utility as detectors of the more or less valuable manifestations which have appeared. The arrangement of the diagnostic tests shown
in Fig. 6, ordered according to their diagnostic value in the given diagnostic situation, distinguishes "nervousness" as one of the two tests which it
is suggested should be made next. It appears quite well justified when
looking at Fig. 2, presenting an almost linearly ordered set of hypotheses
in the light of "nervousness observed". This way of ordering tests has been
described in a more detailed manner in Ruszkowski (1988) and Ruszkowski and
Milewski (1980). There could also be found the principles of diagnostic
test selection according to their ability to differentiate between the
chosen groups of hypotheses. This method has succeeded in application to
computer-assisted consultation in aiding the early detection of neoplastic
changes in the body of the uterus (Ruszkowski, 1987b; Ruszkowski and
Skarzynski, 1985).

     As to the similarity between this and other consultation systems, in
view of their comprehensiveness in the given branch of medicine, it seems
that INTERNIST (Miller, 1984) could be seen as being one of comparable
value. That system includes the majority of problems of internal medicine
which are usually met in either in- or out-patient clinics.

OPERATION OF THE SYSTEM

     Following on from the assumptions previously mentioned and the methodo-
logical background, it is now time to show how the system operates. Fig. 7
presents the sequence of the actions of the user as well as the system's

reactions to them. Although the chart simplifies the course of the consultation, it does reflect sufficiently well the basic steps of the system-user dialogue. The user's role is to inform the system regarding the patient's initial manifestations and further findings acquired, as well as on his diagnostic concepts defined through the group of working hypotheses selected. All these data are on each occasion confronted with the medical knowledge stored in the system's files and then the respective suggestions are given out; those concerning the conceivable reasons for the patient's complaints as well as those aiding selection of suitable steps to explain them in an efficient manner. An unlimited number of iterations is possible, dependent only on the user's decisions as to how far and to what extent the system's suggestions should be taken into account.

A module shown in this chapter has been generated from the system shell MEDIATOR (Ruszkowski, 1985). It is called AMIGO.ivb (Aiding Medical Inferences in Gynaecology and Obstetrics) through its application to this branch of medicine. The abbreviation "ivb" denotes the diagnostic problem in the module considered. The medical knowledge database currently contains about twelve thousand answers acquired from experts, and this can be developed further according to the rules of the ELSA-method.

The system operates on an IBM-PC/XT or compatible micros, equipped with DOS 3.1, 256 kB RAM and a 360 kB floppy drive.

CLINICAL EXAMPLE: CLOSING REMARKS

In this study each case of bleeding from the genital organs which has been assessed as abnormal by the patient herself has been assumed to be "irregular vaginal bleeding". Particular features included:

- a prolonged menstrual period,
- hypermenorrhea,
- algomenorrhea, and
- vaginal bleeding after the menopause.

The bleeding caused by senile changes in the genital organs has been omitted. The notion of "irregular vaginal bleeding" defined above is not of an aetiological character; it relies only upon symptoms assessed subjectively by the patient herself as being sufficiently abnormal to consult a gynaecologist.

Let us now consider a case of a patient presenting with "irregular vaginal bleeding" who has been manifesting the following signs and symptoms:

- lower hypogastric pain,
- sacrodynia,
- sub-febrile conditions,
- no general weakness,
- no diminished appetite,
- nervousness observed,
- no sleeplessness mentioned,
- no neoplastic diseases in the family,
- age: 18 - 40 years,
- good welfare conditions,
- one childbirth,
- secondary amenorrhea,
- no hypermenorrhea,
- no algomenorrhea,
- no intermenstrual bleeding,
- subjective symptoms of pregnancy,
- normal weight,

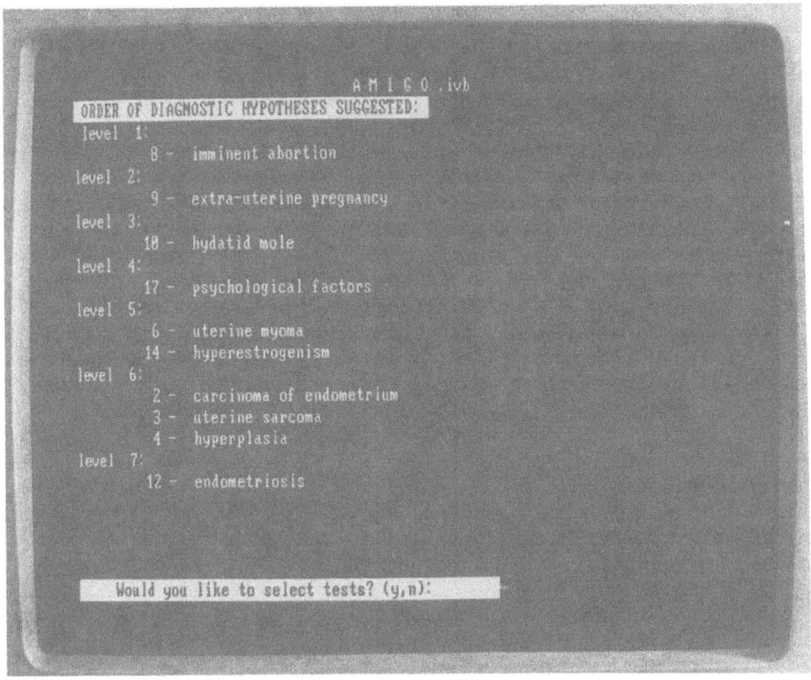

Fig. 8.   An example of consultation in the case of a patient with
suspected early pregnancy disorders.

- no diabetes,
- no hypertension,
- digestive tract disorders observed,
- no urinary system disorders,
- no rectal tenesmus,
- no lower limbs oedema,
- no ischialgia,
- no muscular defence,
- no hirsutism,
- no sexual precocity,
- increased erythrocyte sedimentation rate,
- no leukocytosis.

Fig. 8 shows the result of the consultation performed by the system, sug-
gesting an order within the group of working hypotheses previously selected
by the user.   Ten hypotheses (from the set of seventeen) have been considered, but upon the system's suggestions it seems that a careful differentiation within the disorders of early pregnancy should be performed in the
next diagnostic steps.

The system also has several other mechanisms built-in, making the consultation easy and efficient for the user;   the explanations of the suggestions of the system concerning the arrangement of hypotheses are provided
on request.   From our ongoing experience it follows that AMIGO could be a
proper facility for a tutorial aid in medicine, for example in developing
countries.

REFERENCES

Doroszewski, J., Bolc, L., Rózańska, L., and Milewski, J., 1981, Natural
    language question-answering medical system MEDLING, in: "Proc. Int.
    Congress on Medical Informatics", Strasbourg.
Miller, R. A., 1984, INTERNIST/CADUCEUS: Problems facing expert consultant
    programs, Meth. Inform. Med., 23 : 9.
Ruszkowski, J., 1985, Computer-aided medical consultation based on experts'
    knowledge, in: "Medical Decision Making: Diagnostic Strategies and
    Expert Systems", J. H. van Bemmel, F. Grémy and J. Zvárová, eds.,
    North Holland Elsevier, Amsterdam : 65.
Ruszkowski, J., 1986, ELSA - A method for medical knowledge acquisition
    and processing to aid medical reasoning, Meth. Inform. Med., 25 : 79.
Ruszkowski, J., 1987a, "Vaginal Bleeding in Early Pregnancy: Computer-
    aided Consultation Based on Medical Expert Knowledge", in: Journal
    of Perinatal Medicine, 15, suppl. 1, de Gruyter, Berlin.
Ruszkowski, J., 1987b, Expert knowledge-based system to aid an early detec-
    tion of uterus body neoplastic changes, in: "Proc. Medical Informatics
    Europe 87", Rome.
Ruszkowski, J., 1988, "Clinical Significance of Symptoms and Signs Assessed
    upon Medical Experts' Knowledge. Application to the Common Gynecologic
    Problems", Wissenschaftliche Zeitschrift der Tech. Hochschule,
    Ilmenau, 1 (in press).
Ruszkowski, J., and Milewski, J., 1980, Relational model for the evaluation
    of medical diagnostic tests based on expert opinions. Differentiation
    of jaundice etiology, in: "Progress in Cybernetics and System Research,
    Vol. 9", R. Trappi, ed., Hemisphere Publ. Co., Washington.
Ruszkowski, J., and Milewski, J., 1987, Pancreatic carcinoma versus chronic
    pancreatitis; computer-aided diagnosis, Materia Medica Polona, 19(3).
Ruszkowski, J., and Skarzyński, B., 1985, The evaluations of indications
    to the biopsy of endometrium performance in case of vaginal bleeding,
    J. Tech. Probl. Med., 15(3) : 198 (in Polish).
Wagner, G., Tautu, P., and Wolber, V., 1978, Problems of medical diagnosis -
    a bibliography, Meth. Inform. Med., 17(3) : 55.

SEQUENTIAL METHODS IN MEDICAL DECISION MAKING

M. Csukás, A. Krámli and J. Soltész

National Institute of Cardiology, and
Computer and Automation Institute, Hungarian Academy of Sciences
Budapest, Hungary

INTRODUCTION

The basic problem of statistical discrimination can be formulated as follows: there are several populations and a sample of individuals drawn from each. Using measurements from these individuals a certain rule should then be set up which enables us to place a new individual, on the basis of his measurement values, into the population to which he belongs. When the results of these measurements can be regarded as random vectors having joint normal distributions with common covariance matrix and different mean vectors for different populations, then the well-known method of linear statistics can be used.

In this chapter two types of stepwise non-parametric statistical decision procedures are presented. For the data which are the basic subjects of our investigations standard linear analysis was carried out. However, using the first method presented here the same effectiveness of discrimination can be reached using the same number of parameters, but our experiences with new data show that the non-parametric method is more robust than the linear one.

Some words about the data are appropriate. In nine medical centres from different countries 297 + 73 + 93 patients (suffering from three different diseases) were investigated by a WHO working group. Approximately 25 relevant characteristics of these patients were measured by non-invasive methods (such as ECG waves and blood pressures). In addition, the so-called PAP value was determined for each patient by an invasive method. The PAP values are divided into groups determined by the WHO working group. Our statistical methods are intended to predict only whether a patient belongs to a given PAP group, and always deal with a pair of PAP groups consisting of a lower and an upper PAP group (PAP_L and PAP_U).

THE METHOD BASED ON KOLMOGOROFF-SMIRNOFF STATISTICS

The objective of the method described below is to minimise the probability of misclassification using not more than 3 explanatory variables. The method determines thresholds of the explanatory variables by which the patient can be classified into the correct PAP group with minimal misclassification error. In order to determine the threshold of a given variable, we compute the empirical distribution function of the variable in the groups PAP_L and PAP_U separately.

Let $F_L(x_i)$ and $F_U(x_i)$ denote the empirical distribution functions of the variable x_i within the PAP_L and PAP_U, respectively. Comparing these two distribution functions, a cutting point (threshold value) q_i can be determined at which the difference of the two distribution functions is maximal and the sum of the false negative (α) and false positive (β) errors is minimal where: α = (number of patients belonging to the group PAP_U and classified into the group PAP_L)/(number of patients belonging to PAP_U); and β = (number of patients belonging to PAP_L and classified into PAP_U)/(number of patients belonging to PAP_L).

A Kolmogoroff-Smirnoff statistic is used to test whether the difference between the distribution functions $F_L(x_i)$ and $F_U(x_i)$ is significant. The formula for this statistic is:

$$d_i = \left(\frac{n_L\, n_U}{n_L + n_U}\right)^{1/2} \max_{x_i} \left|F_L(x_i) - F_U(x_i)\right| \tag{1}$$

where n_L and n_U are the number of cases in the group PAP_L and PAP_U, respectively (see, for example, Hoeffding, 1951). If the value of the Kolmogoroff-Smirnoff statistic is greater than 1.36 (1.63), then the difference between the distribution functions $F_L(x_i)$ and $F_U(x_i)$ is significant at the 5% (1%) level. The Kolmogoroff-Smirnoff test values were calculated for each continuous and discrete variable given by the WHO working group (note that the use of the Kolmogoroff-Smirnoff test for discrete variables is not quite correct). The threshold value q_i of the variable x_i with the greatest value d_i was selected for the first step of the stepwise decision procedure.

The threshold value q_i divides all cases belonging to the groups PAP_L and PAP_U into two sub-populations. The patients with value x_i less than or equal to the threshold value q_i are put into the group PAP_L, the others into PAP_U. Then the false negative and false positive errors can be calculated.

Remark: As for variables whose values decrease when the value of PAP increases, the patients with value $x_i > q_i$ are classified to the group PAP_L, and vice versa.

In the second step of our stepwise decision method the above procedure is repeated for each of the two sub-populations above by selecting new classification variables with new threshold values.

The procedure in principle can be continued until all d_i values become less than 1.36 (significance level) or the size of the respective sub-population is too small and so any further division would be meaningless. After finishing the sequential decision procedure, the errors α and β can be recalculated for the total classification.

An example for this method can be seen in Fig. 1.

Remark: On Figs. 1 and 2 the thick (thin) arrows point to the rectangles containing the number of patients classified into the group PAP_U (PAP_L).

The second row in the rectangle marked with ▲ on Fig. 1 contains the number of patients whose PAD values were measured. The PAP $\leqslant$ 29 and PAP $\geqslant$ 30 groups (PAP_L and PAP_U, respectively) were to be compared. At the first classification step the variable SWD1 had the greatest value d_i of the Kolmogoroff-Smirnoff test, so SWD1 and its threshold value q_i were

428

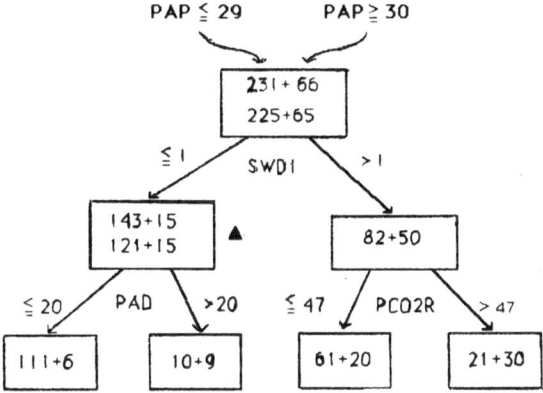

Fig. 1. An example of the sequential decision procedure.

entered in the first classification step. The patients with SWD1 > 1 have
been classified as belonging to the group PAP_U and the others came in the
group PAP_L (see Fig. 1). After the first step the error figures α = 15/65
= 0.231 and β = 82/225 = 0.364 were found.

At the second step the variable PAD (q_i = 20) was involved in the classification on the branch PAP_L, while for PAP_U the variable PCO2 gave the greatest d_i. From the decision tree the errors were computed: α = (6 + 20)/
65 = 0.4 and β = (10 + 21)/(111 + 10 + 61 + 21) = 31/203 = 0.153. If we
want to minimise the false negative error α then it is better to stop after
the first classification step on the branch PAP_U. In this case α = 6/65
= 0.092 and β = (82 + 10)/(82 + 10 + 111) = 92/203 = 0.453.

THE METHOD BASED ON FISHER'S EXACT TEST

The objective of this method is to select a relatively large sub-
population where the probability of misclassification is very low. The
restriction that the decision graph should be a tree is dropped, and the
number of explanatory variables can exceed 3. For the sake of simplicity
in the description of the procedure we assume that all explanatory variables
are positively correlated with the PAP value.

In the example presented here we intend to separate the sub-population
PAP_U. The procedure can be described inductively as follows. We are looking
for the explanatory variable and its threshold value such that the sub-
population having a value of the explanatory variable greater than the
threshold is at least 10% of the whole population; and the probability of
the event (computed using Fisher's exact test) that in this sub-population
the number of patients from PAP_L is less than or equal to its actual value
is minimal. If the number of patients belonging to PAP_L and classified to
PAP_U is still large, the procedure should be continued for the sub-population separated at the preceding step.

At the next step the remaining populations are joined together. For
this union we can repeat the former procedure, and so we get a new sub-
population with a small "one-sided" misclassification rate β. At the end
the separated sub-populations with low misclassification rate can be joined,
and as a result we obtain a relatively large sub-population in which the
probability of lower PAP value is small.

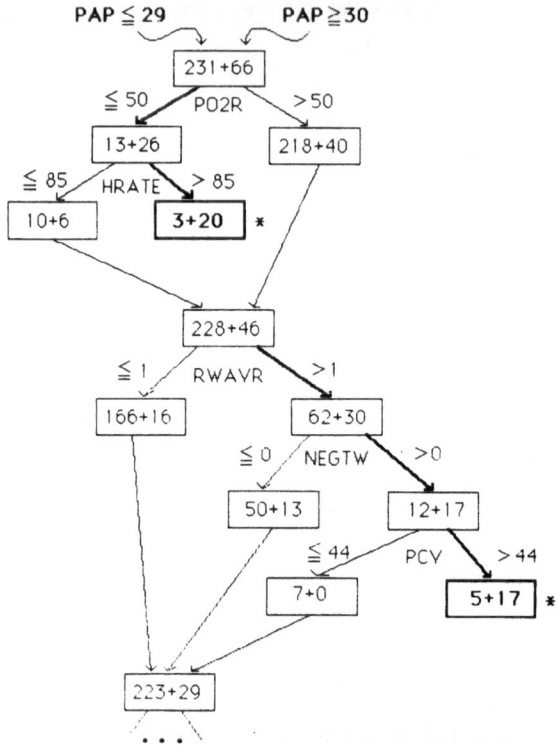

Fig. 2. An example of the second procedure.

An example of the second procedure is shown in Fig. 2. The rectangles marked with * contain the number of patients finally classified into the group PAP_U. The procedure was continued for the sub-population in the lowest rectangle (223 patients from PAP_L and 29 from PAP_U). Finally, 20 patients from PAP_L and 55 from PAP_U were classified into the PAP_U group. This seems to be a rather good result compared with the initial distribution of patients between PAP_L and PAP_U groups: 231 and 66.

REFERENCE

Hoeffding, W., 1951, Optimum nonparametric tests, in: "Proc. Second Berkeley Symp. on Math. Stat. and Probab.", J. Neyman, ed., University of California Press, Berkeley and Los Angeles: 83.

APPLICATION OF FUZZY SETS IN RHEUMATOLOGY

P. Masaryk, P. Píš, D. Žitňan, J. Lukáč and J. Rovenský

Research Institute of Rheumatic Diseases, Piešťany, and
State Sanatorium
Bratislava, Czechoslovakia

INTRODUCTION

Up to the present time, data on the application of mathematical models in rheumatology are very rare. The isolated attempts to apply mathematics in rheumatology remain on the level of experiment. This results from the fact that rheumatology deals mostly with qualitative signs. Some applications of discriminant analysis and cluster analysis have occurred, mainly in the analysis of clinical data (Adlassnig et al., 1985; Alpert, 1984; Inoue et al., 1986; Kingsland et al., 1983).

This chapter describes an attempt to apply fuzzy set theory in rheumatology and, in particular, in the process of the choice of suitable treatment. For this purpose one nosological entity has been selected - systemic lupus erythematosus (SLE). SLE is a systemic inflammatory chronic disease marked by temporary or permanent involvement of various organs (for instance, joints, kidneys, heart, lungs, skin and brain) and by the production of autoreactive antibodies against nuclear and other antigens, owing to abnormal function of the immune system and its effector mechanism. Before corticotherapy was discovered the average survival of patients had been 3 - 5 years; today it is extended threefold. In patients with SLE, the focus has been on the selection of suitable treatment. Four types of treatment may be considered according to the severity of the disease:

  (i)   basal (that is, non-steroidal antirheumatic drugs, antimalarials);
 (ii)   monotherapy with glucocorticosteroids (prednison, triamcinolon, hydrocortisone);
(iii)   combined therapy with cytostatics (glucocorticosteroids + cyclophosphamide); and
 (iv)   immunomodulatory therapy (glucocorticosteroids + cyclophosphamide + levamisole).

Next, the prognosis for a patient depends on the decision regarding the type of treatment. The physician must take into account a number of parameters. It is obviously very difficult to orientate oneself given the large quantity of data and so the physician's decision may also sometimes be incorrect.

The aim of this chapter is to ascertain whether there exists the possibility to model the decision-making of the physician in the real situation of choosing treatment for SLE by the assistance of fuzzy sets and, if yes, then to study the role of simple parameters and their combinations on the decision-making process of the physician.

MATERIAL AND METHODS

For this purpose the classification system has been used which is a part of the expert system with inductive learning created in the State Sanatorium in Bratislava. One of the possibilities of this system is also to make use of the fuzzy principle. Considering that the system can deal with a maximum of 16 parameters at once, the expert rheumatologist selected the parameters shown in Table 1 which he assumed to have a bearing on the choice of treatment for SLE.

The work had two phases. In the first one (learning with the teacher) the system learned to classify the patients in the single types of treatment (fuzzy sets) on the basis of the parameters and the type of treatment administered by the expert. In the second phase, the system alone classified the patients regarding the type of treatment. The degree of success of the system was then evaluated by examining the agreement of classification of the system with that of the expert.

RESULTS AND DISCUSSION

To begin with various haphazard combinations of parameters were chosen and the degree of success was observed. Eighty per cent success was achieved using combinations of all 16 parameters. Since all the valid parameters could not be selected when choosing the treatment (they were not available in all the cases considered in the training set), it can be assumed that the percentage success would be even higher. Therefore, it is possible to extrapolate that this system could be used as a model of the decision-making process of the physician in choosing treatment in SLE.

Next the role played by single parameters in the success of the decision was studied, that is the system knew the value of one parameter only. In this part of our experiment the most successful parameters were:

S-creatinine	58%
duration of disease	45%
anti-ds-DNA	44%
leucocytes in blood	41%
S-gamma-globulins	40%

These findings have a clinical correlation, as was confirmed retrospectively by the expert: S-creatinine is an indirect indicator of the function

Table 1. Sixteen Parameters Influencing the Choice of Treatment for SLE

1.	duration of disease	9.	glomerular filtration
2.	serositis	10.	resorption
3.	haemoglobin in blood	11.	U-erythrocytes
4.	E.S.R.	12.	U-leucocytes
5.	leucocytes in blood	13.	U-cylinders
6.	S-alpha-2-globulins	14.	U-proteins
7.	S-gamma-globulins	15.	anti-ds-DNA
8.	S-creatinine	16.	CH_{50}

of the kidneys (involvement of the kidneys is the most frequent and most severe finding in SLE); duration of disease (in cases of shorter duration a less aggressive therapy is applied); anti-ds-DNA (the sign of the activity of the disease and a sign of poor prognosis); likewise, leucocytes and gamma-globulins.

On the other hand, surprisingly less successful were parameters such as glomerular filtration (25%), resorption (23%), U-erythrocytes (21%) and serositis (17%). A "quasi-experiment" brought a partial explanation: parameters such as creatinine (which alone had the greatest degree of success) were combined with other parameters of renal function (having little success). It was assumed that the total success would increase, that they would together carry more information regarding renal function. However, in the experiment, the degree of success did not increase even by as little as 1% (that is to say, 58% either with creatinine alone or with the other parameters). S-creatinine alone yielded as much success as the other 6 parameters. This is correct. It reflects Cockcroft's pattern in nephrology.

Next, combinations of single parameters belonging together were tried (blood count, electrophoresis, immunological examinations, examinations of renal function). The best was the last; the others were approximately equal.

A trial was also carried out combining the three most successful parameters (S-creatinine, duration of disease and anti-ds-DNA). However, the degree of success did not increase in this case. On the contrary, it decreased by 20%. This could not be explained even by the expert. It was assumed that these parameters must have some mutual antagonistic validity in decision-making - they eliminate each other and so they confuse the model as well as the physician.

CONCLUSIONS

It is concluded that the model of decision-making in the choice of treatment for SLE presented here could be a very effective tool for studying the physician's decision-making process in this area of rheumatology. At present the model is able to classify in conformity with the physician's decision - with the expert. In future it is planned "to teach the model" to select the best therapy in such a way that the training set will be selected from patients with the best effect from a particular type of therapy. Such a model has considerable potential in rheumatology.

REFERENCES

Adlassnig, K. P., Kolarz, G., Scheithauer, W., Effenberger, H., and Grabner, G., 1985, CADIAG: Approaches to computer-assisted medical diagnosis, Comput. Biol. Med., 15:315.

Alpert, E. J., 1984, Computers in rheumatology practice; implications for clinical research, J Rheumatol., 11:697.

Inoue, T., Takeda, T., Koda, S., Negoro, N., Okamura, M., Amatsu, K., Kohno, M., Horiguchi, T., and Kanayama, Y., 1986, Differential diagnosis of fever in systemic lupus erythematosus using discriminant analysis, Rheum. Int., 6:69.

Kingsland, L. C., Lindberg, D. A. B., Sharp, G. C., 1983, AI/RHEUM, A consultant system for rheumatology, J. Med. Syst., 7:221.

EXPERIENCE FROM RUNNING THE CLINICAL INFORMATION SYSTEM CIS 1.T

A. Sušil, J. Münz, M. Dvořák, O. Gotfrýd and O. Volejníček

Research Institute of Traumatology
Brno
Czechoslovakia

INTRODUCTION

The main goal of clinical information systems is storing various kinds of information concerning the state of the patient's condition and processing them from various points of view. By "clinical information system" we understand, in most cases, the local information system of the clinical wards with its patient-oriented database (Dvořák and Gotfrýd, 1985).

An important characteristic of clinical information systems is that they are used almost exclusively by medical or paramedical staff. Nevertheless, their main working activity is not to collect information about patients and to store it in the database, but first of all to provide diagnostic and therapeutic care. The user's interactive dialogue with the clinical information system performed via a display has to be designed in such a way as to be simple and understandable for the non-technical user. Consider the requirements necessary for a non-technical user to have dialogue with a clinical information system. The dialogue should be

- sufficiently fast;
- non-demanding of the user's knowledge;
- easily interruptable and modifiable;
- resistant to the user's errors;
- unified and easy to survey; and
- individually directed.

The data collection process is, given the current state of accessible technology, time-consuming and as such is not attractive for medical and paramedical staff. Software has to control the user's dialogue and immediately to perform a great number of checks regarding the state of the input information.

For a clinical information system to function correctly it is necessary to store in the database structures a great number of data representing the medical knowledge base, the knowledge base concerning the clinical setting and the list of persons entitled to communicate with the system. From the point of view of a dynamic variable patient database, these structures and data files are usually termed "fixed structures".

In addition to fixed data regarding clinics, departments, patient rooms, relation between doctors and their patients and the technical equipment of the system, the medical knowledge bases create the most important part of the fixed structures. The extent of the medical knowledge base is proportional to the application of medical decision processes. Medical knowledge bases are also relatively extensive in the case when the clinical information system is reduced to minimum. They have to contain all the sentence fragments necessary for generating and simultaneously coding sentences which provide patient history and diagnostic information. Medical knowledge bases also contain all the necessary data about infusions, medication and other therapeutic preparations. The data in these databases yield information about the name, dosage, method of application, composition and possible interactions and contra-indications.

THE CLINICAL INFORMATION SYSTEM CIS 1.T

The Research Institute of Traumatology has developed a clinical information system, CIS 1.T, which has been used routinely since the beginning of 1985 (Münz et al., 1985; 1986). Although the present model cannot be considered ideal, the aim of its authors was to preserve all basic principles of the clinical information system that had been built up. We therefore assumed that by describing the modules of the CIS 1.T information system, which had been implemented and verified in practice, the concrete modules, system of communication, data storing and data processing could be made more understandable for the reader. The types of input information that have been implemented include:

1. Data input at the time of the patient's admission (IPAD);
2. Diet prescribing (IDIE);
3. Infusion therapy (IITO);
4. Drug therapy (IDRT);
5. Laboratory test ordering (ILVO);
6. Nursing care (INCR);
7. Monitored signals and measured values (IMMV);
8. Fluid balance (IFBL);
9. Laboratory test results (ILTR);
10. Patient's history (IANA);
11. Diagnoses (IDIA);
12. Subjective complaints (ISCM);
13. Operations performed and anaesthesia (IPOA);
14. Special analyses (ISAN);
15. Physiotherapy (IPHT);
16. Consultant examinations (ICON);
17. Updating patient's location (IUPL); and
18. Discharge summary (IDSM).

Programs can be chosen by number or mnemonic codes. Mnemonic codes also serve for fast connection with the chosen program in the case when the other list of programs is displayed. There are six similar groups of programs - for information output, for patient identification, for system programmers and so forth. The following paragraphs describe selected interesting CIS user modules.

Patient Identification

This module ensures an unambiguous identification of the patient. The patient is identified by his citizen's card-index number (or by a substituted citizen's card-index number as generated by the system), further by his surname, Christian name, and the number of his patient record. The identification entry data include mainly items from the front page of the classical patient record. The sub-system of identification includes programs enabling the retrieval of a patient defined earlier. The data in the patient identification structures can be used to output printed reports about the patient's admission to and discharge from the hospital as well as for statistical reports of all kinds.

The patient identification module also incorporates programs that enable data about the patient's location in the system to be updated or displayed. During the input of information about a new patient at the time of his admission, the patient is located in the CIS in the so-called (fictitious) admission ward. As soon as the patient is transferred to a particular clinical ward, the medical staff inform CIS about the patient's transfer (to the particular ward) with the help of a special program "updating the location of the patient". When the patient's stay in hospital is over, the patient's location in the CIS automatically changes and the patient is newly located in a (fictitious) discharge ward. CIS also has a program which provides a clear image about the actual location of the patient in the ward.

Patient History

The module for Patient History enables the patient's family and personal history to be input together with a description of the history of the injury and the condition of the patient.

Diagnoses

Diagnoses rank among the most important data on the patient's condition. These data are used both in the diagnostic and therapeutic processes and in nearly all types of actual and archive database processing. Great attention was therefore given to the method of recording the patient's diagnoses in the CIS database. Since automatic coding of a diagnosis is necessary at the moment of its input into the system, we rejected the simple storing of the diagnostic text in the record. The use of standard numerical codes (for example WHO) was not found suitable for clinical practice either.

There are several ways of inputting diagnoses into the computer. The first involves a successive choice from a list displayed on the terminal. These lists are sorted and contain diagnostic and location units which are linked logically.

The second method of diagnosis input demands observing certain criteria. The diagnosis is written in the form of Latin text, for example "FRACTURA APERTA SPIRALIS FEMORIS DX.". The input is followed by automatic analysis and simultaneous coding of the text (Gotfrýd and Dvořák, 1985). If the analysis of the submitted text is not completely successful, the system offers to complete the diagnosis by listing the items on the terminal followed by a choice of a particular alternative and the communicator continues to compile the required diagnosis.

Subjective Complaints

For the medical report to be complete, parallel data providing information about the patient's subjective condition are necessary, for example "without complaints", "headache", "insomnia", and so on.

Laboratory Tests

This sub-system includes two, almost mutually independent, modules – laboratory test orders and the entry, display and printing of laboratory test results. The programs concerning entry of the required laboratory tests cancel the order and also print a summary application form for the laboratory tests. This summary application form is used by paramedical personnel as the authority necessary for the taking of samples. The laboratory technicians in the Clinical Biochemistry Department use it when checking the completeness of the samples required and the samples delivered. This program module does not handle the requirements for statim tests.

The CIS does not participate in the organisation of the work in the Clinical Biochemistry Department. The clinical wards enter requests for individual tests in the form of summary applications after the analyses of samples have been entered with the help of programs for the input of laboratory test results.

The individual laboratory tests are grouped in lists of several types: biochemical tests of serum, urine, urine sediment, blood count and other haematological tests, and so on. All the stored data can be displayed or printed in several ways.

Measured and Monitored Factors

By measured factors we mean basic physiological parameters such as heart rate, blood pressure, respiratory rate and temperature which are typically measured by the nurse. Apart from these factors measured in the classical way, the CIS enables factors about the seriously injured patient, typically in the intensive care unit, obtained from on-line monitors, to be stored automatically in the CIS database. The set of monitored factors is given by the needs of our Institute and can be modified when necessary.

The unit processing the monitored signals is implemented on a micro-computer and it is able, from the given parameters, to compute the following results: heart rate, systolic and diastolic pressures, respiratory rate, inspiratory/expiratory ratio, expiratory volume, oxygen and carbon dioxide content in expired gas, and body temperature. If a signal is out of the given range an alarm is generated by the processing unit.

Fluid Balance

This program module enables the intake and output of fluids during a given day to be registered. The record includes volume values of fluids taken by mouth and infusion, and output by urine, stools, drainage, vomiting and so on. Other computed values include total intake, total output, respiration and the total balance of fluids in 24 h.

Diet

The dietary sub-system in our CIS is quite simple. It enables diets to be administered with additional items being included either by choosing one item from the list of alternatives or by loading in text which is then analysed. The administration of diets for the particular ward is summarised as a meal order for the kitchen. This meal survey is also a key for the distribution of meals.

Drug Therapy

This sub-system is designed for the processes associated with the ordering of drugs and the recording of their use in the management process.

All drugs are stored in the CIS database in coded form. This is necessary to ensure correct processing of the subsequent program modules. The total amount of drugs registered in the CIS is approximately one thousand. It is possible to add any other medication at any time, as required by the medical staff.

During the input of medical therapy orders into the CIS, the doctor first chooses a particular medication either by its full name or by the shortened form of its name. The particular program matches the set text and the name of the medication stored in the CIS database. In case of disagreement, the CIS offers a list of drugs of similar names to choose from. During further communication the doctor sets the mode of administration of the medication and its quantity. The last item to be set is the so-called time course of the administration of the medication. There is a choice of possibilities: once only, regularly at certain intervals, daily at a certain hour, or regularly on certain days. It is assumed that the quantity ordered is equal to the quantity actually administered, and therefore information about the actual quantity of the administered drug is not demanded from the CIS.

The records concerning the ordering of medical therapy serve as underlying data for the program "Time Schedule of Administered Drug Therapy". These printed pages are prepared for the paramedical personnel at given times twice a day, and they represent a schedule for the administration of medication to the patients.

Another possibility for using the records is as a list of drugs administered at the present time, which is a part of the printed page "Report of Patient Condition for the Doctor's Morning Rounds". The records can be further used for summarising - from various points of view - the total amount of drugs applied.

Infusion Therapy

This sub-system enables infusions to be ordered for patients and also provides for the storage of information about the composition of each infusion. Transfusions and blood derivatives are stored in the same way as infusions. The record about the infusion therapy contains: date, name of infusion, applied volume; further non-obligatory items are special admixtures (and their quantities) which are recorded in coded form (similarly to the main infusion), and the last item is a remark in the form of free non-coded text. The manner of choosing an infusion is similar to the choosing of drugs in the drug therapy sub-system. The sub-system also incorporates a program which helps the doctor to order the infusion therapy with regard to considerations of acid-base, ionic, energetic and water balance.

Operations and Anaesthesia

This sub-system is for the recording of operations and the subsequent processing of this information. The programs for the input of data about operations performed enable the date, time and place of the performance to be set, together with the members of the operating staff, operation diagnosis and features of the operations, which are coded for subsequent processing. The operation record is completed by free non-coded text describing the operation which is necessary for the printing of the whole operation report at a later time.

The records about the operations performed, taken from the patient database, can be used for printing summary reports giving the frequency of various types of operation over a certain period and about the number of operations performed by a certain surgeon. The chief of the department thus obtains the information needed for the planning of operations and the medical and support teams needed.

Special Analyses

An extensive CIS database enables a number of special programs to be worked out, which result in summary information about the functions of separate organs or physiological functions of the observed patient. Programs of this type are commonly called "Special Analyses". The proposed system allows for the gradual expansion of this group of programs. At present the module incorporates a frequency analysis of EEG signals, calculations of circulatory and respiratory parameters, and calculations of metabolic balance.

EEG signal analysis. The aim of this module is to obtain the efficiency spectrum of the EEG signal which was calculated with the help of a Fast Fourier Transformation, and to define the delta, theta, alpha and beta activities, and the frequency of maximal activity.

The calculation of circulatory and respiratory parameters. This sub-system facilitates the calculation of the following parameters: respiratory quotient, acid-base balance and pulmonary shunt, cardiac output (according to Fick), and dead space. The calculated parameters are used in the diagnostic and therapeutic processes, especially when it is necessary to determine the stage of pulmonary insufficiency, or when it is necessary to start, regulate or finish mechanical ventilation. Apart from this, by processing the results of acid-base balance and other biochemical tests, it is possible to diagnose miscellaneous disturbances in acid-base balance, the degree of kalaemia and to calculate the infusion volume needed to restore the "internal milieu" to normal.

The calculation of metabolic balance. From the data stored in the patient CIS database on applied infusion therapy, laboratory test results, temperature and fluid balance it is possible to calculate the metabolic balance including fluid balance, energetic, nitrogen, and ionic balance and also in patients receiving parenteral feeding. Calculations of metabolic balance are often of key importance in planning further parenteral feeding in patients who cannot be fed by mouth.

"Patient Passing" Through the CIS and the Role of the CIS in Ward Management

"Patient passing" means the patient's admission to the CIS, everyday checks of his location in the CIS and the occasional updating of information on the diagnostic and therapeutic processes, data processing from the patient-oriented database to produce printouts giving information about the patient's condition, and printouts for paramedical staff (time schedules), the semi-automatic generation of the discharge summary and the generation of "shadow structures" for further statistical processing. This activity survey can be extended by data archiving on, and reactivation from, magnetic tapes.

The input of patient data is performed in the central admission ward. The following data are gradually put into the CIS: basic identification data, patient's history, patient's weight and height, diagnoses on admission, subjective complaints, nursing care ordered, laboratory tests ordered, consultant examinations, prescribed diet, drugs and infusions, and non-coded text about the patient's admission. These data are immediately printed and serve as a primary source of information about the patient.

During admission, the patient is located in a fictitious admission ward. As soon as the patient is transferred to a clinical ward, the paramedical staff immediately inform the CIS about the patient's actual location. The CIS must be informed about any transfer of the patient.

Most data are input into the CIS after the doctor's morning rounds by clerks. Technicians on night duty print a "Data for the doctor's morning round" paper for each patient. This paper contains: subjective complaints, a survey of diagnoses, fluid balance, measured values in the last 24 h, treatments ordered and performed, laboratory test orders and results, current diet, current drug and infusion therapy, and a survey of rehabilitation (for example, physiotherapy) performed. On this page the doctor doing the ward round writes new data about the patient or new orders, or cancels some of the old orders. These new data are then input into the CIS by clerks. •

New time-schedules for paramedical staff are printed after up-dating the database. These printed reports are sorted by departments and for each department by time, patient rooms and beds. The CIS enables a discharge summary to be printed semi-automatically after the end of the treatment. The discharge summary contains the most important data stored in the database during the patient's treatment (identification, history, diagnoses, last laboratory results, last measured factors, drugs, and so on). These data are completed by a doctor before discharging the patient, including some important facts that were not yet among the patient data (for instance a verbal description of the development of the illness and recommendations to the general practitioner).

The information stored in the database is used for the automatic generation of several administrative and statistical reports. The "shadow structures" mentioned above (for example, diagnosis-oriented, operation-oriented) serve for the rapid selection of data for research purposes.

CONCLUSION

The clinical information system CIS 1.T has been working for nearly three years in our Institute, 24 hours a day, 7 days a week. This demonstrates its vitality and also indicates the fact that it is possible to extend it by the inclusion of new functions and modules.

REFERENCES

Dvořák, M., and Gotfrýd, O., 1985, Descriptor - a software tool for construction of flexible database, in: "Proceedings of Medical Informatics Europe 1985" (Lecture Notes in Medical Informatics 25), F. H. Roger, P. Grönroos, R. Tervo-Pellikka and R. O'Moore, eds., Springer-Verlag, Berlin : 791.

Gotfrýd, O., and Dvořák, M., 1985, An approach to free text syntactic analysis, in: "Medical Informatics Europe 1985" (Lecture Notes in Medical Informatics 25), F. H. Roger, P. Grönroos, R. Tervo-Pellikka and R. O'Moore, eds., Springer-Verlag, Berlin : 216.

Münz, J., Dvořák, M., Gotfrýd, O., and Peštál, M., 1985, Clinical information system for urgent surgery and traumatology, in: "Medical Informatics Europe 1985" (Lecture Notes in Medical Informatics 25), F. H. Roger, P. Grönroos, R. Tervo-Pellikka and R. O'Moore, eds., Springer-Verlag, Berlin : 101.

Münz, J., Dvořák, M., Gotfrýd, O., and Volejníček, O., 1986, Patient oriented database structure and data processing for departmental IS, in: "Medinfo 86", R. Salamon, B. Blum and M. Jørgensen, eds., North Holland, Amsterdam : 1130.

DPS - A DIAGNOSTIC PREVENTIVE SYSTEM

I. Horský, L. Scheidová and I. Seress

Medical Bionics Research Institute
Bratislava
Czechoslovakia

INTRODUCTION

This chapter describes work which is being carried out, directed
towards the development of automatic diagnostic preventive systems (DPS).
Both technical and medical dimensions of the work are considered in terms
of the technical processes associated with the development of the equipment and its possible use in medical practice.

TECHNICAL ASPECTS OF DPS

Medical instruments and the associated measuring techniques constitute
that part of the screening system with which the person being examined has
direct contact. This fact significantly influences its choice, construction and the technological demands. An analysis confirmed the need for
the application of digital computing techniques and microelectronics with
the aim of creating rational, high-effective automatic diagnostic instruments and instrument complexes. The functional and constructional compatibility of self-contained medical instruments, nodes and blocks is the un-
conditional prerequisite for the complete delivery of the individual sub-
systems as well as the whole complex for the system of mass screening.

The measurement techniques proposed for such diagnostic centres must
take into account the generally valid demands of the system, especially
with regard to: functionality, long-term reliability, computer-compati-
bility, accuracy, acceptability to the person being examined, operational
simplicity and modularity.

All the blocks of the measuring and instrument complexes must fulfil
criteria relating to information transfer, ergonomic considerations and
their construction, together with metrological and energy criteria as well
as considerations of user compatibility. Other factors include: unification of input and output signals, selection of metrological characteristics
and methods, hierarchical compatibility in relation to lower level and
higher level complexes, unification of conditions for compatibility co-
operation with computing techniques and the possibility of connecting the
sub-system to the central computer; also elaboration of the user's
characteristics as an indicator of reliability, unification of power supply
parameters - maintaining electric security, the possibility and realisation

of rationalisation tests on the single function sub-systems as well as on the system as a whole.

The measuring technique adopted, with the exception of basic functional demands used in the automatic sub-systems, must enable the measured data and other discrete information to be transmitted in ASCII code. For continuous signals an output in analog form with output levels ±5V, ±1V respectively was considered. The transfer of continuous signals is realised in analog form by means of a transfer set consisting of modulators and demodulators. Furthermore, the basic functions in such a system must be controllable by electronic means. From a detailed analysis it was concluded that for most of the sub-systems being incorporated the following set of electronically controllable functions would be sufficient: scanning, with the continuous output of an analog signal, calibration, zero level, shift of the recorder, measurement, discrete data emission, automatic cycle start up, control of auxiliary equipment, special functions, sensitivity switching and filtering. Communication demands between the central computer and the individual sub-systems has been considered in detail and published on a number of occasions (Seress, 1986; Seress et al., 1984). To achieve this linkage and further necessary associated functions, a communications module with the working name "Computer connection unit" was designed and developed.

Experience gained during the development of a system for mass preventive screening of the population in the District Hospital of Trencín, as well as from literature on this subject over the whole world (Anggard et al., 1986), has become the point of departure for the DPS - diagnostic preventive systems project in the Medical Bionics Research Institute. From the technical point of view, the following are important:

  (i)   the mode of connection of medical equipment to the computer;
 (ii)   the modular nature of the system and resultant flexibility; and
(iii)   the maximum reliability of the computing techniques employed in
        the system.

Hence, in developing the DPS in the context of the prevention of cardiovascular disease, the medical equipment is linked in an on-line manner to an SM 50/50 M1 microcomputer in extended configuration, or to an SM 4/20 minicomputer. In the first stage of DPS development, a TK automatic blood pressure instrument is used in the tonometry sub-system, a Chirastar 32U in the ECG sub-system, and a Chiradat in the spirometry sub-system. The anamnesis and anthropometric sub-systems communicate with the computer by means of video-terminals.

MEDICAL ASPECTS OF DPS

Preventive systems may be oriented to general or special medical problems such as to oncological, cardiopulmonary or other diseases. Accordingly it is necessary to define the specific examination methods to be adopted in seeking to identify a selected type of disease, as well as the risk factors which result in a predisposition towards such a disease. This is an important and serious medical problem.

The choice of methods may be considered at two levels:

 (i)   first, there are the obligatory methods representing the very basal
       and most necessary examination methods in order that the system
       should be capable of realising its fundamental purpose; and
(ii)   at a second level there are facilitative methods which are capable
       of extending the spectrum of examinations. The choice is secured

444

by the decision-making algorithm software according to the results
of the first level of examinations.

In our Institute such a system is currently being evaluated in the
context of cardiopulmonary diseases, and it is this which will be described
in more detail.

The person to be examined receives an invitation by post at home, together with some instructions. He brings with him a completed medical
questionnaire, identification data, and early-morning urine sample, and,
after changing his dress, is examined in terms of the following sub-systems
which are depicted in Fig. 1:

Anamnesis (history taking) sub-system. This is realised by a questionnaire form with 50 questions, which the person being examined completed at
home; he marks the positively-answered questions. The relevant numbers of
the questions are fed into the computer by a staff member via a video-
terminal.

The laboratory sub-system. Here urine and blood samples are examined
for their haematological and biochemical profiles. After the usual withdrawal, the biological material is analysed and the results are fed into the
computer by means of a video-terminal.

The spirometry sub-system. After relaxation and instruction regarding
the nature of the test, the person being examined performs forced expiration
following a maximal inspiration. The measured values are recorded twice.

The tonometry sub-system. After the person being examined has relaxed,
the automatic measuring equipment secures the automatic blowing up of the
tonometer cuff; measurement of systolic and diastolic blood pressure is
performed twice, and the measured values are transmitted to the computer.

The ECG/VCT sub-system. The electrodes are placed in defined positions
in the usual way for the orthogonal conducting system according to Frank.
The sensed ECG signal is both recorded on a chart recorder and transmitted
via an A/D converter to the computer. The Pipberger software performs an
evaluation of the ECG signal.

The anthropometry sub-system. In this sub-system, height and weight
are measured in the traditional way, together with measurements of 6 fat
folds and the defined circumferences. The measured parameters are put into
the computer by means of a video-terminal.

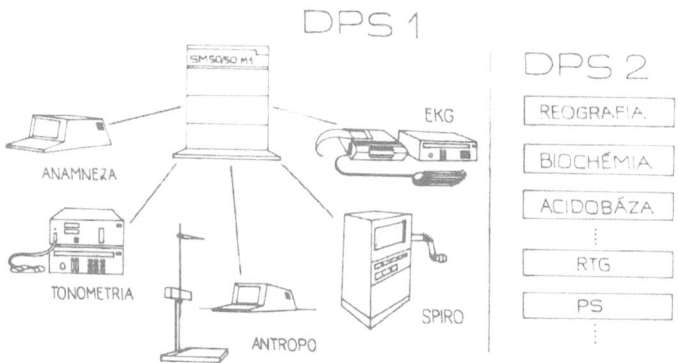

Fig. 1. Sub-systems of the DPS diagnostic preventive
system.

The application software secures a recording of the measured values of the defined parameters, as well as averaging them and comparing them with normal values. The evaluation process considers sex, age, weight, height and other physical measurements. The software also prints the listings of the measured and normal values, pointing out pathological findings in a logical well-structured table, as well as drawing possible diagnostic conclusions. In parallel, decision-making algorithms on each sub-system have been established that enable risk factors to be selected, an indication to be given of the need for blood withdrawal for biochemical analysis as well as the necessity of proceeding to the second level of examination (DPS 2). This second level includes: examination of present status, chest X-ray, impedance plethysmography of the extremities, and acid-base and blood gas analyses.

Up to now 550 subjects have been examined in this way, subjects whose state has been correlated by the physicians in the traditional way. At present this group is being evaluated statistically. Although the system is currently oriented towards cardiopulmonary diseases, its flexibility enables it to be adapted to other diseases dependent on organisational, technical and economic possibilities, as well as the defined main goals of the users.

REFERENCES

Seress, I., Horský, I., and Mikulás, M., 1984, Komunikácia zdravotníckej techniky s pocítacom (Communication of medical equipment with the computer), Lékar a Technika, 15(1) : 4.

Seress, I., 1986, Ergonómia zdravotníckych prístrojov pre ASR (Ergonomics of medical instruments for automatic control systems), in: "Elektrotech. Rocenka 1986", ALFA, Bratislava : 305.

Anggard, E. E., Land, J. M., Lenihan, C. J., Packard, C. J., Percy, M. J., Ritchie, L. D., and Shepherd, J., 1986, Prevention of cardiovascular disease in general practice: a proposed model, Br. Med. J., 293 : 177.

TRANSFORMATION OF DATA TO FINDINGS

O. Gotfrýd, M. Dvořák, J. Münz and L. Popelínský

Research Institute of Traumatology
Brno
Czechoslovakia

CLINICAL INFORMATION SYSTEMS

There are several institutes in Czechoslovakia which have been developing relatively extensive software for clinical information systems that
have been realised over the last few years. These systems are designed as
interactive database information systems and make possible the automation
of the following fields:

  (i)   display and printout of information concerning patient state and
         his therapy hitherto;
 (ii)   clinical ward control and organisation including working documen-
         tation;
(iii)   summary and specialised printouts for every patient;
 (iv)   statistical and summary printouts concerning clinical ward and
         staff activities;   and
  (v)   retrospective statistics and the processing of research archive
         data.

It is obvious that the majority of processed data must be stored in
the database in a coded form, that is in a formalised structure. The coding
is performed semi-automatically as part of the user-computer dialogue during
data input.

The basic features (rather negative) of such clinical information systems may be summarised as follows:

  (i)   neither diagnostic nor therapeutic procedure is influenced automati-
         cally by the contents of the database;
 (ii)   the extent and contents of clinical documentation (and also of the
         patient-oriented database) are defined by a head of a department,
         and they are limited by the feasibility and working regime of the
         clinical ward;   and
(iii)   data are presented mainly in the extent and form in which they have
         been entered into the system;   an exception to this is a relatively
         small number of calculations of derived parameters and a summary of
         selected information.

Experience gained during running the information system in our Institute over the last three years may be summarised as follows:

(i)   there were no problems with data coding and with the extension of
data types due to a flexible database structure;

(ii)   the relative independence of the contents of the patient database
became evident in a negative manner quite soon; and

(iii)   the system has relatively limited possibilities for the medical
interpretation of data and their ambiguity.

LOGICAL DATA REDUCTION

After analysis of the properties and experiences gained from running
the clinical information system, we came to the conclusion that it was
necessary to transform the "rough" primary data stored in the database to
logically higher entities that may be marked as <u>findings</u>. The automatic
generation of entities and statements is called data transformation to
findings. This generation is based on data processing and interpretation.
Data may be both in a quantitative form and also in a descriptive form and
are stored in the patient-oriented database.

Let us suppose that existing findings, entities or statements may be
separated to the three basic categories of: event, phenomenon, and parameter course.

<u>Event</u> is a term indicating the occurrence of a fact concerning patient
state or his therapy. The facts must be previously defined and should significantly influence data interpretation or should delimit different phases
of disease or therapy (for example, road accident, surgical operation,
start of oxygen therapy).

<u>Phenomenon</u> is a process characterised by the simultaneous satisfaction
of one or more conditions previously defined. These conditions must be
satisfied during the whole time interval (for example, specified drug administration, assisted ventilation).

<u>Parameter course</u> is also previously defined. It is a classification
of the sample sequence of quantitatively measured data in a given time
interval (for example, a parameter course alternating, constantly low).

The proposition of these definitions satisfies the interpretation requirements for the long-term care of patients. The basic time quantification is
one day.

There is considerable difficulty in defining the necessary set of
entities even for the description of a relatively well-specified disease.
There are two possibilities:

(i)   the construction of a necessary entities structure in a manner
common with the getting together of the medical knowledge base for
an expert system; and

(ii)   the entity structure may be obtained by inter-personal communications
among the working team of physicians.

Search for Event Occurrence

The search for event occurrence in patient data is realised on a day
by day basis from the beginning of the patient's stay in hospital. An
event can be:

(i)   the existence of a certain type of information record in the patient
file in a corresponding day (for example, surgical operation record);

(ii)   the existence of a given daily finding of a certain parameter in a patient file in the corresponding day (for example, when a certain parameter exceeds previously defined limits; or

(iii) the existence of a given item in a patient record (for example, the start of oxygen therapy).

The result of the event occurrence search is stored in coded form in a patient data file.

Search for the Occurrence of Phenomena

The search for phenomenon occurrence in patient data is realised day by day from the beginning of the patient's stay in hospital. The difference between an event and a phenomenon relates to the continuous existence of a given fact for more than one day (for example, antibiotic treatment). The result of the search for the occurrence of phenomena is stored in a coded form in a patient data file.

Search for Typical Trends in Quantitative Data

Those data that may be expressed by a number related to the time of occurrence or detection of the value are called quantitative data. We may consider the sequence of samples of such a real data type to be a time sequence and typical trends in such a sequence may be searched.

The first step in the processing of a numeric sequence is the representation of the numeric sequence corresponding to one day by a symbol. The symbol will be called a daily finding. A time sequence of quantitative data indicated as a concrete daily finding is determined by the physician in accordance with his opinion (for example, low alterating, extremely high).

Reduced trends in the data are obtained by lining up the daily findings of concrete quantitative data as a time sequence. It is necessary to establish in advance a set of pre-defined typical trends of quantitative data. This allows typical trends in the daily finding sequence (for example, temporary increase over the normal limit, immediate decrease under the normal limit) to be searched. The result of a typical trend search is stored in a database in coded form.

RETROSPECTIVE PATIENT DATABASE PROCESSING

It is necessary to perform retrospective database processing to be able to establish which findings and entities pre-defined for the specified disease description are significant. The processing must be performed with data from a specialised set of patients with a given disease. Statistical evaluation of finding occurrences and of automatically generated relations between each of them verifies which events, phenomena or typical trends are acceptable and which newly-found correlations between them are valid.

PATIENT DATABASE REDUCTION

Effective reduction of stored patient data may be applied when we have a sufficiently extensive "dictionary" of events, phenomena and typical trends for certain disease types. Instead of storing primary patient data it is possible to archive the data transformed to findings, that is, the statements characterising the patient's state and the treatment of the patient. The volume of stored data will be significantly reduced with minimal loss of information by eliminating banal and redundant data.

The database of findings automatically generated from the primary

patient data, sufficiently extended and with acceptable success, will
accelerate a dialogue between users and the clinical information systems.

VERBAL PATIENT STATE DESCRIPTION

Standard situations may be described using schematic sentences consisting of text constants and text variables. To allow such a description, a
verbal description of pre-defined typical trends, events and phenomena and
relations between them must be known. Text variables take on the values of
the verbal description of typical trends, events and phenomena when a concrete situation is described.

The length of the daily finding sequence specifies the number of days
D on which quantitative data Q of a given parameter are reduced. An output
sequence of typical trends is composed of symbols $T1$, $T2$, ... The parameter
is investigated between two events $E1$ and $E2$. The schematic sequence may
then have a structure like, for example:

The trend of Q was between $E1$ and $E2$ (that is D days) first $T1$, later
$T2$, ...

Example:

The trend of the arterial blood pressure was, between the surgical operation
and the start of physiotherapy (that is 5 days), first temporarily decreased
to normal value, later high alternating.

The substituted text variables are typed in this expressive manner.

A description of phenomenon P may be substituted by a schematic sentence
in a similar fashion. Such a sentence may have a shape of the form, for
example:

P has continued D days after E.

Example:

Antibiotic treatment has continued 7 days after the surgical operation.

Similar schematic sentences describing a patient's state and his therapy may
be generated. Verbal descriptions of previously determined facts searched
in a patient's data will be substituted into such sentences.

EXAMPLE OF REDUCTION AND VERBAL DESCRIPTION

In this section there is shown an example of the automatic interpretation of physiological patient data collected during a period of five days
after a surgical operation and after the beginning of antibiotic treatment.
The data concerning the facts mentioned above are dispersed in the database
in records of different types. They are searched, found and reduced to
findings that may be described verbally. The user of the information system,
following his query, is presented not only with columns of numbers but also
with a text describing facts important for the patient's state and its
dynamics.

In Table 1 there is shown a survey of real patient physiological data.
A description of these facts looks as follows:

The heart rate at first remained high, later decreasing to normal values
over the 5 days.

Table 1. Example of Measured Physiological Parameters Over the Period
12.12.1986 to 16.12.1986

DATE:	12.12	12.12	13.12	13.12	13.12
TIME:	0:00	20:00	0:00	8:00	12:00
HEART RATE:	102	100	96	76	68
ART. PRESSURE:	110/80	130/90	150/90	150/80	170/110
RESP. RATE:	24	24	20	24	20
TEMPERATURE:	36.4	36.0	37.0	37.0	36.5
DATE:	13.12	14.12	14.12	14.12	15.12
TIME:	20:00	8:00	12:00	20:00	0:00
HEART RATE:	96	80	84	92	84
ART. PRESSURE:	180/110	160/90	140/80	150/90	150/100
RESP. RATE:	22	20	20	20	20
TEMPERATURE:	37.0	36.4	36.4	37.2	36.3
DATE:	15.12	15.12	15.12	16.12	16.12
TIME:	8:00	12:00	20:00	8:00	20:00
HEART RATE:	82	78	90	98	100
ART. PRESSURE:	140/80	140/80	150/100	150/90	120/80
RESP. RATE:	20	20	20	24	24
TEMPERATURE:	36.3	36.5	36.3	37.2	36.8

The systolic pressure at first remained normal, later increasing above normal
values over the 5 days.

The diastolic pressure at first remained normal, later increasing above
normal values over the 5 days.

The respiratory rate remained normal during the 5 days.

The temperature remained normal during the 5 days. Antibiotic treatment
was continued for 5 days after the surgical operation.

The example shown illustrates the difference in the form of presentation of the same information concerning the development of the patient's
state. When comparing the volume of coded findings (in our example mostly
typical trends) with the primary data volume we can appreciate the great
reduction of the stored volume of data.

CONCLUSION

The approach presented for the processing of patient data stored in
the clinical information system is an attempt to create algorithms for the
formation of higher logical entities with a larger volume of important
information. This importance refers to the general situation where the
particular data are evaluated. The difficult definition of the description
of evaluated facts so that they could be automatically searched and transformed into findings remains a relatively large problem. The considerable
reduction in the volume of stored data and the possibility of automatic
verbal description of patient state at the level of communication usual
among physicians make the tackling of this difficulty worthwhile.

BIBLIOGRAPHY

Duisterhout, J., Franken, B., and Schoemaker, J., 1986, Constructing departmental systems by using the 4th generation software package AIDA, in: "Medinfo 86", R. Salamon, B. Blum and M. Jørgensen, eds., North Holland, Amsterdam : 1081.

Dvořák, M., and Gotfrýd, O., 1985, Descriptor - a software tool for construction of flexible database, in: "Medical Informatics Europe 1985" (Lecture Notes in Medical Informatics 25), F. H. Roger, P. Grönroos, R. Tervo-Pellikka and R. O'Moore, eds., Springer-Verlag, Berlin : 791.

Dvořák, M., Münz, J., Gotfrýd, O., and Sušil, A., 1987, Exploitation of physiologic and laboratory data in a clinical information system, in: "IMIA Conference on Progress in Biological Function Analysis by Computer Technologies", Berlin : 10-04 (Abstract).

Franken, B., and Molenaar, K., 1986, Fifth-generation application generation: AIDA's expert shell AIDEX, in: "Medinfo 86", R. Salamon, B. Blum and M. Jørgensen, eds., North Holland, Amsterdam : 1091.

Münz, J., Gotfrýd, O., and Dvořák, M., 1985, TREND - a high level monitoring system with application of AI methods, in: "Medical Decision Making: Diagnostic Strategies and Expert Systems", J. H. van Bemmel, F. Grémy and J. Zvárová, eds., North Holland, Amsterdam : 57.

Münz, J., Dvořák, M., Gotfrýd, O., and Pestál, M., 1985, Clinical information system for urgent surgery and traumatology, in: "Medical Informatics Europe 1985" (Lecture Notes in Medical Informatics 25), F. H. Roger, P. Grönroos, R. Tervo-Pellikka and R. O'Moore, eds., Springer-Verlag, Berlin : 101.

Münz, J., Dvořák, M., Gotfrýd, O., and Volejníček, O., 1986, Patient oriented database structure and data processing for departmental IS, in: "Medinfo 86", R. Salamon, B. Blum and M. Jørgensen, eds., North Holland, Amsterdam : 1130.

CONCEPTUAL MODELLING IN MEDICAL DATABASE DESIGN

M. Holeňa

SVT ČSAV
Prague
Czechoslovakia

INTRODUCTION

Since the early 1970s, the need for a special design phase at the early stage of the database or information system design process began to emerge. This phase is usually called that of conceptual modelling or conceptual design. The development of database design theory prior to the common acceptance of this design phase was much influenced by the ANSI/SPARC proposal for a three-level architecture for database systems (see, for example, Tsichritzis and Klug, 1978) and the efforts of the ISO TC97/SC5/WG3 (1982). The features of conceptual modelling that contributed most to its acceptance are:

(i)   it postpones the design decisions concerning the hardware configuration and the software components including the decision as to what type of database management system is to be used (for example, relational, Codasyl-like);

(ii) it enables the implementation environment (hardware as well as software) to be changed without the need to repeat the design process completely - however important the change may be, the results of conceptual modelling can always serve as a starting point for the re-design of the database; and

(iii) it leaves the computer-oriented way of thinking in terms of data, records, elements, files and so on - a very important feature from the viewpoint of communication between the database designers and its future users.

In comparison with more traditional areas of database applications, less attention has been paid to conceptual modelling in medical database design so far. Nevertheless, we believe that it is only for historical reasons. At our Institute, we have designed or have been designing various databases, including a medical one - a small clinical database for the Chorioneptiheliom Research Centre in Prague. Consistently carrying out the process of conceptual modelling, we found that this design phase is equally important for medical as for other kinds of databases. Naturally, other aspects must be stressed during conceptual modelling in the case of medical databases rather than, for instance, in the case of ones for business application, to reflect the specific character of the part of reality concerned, particularly its dynamic and often indeterministic nature.

453

This chapter is based on the experience we have with conceptual modelling at our Institute and deals with the first sub-phase, particularly from the viewpoint of medical database design.

OBJECT SYSTEM MODELLING - AN IMPORTANT PART OF CONCEPTUAL DESIGN

Just as database implementation design can be divided into logical design and the physical one according to the relation to data physical storage characterisation, so we can also divide the conceptual modelling phase into two different sub-phases:

Object System Mapping

The subject of this sub-phase is the determination of those elements and classes of elements of reality (for example, things, persons, institutions, properties, decisions, processes, relationship between things and persons) that could be so important for some of the future database users that they would like to be able to obtain information about them from the database. The term object system (universe of discourse) refers to that part of reality, information about which is to be obtainable from the database; hence, the object system mapping means a type of systems analysis of this part of reality. The process of object system mapping requires the close participation of future users. Their requirements serve as a starting point for this process and on the basis of their remarks its results are iteratively corrected and improved.

Database Information Specification

Its subject is specifying the way of identifying and handling the information stored in the database. This sub-phase includes the solving of problems connected with the choice of names and identifiers, synonyms and homonyms, information deducibility, integrity constraints, consistency conditions, specification of transactions and their realisation in algorithmic form, synchronisation, pre- and post-conditions of transactions, exception handling and so on. To solve some of these problems, it may be necessary to make, in advance, formal transformations of the structures representing information, for example, decomposition or different kinds of normalisation. In the case of a database for more users, one of the main problems to be solved in this sub-phase is the consolidation of the views of different users.

Object system mapping is the proper topic of this chapter. This sub-phase has an exceptional position compared with other parts of the database design process. It requires more insight into the problems of the object system, involves more intuition and more often corrections of results already obtained. Furthermore, errors committed during this sub-phase are very expensive to remove at a later time. If we may, for guidance, use the results of Boehm (1981) (concerning other kinds of software systems than databases), then the increase in cost of removing an error committed during object system mapping is about 2 - 5 times if it occurs during the database information specification, about 5 - 15 times if it occurs during the implementation design, and if it occurs during the database implementation, about 15 - 35 times compared with its removal directly during the object system mapping itself (Brodie and Ridjanovic, 1984).

The basic and most widely used method of object system mapping is the representation of the object system by one or more abstract structures built up by means of several formally distinguishable constructs, each having a specific interpretation in reality. Hereafter, we shall call such abstract structures object system models and the object system mapping by means of them object system modelling. Specificity of construct interpretation differentiates object system models from traditional data models (that is, relational, hierarchical and network models).

OBJECT SYSTEM STRUCTURE AND DYNAMICS MODELS

As far as medical databases are concerned, of particular importance is usually the dynamic component of the object system, which comprises, for example, disease occurrence, progression, therapeutic decisions, surgical actions, response, relapse, recurrence, death, arrival of a patient for a consultation, admission to hospital (Hammer and McLeod, 1981; Peimann, 1983; Rolland et al., 1982; Rothemund and Ellsässer, 1984). More generally, we shall give the name dynamic component to that part of the object system formed by changes of things, properties, abstract concepts, relationship and so on, different events, actions, activities, decisions, evolution, but also, for instance, the passage of time. The importance of the dynamic component of the object system is twofold:

(i)   Some elements or classes from the object system dynamic component belong to the ones in which the future users are most interested. In the case of a medical database, a very important purpose for which it is designed is usually that of keeping track of the development of each patient's disease and therapy. This information is needed for the physician's decision making, for statistical analysis and for prediction.

(ii) Many elements of the dynamic component during the use of the database regularly result in certain operations or successions of operations with information and, consequently, also with data (data insertion, deletion, modification and retrieval, transfer of data between main and external storage or between different storage units, and so on). On the other hand, for some of the operations with data one or more stimulating elements of the dynamic component may be found. That is why this component plays a very important role during the specification of database information (for example, for specification of transactions) and during the implementation design (it serves as a basis for the estimation of database performance characteristics and for the design optimisation accordingly).

Some of the object system models enable the fact that an element of the model represents an element or a class from the object system dynamic component to be caught explicitly - that is, in such a way that one or more of the model constructs is applied only to represent such elements or classes. We shall call such models object system dynamics models. The other models, in turn, enable this fact to be caught only implicitly, that is, by means of the semantics of each single element forming the model. In some cases, for instance if we want to specify the operations with information during the database information specification, such a way of representation may be very cumbersome and awkward. These models, for the sake of convention, will be called object system structure models.

So far, object system structure models have been used far more often during the conceptual modelling. Also at our Institute, we used to confine ourselves to an object system structure model, namely, to the entity-relationship (ER) model (see, for example, Elmasri et al., 1985; Chen, 1976). This was chosen because of its most common use as an object system model. In this respect, it surpasses both the semantically richer models - for instance, SAM (Su, 1986), SHM (Brodie and Ridjanovic, 1984), SDM (Hammer and McLeod, 1981), or the infological model (Sundgren, 1975; Tsichritzis and Lochovsky, 1982), and models enabling easier communication with future users - for example, binary (Nijssen, 1979; Tsichritzis and Lochovsky,

1982), HIT (Staníček and Krejčí, 1984; Zlatuška, 1984), and other functional models (Kulkarni and Atkinson, 1986; Manola and Orenstein, 1986). Also, several international conferences held on the ER approach since 1979 have documented the attention that the ER model attracts (the 6th of them was held in New York in 1987).

The ER model is described in more detail in the next section. The final section contains a very brief review of 12 object system dynamic models. This review is based on results of a comprehensive study and comparison of these 12 models carried out at our Institute in connection with the object system modelling of our medical database.

THE ENTITY-RELATIONSHIP MODEL

An object system model, being an abstract structure, can be defined formally. The version of the ER model we used during the object system modelling of our medical database was defined as follows:

An <u>ENTITY-RELATIONSHIP MODEL</u> is a triple $\left((E_i)_{i \in \hat{p}}, (R_j)_{j \in \hat{q}}, (A_k)_{k \in \hat{t}} \right)$ where

$p, q, t \in N \& \left(\forall i \in \hat{p} \right) \left(\forall j \in \hat{q} \right) \left(\forall k \in \hat{t} \right) E_i, R_j, A_k$ are sets $\& E_i \cap R_j = \emptyset \&$

$E_i \cap A_k = \emptyset \& R_j \cap A_k = \emptyset \& \left(\forall j \in \hat{q} \right) \left(\exists s_j \in N \setminus \{1\} \right) \left(\exists p_{j,1} \ldots, p_{j,s_j} \in \hat{p} \right)$

$R_j \subset E_{p_{j,1}} \times \ldots \times E_{p_{j,s_j}} \& \left(\forall k \in \hat{t} \right) \left(\exists V_k - a \text{ set} \right) \left(\exists D_k \in \{ E_1, \ldots, E_p \} \cup \right.$

$\{ R_1, \ldots, R_q \} \right) \left(\forall f \in A_k \right)$ f is a mapping $\&$ def$(f) = D_k \&$ val$(f) \subset$

$2^{V_k} \& \left(\forall d \in D_k \right)$ card $f(d) < \infty$.

Here $\left(\forall i \in \hat{p} \right) E_i$ is called an ENTITY TYPE, $\left(\forall e \in E_i \right) e$ is called an ENTITY, $\left(\forall j \in \hat{q} \right) R_j$ is called a RELATIONSHIP TYPE, $\left(\forall r = (e_1, \ldots e_{s_j}) \right.$ $\in R_j) r$ is called a RELATIONSHIP, $\left(\forall k \in \hat{t} \right) A_k$ is called an ATTRIBUTE, $\left(\forall f \in A_k \right) f$ is called a CONSTELLATION of A_k. In particular, if $k \in \hat{t} \& \left(\exists u \in N \right) \left(\exists t_1, \ldots, t_u \in \hat{t} \setminus \{k\} \right) V_k \subset V_{t_1} \times \ldots \times V_{t_u} \& \left(\forall h \in \hat{u} \right) D_{t_h} =$ $D_k \& \left(\forall f \in A_k \right) \left(\exists (f_1, \ldots, f_u) \in A_{t_1} \times \ldots \times A_{t_u} \right) \left(\forall d \in D_k \right) f(d) \subset f_1(d)$ $\times \ldots \times f_u(d)$, then A_k is called COMPOSED and $\left(\forall h \in \hat{u} \right) A_{t_h}$ is called a COMPONENT of A_k.

Furthermore, we formally defined the cardinality of relationship types and multiplicity of attributes, but the definitions will not be given here due to lack of space.

In the case of an object system model, the formal definition plays only a secondary role - it enables different versions of the model to be differentiated from each other. Much more important is the interpetation of the constructs used to build up the model - the same abstract structure may be interpreted in various ways, while constructs of different object system models often have the same interpretation in reality. The interpretation of the constructs used to build up an ER model is briefly sketched below.

<u>Entity</u> - a distinguishable thing, person, property, abstract concept, process, action, event and the like that is in itself important for future users of the database.

<u>Entity type</u> - the class of all the entities that have certain features, important for future users.

Relationship - an ordered set of a given number of entities with the property that the interconnection, mutual relations, interaction, and the like of all of them together, is important for future users.

Relationship type - the class of all the relationships between entities of given types that have certain features, important for future users.

Attribute - a property, characterisation and the like that is important for future users, however, not in itself but only as a means of characterising entities or relationships of a given type.

Constellation - the distribution of a property, characterisation and the like on all or some of the entities or relationships of a given type, occurring during the database utilisation period and important for future users.

To make the object system modelling results understandable for future users, a convenient graphical representation must be used. In the case of the ER model, such graphical representations are called ER diagrams. The symbolism we use in ER diagrams is depicted in Fig. 1. The individual symbols in this figure have the following meaning:

(a) the entity type A;
(b) a relationship type between the entity types A and B;
(c) the attribute a_1 for the entity type A;
(d) the composed attribute with components b_1, b_2 for the entity type B;
(e) the composed attribute with components u_1, u_2, u_3 for the relationship type U;
(f) - (j) several different possibilities for cardinalities of relationship types; and
(k) - (n) several different possibilities for multiplicities of attributes and components of composed attributes.

Fig. 2 offers a "bird's eye view" of an ER diagram representing the entity types, relationship types and attributes for the ER model of the object system of our medical database.

A BRIEF REVIEW OF THE OBJECT SYSTEM DYNAMICS MODELS

In the literature we have found the following models complying with the definition of object system dynamics models stated above:

1. the models considered in the papers of Bubenko (1977; 1980);
2. the entity/event model (Robinson, 1979), later used in the methodology JSD (Cameron, 1986);
3. the object system model included in the metamodel PIOCO (Iivari, 1983);
4. the object system model used in the methodology Remora (Huet et al., 1983; Rolland and Richard, 1982; Rolland et al., 1982);
5. the object system model used in the methodology ISAC (Lundeberg et al., 1981; Stübel, 1979);
6. the object system model used in the method ABRAHAM (Godbersen, 1983; Karagiannis and Schneider, 1985);
7. the object system model used in the methodology IPSO (Moulin, 1983), and a related model used in the methodology of Roussopoulos and Yeh (1984);
8. the object system model used in the methodology Insyde (King and McLeod, 1985);
9. the object system model included in the Conceptual Knowledge Model (CKM) for the design of data- and knowledge-bases of expert systems (Yasdi, 1985);
10. object system models using Petri nets (see, for example, Antonellis and Leva, 1985; Sakai, 1984; Studer and Horndasch, 1986);

457

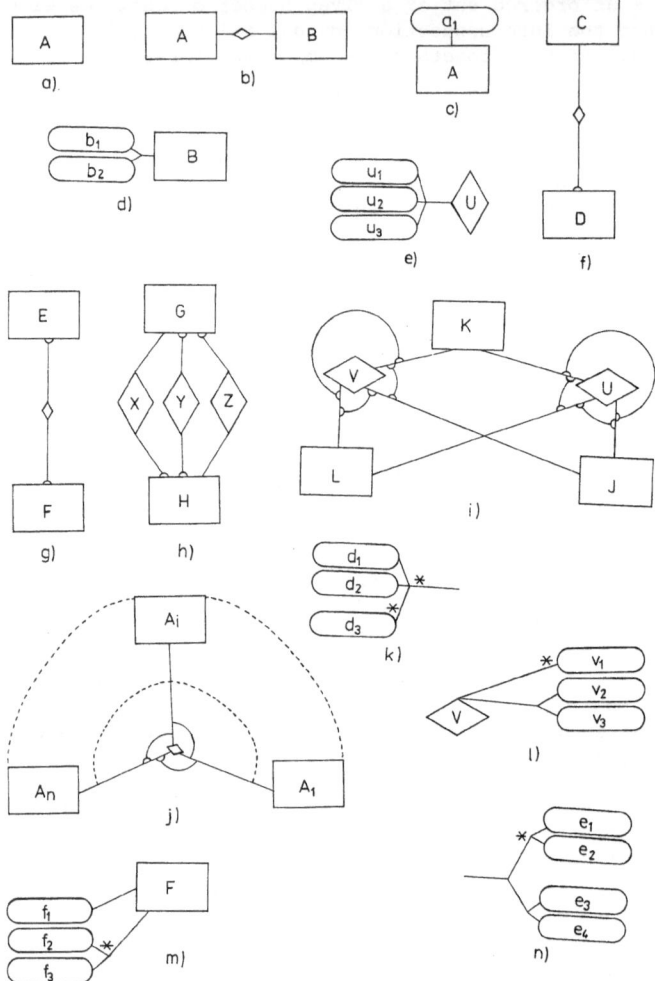

Fig. 1. Symbolism used in ER diagrams.

11. an object system model using finite state machines (Rosenquist, 1982); and

12. an object system model using homogeneous Markov chains (Oftedal and Sølvberg, 1981).

The models 1 - 9 have been proposed directly for the purpose of database or information system design, while 10 - 12 are models from outside the area of databases and information systems. The latter have been used either independently or together with an object system structure model - amongst the quoted papers the use of Petri nets and finite state machines, together with the ER model, is reported in Antonellis and Leva (1985), Rosenquist (1982) and Sakai (1984). As far as medical databases and information systems are concerned, probably the only object system dynamic models that have ever been used during their design so far are the model of the methodology Remora and Petri nets.

A very brief overview of some features of the above-mentioned models and modelling with them is given in Table 1. The individual columns of this table have the following meaning:

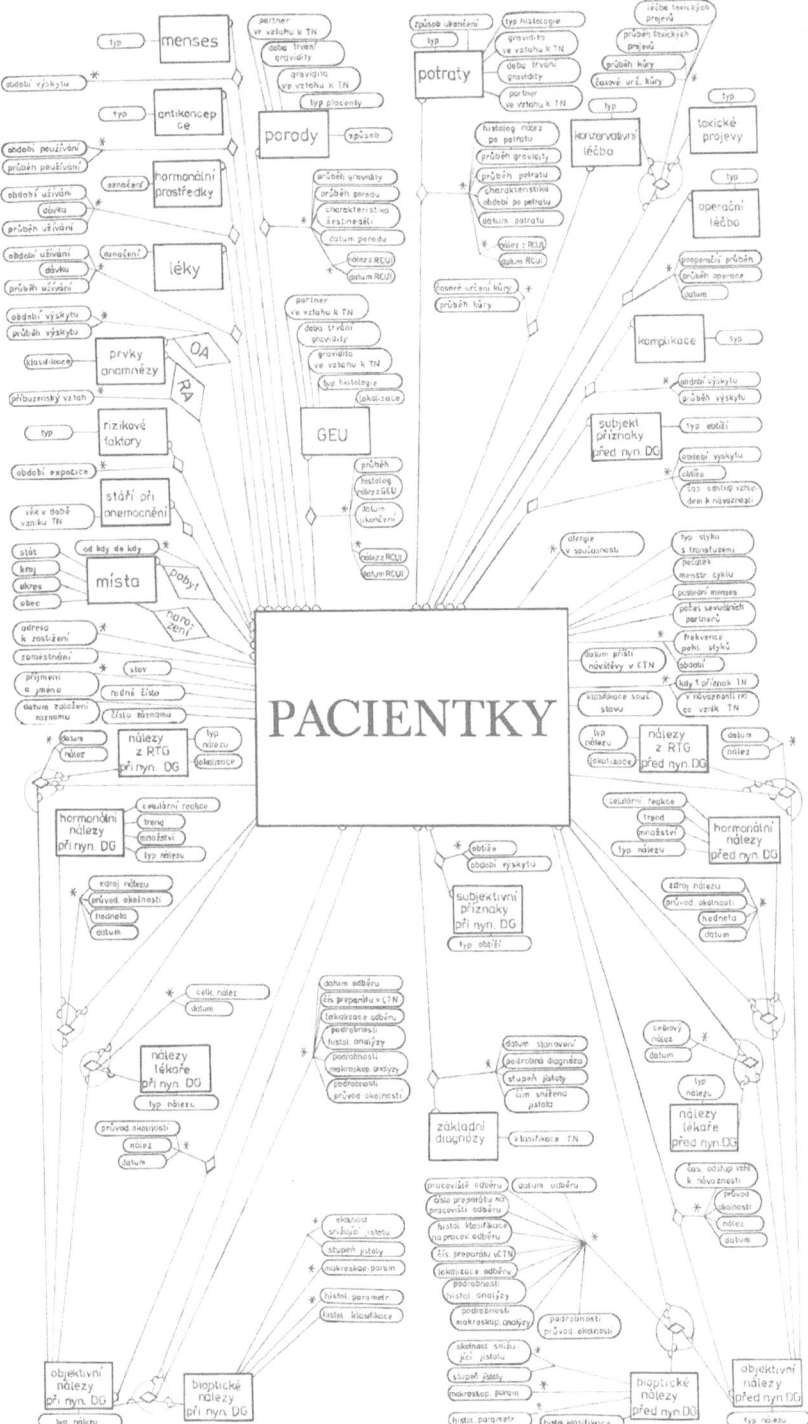

Fig. 2. Bird's eye view of an ER diagram.

Table 1. Principal Features of the Models Being Reviewed.

Model	Included in	Constructs for the dynamic component	Related constructs	Graphical representation	Results utilised for
Bubenko	conceptual information model, abstract model realm	event, event type	attribute function, time of occurrence	nets	entity and event types specifications, information model
Entity/event + JSD	modelling phase	event, sequence, selection, iteration		behaviour diagrams	network phase, model processes (life history programs), synchronisation (messages, state vectors)
PIOCO	I/O model	event, event type, event generalisation, event sub-set, event classification, action process	association, situation	OS (object system) graphs, OA (object system event action) graphs	information model, information process model, interaction model, IS graphs, IT graphs, IP graphs, IA graphs
Remora	conceptual schema design	event, operation, modify, induce trigger, t-event, (ascertain), t-operation, t-modify, t-trigger t-induce		only for the database information specification	static sub-schema (c-objects), dynamic sub-schema (c-operations, c-events), elementary transactions, transactions sequencing graph, design of the ISMS
ISAC	change analysis, activity studies	activity	real flow, message flow	A-graphs (activity graphs)	co-ordination of information sub-systems, precedence and component analysis, information analysis, I-graphs
ABRAHAM	"Istanalyse", "Sollkonzept"	activity	access path	ABRAHAM symbolism	system design, activity trees, information trees

Method					
IPSO + Roussopoulos-Yeh	information system analysis, data conceptual analysis, applications analysis, environment and requirements analysis	process, sub-process, synchronisation, task, sub-task	channel, flow	information flow diagrams, data conceptual schema, process diagrams, task flow diagrams	data logical schema design, module design, external data schemata design, application structured diagrams, data logical schemata, specification of tasks and sub-tasks.
Insyde	phase one	process event, function link, communication link		information flow schema (design schema)	semantic schema (conceptual schema), application events definition
CKM	graphical representation	function, event, trigger	information flow	conceptual requirement graphs, conceptual structure graphs	functional decomposition, conceptual behaviour graphs (actions), access table generation
Petri nets	first overview net, second overview net, conceptual schema design, views conceptual design	transition (agency)	flow (arc)	Petri nets, behaviour diagrams	high-concurrency net, low-concurrency net, behaviour description (transactions), views integration, dynamics analysis (operation frequencies)
Finite state machines	conceptual analysis, detailed analysis	next stage function	life cycle history	ER diagrams, entity life cycle diagrams	data structure and file design, program design, state/event matrix hierarchy
Markov chains	requirements definition	Markov chain, state of a Markov chain		only for the database information specification	decomposition of transactions into processes, estimation of traffic load on the global information structure

<u>model</u> - indicates the object system model in question;

<u>included in</u> - a more comprehensive model or a design phase in which the model in question is included or used;

<u>constructs for the dynamic component</u> - model constructs applied only to model elements or classes from the dynamic component of the object system;

<u>related constructs</u> - model constructs applied to model interconnections, mutual relations and the like between elements or classes from the dynamic component and elements or classes from outside this component;

<u>graphical representation</u> - the graphical representation or representations used for the model; and

<u>results utilised for</u> - parts of the database information specification or parts of the implementation design where the results of object system modelling with the model in question are particularly well utilised or particularly necessary, or results of such parts.

On the basis of the comparison of the models mentioned above, we have chosen four of them - Petri nets, the model of the methodology Remora, finite state machines and Markov chains - by means of which we are trying to model some parts of the object system of our medical database.

CONCLUDING REMARKS

Some readers may find two topics missing from this chapter - the problems of database information specification (particularly the ones of specification of dynamics), and the supporting software tools for object system modelling, or for conceptual modelling generally (that is, the topic of computer-aided conceptual modelling). There are three reasons why they were not included:

 (i) the limited extent of this chapter;
 (ii) the fact that there is ample literature dealing with these problems; and
(iii) the fact that neither the database information specification nor the supporting tools are so closely connected with the object system itself and its specificity as the object system modelling; that is why we do not consider it so important for database users to know about them.

On the other hand, we assume that it would be worth dealing with possible stochastic object system dynamic models, of which the only one presented so far is that using homogeneous Markov chains. The important role stochastic models play in medicine generally seems to indicate that they could also be important for object system modelling in medical informatics. We are going to pursue research on stochastic object system dynamic models in the future.

REFERENCES

Antonellis, V. De, and Leva, A. Di, 1985, DATAID-1: A database design methodology, <u>Infor. Sys.</u>, 10 : 181.

Boehm, B. W., 1981, "Software Engineering Economics", Prentice Hall, New Jersey.

Brodie, M. L., and Ridjanovic, D., 1984, On the design and specification of database transactions, <u>in</u>: "On Conceptual Modelling. Perspectives from Artificial Intelligence, Databases and Programming Languages", Springer Verlag, Berlin : 277.

Bubenko, J. A., 1977, The temporal dimension in information modelling, <u>in</u>: "Architecture and Models in Data Base Management Systems", North Holland, Amsterdam : 93.

Bubenko, J. A., 1980, Information modeling in the context of system development, in: "Information Processing 80", North Holland, Amsterdam : 395.

Cameron, J. R., 1986, An overview of JSD, IEEE Trans. Software Eng., SE-12 : 222.

Chen, P. P. S., 1976, The entity-relationship model: towards a unified view of data, ACM Trans. Datdabase Sys., 1 : 9.

Elmasri, R., Weeldreyer, J., and Hevner, A., 1985, The category concept: an extension to the entity-relationship model, Data Knowl. Eng., 1 : 75.

Godbersen, H. P., 1983, "Funktionsnetze Eine Modellierungskonzeption zur Entwurfs- und Entscheidungsunterstützung", Ladewig, Birkach.

Hammer, M., and McLeod, D., 1981, Database description with SDM: a semantic database model, ACM Trans. Database Sys., 6 : 351.

Huet, B., Rolland, C., and Martin, J., 1983, A methodology for information analysis of emergency and triage units in hospital information systems, Med. Infor., 8 : 255.

Iivari, J., 1983, "Contributions to the Theoretical Foundations of Systemeering Research and the PIOCO Model", Dissertation, University of Oulu, Institute of Data Processing.

International Organization for Standardization/Technical Committee 97 Computers and Information Processing, 1982, "Concepts and Terminology for the Conceptual Schema and the Information Base", ISO TC97/SC5/WG 3 Technical Report, ISO, New York.

Karagiannis, D., and Schneider, H. J., 1985, Knowledge based systems for information system development methods: a case study with ISAC-prototypes, in: "Proc. 2nd Baghdad Conference on Information Systems", Baghdad : 1.

King, R., and McLeod, D., 1985, A database design methodology and tool for information systems, ACM Trans. Office Infor. Sys., 3 : 2.

Kulkarni, K. G., and Atkinson, M. P., 1986, EFDM: extended functional data model, Comput. J., 29 : 38.

Lundeberg, M., Goldkuhl, G., and Nilsson, A., 1981, "Information Systems Development. A Systematic Approach", Prentice Hall, New Jersey.

Manola, F., and Orenstein, J. A., 1986, Toward a general spatial data model for an object-oriented DBMS, in: "Proc. 12th Int. Conf. on Very Large Data Bases", Morgan Kaufmann, Los Altos : 328.

Moulin, B., 1983, The use of EPAS/IPSO approach for integrating entity relationship concepts and software engineering techniques, in: "Entity-Relationship Approach to Software Engineering", North Holland, Amsterdam : 671.

Nijssen, G. M., 1979, Modelling in data base management systems, in: "EUROIFIP 79", North Holland, Amsterdam : 39.

Oftedal, H., and Sølvberg, A., 1981, Data base design constrained by traffic load estimates, Infor. Sys., 6 : 267.

Peimann, C. J., 1983, Über den Einsatz von Petrinetzen zur Beschreibung und Analyse von Informationssystemen im Krankenhaus, in: "Methoden der Statistik und Informatik in Epidemiologie und Diagnostik", Springer-Verlag, Berlin : 192.

Robinson, K. A., 1979, An entity/event data modelling method, Comput. J., 22 : 270.

Rolland, C., Huet, B., and Battesti, J. P., 1982, Medical and administrative information modelling of a hospital university centre pneumology department, Int. J. Bio-Med. Comput., 13 : 457.

Rolland, C., and Richard, C., 1984, The Remora methodology for information systems design and management, in: "Information Systems Design Methodologies: A Comparative Review", North Holland, Amsterdam : 369.

Rosenquist, C. J., 1982, Entity life cycle models and their applicability to information systems development life cycles. A framework for information systems design and implementation, Comput. J., 25 : 307.

Rothemund, M., and Ellsässer, K. H., 1984, Darstellung der Informationsflüsse in der Tumornachsorge mit Petri-Netzen, in: "Der Beitrag der Informationsverarbeitung zum Fortschritt der Medizin", Springer-Verlag, Berlin : 513.

Roussopoulos, N., and Yeh, R. T., 1984, An adaptable methodology for data-
    base design, Computer, 17(5) : 69.
Sakai, H., 1984, Entity-relationship behaviour modeling in conceptual
    schema design, J. Sys. Software, 4 : 135.
Staníček, Z., and Krejčí, F., 1984, The contribution to the IDMS data base
    design based on the HIT data model, in: "Proc. 7th Int. Seminar on
    Database Management Systems", Bulgarian Academy of Sciences, Varna : 34.
Stübel, G., 1979, ISAC - eine formale Methode zur rechnergestützten
    Beschreibung von Betriebsabläufen, in: "Formale Modelle für Infor-
    mationssysteme", Springer-Verlag, Berlin : 25.
Studer, R., and Horndasch, A., 1986, Modeling static and dynamic aspects
    of information systems, in: "Data Semantics (DS-1). Proc. IFIP WG
    2.6 Working Conference on Data Semantics", North Holland, Amsterdam :
    13.
Su, S. Y. W., 1986, Modeling integrated manufacturing data with SAM*,
    Computer, 19(1) : 34.
Sundgren, B., 1975, "Theory of Data Bases", Mason/Charter, New York.
Tsichritzis, D. C., and Klug, A., 1978, The ANSI/X3/SPARC DBMS framework
    report of the study group on database management systems, Infor. Sys.,
    3 : 173.
Tsichritzis, D. C., and Lochovsky, F. H., 1982, "Data Models", Prentice
    Hall, New Jersey.
Yasdi, R., 1985, A conceptual design aid environment for expert-data-
    base systems, Data Knowl. Eng., 1 : 31.
Zlatuška, J., 1984, HIT data model. A functional approach to data bases,
    in: "Proc. 7th Int. Seminar on Database Management Systems", Bulgarian
    Academy of Sciences, Varna : 21.

APPLICATION OF THE FEL-EXPERT SYSTEM IN THE DIAGNOSIS OF GENETIC DISEASES

V. Mařík, I. Šedivá, P. Rajchl, M. Sláma, Z. Zdráhal,
T. Maříková, J. Hyánek, M. Kubík and V. Kožich

Faculty of Electrical Engineering
Czech Technical University
Prague; and
Faculties of Paediatrics and Medicine
Charles University
Prague, Czechoslovakia

INTRODUCTION

The FEL-EXPERT shell and its versions are briefly described in the first part of the chapter. The main focus is clinical applications developed under the FEL-EXPERT project which are oriented towards genetic counselling.

The first case study concerns the differential diagnosis of three of the most common muscular dystrophies. The second case study deals with the differential diagnosis of inherited metabolic disorders. Other applications are briefly presented to demonstrate the flexibility and wide applicability of the shell.

FEL-EXPERT EXPERT SYSTEM

Basic Features

FEL-EXPERT, an empty rule-based expert system (shell), has the following characteristics:

- Domain independence: The application area can be changed by replacing the knowledge base, with no modification required to the empty FEL-EXPERT system.
- Machine independence: FEL-EXPERT is written in the standard Pascal programming language. It is conservative of memory, permitting its use on a personal computer.
- Diagnostic character: FEL-EXPERT is a suitable tool for solving diagnostic tasks. A finite set of goal hypotheses is considered, and evaluated and re-evaluated during the consultation run.
- Ability to handle uncertainty: Uncertainty of both knowledge and data is considered and accepted. A built-in model for handling uncertainty is based on ideas previously used by PROSPECTOR (Duda et al., 1976).
- Explanation capabilities: A wide spectrum of explaining abilities (including the answering of "What?" and "Why?" questions) makes possible

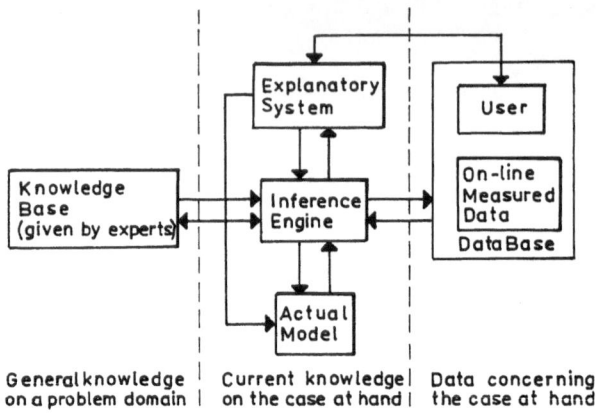

Fig. 1.   Architecture of the FEL-EXPERT

very detailed, user friendly, explanations of the decision making process
and also of the actual model state of the case in hand.

The FEL-EXPERT Architecture

The principle of the FEL-EXPERT architecture is shown in Fig. 1.
The user's data (the data from the database) are provided in a sequential
manner, in the form of a dialogue between the expert system and the user.
Making use of both knowledge base (general rules, given by the expert) and
the user's particular information, the actual model becomes tailored for
the case in hand and the resulting conclusion is formed. The explanatory
system serves for the user's better understanding of the reasoning process.

Knowledge Representation

Basically, three types of knowledge representation in the knowledge
base are used: production rules, logical functions, and context links.

A. The production rules have the following form:

If {evidence E}   THEN   {hypothesis H}   WITH   {probability P_1}
                  ELSE   {hypothesis H}   WITH   {probability P_2} ,

where {evidence E} and {hypothesis H} are propositions, {probability P_1},
{probability P_2} are subjective uncertainty measures (not probabilities
in an exact mathematical sense!), called, as in Duda et al. (1976), sufficiency and necessity measures, respectively. They can be expressed, in
terms of PROSPECTOR, as subjective conditional probabilities $P(H/E)$ and
$P(H/\bar{E})$, respectively. Their values are given by the expert. The model
for uncertainty handling requires assigning a prior probability to each
proposition. This value is also given by the expert.

Hypothesis H can serve as evidence for the other production rule.
Proposition E could be a hypothesis with respect to quite another production rule, and so on. In this way the knowledge base represented only by
the production rules may be expressed graphically as an oriented graph,
usually called an inference net. Each proposition is represented as a node
of the graph, and each production rule as an oriented edge.

The nodes with no outcoming edges are called the top hypotheses; the
nodes with no incoming edges are called the leaves. The nodes that are
neither the leaves nor the top hypotheses are marked as inner ones. The
aim of the particular consultation is to prove all the goal hypotheses

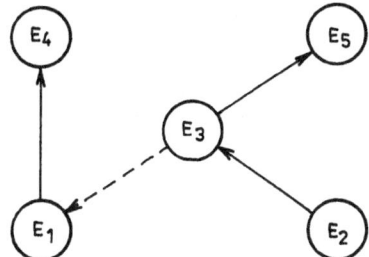

Fig. 2.   Logical connections
and context link of
the FEL-EXPERT.

(the goal hypotheses are the top ones and, sometimes, selected inner ones
too) or at least some of them.

B.   The logical function makes the expression of composed propositions possible.   Three types of logical functions are considered in the FEL-EXPERT
system: AND, OR, NOT.

C.   The context links are used in cases where, before one proposition can
be investigated, the complete checking of the other one is necessary, see
Fig. 2;   the context link is expressed by a dashed line.   The proposition
E_3 is a context of E_1, that means that before investigating E_1 the proposition E_3 has to be proved in a sufficient way (the context must be satisfied).
In the opposite case E_1 is excluded from the investigation.   To each context link two numbers, α_1 and α_2, are assigned by the expert;   they represent the range within which the probability (validity of the context) has
to lie.

The context links are, as a matter of fact, meta-rules that ensure
the common sense ordering of questions.   These links are very often used
to remove dynamically that part of the knowledge base which is not relevant
to the case at hand.

The knowledge representation language is simple, and allows construction and change in the knowledge bases, without any knowledge of programming languages.   The syntax of the knowledge representation language is described in Mařík and Zdráhal (1985).

Control Strategy

The control strategy can be described briefly in the following way:

(1)   There is one investigated hypothesis at each step of a consultation.
This hypothesis is either (a) chosen by the system, but its investigation has not been completed until now (starting the consultation the
goal hypothesis with the highest prior probability is chosen to be
completely investigated;   later the goal hypothesis with the highest
actual probability is always chosen to be completely investigated,
and so on), or (b) chosen by the user but its investigation has not
been finished until now.

(2)   The following two control strategy modes are permanently alternated
during the course of the consultation:

(a)   During a question-selection mode a backward chaining strategy in the
inference net is used: starting from the investigated goal hypothesis,

467

the suitable node (the question for the user) is chosen and a special <u>heuristic criterion function</u> is recursively utilised at each step of the backward chaining process to select the proper rule.

(b) During a model-updating mode the information obtained from the user (from the database) is propagated along all oriented paths leading from the node being answered to all the top hypotheses: the probabilities of the propositions (nodes) on these paths are recomputed, with both the sufficiency and necessity measures being taken into account. The pseudo-Bayesian formulae used by recomputing the probabilities along the edges in the inference net are combined with the fuzzy logic formulae by logical connectives (Mařík and Zdráhal, 1985). The model-updating mode strategy ensures that top/goal hypotheses are evaluated in parallel.

(3) The automatic run of a consultation can be interrupted by the user by means of special user's instructions ($-instructions).

(4) The consultation is finished (a) when there are no questions to be answered (all the relevant questions have been exhausted), or (b) at the user's request.

Actual Model

The actual model is composed of a set of all posterior node (proposition) probabilities. Starting the consultation it consists of the set of the prior probabilities. With any user's particular information the node probabilities change to posterior ones and thus the actual model becomes - step by step - tailored for the case in hand.

Communication Module

The communication module ensures that the dialogue between the user/knowledge engineer and the system takes place in a (restricted) natural language. It is responsible for:

(a) putting the question and understanding the user's answer;
(b) the creation of the protocol of the consultation (as a separate computer file); and
(c) the $-instruction execution. Some of the $-instructions serve as explaining/inspection facilities (explanatory system); the remaining ones are used to make the user's influence on the consultation run possible.

The communication sub-system has a highly modular structure. At a customer's request further supporting facilities can be added. For instance, a WHAT-IF module for hypothetical situation modelling, and also a module for the automatic creation of a patient's data list (filling in of pre-specified forms in medical applications) were developed with respect to the particular request.

Versions of FEL-EXPERT

Version FEL-EXPERT 1.5: This is the basic version completed in 1983 and showing all the features described above. A slightly modified version, with the communication module in English, was distributed as the system MEXEXP 1.0 (Mařík et al., 1984).

Version FEL-EXPERT 2.5: facilitates - in comparison with version 1.5 from which it was derived - a simple representation and processing of quantitative information. Two types of the so-called quantitative nodes have been involved in the knowledge representation language.

Version FEL-EXPERT 2.6: derived from version 2.5, makes use of a modified question-selection strategy. The new criterion function enables both (a) the actual model parameters and (b) the cost of the answers to be combined properly.

Version FEL-EXPERT 2.7: has been derived from version 2.5 by adding a blackboard control structure which allows the evocation and use of different knowledge bases in the course of a consultation. From the point of view of the blackboard philosophy, the knowledge base set may be considered as a distributed set of knowledge sources.

The blackboard consists of two sections: one for control and one for data. The control section is used for the accomplishing of different "global" actions over the knowledge source set (switching-over of the knowledge sources, changing the hypothesis and so on). The data section serves as an information transfer medium among different knowledge sources.

At the blackboard level two control strategies are properly combined: demons and agenda.

Demons. The control section contains the WHEN rules having the form:

WHEN {logical combination of elementary conditions}
THEN {action}

where an elementary condition is expressed as

$$P(U_i) \lesseqgtr t_i \, ,$$

where $P(U_i)$ is the actual posterior probability of the node U_i, and t_i is a threshold within the interval <0;1>.

The FEL-EXPERT is primarily a goal-driven, backward-chaining system. In addition to goal-driven inferencing, the blackboard structure allows data-driven, forward-chaining reasoning through the use of WHEN rules. These rules act as demons that come alive when certain events occur (that is, when actual posterior probabilities acquire values within a specified range). This is why the blackboard structure may be applied conveniently, even in the case of a single knowledge base such as a meta-level reasoning tool. In a manner similar to Laffey et al. (1986), we have also proved that the WHEN rules, when added to the standard IF-THEN production rules, bring a higher expert system efficiency. Moreover, through the use of the WHEN rules, FEL-EXPERT allows the knowledge engineer to state explicitly the expert's reasoning process in a more natural way.

Agenda. To make possible a certain planning procedure during the problem-solving process, the method of agenda has been applied. The sequences of partial tasks to be executed (the actual plan) are contained in special stacks. The following two stacks are considered for control purposes: the A-stack and the B-stack.

The A-stack (associative stack) contains the names of the partial nodes which should be completely investigated. The content of the A-stack is formed/changed by means of some of the WHEN rules expressing associative links among the different pieces of knowledge.

The B-stack (knowledge base stack) expresses the natural ordering of the knowledge base (= source) invocation. The initial priority of the sources is given by the expert. The WHEN rules may be used to change both the content of the B-stack and the priority of the bases in it.

The B-stack has lower priority than the A-stack; it is used only in the case when the A-stack is empty. The system makes use of the A-stack whenever the current goal has been completely investigated and it is necessary to choose the other one.

The implementation of the blackboard structure led to many difficulties with the explanation module because some user's commands could be understood from two points of view: either from the global point of view or from a local one (that is in the frame of the current knowledge source). The third stack (F-stack) has been implemented to store/order the subgoals during the global explanation activity of the system.

Implementation

The FEL-EXPERT shell has been written in Pascal, and at present it is available for PDP 11 - RT 11, PDP 11 - RSX, HP 1000, IBM-PC (Turbo Pascal) and other machines.

The FEL-EXPERT system is conservative of memory; the shell (IBM-PC version) occupies about 60 kbytes of core. The "medium-size" knowledge base needs about 20 - 50 kbytes (for example, the MYOPAT II knowledge base described below occupies 26 kbytes of core). Making use of version 2.7, the "size" of a knowledge base is - in principle - not restricted. To make the construction of the knowledge base much more comfortable, user-friendly knowledge base editors are now available.

APPLICATIONS

The FEL-EXPERT shell is a flexible and effective tool enabling the build-up of problem-oriented expert systems for solving different problems in a wide spectrum of application areas (Mařík and Zdráhal, 1986). We find that good technical people can become knowledge engineers for FEL-EXPERT in a couple of months. In particular, when the knowledge engineer sees not only the knowledge representation language but also examples of how this language has been applied, a good technical person can catch on rather quickly. By making use of FEL-EXPERT it is possible for a non-technical person, with no experience in programming, to become an expert in domain-oriented knowledge engineering.

Various knowledge bases for the FEL-EXPERT shell have been prepared and are tested or used in over 40 hospitals, factories and research institutes in Czechoslovakia and abroad. Over 30 knowledge bases have been developed within the frame of the FEL-EXPERT project. There have been several reasons for developing them: to test the FEL-EXPERT versions by solving real practical tasks, to gather experience from knowledge engineering activity, and to prepare examples for the education of knowledge-engineering groups - FEL-EXPERT users. Most of the applications have been aimed at the area of medical diagnosis, especially in relation to genetic counselling.

MYOPAT Knowledge Base - Case Study I

The first real knowledge base developed under the FEL-EXPERT project was oriented to genetic counselling. This part of the project resulted from collaboration with the Genetic Counselling Centre, Faculty of Paediatrics, Charles University, Prague. The task was to diagnose three genetic syndromes: Duchenne's progressive muscular dystrophy, Becker's progressive muscular dystrophy and limb girdle dystrophy. All these syndromes are concerned with the muscular dystrophy of the lower and upper extremities. They are rare in the population (less than 0.02% in Central Europe). Therefore, with the exception of several genetic laboratories, the experience of physicians in this area is rather limited. Usually, the clinical findings, as well as results of laboratory tests, are very similar. The principal difference is in the type of heredity and in the prognosis. In the early stage of defect manifestations (age 2 - 7 years) the symptoms are not distinct enough and it is difficult to produce the correct diagnosis, even for experienced geneticists. However, during these early stages the diagnosis is very important, since it is possible to estimate the risk of the birth of another affected child in the family and reconsider the consequences of another pregnancy. (At present no therapy is known.) Later on, the syndromes are easier to distinguish, since their symptoms become more evident and the progress of the disease provides additional information.

In the early stage, the clinical findings, results of laboratory tests and the pedigree information are the only data available for producing the diagnosis. Their mutual relations and their relations to the syndromes are described mostly by heuristic rules. The problem is well structured and, in our opinion, it was suitable for an expert system application.

For this task two knowledge bases (MYOPAT I and MYOPAT II) have been developed. MYOPAT I has been prepared for the FEL-EXPERT version 1.5 and contains 120 nodes, 276 rules and 68 context links. The revised version MYOPAT II has been developed for the FEL-EXPERT 2.5. The more efficient representation decreased the size of the knowledge base to 97 nodes, 177 rules and 24 context links and significantly improved results. The maximum depth of the inference network (number of nodes from the current goal to the node associated with the question) is 6 for both the knowledge bases.

The set of 35 cases has been used for debugging the knowledge base MYOPAT II and then the base has been tested on a set of 113 other cases (about 70% of all the cases registered at the laboratory since 1970). In 112 cases the results achieved from the MYOPAT II were identical with those provided by the panel of geneticists. The distribution of syndromes and results is shown in Table 1.

In the test set there were 7 pairs of brothers and/or sisters. They were investigated independently and the diagnosis of MYOPAT II was completely approved by the geneticists. The possibility of explanation of the conclusions in terms of the most supporting and the most opposing symptoms has also been greatly appreciated. The knowledge base MYOPAT II has been used as a regular supporting tool for geneticists since 1984.

The MYOPAT II knowledge base has been translated into English and Spanish and tested by Mexican geneticists. The results have also been found adequate in Mexican conditions (the relevance of some symptoms differs in various geographical conditions and for various ethnic groups).

Table 1. Distribution of Syndromes and Results Obtained
Using MYOPAT II

EXPERTS:	MYOPAT II		
	Duchenne	Becker	L.G.D.
Duchenne	80	1	0
Becker	0	17	0
L.G.D.	0	0	15

Diagnosis of Inherited Metabolic Disorders - Case Study II

The first problem to be solved in collaboration with the Diagnostic
Centre for Inherited Metabolic Disorders, Faculty of Medicine, Charles
University of Prague, was the differential diagnosis of heterozygotes for
hyperphenylalaninaemia I (classical phenylketonuria - PKU). This, with an
incidence of about 1 in 6,000, is one of the most common inherited metabolic disorders in our region. High phenylalanine blood levels due to
lack of hydroxylation lead, in untreated cases, to severe mental defects.
For genetic counselling purposes it is necessary to know which persons in
the families with the occurrence of PKU are the heterozygotes. One of the
most reasonable methods to detect the carrier status is the oral load test
with L-phenylalanine. This test is based on an impaired hydroxylation of
phenylalanine to tyrosine in heterozygotes as compared to healthy individuals. Carrying out this test is simple, but the evaluation needs great
experience. Many discriminants derived from the phenylalanine and tyrosine
levels have been used all over the world. None of them is "strong" enough
to distinguish completely the heterozygotes from healthy persons.

The knowledge base METABOL-A which we have prepared makes use of the
experience of two experts and of the evaluation of 353 load tests which
were previously performed. The knowledge base is a comparatively simple
one (see Table 2). The rules express the interdependencies among the data
(results of laboratory tests, for instance), the values of the discriminants and the final decisions.

To make the decisions more precise, the knowledge base has been improved by learning (the METABOL-AC knowledge base). The Fukunaga-Koontze
method of feature extraction (originally developed in the area of pattern
recognition) has been used for this purpose with advantage. As a result,
new, far more efficient discriminants have been induced (they are of a
syntactic nature) and, on the other hand, some of the rules have been excluded.

The 353 loading tests mentioned above have been evaluated by means of
the METABOL-AC to test the reliability. The results were always thoroughly
checked from other points of view by a panel of physicians. No substantial
discrepancy was revealed. This system of evaluation of the loading tests
with L-phenylalanine has been routinely used since August, 1986 (approximately 70 loading tests have been consulted).

The other and more serious and complicated problem can be found in
deciding which laboratory tests should be used to establish the diagnosis
of an inherited metabolic disease. The Diagnostic Centre receives about
7,000 samples per year from the high risk population from all over Czechoslovakia. Only a small number of the samples arrive with a diagnosis
designated; most of the samples are designated as being suspected of indicating metabolic disease without precise specification. In 1986 about 300
nosologic entities with primary metabolic defects were known. The clinical
findings and the course and time of manifestation are very different and
are overlapping. It is usually impossible to establish the diagnosis of
any inherited metabolic disorder from the clinical findings alone. The
clinical findings enable the preliminary diagnosis (usually several diagnoses) to be determined. According to this preliminary diagnosis further
investigations are performed to confirm or refute the diagnosis from the
set of the suspected ones. There are hundreds of methods, but the choice
must be correct. Inherited metabolic diseases are not rare as a group
(about 1 - 2 per 1,000 newborns). However, each of them is sufficiently
rare that many clinicians may not have had considerable personal experience
of dealing with it. Decision making is mainly based on the knowledge contained in very specialised handbooks.

Table 2. The Structure of the Knowledge Bases (the Third Column in the Table is Expressed in the Form: Number of Nodes/Rules/Context Links/Top Nodes/Goals/Inference Net Depth)

Knowledge Base Name	FEL-EXPERT Version	Structure	State of Exploration
MYOPAT II	2.5	97/177/24/3/3/6	in use
METABOL-A	2.5	24/28/6/2/2/3	no longer in use
METABOL-AC	2.6	26/31/4/2/2/3	in use
OB	2.7	87/135/31/15/15/6	in use
MPK	2.7	82/130/27/8/8/5	in use
OK-2	2.7	68/87/17/15/15/6	undergoing experimental testing
PK	2.7	78/82/22/17/17/5	undergoing experimental testing
POLYD	1.5	134/234/28/21/21/3	in use
FRA(X)	2.5	27/33/1/1/1/3	undergoing experimental testing
TUMOR	1.5	218/391/89/50/50/6	undergoing experimental testing
PSYCH	2.5	43/72/7/2/5/8	in experimental use
GTIS	1.5	42/59/22/2/2/5	in use
MOTOR	1.5	514/605/63/167/154/6	undergoing experimental testing
FERPLAN	2.5	35/126/14/7/7/3	in experimental use

There were two reasons for splitting the problem-oriented knowledge into several knowledge bases:

(i) there are many nosologic entities and a vast volume of knowledge considered; and

(ii) the type of disease can be fully confirmed only by biochemical tests, but the clinical findings have to be taken into account before these tests are recommended.

That is why the blackboard control structure has been applied and the structure of the knowledge base set had to be designed. The knowledge bases are arranged in two levels (see Fig. 3). The bases for the evaluation of clinical findings have to be used first. The preliminary results obtained by them are exploited for the more precise diagnosis at the "deeper", second level. This level consists of knowledge bases for the evaluation of biochemical tests. Each of the "second level" knowledge bases has been constructed for precise decision making in a group of "related" diseases.

The total number of knowledge bases is not yet known (it will be about twenty), but only four of them plus the blackboard knowledge base METABOL have been developed until now. There is one knowledge base on the first level (OB for the evaluation of biochemical findings) and three bases on the second one (MPK for the differential diagnosis of mucopolysacharidoses, OK-2 for disorders of the metabolism of organic acids, and PK for disorders of the metabolism of purines and pyrimidines). Their structure is presented

473

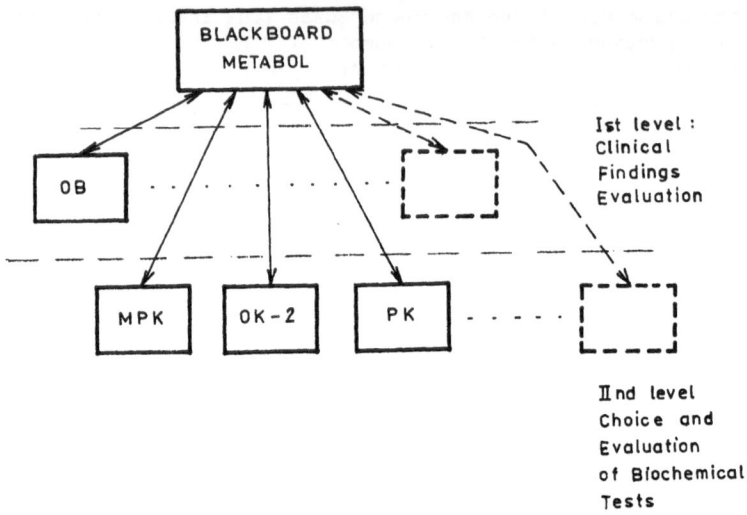

Fig. 3. METABOL blackboard with two levels of knowledge base

in Table 2. Each partial knowledge base has been developed and tested
separately.

The METABOL blackboard knowledge base observes 26 partial hypotheses
and contains 20 rules, for instance the following ones:

WHEN   {The probability of the proposition "Changes of skeleton" < 0.2}

THEN   {Start to investigate the knowledge base PK immediately},

WHEN   {The probability of the proposition "Changes of skeleton" < 0.1
       AND the probability of the proposition "It deals with the syn-
       drome Sanfilippo" < 0.005}

THEN   {The utilisation of the MPK knowledge base is not relevant},
       and so on.

The development of the other knowledge bases involved in the overall
complex structure continues.

Other Medical Applications

Genetic counselling. The knowledge base POLYD for the differential diagnosis of 21 syndromes with a symptom of polydactyly has been developed and
used in genetic practice. A part of the POLYD knowledge base has been induced from examples by an original learning method (Mařík et al., 1987).
This method is based both on the inductive construction of a decision tree
and on the conversion of this tree into the knowledge base notation by
making use of the intentional approach (Perez and Jiroušek, 1985).

A knowledge base FRA(X) was prepared for the screening of "fragile-X
syndrome" patients among mentally-retarded boys. It contains only one goal
hypothesis ("Suspicion of the fragile-X syndrome") and makes use of simple
clinical findings. (The more demanding chromosomal examination can only be
performed on the suspected boys.)

Peri-operative diagnosis of brain tumours. A knowledge base TUMOR, for the
peri-operative diagnosis of cerebral tumours was prepared, in collaboration
with the Central Military Hospital and the Institute of Molecular Genetics

of the Czechoslovak Academy of Sciences. The knowledge base was corrobora-
ted on a testing set of seventy cases of tumours where the diagnosis was
correctly established in more than 90% of the test cases. The explanatory
capacity of the expert system makes it possible to use it also for learning
purposes; therefore, its use for better under- and postgraduate teaching
is foreseen.

Recognition of long-term psychophysiological states. A knowledge base PSYCH
serves for the classification of long-term psychophysiological states. The
parameter of the preprocessed EEG-signals (the original EEG-signal is fil-
tered by α, β, δ and μ-activity band-pass filters, integrated and statistically tested), as well as the results of several mental tests, are used as
the database. 97% correct decisions were obtained in the course of experiments. These experiments were prepared in co-operation with the Research
Institute of Psychiatry, Prague.

REFERENCES

Duda, R. O., Hart, P. E., and Nilsson, N. J., 1976, "Subjective Bayesian
    Methods for Rule-based Inference Systems", Report TN 124, SRI Inter-
    national, Stanford.
Laffey, T. J., Perkins, W. A., and Nguyen, T. A., 1986, Reasoning about
    fault diagnosis with LES, IEEE Expert, Spring 1986 : 13.
Mařík, V., and Zdráhal, Z., 1985, Expertní systémy FEL-EXPERT 2.5 a
    MIFELEX - prírucka uzivatele, in: "Metody Umělé Inteligence a Expertní
    Systémy II", V. Mařík and Z. Zdráhal, eds., CSVTS FEL, Czech Technical
    University, Prague : 175 (in Czech).
Mařík, V., and Zdráhal, Z., 1986, "The FEL-EXPERT and MIFELEX Empty Expert
    Systems and Their Applications", Working Paper WP-86-61, IIASA, Laxen-
    burg, Austria : 47.
Mařík, V., Zdráhal, Z., and Mendez, R. A., 1984, "A Skeleton for Expert
    Systems: MEXEXP 1.O", Technical Report AHR-84-29, C.I.E.A. del Insti-
    tuto Politécnico Nacional, Mexico City.
Mařík, V., Kouba, Z., and Zdráhal, Z., 1987, Knowledge acquisition experi-
    ments under the FEL-EXPERT project, in: "Artificial Intelligence and
    Information-Control Systems of Robots", North-Holland, Amsterdam : 327.
Perez, A., and Jiroušek, R., 1985, Constructing an intentional expert system
    (INES), in: "Medical Decision Making: Diagnostic Strategies and
    Expert Systems", J. H. van Bemmel, F. Grémy and J. Zvárová, eds.,
    North-Holland, Amsterdam : 307.

METHODS FOR THE EVALUATION OF EXPERT SYSTEMS

P. PÍŠ

State Hospital
Bratislava
Czechoslovakia

INTRODUCTION

Expert systems have their origins in traditional data processing,
which is based on the definition of the appropriate representation for the
sequence of operations and for the data. In conventional systems, the
classification of sets of similar entities is usually achieved by providing
a variable name and indexing. Arrays are normally used to define relationships. In a conventional control structure what heppens next is predefined
by the program. The knowledge processing of a human expert can be presented
in different forms, because traditional representation techniques are
inadequate to cope with the problems arising in such systems. One approach
to representing classes and relations can be achieved, for example, through
the use of a predicate calculus and the control structure can be the production system. The productions are the set of rules and the control structure
determines what rule is tried next. If the control structure is domain-
independent, it could be very simple. In many problems it is convenient to
allow a rule to be considered only in certain circumstances. The control
structure of such a system is more complex and conventional algorithms are
not well suited because in real situations they have a long response time.

For these reasons many expert systems have come into use at the present
time, and in medicine too. This has come about also as a result of the
fact that a number of problem-oriented expert systems has been developed.
These systems can be used in various applications so that experts create
problem-oriented knowledge bases. Some of them are able to work with imperfect information. The imperfection is usually expressed through weights
(uncertainty factors) for every fact and piece of knowledge. Simultaneously,
by using the expert systems, an effort has been made to evaluate their performance (see, for example, Yu et al., 1979).

In this chapter we present two methods of expert system evaluation.
The first method looks at the activity of the expert as a whole. To illustrate this point of view we consider a diagnostic expert system in medicine.
In this sphere the user has a certain set of information about the patient
which he delivers to the expert system. On the basis of this information,
the expert system provides a recommendation. Of course, we do not know the
real diagnosis of the patient with certainty and the recommended diagnosis
may be correct or incorrect. In this way the expert system can be understood as a method of estimating reality - the real diagnosis of the patient.

The recommended diagnosis can be understood as a concrete estimate-realisation of response statistic. This point of view on the activity of the expert system can be used in its evaluation, adopting methods of mathematical statistics.

It must be taken into account that suitable methods of mathematical statistics, for example decision theory (see Johnson and Leone, 1977), require complete knowledge of the problem. However, more frequently the knowledge is fragmentary and uneven. In the first method we try to overcome these difficulties.

The second method of evaluation takes a more detailed view of the input information and the internal structure of the knowledge base. Therefore we restrict our attention to the class of system which is known as a goal-oriented expert system based on rules. For the purpose of this chapter we present a formalised notion of the expert system. Such a notion does not reflect all aspects which are important in the construction of an expert system, but on the basis of the notion presented it is possible to define the goal of the expert system, the structure of the knowledge base in the form of rules and the basis of the control structure.

For detailed analysis of the knowledge base we use Petri nets. An expert system capable of solving real problems must have a large knowledge base with a complex structure, otherwise the user can replace the system by himself. Davis (1982) calls such a system "robust". The robustness of the system works to some extent against the user and can be a serious problem in the creation of a useful expert system. To create a useful knowledge base, an expert must know if the set of rules and their structure are sufficient for the expert system to make a decision and if a set of input information for the given set of rules is sufficient for the system too. Advice about the input information needed for the expert system to make a decision is very useful for the user. Petri nets prove to be a useful model for this range of problems.

METHOD I

In this method information about every patient can be regarded as a population from which the sample is drawn. The information supplied to the expert system corresponds to the set of training samples. The system recommends, on the basis of the information supplied, one of the possible diagnoses. Then the system itself can be taken as a sample-based estimator and its recommendation as realisation of the estimator.

Let us assume that there exist data about m patients with their diagnoses. Let 1, 2, ..., k be possible diagnoses, where

$d_i = j$ denotes ith patient with jth diagnosis
$e_i = j$ denotes that the expert system for the ith patient recommended the jth diagnosis.

Let $D(d_{ij})$ be a matrix of diagnoses, where

$$d_{ij} = \begin{cases} 0 & \text{if } d_i \neq j \\ \\ 1 & \text{if } d_i = j \end{cases} \qquad i = 1, 2, \ldots, m \quad j = 1, 2, \ldots, k \qquad (1)$$

$E(e_{ij})$ is a matrix of estimators, where

$$e_{ij} = \begin{cases} 0 & \text{if } e_i \neq j \\ \\ 1 & \text{if } e_i = j \end{cases} \qquad 1 = 1, 2, \ldots, m \quad j = 1, 2, \ldots, k \qquad (2)$$

and $S(s_{ij})$ is a loss matrix where s_{ij} is the loss when the expert system recommended ith diagnosis, but the true diagnosis is j.

Then the loss for the m patients for the set of input information and for the knowledge base can be written as

$$LS_m = Tr(D.S.E^T) \qquad (3)$$

and average loss is

$$LS = LS_m/m \qquad (4)$$

To illustrate (3) we give a very simple example:

$$\text{Let } s_{ij} = 1 - \delta_{ij}, \text{ where } \delta_{ij} = \begin{cases} 1 & \text{if } i = j \\ \\ 0 & \text{if } i \neq j \end{cases} \qquad (5)$$

Then the loss matrix is

$$S = A_k - I_k, \qquad (6)$$

A_k is the matrix where $a_{ij} = 1$ $\quad i = 1, 2, \ldots, k \quad j = 1, 2, \ldots, k$ and I_k is the identity matrix.

Then $D.S.E^T = D.A_k.E^T - D.I_k.E^T = I_m - D.E^T \qquad (7)$

and

$$LS = Tr(I_m - D.E^T) = m - \sum_{i=1}^{m} \delta_{d_i, e_i} = \sum_{i=1}^{m} \left(1 - \delta_{d_i, e_i}\right) \qquad (8)$$

In this case LS is equal to the number of patients in which the expert system determines an incorrect diagnosis.

The method described above is based on the assumption that every patient has only one diagnosis and the expert system recommended only one diagnosis too. Generally, the ith patient can be represented by the set of k diagnoses ($j = 1, 2, \ldots, k$) with weights d_{ij} and the set of recommendations with the weight e_{ij}. We suppose that

$$\sum_{i=1}^{k} d_{ij} = 1 \quad , \quad d_{ij} \in <0, 1> \qquad (9)$$
$$i = 1, 2, \ldots, m$$
$$j = 1, 2, \ldots, k$$
$$\sum_{i=1}^{k} e_{ij} = 1 \quad , \quad e_{ij} \in <0, 1> \qquad (10)$$

In this case it is suggested that LS_m be computed as follows:

$$LS_m = Tr\left(R.G.A_{k,m} + D.S.(E - R)^T\right) \qquad (11)$$

R is a matrix, where $r_{ij} = \min(d_{ij}, e_{ij})$

G is a matrix, where $g_{ii} = s_{ii}$, $g_{ij} = 0$ and $i \neq j$

479

$A_{k,m}$ is a matrix $k \times m$, where $a_{ij} = 1$

If $s_{ii} = 0$, $I = 1, 2, \ldots, k$, then $G = 0$ and (11) can be written as

$$LS_m = Tr\left(D.S.(E - R)^T\right) \quad \text{and} \quad LS = LS_m/m . \tag{12}$$

In the general case it is suggested that (12) be used. Note that (3) is a special case of (12).

So far it has been assumed that the set of input information is constant. We shall next consider the case of a constant knowledge base and a variable set of input information. We shall also suppose that for a set of input information the cost C(II) is known. Then for the knowledge base and for the set of input information it is suggested that the suitability H of the set of input information II be evaluated as follows:

$$H(II) = \left(C(II) + LS(II)\right)^{-1} \tag{13}$$

where LS(II) is computed from (12).

The evaluation of the input information mentioned above is based on the assumption that the user knows sufficient amount of it, that is, the set of input information is sufficient for the expert system to make a decision. The next assumption is that the set of input information must have the property of monotony, that is, if the set of information J contains the set of information I, then the loss S(J) must not exceed the loss S(I), then

$$I \subset J \to S(I) \geqslant S(J). \tag{14}$$

On the other hand, if the number of pieces of input information is increasing, then their cost is growing too. In such a case

$$I \subset J \to C(I) \leqslant C(J). \tag{15}$$

Therefore we can restrict ourselves to the following conclusion:

From two sets of input information we shall consider as better that one which has the greater suitability H.

In connection with the evaluation of the suitability of a set of input information, there arises a natural question: how to optimise a set of input information?, that is, how to find the most suitable set of input information? In our opinion it is not possible to propose a general method for this purpose, and the one to be adopted would depend on:

  (i)   the internal structure of the knowledge base;
  (ii) the cost of information C; and
 (iii) the matrix of loss S.

We give a certain view on this problem in the next method. This method has to do with knowledge based on rules.

METHOD II

In our context the expert system is a goal-oriented system, and the knowledge about what deduction can be made is contained in the static database (knowledge base) in the form of rules. In the system, rules are invoked in a recursive fashion and thus claimed together. This is a general strategy in artificial intelligence approaches to problem solution by goal decomposition with the definition of intermediate states.

Formalised Model

Next we present a formalised notion of the expert system, where entities are designated as parameters (pm) and characteristics of a parameter are called values of the parameter (vp). It is necessary to choose those ones which are important in information processing by the expert system.

Let the parameters $(pm_1, pm_2, \ldots, pm_k)$, $pm_i \in gp$ be in some relationship to each other. On the basis of this relationship we can consider $(pm_1, pm_2, \ldots, pm_k)$ as another entity, which we call a group of parameters (gp). The entities can be expressed as (gp_i, pm_j, vp_{jk}). In our formalised notion we shall assume that entities are expressed as

$$(pm_i, vp_{ik}) \tag{16}$$

where $pm_i \in PM$ and $vp_{ik} \in pm_i$.

Let us suppose that an elementary proposition is a proposition in which each triple occurs at most once with the weight $w(pm_i, vp_{ik})$, $w \in <-1,1>$. A proposition in the normal conjunctive form, composed of elementary propositions, is called a conjunction. A rule r is the expression of the form:

$$r:w(P) \rightarrow w(C) \tag{17}$$

where P is the premise of a rule with the weight $w(P)$, and C is the conclusion of the rule with the weight $w(C)$. The rule can be interpreted in the following way:

If the premise P has the weight $w(P)$, then the conclusion C has the weight $w(C)$.

Next we shall consider the following form of $w(P)$ and $w(C)$:

$$w(P) = f_P\bigl(w(L_1), w(L_2), \ldots, w(L_n)\bigr) \tag{18}$$

where L_i is an elementary premise of the form of the elementary proposition with weight $w(L_i)$.

$W(P)$ is a premise of the form of the conjuction, and
f_P is a weighting function of the premise P.

(In this chapter we do not deal with weighting functions in detail. This aspect of our work is presented in Pís and Mesiar (1986).)

$$w(C) = \bigl(w(C_1), w(C_2), \ldots, w(C_n)\bigr) \tag{19}$$

where C_i is an elementary conclusion of the form of the elementary proposition with the weight

$$w(C_i) = f_{C_i}(w(P)) \tag{20}$$

where f_{C_i} is a weighting function of the elementary conclusion C_i.

For our purpose we shall consider that

$$w(C_i) = f_C(K_i, w(P)) \tag{21}$$

where f_C is the weighting function equal for all the rules, and K_i is the constant characterising the contribution to the weight of the C_i by the rule r.

Suppose that R is not an empty set of rules; pm_i is an internal parameter $pm_i \in IP$, if it occurs both in the premise of a rule and in the conclusion of another rule as well; pm_j is an axiomatic parameter $pm_j \in AP$, if it occurs in the premise of a rule but it is not in the conclusion of any rule (the set AP we call the set of input information); and $pm_g \in IP$ is the goal parameter if it occurs only in the conclusion of a rule.

Then the set of rules R is the system of rules if the set $Z(i)$ is defined for every parameter $pm_i \in IP$ by:

$$Z(i) = \{r^{1i}, r^{2i}, \ldots, r^{ni}\} \tag{22}$$

where $r^{ji} \in R$ so that pm_i occurs in the conclusion of the rule r^{ji}.

Let R be the system of rules. Then the rule r^j in the set $Z(i)$ can be described as follows:

$$r^{ji} : w(p^{ji}) \to w(c^{ji}) \tag{23}$$

where $w(P^{ji}) = f_P\big(w(L_1^{ji}), w(L_2^{ji}), \ldots, w(L_n^{ji})\big)$ and

$$w(C^{ji}) = \big(w(C_1^{ji}), w(C_2^{ji}), \ldots, w(C_n^{ji})\big).$$

Next, let us suppose that the parameter pm_i is in the elementary conclusions $C_k^{1i}, C_k^{2i}, \ldots, C_k^{ni}$. Then, using the set of rules $Z(i)$, we obtain:

$$w(C_k^{1i}) = f_C\big(K_k^{1i}, w(P^{1i})\big)$$

$$\vdots \tag{24}$$

$$w(C_k^i) = w(C_k^{ni}) = f_V\big(f_C(K_k^{ni}, w(P^{ni}), w(C_k^{(n-1)i})\big)$$

where f_V is the global weighting function and

$$w\big(L_v^{ji}\big) = \begin{cases} w(pm, vp) & \text{if } pm \in AP \tag{25} \\ \\ w\big(C_z^{ns}\big) \tag{26} \end{cases}$$

if L_v^{ji} possesses $pm \in IP$ and occurs at C_z^{1s}

$$C_z^{2s}, \ldots, C_z^{ns} \text{ of the rules in } Z(s) = \{r^{1s}, r^{2s}, \ldots, r^{ns}\}.$$

Then $w\big(C_k^i\big)$ may be written as follows:

$$w\big(C_k^i\big) = f_k^i\big(w(pm_1, vp_1), \ldots, w(pm_n, vp_n), w(C_1^{n1}), \ldots, w(C_m^{nm})\big) \tag{27}$$

where

$f_k^i = f_k^i(f_C, f_V, f_P, K)$
K is the set of constants K_k^{ij} of the rules used,
$w(pm_i, vp_i)$ are weights for which (25) holds, and
$w\big(C_i^{nj}\big)$ are weights for which (26) holds.

Equation (27) shows that the weight of an elementary conclusion of the rule r can be computed from the weights of the couples (pm_i, vp_i) $pm_i \in AP$ and from the weights of the elementary conclusions of rules simultaneously occurring in the elementary premises of the rule r. An algorithm to be developed on the basis of the equations (17) - (27) could be simple and domain-independent. A more complex domain-dependent algorithm has been developed and has been used in an empty expert system, see, for example, Píš and Mesiar (1987a,b). Domain-dependence of the algorithm has been reached to group the rules together by gp and it allowed a rule to be considered if its appropriate gp had been established. The system developed has been used in various spheres of medicine, see, for example, Masaryk et al. (1987) and Píš and Trupl (1987).

Now we are ready to present our definition of the goal of the expert system:

Let R be the system of the rules; $w(pm, vp)$, $pm \in AP$ are the weights of axiomatic parameters (we describe them as $(w_1, w_2, ..., w_m)$); f_V, f_C, f_P are weighting functions; $pm_g \in IP$ is the goal parameter which occurs in the goal elementary conclusions C_k^{1g}, C_k^{2g}, ..., C_k^{ng} of the rules in $Z(g)$. Then the goal of the expert system is to create the function $f = f(f_V, f_P, f_C, K)$ for which the following holds:

$$w(C_k^g) = f(w_1, w_2, ..., w_z) , z \leqslant m \tag{28}$$

Three cases may be noted:

(i) $w(C_k^g) \geqslant$ const.

(ii) $w(C_k^g) \leqslant -$ const.

(iii) $-$ const. $< w(C_k^g) <$ const.

where $0 \leqslant$ const. $\leqslant 1$

The first two possibilities are those in which the expert system makes the decision. In case (iii) the user really derives no benefit from the fact that the expert system is not capable of ascertaining the validity of the goal conclusion, but in that case he cannot expect more information from the expert system. A more detailed analysis of this case is described in Píš et al. (1985).

For the purpose of Method II we define a number of additional concepts.

A set of parameters can be called sufficient if the parameters can acquire the values on the basis on which (from successive use of rules) the expert system is able to make a decision. The meaning of the concept "sufficient set" is to find such a set of parameters on the basis of which the user has, theoretically, the possibility (if the parameters take on suitable values) of receiving a positive or negative answer to the basic question from the expert system.

It is clear that if a set of parameters is sufficient, then each of its supersets has this characteristic too. Therefore the entire information about the system of all sufficient sets is given by minimal sets. (A permitted set of parameters A is termed minimal if $pm \in A$ and set A-pm are not sufficient.) If the expert system is robust there exist several minimal sets. The concept "cost" of a parameter can help the user to distinguish between them. Let $e(pm)$ be the elementary cost of a parameter, that is, the means needed for finding the value of the parameter pm without using the rules of the expert system. Then the cost $C(pm)$ of the parameter pm is

$$C(pm) = \min\{e(pm), \sum_{x \in A} e(x)\} \tag{29}$$

where A is a sufficient set of the parameters to ascertain the value of the parameter pm using the expert system rules.

Petri Nets

Petri nets are a tool for the study of systems. They were designed and are used mainly for modelling. Many systems, especially those with independent components, can be modelled by means of Petri nets. In particular, Petri nets may model the flow of information or other resources within a system. They prove to be a useful model for a certain kind of expert system and for investigating certain kinds of knowledge base, see, for example, Píš et al. (1986).

A Petri net is a triple $C = (P, Z, D)$, where P is a finite set of places, Z is a finite set of transitions, and D is a set of couples in the form (p, z) or (z, p), $z \in Z$, $p \in P$.

The function $f : P \rightarrow Z^+$ is a marking of a Petri net. We note that a transition is facilitated if for each input place (a place such that $(p, z) \in D$) there exists a positive marking. A facilitated transition fires by adding (subtracting) one to (from) every output place (input place). A graphical representation of the Petri net structure is useful in its application. A Petri net graph is a presentation of a Petri net structure as a bipartite directed multigraph. A Petri net structure consists of places and transitions. Corresponding to these a Petri net graph has two types of node. Direct arcs connect the places and the transitions. An arc directed from a place p_i to a transition z_j defines the place which is to be an input of the transition. An output place is indicated by an arc from the transition to the place. For more detailed information see, for example, Peterson (1981).

Petri Nets and Expert Systems

Method II can be used for detailed analysis both of the structure of a knowledge base and of a set of input information. The objective of the first possibility is to obtain a view regarding the structure of the set of parameters. The concept "structure of parameters" means dependence of parameters, that is, by using rules to establish the value of one parameter on the basis of other parameters; especially to determine the logical circle in the knowledge base of the expert system, that is, in the set of rules. The objective of the second possibility is to investigate the sufficiency of a set of parameters, especially that of a set of axiomatic parameters AP.

In our context the knowledge base of an expert system is in the form of a set of rules. The set of rules can then be modelled through Petri nets. We can design several Petri nets to model the set of rules with different levels of differential ability. A model based on this ability is an abstraction of the modelled system. As such it ignores the specific details as much as possible. If all the details were to be modelled, then the model would be a duplicate of the modelled system and not an abstraction. Next we describe three Petri nets:

(i) Let $C = (P, Z, D)$ be a Petri net, where
$P = \{x;$ is a parameter in some rule $r \in R\}$
$Z = \{r ; r$ is the rule $r \in R\}$
$D = \{(x, r) ; x$ is a parameter in an elementary premise of the rule $r\} \cup$
$\{(r, x) ; x$ is a parameter in an elementary conclusion of the rule $r\}$

(ii) Let $C_1 = (P_1, Z_1, D_1)$ be a Petri net, where
$P_1 = \{x \,; x \text{ is a value of the parameter } x \text{ in some rule } r \in R\}$
$Z_1 = Z$
$D_1 = \{(x, r) \,; x \text{ is a value of the parameter in an elementary premise}$
      $\text{of the rule } r\} \cup$
      $\{(r, x) \,; x \text{ is a value of the parameter in an elementary conclu-}$
      $\text{sion of the rule } r\}$

(iii) Let $C_2 = (P_2, Z_2, D_2)$ be a Petri net, where
$P_2 = \{t^+, t^- \,; t \text{ is a value of the parameter in some rule } r \in R\}$
$Z_2 = Z$

$D_2 = \{(t^+, r), \text{ resp. } (t^-, r); \text{ if } t \text{ is a value of the parameter } x$
      $\text{in an elementary premise of the rule } r \text{ such that}$
      $w(x, t) = 1, \text{ resp. } w(x, t) = -1\} \cup$
      $\{(r, t^+), \text{ resp. } (r, t^-); \text{ analogically as in the elementary}$
      $\text{premise, but now in an elementary conclusion of the rule } r\}$

Now we are ready to formulate some problems connected with Method II by means of Petri nets and outline the solutions also.

Problem 1. To determine a logical circle in the set of rules. In our context the logical circle in the set of rules is the following situation: $A_1 \rightarrow A_2 \rightarrow , \ldots, \rightarrow A_n \rightarrow A_1$, where $A_i \rightarrow A_{i+1}$ designates the existence of the rule which has in its elementary premise the parameter A_i and in its elementary conclusion the parameter A_{i+1}. The logical circle in the set of rules corresponds to the oriented circuit in Petri nets C, C_1, C_2 of length $2n$. We can use Petri nets C, C_1, C_2 to find the logical circle in a set of rules. The application of Petri net C, C_1, C_2 depends on the level of detailed analysis of a set of rules being reached. Algorithms for finding oriented circuits in oriented graphs are simple and are described in most of the standard works on graph theory.

Problem 2. To evaluate the structural dependence of parameters. Next we assume that the nets C, C_1, C_2 do not contain any oriented circuit, that is the set of rules does not contain any logical circle. Furthermore, in these nets there is at least one transition where input places are not output places of another transition, that is, there exists at least one rule in which the parameters of the premise are not in the conclusion of other rules. Further, we define the number $d(x)$ for every place of the Petri net as follows:

$$d(x) = \begin{cases} 0, \text{ if } x \text{ is only an input place} \\ \\ k, \text{ otherwise.} \quad 2k \text{ is the length of the most length-} \\ \quad \text{oriented path with starting place } y, \text{ for which } d(y) = 0 \\ \quad \text{and with the terminal place } x. \end{cases}$$

We find all the oriented paths starting in places which are only input places. The length of every one of these oriented paths is an even number because every path contains, alternately, places and transitions. Half of the path of greatest length is the number $d(x)$.

Then the problem of structural dependency of parameters can be formulated as affixing to every place in the Petri nets C, C_1, C_2 the value $d(x)$. A simple algorithm which does this is as follows:

1. $d(x) = 0$ for every place x.
2. Find the transition r in which input places are not output places of other transitions.

3. M = max d(x) ; x is input place of transition r.
4. d(x) = M + 1 for every input place of the transition r.
5. Delete the transition r and all its input and output places.
6. If the net has a transition, go to 2.
7. End.

If we use this algorithm in the net C, it is evident that the greatest value of d has the place corresponding to the goal parameter pm_g and the goal parameter is the only one which has this value. In the case of the nets C_1 or C_2 there exist more of these places, but all these places correspond to the same values of the goal parameter.

Problem 3. To investigate the adequacy of a set of parameters. This problem can be formulated through cost of parameters, as follows:

(i) To find the costs of the parameters, especially to find the cost of the goal parameter.

The following problem is a generalisation of the previous one:

(ii) To find all minimal sufficient sets.

For this purpose we shall use the marked Petri net C_1, where the marking is as follows: $f(x) = 1$ if the parameter's value is x, and $f(x) = 0$ if the parameter's value is not known or the parameter is not x. Then the expert system can make the decision only if there exists the sequence of firing which transforms marking f to the marking in which the place corresponding to the value of the goal parameter has the value 1.

Generally, the firing of a transition represents a change in the state of the Petri net. A state is reachable if there exist some sequences of firings which transform the start state into the desired one. To date it has not been proved whether the problem of reachability can be solved algorithmically. Our net C_1 has the specificity which enables it to solve the problem of finding all the minimal sufficient sets. In Pís et al. (1986) the algorithm which gives an upper estimate of the cost of the parameters is described. Its complexity is linear with regard to the number of transitions and it can be used for the first judgement of a situation. In this work another algorithm which computes the cost of the parameters is described and this finds all the minimal sufficient sets regarding all the parameters. These algorithms are suitable for expert systems which work with certain information. They can be changed easily and then used in a situation where every piece of information is given a weight in the interval <-1, 1>, but in this case the complexity of the algorithm is considerably increased. It seems that this complexity is not the consequence of our approach, but rather that the problem itself is complicated. Restricting one's attention only to some minimal sets of parameters represents one possible solution.

The list of minimal and sufficient sets of parameters yields useful information. The user can use this information to prepare for the consultation. On the other hand, the Petri net can be used in the case when the expert system is not capable of offering an explanation because of the lack of input information. This model can be used to find the additional input information required by the expert system in order to make a decision. Our algorithm provides minimal sets in terms of the values of parameters. This detailed information is redundant for the user, but for the expert who creates a knowledge base it provides a detailed view on the internal structure of the parameters.

CONCLUSION

In this chapter we have briefly described two methods of expert system evaluation. In the first method, the suggestion is to evaluate the expert system as a whole on the basis of its results in practice. The second method gives a detailed view on the structure of knowledge which is based on rules. Future work should be directed towards those methods which, from a particular point of view, optimise the structure of the knowledge base.

REFERENCES

Davis, R., 1982, "Expert Systems: Where Are We and Where Do We Go From Here?" A.I. Memorandum 665, Artificial Intelligence Laboratory, MIT.

Johnson, N. L., and Leone, C. L., 1977, "Statistics and Experimental Design, vol. 1", Wiley, New York.

Masaryk, P., Píš, P., Urbánek, T., and Tkáčik, J., 1987, The consulting system in rheumatology, in: "13th SCS Symposium, Piešťany, Czechoslovakia" : 107.

Peterson, J. L., 1981, "Petri Nets Theory and the Modeling of Systems", Prentice-Hall, Englewood Cliffs, NJ.

Píš, P., Mesiar, R., and Horák, P., 1985, "Expert Systems' Evaluation", Technical Report 04-70-85, SF SVŠT, ŠÚNZ, Bratislava (in Slovak).

Píš, P., Horák, P., and Mesiar, R., 1986, "Expert Systems and Petri Nets", Technical Report 04-50-86, SF SVŠT, ŠÚNZ, Bratislava (in Slovak).

Píš, P., and Mesiar, R., 1987a, Fuzzy model of inexact reasoning in medical consulting systems, in: "Proc. 1st Joint IFSA-EC and Euro-WG Workshop on Progress in Fuzzy Sets in Europe, Warsaw, 1986".

Píš, P., and Mesiar, R., 1987b, Expert consulting system JANO, in: "XI Internationaler Congress über Anwendungen der Mathematik in dem Ingenierwissenschaften, Berichte 2, Weimar" : 70.

Píš, P., and Trupl, 1987, The expert system for antibiotic therapy, in: "3rd European Congress on Clinical Microbiology, The Hague", abstract 499.

Yu, V. L., Buchanan, B. G., Shortliffe, E. H., Wraith, S. M., Davis, R., Scott, A. C., and Cohen, E. N., 1979, Evaluating the performance of a computer-based consultant, Comput. Prog. Biomed., 9 : 95.

CONTRIBUTORS

ALBANI, C., Neurology Department, University Hospital Zurich,
    Raemistrasse 100, CH-8091 Zurich, Switzerland.
ALBRECHT, G., PTI Jena, Academy of Sciences of the GDR, Helmholtzweg 4,
    6900 Jena, GDR.
ALLINEY, S., Department of Mathematics and Informatics, University of
    Udine, Via Zanon 6, 33100 Udine, Italy.
ANDĚL, J., Department of Statistics, Charles University, Sokolovská 83,
    186 00 Prague 8, Czechoslovakia.
ANDĚL, M., First Research Department, Institute of Clinical and Experi-
    mental Medicine, Vídeňská 800, 146 22 Prague 4, Czechoslovakia.
ANTALÓCZY, Z., Postgraduate Medical School, 2nd Medical Department
    (Cardiology), P.O.B. 112, Szabolcs str. 33, H-1389 Budapest, Hungary.
ASZALOS, J., Computer Applications and Service Co., Expert System Depart-
    ment, Csalogány str. 30/32, H-1015 Budapest, Hungary.
BAJLA, I., Institute of Measurement and Measuring Technique, Electro-
    Physical Research Centre, Slovak Academy of Science, Dúbravská cesta 9,
    84219 Bratislava, Czechoslovakia.
BARCSÁK, J., Postgraduate Medical School, 2nd Medical Department (Cardio-
    logy), P.O.B. 112, Szabolcs str. 33, H-1389 Budapest, Hungary.
BARTKOWIAK, A., Institute of Computer Science, University of Wrocław,
    Przesmyckiego 20, 51-151 Wrocław, Poland.
BATORSKI, L., Clinic of Neurosurgery, Child's Health Centre, Al. Dzieci
    Polskich 20, 04-736 Warsaw, Poland.
BERTHEL, K.-H., FSU Jena, Department of Physics, Max-Wien-Platz 1,
    6900 Jena, GDR.
BOGÁNYI, G., "Kandó Kálmán" College MITI, P.O.B. 112, H-1431, Budapest,
    Hungary.
BOGNÁROVÁ, M., Institute of Measurement and Measuring Technique, Electro-
    Physical Research Centre, Slovak Academy of Science, Dúbravská
    cesta 9, 842 19 Bratislava, Czechoslovakia.
BRELIDZE, Z., Central Science Laboratory, Ministry of Health Georgia,
    Ulica Klara Zetkin 126, Tbilisi, Georgia, USSR.
BUHSS, U., Academy of Sciences of the GDR, Central Institute for Cardio-
    vascular Research, Wiltbergstrasse 50, Berlin 1115, GDR.
BURGHOFF, M., PTI Jena, Academy of Sciences of the GDR, Helmholtzweg 4,
    6900 Jena, GDR.
CAMM, A. J., Department of Cardiological Sciences, St. George's Hospital
    Medical School, Cranmer Terrace, London SW17 ORE, England.
CARSON, E. R., Centre for Measurement and Information in Medicine, City
    University, Northampton Square, London EC1V OHB, England.
CHWISTECKI, K., Department of Cardiology, Medical Academy, Pasteura 4,
    Wrocław, Poland.
CIPRA, T., Department of Statistics, Charles University, Sokolovská 83,
    186 00 Prague 8, Czechoslovakia.
COBELLI, C., Department of Electronics and Informatics, University of
    Padova, Via Gradenigo 6/A, 35131 Padova, Italy.

CRAMP, D. G., Departments of Chemical Pathology and Medical Informatics,
Royal Free Hospital School of Medicine, Rowland Hill Street,
London NW3 2PF, England.

CSUKÁS, M., Hungarian Institute of Cardiology, P.O.B. 88, IX Hámán Kató
ut 29, H-1450 Budapest, Hungary.

CZOSNYKA, M., Institute of Electronics Fundamentals, Warsaw University of
Technology, Nowowiejska 15/19, OO-665 Warsaw, Poland.

DANKO, S. G., Department of Human Neurophysiology, Institute for Experimental Medicine, AMS USSR, Pavlov's Str. 12, 197022 Leningrad, USSR.

DASKALOV, I., Institute of Biomedical Engineering, Medical Academy,
G. Sofiysky str. 1, Sofia 1431, Bulgaria.

DIMITROVA, M., Institute of Biomedical Engineering, Medical Academy,
G. Sofiysky str. 1, Sofia 1431, Bulgaria.

DITTERT, I., Institute of Physiology, Czechoslovak Academy of Sciences,
Vídeňská 1083, 142 20 Prague 4, Czechoslovakia.

DOLEŽEL, S., Institute of Pathological Physiology, Department of Artificial
Heart, Faculty of Medicine, University of J. E. Purkyně, Komenského
nám. 2, 662 43 Brno, Czechoslovakia.

DOSKOČIL, M., Department of Anatomy, Faculty of Medicine, Charles University,
U nemocnice 3, 120 00 Prague, Czechoslovakia.

DOSTÁL, M., Institute of Pathological Physiology, Department of Artificial
Heart, Faculty of Medicine, University of J. E. Purkyně, Komenského
nám. 2, 662 43 Brno, Czechoslovakia.

DOSTÁLEK, C., Czechoslovak Academy of Sciences, Institute of Physiological
Regulations, Bulovka pav. 11, 180 85 Prague, Czechoslovakia.

DUCHÁČ, V., Department of Pathological Physiology, Faculty of General
Medicine, Charles University, U nemocnice 5, 128 53 Prague, Czechoslovakia.

DVOŘÁK, J., Institute of Hygiene and Epidemiology, Šrobárova 48, Prague 10,
Czechoslovakia.

DVOŘÁK, M., Department of Histology and Embryology, Faculty of Medicine,
University of J. E. Purkyně, Tř. Obráncú 10, 662 43 Brno, Czechoslovakia.

DVORÁK, M., Institute of Meaical Research, Centre for Medical Informatics,
Ponávka 6, 662 50 Brno, Czechoslovakia.

FIEHRING, H., Academy of Sciences of the GDR, Central Institute for Cardiovascular Research, Wiltbergstrasse 50, Berlin 1115, GDR.

FORMÁNEK, J., Institute of Hygiene and Epidemiology, Šrobárova 48,
Prague 10, Czechoslovakia.

FUJII, M., Department of Electrical and Electronic Engineering, Sophia
University, 7-1 Kioi-cho, Chiyoda-ku, Tokyo 102, Japan.

FUSEK, M., 2nd Department of Medicine, Faculty of General Medicine, Charles
University, U nemocnice 2, 128 08 Prague, Czechoslovakia.

GODIK, E. E., Institute of Radio Engineering and Electronics, USSR Academy
of Sciences, GSP-3, Marx Av. 18, 103907 Moscow, USSR.

GONDZHILASVILI, J., Central Science Laboratory, Ministry of Health Georgia,
Ulica Klara Zetkin 126, Tbilisi, Georgia, USSR.

GOTFRÝD, O., Institute of Medical Research, Centre for Medical Informatics,
Ponávka 6, 662 50 Brno, Czechoslovakia.

GROTH, T., Biomaterials Research Unit, Humboldt University School of
Medicine (Charite), Tucholskystrasse 2, 1040, Berlin, GDR.

GULJAEV, Y. V., Institute of Radio Engineering and Electronics, USSR
Academy of Sciences, GSP-3, Marx Av. 18, 103907 Moscow, USSR.

HABERKORN, W., ZWG Berlin, Academy of the Sciences of the GDR, Rudower
Chaussee 6, 1199 Berlin, GDR.

HACISALIHZADE, S. S., Institute of Automatic Control, Swiss Federal
Institute of Technology (ETH), Physikstrasse 3, CH-8092, Zurich,
Switzerland.

HAENO, M., Department of Electrical and Electronic Engineering, Sophia
University, 7-1 Kioi-cho, Chiyoda-ku, Tokyo 102, Japan.

HAUSER, F., Hybrid Computation Laboratory, Institute of Social Medicine
and Health Services Organisation, Vítězného Února 54, 121 39 Prague 2,
Czechoslovakia.

KRÁL, V., Institute of Physiological Regulations, Czechoslovak Academy of
Sciences, Bulovka, Pav. 11, 180 85 Prague 8, Czechoslovakia.
KRÁMLI, A., Computer and Automation Institute, Hungarian Academy of
Sciences, P.O.B. 63, XI Kende utca 13-17, H-1502 Budapest, Hungary.
KRECHNAKOVÁ, A., Medical Bionics Research Institute, Jedĺová 6,
833 08 Bratislava, Czechoslovakia.
KREKULE, I., Institute of Physiology, Czechoslovak Academy of Sciences,
Vídeňská 1083, 142 20 Prague 4, Czechoslovakia.
KUBÁK, R., Department of Medical Electronics, Technical University,
Purkyňova 95B, 612 00 Brno, Czechoslovakia.
KUBÁT, J., Institute of Hygiene and Epidemiology, Šrobárova 48, Prague 10,
Czechoslovakia.
KUBÍK, M., Faculty of Medicine, Charles University, Karlovo nám 32,
120 00 Prague 2, Czechoslovakia.
KÜCHLER, G., Zentralinstitut für Arbeitsmedizin der DDR, Direktions-
bereich Arbeitsphysiologie, Nölderstrasse 40-42, 1134 Berlin, GDR.
LEANING, M. S., Department of Medical Informatics, Royal Free Hospital
School of Medicine, Rowland Hill Street, London NW3 2PF, England.
LEVKOV, C., Institute for Biomedical Instrumentation, Medical Academy,
G. Sofiysky Str. 1, 1431 Sofia, Bulgaria.
LUKÁČ, J., Výzkumný ústav Reumatichých Choròb, 921 01 Piešťany, Czecho-
slovakia.
ŁUKASIK, S., Department of Cardiology, Medical Academy, Pasteura 4,
Wrocław, Poland,
MADER, R., Soběslavská 1, 130 00 Prague 3, Czechoslovakia.
MAKIE, K., Department of Electrical and Electronic Engineering, Sophia
University, 7-1 Kioi-cho, Chiyoda-ku, Tokyo 102, Japan.
MALIK, M., Department of Cardiological Sciences, St. George's Hospital
Medical School, Cranmer Terrace, London SW17 ORE, England.
MALÍK, M., Department of Computer Science, Charles University, Malostranské
náměstí 25, 118 00 Prague 1, Czechoslovakia.
MANSOUR, M., Institute of Automatic Control, Swiss Federal Institute of
Technology (ETH), Physikstrasse 3, CH-8092 Zurich, Switzerland.
MAŘÍK, V., Faculty of Electrical Engineering, K335, Czech Technical
University, Suchbátarova 2, 166 27 Prague 6, Czechoslovakia.
MAŘÍKOVÁ, T., Faculty of Paediatrics, Charles University, V úvalu 84,
150 00 Prague 5, Czechoslovakia.
MASARYK, P., Výzkumný ústav Reumatichých Choròb, 921 01 Piešťany, Czecho-
slovakia.
MATEJ, S., Institute of Measurement and Measuring Technique, Electro-
Physical Research Centre, Slovak Academy of Science, Dúbravská
cesta 9, 842 19 Bratislava, Czechoslovakia.
MATVEJEV, M., Institute of Biomedical Engineering, Medical Academy,
C. Sofiysky str. 1, Sofia 1431, Bulgaria.
MIERTUSOVÁ, J., Elektrotechnická Fakulta SVST, Mlynská Dolina,
815 19 Bratislava, Czechoslovakia.
MIKULÁŠ, M., Elektrotechnická Fakulta SVST, Mlynská Dolina, 815 19 Bratis-
lava, Czechoslovakia.
MIYAMOTO, A., Department of Hygiene, Nihon University, 30-1 Kamimachi,
Ohoyaguchi, Itabashi-ku, Tokyo, Japan.
MRUKOWICZ, M., Department of Cardiology, Medical Academy, Pasteura 4,
Wrocław, Poland.
MUCKE, R., Zentralinstitut für Arbeitsmedizin der DDR, Direktionsbereich
Arbeitsphysiologie, Nölderstrasse 40-42, 1134 Berlin, GDR.
MÜLLER, W., Medical Academy Erfurt, Clinic for Eye Diseases, Nordhäuser
Str. 74, 5010 Erfurt, GDR.
MUNCLINGER, M., 2nd Department of Medicine, Faculty of General Medicine,
Charles University, U nemocnice 2, 128 08 Prague, Czechoslovakia.
MÜNZ, J., Institute of Medical Research, Centre for Medical Informatics,
Ponávka 6, 662 50 Brno, Czechoslovakia.
MUSHINSKI, A., Zakłady Techniki Medycznej, ul. Słowiańska 15, Koszalin,
Poland.

NAKAYAMA, K., Department of Electrical and Electronic Engineering, Sophia
    University, 7-1 Kioi-cho, Chiyoda-ku, Tokyo 102, Japan.
NOVAKOV, E., Institute of Biomedical Engineering, Medical Academy,
    G. Sofiysky str. 1, Sofia 1431, Bulgaria.
NOWAK, H., PTI Jena, Academy of Sciences of the GDR, Helmholtzweg 4,
    6900 Jena, GDR.
ODEHNAL, M., Institute of Physics, Czechoslovak Academy of Sciences,
    Na Slovance 2, 180 40 Prague, Czechoslovakia.
ORANIEN, S., Department of Biomedical Engineering, Ilmenau Intitute of
    Technology, P.O.B. 327, 6300 Ilmenau, GDR.
ORAVCOVÁ, M., Medical Bionics Research Institute, Jedlová 6, 833 08 Bratis-
    lava, Czechoslovakia.
PACINI, G., Bioengineering, LADSEB-CNR, Corso Stati Uniti 4, 35020 Padova,
    Italy.
PENEV, H., Institute of Cybernetics and Robotics, Bulgarian Academy of
    Sciences, Sofia 1113, Bulgaria.
PETRÁNEK, S., Neurologická Klinika ILF, Vídeňská 800, 140 59 Prague 4,
    Czechoslovakia.
PÍŠ, P., Štátne Sanatorium, Klenova 1, 833 10 Bratislava, Czechoslovakia.
POKORNÝ, Z., Department of Pathological Physiology, Faculty of General
    Medicine, Charles University, U nemocnice 5, 128 53 Prague, Czecho-
    slovakia.
POPELÍNSKÝ, L., Brožíkova 13, 638 00, Brno, Czechoslovakia.
POTUCEK, J., Krskova 788, 152 00 Prague 5-Hlubocepy, Czechoslovakia.
PRÁŠKOVÁ, Z., Department of Statistics, Charles University, Sokolovská 83,
    186 00 Prague 8, Czechoslovakia.
PRÉDA, I., Postgraduate Medical School, 2nd Medical Department (Cardiology),
    P.O.B. 112, Szabolcs str. 33, H-1389 Budapest, Hungary.
RAJCHL, P., Faculty of Electrical Engineering, K335, Czech Technical
    University, Suchbátarova 2, 166 27 Prague 6, Czechoslovakia.
RECH, F., Institute of Physiology, Czechoslovak Academy of Sciences,
    Vídeňská 1083, 142 20 Prague 4, Czechoslovakia.
REIME, B., Academy of Sciences of the GDR, Central Institute for Cardio-
    vascular Research, Wiltbergstrasse 50, Berlin 1115, GDR.
ROSÍK, V., Institute of Measurement and Measuring Technique, Electro-
    Physical Research Centre, Slovak Academy of Sciences, Dúbravská
    cesta 9, 842 19 Bratislava, Czechoslovakia
ROTTOVÁ, I., Neurologická Klinika ILF, Vídeňská 800, 140 59 Prague 4,
    Czechoslovakia.
ROVENSKÝ, J., Štátny ústav pre Kontrolu Liečiv, Kvetná 11, 825 08 Bratis-
    lava, Czechoslovakia.
RUSZKOWSKI, J., Medical Computer Laboratory, Department of Biophysics and
    Biomathematics, Medical Centre of Postgraduate Education, ul. Mary-
    moncka 99, 01-813 Warsaw, Poland.
SAKAMOTO, K., Department of Electrical and Electronic Engineering, Sophia
    University, 7-1 Kioi-cho, Chiyoda-ku, Tokyo 102, Japan.
SCHEIDOVA, L., Medical Bionics Research Institute, Jedlová 6, 833 08 Bratis-
    lava, Czechoslovakia.
ŠEDIVÁ, I., Faculty of Electrical Engineering, K335, Czech Technical
    University, Suchbátarova 2, 166 27 Prague 6, Czechoslovakia.
SERESS, I., Medical Bionics Research Institute, Jedlová 5, 833 08 Bratis-
    lava, Czechoslovakia.
SERF, B., 2nd Department of Medicine, Faculty of General Medicine, Charles
    University, U nemocnice 2, 128 08 Prague, Czechoslovakia.
SHAMSOLMAALI, A., Department of Medical Informatics, Royal Free Hospital
    School of Medicine, Rowland Hill Street, London NW3 2PF, England.
SLÁMA, M., Faculty of Electrical Engineering, K335, Czech Technical
    University, Suchbátarova 2, 166 27 Prague 6, Czechoslovakia.
SOLTÉSZ, J., Computer and Automation Institute, Hungarian Academy of
    Sciences, P.O.B. 63, XI Kende utca 13-17, H-1502, Budapest, Hungary.
ŠRAMKA, M., Research Laboratory for Clinical Stereotactics, Research Insti-
    tute for Medical Bionics, Jedlová 6, 833 05, Bratislava, Czechoslovakia.

STAMBOLIEV, I., Department of Electronics, Higher Institute of Mechanical
    and Electrical Engineering, Sofia 1156, Bulgaria.
ŠŤASTNÁ, J., Department of Histology and Embryology, Faculty of Medicine,
    Purkyně University, Tr. Obráncú 10, 662 43 Brno, Czechoslovakia.
SUMMERS, R., Centre for Measurement and Information in Medicine, City
    University, Northampton Square, London EC1V OHB, England.
SUŠIL, A., Institute of Medical Research, Centre for Medical Informatics,
    Ponávka 6, 662 50 Brno, Czechoslovakia.
SVAČINA, S., Third Medical Department, Charles University, U nemocnice 1,
    128 21 Prague 2, Czechoslovakia.
SVETLOVSKÁ, E., Medical Bionics Research Institute, Jedlová 6, 833 08 Bratis-
    lava, Czechoslovakia.
TAGAWA, H., Department of Internal Medicine, Mitsui Memorial Hospital,
    1 Izumi-cho Kanda Chiyoda-ku, Tokyo 101, Japan.
TARATORIN, A. M., Institute of Radio Engineering and Electronics, USSR
    Academy of Sciences, GSP-3, Marx Av. 18, 103907 Moscow, USSR.
TITOMIR, L. I., Institute of Problems of Information Transmission, USSR
    Academy of Sciences, Jermolovoj 19, 102051 Moscow, USSR.
TKACZ, E., Department of Medical Electronics, Technical University of
    Silesia, W. Pstrowskiego Str. 16, 44-100 Gliwice, Poland.
TURZOVÁ, M., Institute of Measurement and Measuring Technique, Electro-
    Physical Research Centre, Slovak Academy of Sciences, Dúbravská
    cesta 9, 842 19 Bratislava, Czechoslovakia.
TYŠLER, M., Institute of Measurement and Measuring Technique, Electro-
    Physical Research Centre, Slovak Academy of Sciences, Dúbravská
    cesta 9, 842 19 Bratislava, Czechoslovakia.
VAŠKŮ, Jan, Institute of Pathological Physiology, Department of Artificial
    Heart, Faculty of Medicine, University of J. E. Purkyně, Komenského
    nám. 2, 662 43 Brno, Czechoslovakia.
VAŠKŮ, Jaromír, Institute of Pathological Physiology, Department of
    Artificial Heart, Faculty of Medicine, University of J. E. Purkyně,
    Komenského, nám. 2, 662 43 Brno, Czechoslovakia.
VOLEJNÍČEK, O., Institute of Medical Research, Centre for Medical Infor-
    matics, Ponávka 6, 662 50 Brno, Czechoslovakia.
VOSS, A., Academy of Sciences of the GDR, Central Institute for Cardio-
    vascular Research, Wiltbergstrasse 50, Berlin 1115, GDR.
WACKERMAN, J., Psychiatric Research Institute, Ústavní 91, 181 03 Prague,
    Czechoslovakia.
WOLF, H., Biomaterials Research Unit, Humboldt University School of
    Medicine (Charite), Tucholskystrasse 2, 1040 Berlin, GDR.
WOLLK-LANIEWSKI, P., Department of Anaesthesiology and Intensive Care,
    Child's Health Centre, Al. Dzieci Polskich 20, 04-736 Warsaw, Poland.
YAGI, S., Department of Electrical and Electronic Engineering, Sophia
    University, 7-1 Kioi-cho, Chiyoda-ku, Tokyo 102, Japan.
YAGIMUMA, T., Department of Electrical and Electronic Engineering, Sophia
    University, 7-1 Kioi-cho, Chiyoda-ku, Tokyo 102, Japan.
YAJIMA, K., Department of Hygiene, Nihon University, 30-1 Kamimachi,
    Ohoyaguchi, Itabashi-ku, Tokyo, Japan.
ZACH, H.-G., PTI Jena, Academy of Sciences of the GDR, Helmholtzweg 4,
    6900 Jena, GDR.
ZAHLMANN, G., Department of Biomedical Engineering, Ilmenau Institute of
    Technology, P.O.B. 327, 6300 Ilmenau, GDR.
ZAWORSKI, W., Institute of Electronics Fundamentals, Warsaw University of
    Technology, Nowowiejska 15/19, 00-665 Warsaw, Poland.
ZDRÁHAL, Z., Faculty of Electrical Engineering, K335, Czech Technical
    University, Suchbátarova 2, 166 27 Prague 6, Czechoslovakia.
ŽITŇAN, D., Výzhumný ústav Reumatichých Chorôb, 921 01 Piešťany, Czecho-
    slovakia.